MATRIX THEORY AND
LINEAR ALGEBRA

Matrix Theory and Linear Algebra

I. N. HERSTEIN
University of Chicago

DAVID J. WINTER
University of Michigan, Ann Arbor

Macmillan Publishing Company
NEW YORK

Collier Macmillan Publishers
LONDON

Copyright © 1988, Macmillan Publishing Company, a division of Macmillan, Inc.

PRINTED IN THE UNITED STATES OF AMERICA

All rights reserved. No part of this book may be reproduced or transmitted in any form or by any means, electronic or mechanical, including photocopying, recording, or any information storage and retrieval system, without permission in writing from the publisher.

Macmillan Publishing Company
866 Third Avenue, New York, New York 10022

Collier Macmillan Canada, Inc.

LIBRARY OF CONGRESS CATALOGING-IN-PUBLICATION DATA

Herstein, I. N.
   Matrix theory and linear algebra.

   Includes index.
   1. Matrices.   2. Algebras, Linear.   I. Winter, David J.   II. Title.
QA188.H47   1988   512.9'434   87-11669
ISBN 0-02-353951-8

Printing: 1 2 3 4 5 6 7 8

Year: 8 9 0 1 2 3 4 5 6 7

To the memory of A. A. Albert
                        I. N. H.

To Alison
                D. J. W.

Preface

Matrix theory and linear algebra is a subject whose material can, and is, taught at a variety of levels of sophistication. These levels depend on the particular needs of students learning the subject matter. For many students it suffices to know matrix theory as a computational tool. For them, a course that stresses manipulation with matrices, a few of the basic theorems, and some applications fits the bill. For others, especially students in engineering, chemistry, physics, and economics, quite a bit more is required. Not only must they acquire a solid control of calculating with matrices, but they also need some of the deeper results, many of the techniques, and a sufficient knowledge and familiarity with the theoretical aspects of the subject. This will allow them to adapt arguments or extend results to the particular problems they are considering. Finally, there are the students of mathematics. Regardless of what branch of mathematics they go into, a thorough mastery not only of matrix theory, but also of linear spaces and transformations, is a must.

We have endeavored to write a book that can be of use to all these groups of students, so that by picking and choosing the material, one can make up a course that would satisfy the needs of each of these groups.

During the writing, we were confronted by a dilemma. On the one hand, we wanted to keep our presentation as simple as possible, and on the other, we wanted to cover the subject fully and without compromise, even when the going got tough. To solve this dilemma, we decided to prepare two versions of the book, this version for the more experienced or ambitious student, and another, *A Primer on Linear Algebra*, for students who desire to get a first look at linear algebra but do not want to go into it in as great a depth the first time around. These two versions are almost identical in many respects, but there are some important differences. Whereas in *A Primer on Linear Algebra* we excluded some advanced topics and simplified some others, in this version we go into more depth for some topics treated in both books (determinants, Markov processes, incidence models, differential equations, least squares methods) and included some others (triangulation of matrices with real entries, the Jordan canonical form). Also, there are some harder exercises, and the grading of the exercises presupposes a somewhat more experienced student.

Toward the end of the preface we lay out some possible programs of study at these various levels. At the same time, the other material is there for them to look through or study if they desire.

Our approach is to start slowly, setting out at the level of 2×2 matrices. These matrices have the great advantage that everything about them is open to the eye—that students can get their hands on the material, experiment with it, see what any specific theorem says about them. Furthermore, all this can be done by performing some simple calculations.

However, in treating the 2×2 matrices, we try to handle them as we would the general $n \times n$ case. In this microcosm of the larger matrix world, virtually every concept that will arise for $n \times n$ matrices, general vector spaces, and linear transformations makes its appearance. This appearance is usually in a form ready for extension to the general situation. Probably the only exception to this is the theory of determinants, for which the 2×2 case is far too simplistic.

With the background acquired in playing with the 2×2 matrices in this general manner, the results for the most general case, as they unfold, are not as surprising, mystifying, or mysterious to the students as they might otherwise be. After all, these results are almost old friends whose acquaintance we made in our earlier 2×2 incarnation. So this simplified context serves both as a laboratory and as a motivation for what is to come.

From the fairly concrete world of the $n \times n$ matrices we pass to the more abstract realm of vector spaces and linear transformations. Here the basic strategy is to prove that an n-dimensional vector space is isomorphic to the space of n-tuples. With this isomorphism established, the whole corpus of concepts and results that we had obtained in the context of n-tuples and $n \times n$ matrices is readily transferred to the setting of arbitrary vector spaces and linear transformations. Moreover, this transfer is accomplished with little or no need of proof. Because of the nature of isomorphism, it is enough merely to cite the proof or result obtained earlier for n-tuples or $n \times n$ matrices.

The vector spaces we treat in the book are only over the fields of real or complex numbers. While a little is lost in imposing this restriction, much is gained. For instance, our vector spaces can always be endowed with an inner product. Using this inner product, we can always decompose the space as a direct sum of any subspace and its orthogonal complement. This direct sum decomposition is then exploited to the hilt to obtain very simple and illuminating proofs of many of the theorems.

There is an attempt made to give some nice, characteristic applications of the material of the book. Some of these can be integrated into a course almost from the beginning, where we discuss 2×2 matrices.

Least squares methods are discussed to show how linear equations having no solutions can always be solved approximately in a very neat and efficient way. These methods are then used to show how to find functions that approximate given data.

Finally, in the last chapter, we discuss how to translate some of our methods into *linear algorithms*, that is, finite-numerical step-by-step versions of methods of linear algebra. The emphasis is on linear algorithms that can be used in writing computer programs for finding exact and approximate solutions of linear equations. We then illustrate how some of these algorithms are used in such a computer program, written in the programming language Pascal.

There are many exercises in the book. These are usually divided into categories entitled *numerical, more theoretical: easier, middle-level,* and *harder.* One even runs across some problems that are downright hard, which we put in the subcategory *very hard.* It goes without saying that the problems are an intrinsic part of any course. They are the best means for checking on one's understanding and mastery of the material. Included are exercises that are treated later in the text itself or that can be solved easily using later results. This gives you, the reader, a chance to try your own hand at developing important tools, and to compare your approach in an early context to our approach in the later context. An answer manual is available from the publisher.

We mentioned earlier that the book can serve as a textbook for several different levels. Of course, how this is done is up to the individual instructor. We present below some possible sample courses.

1. *One-term course emphasizing computational aspects.*

   Chapters 1 through 4, Chapters 5 and 6 with an emphasis on methods and a minimum of proofs. One might merely do determinants in the 3×3 case with a statement that the $n \times n$ case follows similar rules. Sections marked "optional" should be skipped. The sections in Chapter 11 entitled "Fibonacci Numbers" and "Equations of Curves" could be integrated into the course. Problems should primarily be all the numerical ones and a sampling of the easier and middle-level theoretical problems.

2. *One-term course for users of matrix theory and linear algebra in allied fields.*

   Chapters 1 through 6, with a possible deemphasis on the proofs of the properties of determinants and with an emphasis on computing with determinants. Some introduction to vector spaces and linear transformations would be desirable. Chapters 12 and 13, which deal with least squares methods and computing, could play an important role in the course. Each of the applications in Chapter 11 could be touched on. As for problems, again all the numerical ones, most of the middle-level ones, and a few of the harder ones should be appropriate for such a course.

3. *One-term course for mathematics majors.*

   Most of Chapter 1 done very quickly, with much left for the students to read on their own. All of Chapters 2 through 7, including some of the optional topics. Definitely some emphasis should be given to abstract vector spaces and linear transformations, as in Chapters 8 and 9, possibly skipping quotient spaces and invariant subspaces. The whole gamut of problems should be assignable to the students.

4. *Two-term course for users of matrix theory and linear algebra.*

   The entire book, but going easy on proofs for determinants, on the material on abstract vector spaces and linear transformations, totally omitting Chapter 10 on the Jordan canonical form and the discussion of differential equations in Chapter 11, plus a fairly thorough treatment of Chapters 12 and 13. The problems can be chosen from all parts of the problem sections.

5. *Two-term course for mathematics majors.*
   The entire book, with perhaps less emphasis on Chapters 12 and 13.

We should like to thank the many people who have looked at the manuscript, commented on it, and made useful suggestions. We want to thank Bill Blair and Lynne Small for their extensive and valuable analysis of the book at its different stages, which had a very substantial effect on its final form. We should also like to thank Gary Ostedt, Bob Clark, and Elaine Wetterau of the Macmillan Publishing Company for their help in bringing this book into being. We should like to thank Lee Zukowski for the excellent typing job he did on the manuscript. And we should like to thank Pedro Sanchez for his valuable help with the computer program and the last chapter.

I. N. H.
D. J. W.

Contents

Preface ... vii

List of Symbols ... xv–xviii

1 The 2 × 2 Matrices ... 1

1.1 INTRODUCTION   1
1.2 DEFINITIONS AND OPERATIONS   2
1.3 SOME NOTATION   10
1.4 TRACE, TRANSPOSE, AND ODDS AND ENDS   15
1.5 DETERMINANTS   21
1.6 CRAMER'S RULE   26
1.7 MAPPINGS   29
1.8 MATRICES AS MAPPINGS   35
1.9 THE CAYLEY–HAMILTON THEOREM   38
1.10 COMPLEX NUMBERS   45
1.11 $M_2(\mathbb{C})$ 52
1.12 INNER PRODUCTS   55

2 Systems of Linear Equations ... 62

2.1 INTRODUCTION   62
2.2 EQUIVALENT SYSTEMS   74
2.3 ELEMENTARY ROW OPERATIONS. ECHELON MATRICES   79
2.4 SOLVING SYSTEMS OF LINEAR EQUATIONS   85

3 The $n \times n$ Matrices — 90

- 3.1 THE OPENING 90
- 3.2 MATRICES AS OPERATORS 98
- 3.3 TRACE 109
- 3.4 TRANSPOSE AND HERMITIAN ADJOINT 116
- 3.5 INNER PRODUCT SPACES 124
- 3.6 BASES OF $F^{(n)}$ 131
- 3.7 CHANGE OF BASIS OF $F^{(n)}$ 140
- 3.8 INVERTIBLE MATRICES 145
- 3.9 MATRICES AND BASES 148
- 3.10 BASES AND INNER PRODUCTS 157

4 More on $n \times n$ Matrices — 164

- 4.1 SUBSPACES 164
- 4.2 MORE ON SUBSPACES 170
- 4.3 GRAM–SCHMIDT ORTHOGONALIZATION PROCESS 176
- 4.4 RANK AND NULLITY 180
- 4.5 CHARACTERISTIC ROOTS 183
- 4.6 HERMITIAN MATRICES 192
- 4.7 TRIANGULARIZING MATRICES WITH COMPLEX ENTRIES 199
- 4.8 TRIANGULARIZING MATRICES WITH REAL ENTRIES (OPTIONAL) 208

5 Determinants — 214

- 5.1 INTRODUCTION 214
- 5.2 PROPERTIES OF DETERMINANTS: ROW OPERATIONS 221
- 5.3 PROPERTIES OF DETERMINANTS: COLUMN OPERATIONS 233
- 5.4 CRAMER'S RULE 243
- 5.5 PROPERTIES OF DETERMINANTS: OTHER EXPANSIONS 246
- 5.6 THE CLASSICAL ADJOINT (OPTIONAL) 253
- 5.7 ELEMENTARY MATRICES 256
- 5.8 THE DETERMINANT OF THE PRODUCT 262
- 5.9 THE CHARACTERISTIC POLYNOMIAL 266
- 5.10 THE CAYLEY–HAMILTON THEOREM 270

6 Rectangular Matrices. More on Determinants — 273

- 6.1 RECTANGULAR MATRICES 273
- 6.2 BLOCK MULTIPLICATION (OPTIONAL) 275
- 6.3 ELEMENTARY BLOCK MATRICES (OPTIONAL) 278
- 6.4 A CHARACTERIZATION OF THE DETERMINANT FUNCTION (OPTIONAL) 280

7 More on Systems of Linear Equations — 282

7.1 LINEAR TRANSFORMATIONS FROM $F^{(n)}$ to $F^{(m)}$ 282
7.2 THE NULLSPACE AND COLUMN SPACE OF AN $m \times n$ MATRIX   287

8 Abstract Vector Spaces — 292

8.1 INTRODUCTION, DEFINITIONS, AND EXAMPLES   292
8.2 SUBSPACES   296
8.3 HOMOMORPHISMS AND ISOMORPHISMS   300
8.4 ISOMORPHISMS FROM V TO $F^{(n)}$ 307
8.5 LINEAR INDEPENDENCE IN INFINITE-DIMENSIONAL VECTOR SPACES   311
8.6 INNER PRODUCT SPACES   314
8.7 MORE ON INNER PRODUCT SPACES   321

9 Linear Transformations — 330

9.1 INTRODUCTION   330
9.2 DEFINITIONS, EXAMPLES, AND SOME PRELIMINARY RESULTS   331
9.3 PRODUCTS OF LINEAR TRANSFORMATIONS   337
9.4 LINEAR TRANSFORMATIONS AS MATRICES   342
9.5 A DIFFERENT SLANT ON SECTION 9.4 (OPTIONAL)   349
9.6 HERMITIAN IDEAS   351
9.7 QUOTIENT SPACES (OPTIONAL)   357
9.8 INVARIANT SUBSPACES (OPTIONAL)   366
9.9 LINEAR TRANSFORMATIONS FROM ONE SPACE TO ANOTHER   372

10 The Jordan Canonical Form (Optional) — 376

10.1 INTRODUCTION   376
10.2 GENERALIZED NULLSPACES   380
10.3 THE JORDAN CANONICAL FORM   386
10.4 EXPONENTIALS   393
10.5 SOLVING HOMOGENEOUS SYSTEMS OF LINEAR DIFFERENTIAL EQUATIONS   400

11 Applications (Optional) — 408

11.1 FIBONACCI NUMBERS   408
11.2 EQUATIONS OF CURVES   414
11.3 MARKOV PROCESSES   416
11.4 INCIDENCE MODELS   423
11.5 DIFFERENTIAL EQUATIONS   430

12 Least Squares Methods (Optional) — 434

- 12.1 APPROXIMATE SOLUTIONS OF SYSTEMS OF LINEAR EQUATIONS 434
- 12.2 THE APPROXIMATE INVERSE OF AN $m \times n$ MATRIX 443
- 12.3 SOLVING A MATRIX EQUATION USING ITS NORMAL EQUATION 446
- 12.4 FINDING FUNCTIONS THAT APPROXIMATE DATA 450
- 12.5 WEIGHTED APPROXIMATION 455

13 Linear Algorithms (Optional) — 458

- 13.1 INTRODUCTION 458
- 13.2 THE *LDU* FACTORIZATION OF *A* 461
- 13.3 THE ROW REDUCTION ALGORITHM AND ITS INVERSE 467
- 13.4 BACK AND FORWARD SUBSTITUTION. SOLVING $Ax = y$ 476
- 13.5 APPROXIMATE INVERSE AND PROJECTION ALGORITHMS 482
- 13.6 A COMPUTER PROGRAM FOR FINDING EXACT AND APPROXIMATE SOLUTIONS 488

Index — 501

List of Symbols

$s \in S$	the element s is in the set S
$S \subset T$	the set S is contained in the set T
$S \cup T$	the union of the sets S and T
$S_1 \cup \cdots \cup S_n$	the union of the sets $S_1, \ldots, S_n$
$S \cap T$	the intersection of the sets S and T
$S_1 \cap \cdots \cap S_n$	the intersection of the sets $S_1, \ldots, S_n$
$\mathbb{R}$	the set of all real numbers, 2
$M_2(\mathbb{R})$	the set of all 2×2 matrices over $\mathbb{R}$, 2, 90
$A + B$	the sum of matrices A and B, 3, 91
$A - B$	the difference of matrices A and B, 3, 91
0	the zero matrix, 3, 91
$-A$	the negative of the matrix A, 4, 91
uA	the matrix obtained by multiplying the matrix A by the scalar u, 4, 92
AB	the product of matrices A and B, 4, 93
I	the identity matrix, 5, 92
A^{-1}	the inverse of the matrix A, when it exists, 5, 92
aI	the scalar matrix corresponding to the scalar a, 6, 93
$\sum_{r=1}^{n} a_r$	the summation of a_r from 1 to n, 10
$\sum_{r=1}^{m} \sum_{s=1}^{n} a_{rs}$	the double summation of a_{rs}, 10
A^m	the product of the matrix A with itself m times, 12, 97
A^0	the identity matrix, 12, 97
A^{-m}	the product of the matrix A^{-1} with itself m times, when it exists, 12, 97
$\text{tr}(A)$	the trace of the matrix A, 16, 109
A'	the transpose of the matrix A, 18, 116

List of Symbols

$\det(A), |A|$	the determinant of the matrix A, 21, 215
fg	the product or composite of functions f, g, 30
f^{-1}	the inverse of a $1-1$ onto function f, 30
f^n	product of the function f with itself n times, 32
$\mathbb{R}^{(2)}$	the set of all vectors in the Cartesian plane, 35, 98
$v + w$	the sum of vectors in $\mathbb{R}^{(2)}$, 35
dv	the vector v in $\mathbb{R}^{(2)}$ multiplied by the scalar d, 35
Av	the vector obtained by applying $A \in M_2(\mathbb{R})$ to $v \in \mathbb{R}^{(2)}$, 36
$P_A(x)$	the characteristic polynomial of the matrix A, 39, 266
$\alpha = a + bi$	the complex number, α, with real part a and pure imaginery part bi, 46
$\mathbb{C}$	the set of all complex numbers, 46
$\bar{\alpha} = a - bi$	the conjugate of the complex number $\alpha = a + bi$, 49
$|\alpha|$	the absolute value of the complex number α, 49
$M_2(\mathbb{C})$	the set of all 2×2 matrices over $\mathbb{C}$, 52, 90
(v, w)	the inner product of column vectors v, w, 55, 124
A^*	the Hermitian adjoint of A, 57, 116
$\|v\|$	the length of vector v, 58, 127
Add $(r, s; u)$	the operation of adding u times row s to row r, 8, 257
Interchange (r, s)	the operation of interchanging rows r and s, 80, 257
Multiply $(r; u)$	the operation of multiplying row r by the scalar u, 80, 257
F	the set of all scalars: $F = \mathbb{R}$ or $F = \mathbb{C}$, 91
E_{rs}	the matrix whose (r, s) entry is 1 and all of whose other entries are 0, 94
$M_n(F)$	the set of all $n \times n$ matrices over F, 90
$F^{(n)}$	the set of all column vectors with n coordinates from F, 98
$v + w$	the sum of vectors in $F^{(n)}$, 99
0	the zero vector in $F^{(n)}$, 99
$-v$	the negative of the vector v in $F^{(n)}$, 99
tv	the vector v multiplied by the scalar t, 99
Av	the vector obtained by applying the matrix or transformation $A \in M_n(F)$ to the vector $v \in F^{(n)}$, 101
$v^\perp$	the set of all vectors orthogonal to v, 125
e_s	the vector whose r entry is 0 if $r \neq s$ and 1 if $r = s$, 131
$A \sim B$	the matrices A and B are similar, 154
cl(A)	the set of all matrices B such that $A \sim B$, 155
dim (V)	the dimension of V, 167, 310
$V + W$	the sum of subspaces, V, W, 170
$V \oplus W$	the direct sum of subspaces V, W, 172, 225
$W^\perp$	the subspace of vectors orthogonal to W, 173, 325

List of Symbols

$n(A)$	the nullity of the matrix A, 181
$\langle v_1, \ldots, v_r \rangle$	the subspace spanned by the vectors $v_1, \ldots, v_r$, 134
$r(A)$	the rank of the matrix A, 181
$q_A(x)$	the minimum polynomial of the matrix A, 184
V_a	the set of characteristic vectors of A associated with a, 194
M_{rs}	the (r, s) minor of the matrix A, 215
A_{rs}	the (r, s) cofactor of the matrix A, 246
$A^@$	the classical adjoint of the matrix A, 253
$A(r, s; q)$	the elementary matrix corresponding to Add $(r, s; q)$, 257
$M(r; q)$	the elementary matrix corresponding to Multiply $(r; q)$, 257
$I(r, s)$	the elementary matrix corresponding to Interchange (r, s), 257
$v + w$	the sum of vectors v, w in a vector space V, 293
0	the zero vector in a vector space V, 293
av	the vector v multiplied by the scalar a in a vector space V, 186
Ker ϕ	the kernel of the homomorphism ϕ from V to W, 305
$V \cong W$	there exists an isomorphism from V to W, 305
(v, w)	the inner product of elements v, w in an inner product space, 314
$\|v\|$	the length of a vector in an inner product space, 314
$T_1 + T_2$	the sum of linear transformations T_1, T_2, 334
aT	the linear transformation obtained by multiplying the linear transformation T by the scalar a, 335
0	the zero linear transformation, 335
$-T$	the negative of the linear transformation T, 335
$L(V)$	the set of all linear transformations of V over F, 335
$T_1 T_2$	the product of linear transformation T_1, T_2, 337
I	the identity linear transformation, 338
T^{-1}	the inverse of the linear transformation T, when it exists, 339
$m(T)$	the matrix of a linear transformation in a given basis, 342
tr (T)	the trace of the linear transformation T, 346
det (T)	the determinant of the linear transformation T, 346
$P_T(x)$	the characteristic polynomial of the linear transformation T, 346
$[v, w]$	the inner product $(\phi(v), \phi(w))$ corresponding to a given isomorphism ϕ, 352
T^*	the Hermitian adjoint of the linear transformation T, 354
$v + W$	the set of all vectors $v + w$ with w in W, 357
V/W	the quotient space of V by W, 357
$\tilde{T}$	the linear transformation on a subspace W of V induced by the linear transformation T of V when W is invariant under T, 366
$\hat{T}$	the linear transformation on the quotient space V/W induced by the linear transformation T on V when W is invariant under T, 367

List of Symbols

$V_0(T)$	the generalized nullspace of the linear transformation T of V, 380
$V_*(T)$	the intersection of $T^e(V)$ over all positive integers e, 380
$V_a(T)$	the generalized characteristic space of T at a, 381
e^A	the exponential of an $n \times n$ real or complex matrix A, 394
$x'(t)$	the derivative of the vector function $x(t)$, 400
$S^{(k)}$	the kth state in a Markov process, 416
$\text{Proj}_W(y)$	the projection of the vector y on the subspace W, 440
A^-	the approximate inverse of the $m \times n$ matrix A, 444
$\langle v, w \rangle_P$	the inner product $(P(v), P(w))$ corresponding to a given invertible matrix P, 455
$\|v\|_P$	the length of v, given the inner product $(v, w)_P$, 455
$\tilde{A}$	the weighted approximate inverse of an $m \times n$ matrix A, given invertible weighting matrices $P \in M_m(\mathbb{R})$, $Q \in M_n(\mathbb{R})$, 456

MATRIX THEORY AND
LINEAR ALGEBRA

CHAPTER 1

The 2 x 2 Matrices

1.1. INTRODUCTION

The subject whose study we are about to undertake—matrix theory and linear algebra—is one that crops up in a large variety of places. Needless to say, from its very name, it has an important role to play in algebra and, in fact, in virtually every part of mathematics. Not too surprisingly, it finds application in a much wider arena; there are few areas of science and engineering in which it does not make an appearance. What is more surprising is the extent to which results and techniques from this theory are also used increasingly in such fields as economics, psychology, business management, and sociology.

What is the subject all about? Some readers will have made a beginning acquaintance with it in courses in multivariable calculus where 2×2 and 3×3 matrices are frequently introduced. For those readers it is not so essential to explain what the subject is about. For others who have had no exposure to this sort of thing, it might perhaps be helpful if we explain, in a few words, something about the subject matter treated in this book.

We start things off with a very special situation: that of 2×2 matrices whose entries are real numbers. Shortly after that we enlarge our universe slightly by introducing complex numbers as entries.

What will these 2×2 matrices be? At the outset they will be merely some formal symbols, arrays of real numbers in two rows and columns. To these we shall assign methods of combining them, called "addition" and "multiplication," and we show that these combinations are subject to certain rules of behavior. At first, with no experience behind us, we might find these operations and rules to be arbitrary, unmotivated, and even, perhaps, contrived. Later, we shall see that they do have a natural life in geometry and solving equations.

After the initial steps of introducing the formal symbols and how to operate with these, we shall go about experimenting to see what is true for these formal symbols, which we call 2×2 matrices. In doing so, we shall prove a series of results all of which will be very special prototypes of what is to come for the far more general situation of $n \times n$ matrices for any positive integer n.

Why start with the 2×2 case rather than plunge immediately into the general $n \times n$ case? There are several good reasons for this. Because the 2×2 case is small, everything can be done very explicitly and visibly, with rather simple computations. Thus, when we come to the general case, we still have some idea of what may be true there. Hence the explicit case of the 2×2 matrices will be our guideline for the approach to the general case.

But even this passage to the $n \times n$ matrices is still rather formal. It would be nice to be able to see these matrices as living beings. Strangely enough, the best way to do this is to get even more abstract and pass to the abstract notion of a vector space. Matrices will then assume a clearer role—that of objects that do something, that act by transformations on a desirable set in a very attractive way. In making this transition, we are going over from matrix theory to that of linear algebra. In this new framework we shall redo many of the things done before, in a smoother, easier, and more conceptual form. Aside from redoing old things, we shall be able to press on to newer and different notions and results.

With these vague words we have given an even vaguer description of what will transpire. When we come to the more abstract approach—linear algebra—the subject will acquire greater unity and cohesion. It will even, we hope, have a great deal of aesthetic appeal.

1.2. DEFINITIONS AND OPERATIONS

As we mentioned in the introduction, we shall begin everything with a simple case—perhaps even an oversimplified case—of the 2×2 matrices. Even though special, this case offers us the opportunity to be concrete and to do everything with our hands from the ground up.

Let $\mathbb{R}$ be the set of all real numbers. By $M_2(\mathbb{R})$, the set of 2×2 *matrices over* $\mathbb{R}$, we shall mean the set of all square arrays $\begin{bmatrix} a & b \\ c & d \end{bmatrix}$, where a, b, c, d are real numbers. We define how to add and multiply them, and we then single out certain classes of matrices that are of special interest.

Although we shall soon enlarge our setup to matrices over the set $\mathbb{C}$ of complex numbers, we shall restrict our initial efforts to matrices just involving real numbers.

Before doing anything with these matrices, we have to have some criterion for declaring when two of them are equal. Although this definition of equality is most natural to make, it is up to us actually to make it, so that there is no possible ambiguity as to what we mean.

Definition. The two matrices $A = \begin{bmatrix} a & b \\ c & d \end{bmatrix}$ and $B = \begin{bmatrix} e & f \\ g & h \end{bmatrix}$ in $M_2(\mathbb{R})$ are *defined* to be *equal* if and only if $a = e$, $b = f$, $c = g$, and $d = h$.

For instance, in order that the matrix $A = \begin{bmatrix} 1 & 2 \\ -6 & \pi \end{bmatrix}$ be equal to the matrix $B = \begin{bmatrix} 1 & b \\ -6 & d \end{bmatrix}$, it is necessary and sufficient that $b = 2$ and $d = \pi$.

If $A = \begin{bmatrix} a & b \\ c & d \end{bmatrix}$, then a, b, c, and d are called the *entries* of A. In these terms, we declare two matrices to be equal if their *corresponding entries* are equal.

With this definition of equality behind us, we now proceed to introduce ways of combining these matrices. The first operations that we introduce in $M_2(\mathbb{R})$ are *addition* and *subtraction* of two matrices, denoted naturally enough by $+$ and $-$. Here, as in the definition of equality, the definitions strike one as reasonable and natural. We do the obvious in defining these operations as follows, to reflect the corresponding operations on their entries.

Definition. If $A = \begin{bmatrix} a & b \\ c & d \end{bmatrix}$ and $B = \begin{bmatrix} e & f \\ g & h \end{bmatrix}$ are in $M_2(\mathbb{R})$, then $A + B$ and $A - B$ are *defined* by

$$A + B = \begin{bmatrix} a+e & b+f \\ c+g & d+h \end{bmatrix}$$

$$A - B = \begin{bmatrix} a-e & b-f \\ c-g & d-h \end{bmatrix}.$$

Thus, to add or subtract two matrices, simply add or subtract, respectively, their corresponding entries. For instance, if A and B are the matrices

$$A = \begin{bmatrix} 1 & 7 \\ -6 & 3 \end{bmatrix}$$

$$B = \begin{bmatrix} \frac{1}{2} & \frac{2}{3} \\ 0 & 4 \end{bmatrix},$$

then $A + B$ and $A - B$ are the matrices

$$A + B = \begin{bmatrix} 1+\frac{1}{2} & 7+\frac{2}{3} \\ -6+0 & 3+4 \end{bmatrix} = \begin{bmatrix} \frac{3}{2} & \frac{23}{3} \\ -6 & 7 \end{bmatrix}$$

$$A - B = \begin{bmatrix} 1-\frac{1}{2} & 7-\frac{2}{3} \\ -6-0 & 3-4 \end{bmatrix} = \begin{bmatrix} \frac{1}{2} & \frac{19}{3} \\ -6 & -1 \end{bmatrix},$$

both of which are again in $M_2(\mathbb{R})$. In fact, from the definitions themselves, we see that for any A, B, in $M_2(\mathbb{R})$, both $A + B$ and $A - B$ are also in $M_2(\mathbb{R})$.

Note that the particular matrix $\begin{bmatrix} 0 & 0 \\ 0 & 0 \end{bmatrix}$, which we shall simply denote by 0, plays a very special role with respect to the addition operation. As is clear, $A + 0 = 0 + A = A$ for every matrix A in $M_2(\mathbb{R})$. Thus this special matrix 0 plays a role in

$M_2(\mathbb{R})$ very much like that played by the real number 0 in $\mathbb{R}$. It is called the *zero matrix*.

Note, too, that if $A = \begin{bmatrix} a & b \\ c & d \end{bmatrix}$ is in $M_2(\mathbb{R})$, then $B = \begin{bmatrix} -a & -b \\ -c & -d \end{bmatrix}$, which is also in $M_2(\mathbb{R})$, satisfies the equations $A + B = B + A = 0$. So B acts as the "negative" of A. We denote it, simply, by $-A$.

Given $u \in \mathbb{R}$ and $A = \begin{bmatrix} a & b \\ c & d \end{bmatrix}$ in $M_2(\mathbb{R})$, we define the *multiplication of A by the scalar u* as $uA = \begin{bmatrix} ua & ub \\ uc & ud \end{bmatrix}$. Observe that here, too, uA is again in $M_2(\mathbb{R})$.

Before considering the behavior of the addition and subtraction that we have introduced, we want still another operation for the elements of $M_2(\mathbb{R})$, namely the *multiplication* or *product* of two matrices. Unlike the earlier definitions, this multiplication will probably seem somewhat unnatural to those seeing it for the first time. As we proceed we shall see natural patterns emerge, as well as a rationale for choosing this as our desired product.

Definition. If the matrices $A = \begin{bmatrix} a & b \\ c & d \end{bmatrix}$ and $B = \begin{bmatrix} e & f \\ g & h \end{bmatrix}$ are in $M_2(\mathbb{R})$, then their *product AB* is *defined* by the equation

$$AB = \begin{bmatrix} ae + bg & af + bh \\ ce + dg & cf + dh \end{bmatrix}.$$

Let's look at a few examples of this product starting, for example, with:

$$\begin{bmatrix} 1 & 3 \\ -2 & 4 \end{bmatrix} \begin{bmatrix} 4 & 3 \\ 2 & 1 \end{bmatrix} = \begin{bmatrix} 1 \cdot 4 + 3 \cdot 2 & 1 \cdot 3 + 3 \cdot 1 \\ (-2)4 + 4 \cdot 2 & (-2) \cdot 3 + 4 \cdot 1 \end{bmatrix} = \begin{bmatrix} 10 & 6 \\ 0 & -2 \end{bmatrix}.$$

Perhaps more interesting are the products

$$\begin{bmatrix} 0 & 1 \\ 0 & 0 \end{bmatrix} \begin{bmatrix} 1 & 0 \\ 0 & 0 \end{bmatrix} = \begin{bmatrix} 0 & 0 \\ 0 & 0 \end{bmatrix} = 0, \quad \begin{bmatrix} 1 & 0 \\ 0 & 0 \end{bmatrix} \begin{bmatrix} 0 & 1 \\ 0 & 0 \end{bmatrix} = \begin{bmatrix} 0 & 1 \\ 0 & 0 \end{bmatrix} \neq 0.$$

Note several things about the product of matrices:

1. If A, B are in $M_2(\mathbb{R})$, then AB is also in $M_2(\mathbb{R})$.
2. In $M_2(\mathbb{R})$, it is possible that $AB = 0$ with $A \neq 0$ and $B \neq 0$.
3. In $M_2(\mathbb{R})$, it is possible that $AB \neq BA$.

These last two behaviors both run counter to our prior experience with number systems, where we know that

2'. In $\mathbb{R}$, $ab = 0$ if and only if $a = 0$ or $b = 0$.
3'. In $\mathbb{R}$, $ab = ba$ for all a and b.

Here (2′) is, in effect, the *cancellation law of multiplication* for real numbers:

If $ab = 0$ and $a \neq 0$, then $b = 0$.

Thus (2′) says that

The cancellation law of multiplication does not hold in $M_2(\mathbb{R})$.

At the same time, cancellation is possible in $M_2(\mathbb{R})$ under certain circumstances, as we observe in Problem 10. Similarly, (c′) says that real numbers *commute* under multiplication. Thus (3′) says that

Matrices in $M_2(\mathbb{R})$ do not necessarily commute under multiplication.

Matrices in $M_2(\mathbb{R})$ satisfy the *associative law* that $(AB)C = A(BC)$, as you can see by multiplying out the expressions on both sides of the equation. We leave this as an easy, though tedious exercise.

A matrix that plays a very special role in multiplication is the matrix $I = \begin{bmatrix} 1 & 0 \\ 0 & 1 \end{bmatrix}$.

This matrix is called the *identity matrix*, because it has the following properties:

$$AI = \begin{bmatrix} a & b \\ c & d \end{bmatrix}\begin{bmatrix} 1 & 0 \\ 0 & 1 \end{bmatrix} = \begin{bmatrix} a\cdot 1 + b\cdot 0 & a\cdot 0 + b\cdot 1 \\ c\cdot 1 + d\cdot 0 & c\cdot 0 + d\cdot 1 \end{bmatrix} = \begin{bmatrix} a & b \\ c & d \end{bmatrix} = A.$$

Similarly,

$$IA = \begin{bmatrix} 1 & 0 \\ 0 & 1 \end{bmatrix}\begin{bmatrix} a & b \\ c & d \end{bmatrix} = \begin{bmatrix} 1\cdot a + 0\cdot c & 1\cdot b + 0\cdot d \\ 0\cdot a + 1\cdot c & 0\cdot b + 1\cdot d \end{bmatrix} = \begin{bmatrix} a & b \\ c & d \end{bmatrix} = A.$$

Thus, multiplying any matrix A by I on either side does not change A at all. In other words, the matrix I in $M_2(\mathbb{R})$ behaves very much like the number 1 does in $\mathbb{R}$ when one multiplies.

For every nonzero real number a we can find a real number, written as $a^{-1} = 1/a$, such that $aa^{-1} = 1$. Is something similar true here in the system $M_2(\mathbb{R})$? The answer is "no." More specifically, we cannot find, for every nonzero matrix A, a matrix A^{-1} such that $AA^{-1} = I$. Consider, for instance, the matrix $A = \begin{bmatrix} 1 & 0 \\ 0 & 0 \end{bmatrix}$. Can we find a matrix $B = \begin{bmatrix} e & f \\ g & h \end{bmatrix}$ such that $AB = I$? Let's see what is needed. What we require for $AB = I$ is that

$$\begin{bmatrix} 1 & 0 \\ 0 & 1 \end{bmatrix} = AB = \begin{bmatrix} 1 & 0 \\ 0 & 0 \end{bmatrix}\begin{bmatrix} e & f \\ g & h \end{bmatrix} = \begin{bmatrix} e & f \\ 0 & 0 \end{bmatrix}.$$

This would require that $e = 1$, $f = 0$ and the absurdity that $1 = 0$. So no such B exists for this particular A. Let's try another one, the matrix $A = \begin{bmatrix} 1 & 1 \\ 0 & 1 \end{bmatrix}$, where the result is quite different. Again we ask whether we can find a matrix $B = \begin{bmatrix} e & f \\ g & h \end{bmatrix}$ such that $AB = I$. Again, let's see what is needed. What is required in this case is that

$$\begin{bmatrix} 1 & 0 \\ 0 & 1 \end{bmatrix} = AB = \begin{bmatrix} 1 & 1 \\ 0 & 1 \end{bmatrix}\begin{bmatrix} e & f \\ g & h \end{bmatrix} = \begin{bmatrix} e+g & f+h \\ g & h \end{bmatrix}.$$

This requires that $g = 0$, $h = 1$, $e = 1$, $f = -1$, so that the matrix $B = \begin{bmatrix} 1 & -1 \\ 0 & 1 \end{bmatrix}$ does satisfy $AB = I$. Moreover, this matrix B also satisfies the equation $BA = I$, which you can easily verify.

We have seen that for some matrices A we can find a matrix B such that $AB = BA = I$, and that for some A no such B can be found. We single out these "good" ones in our growing terminology.

Definition. A matrix A is said to be *invertible* if we can find a matrix B such that $AB = BA = I$.

A matrix A which is not invertible is called *singular*. If A is invertible, we claim that the B above is *unique*. What exactly does this mean? It means merely that if $AC = CA = I$ for a (possibly) different matrix C, then $B = C$. To see that $AB = BA = I$ and $AC = CA = I$ imply that $B = C$, just equate $AB = AC$ and cancel A by multiplying each side on the left by B. We leave the details as an exercise. You will have to use the associative law for this (see Problem 3).

If A is invertible and $AB = BA = I$ as above, we call B the *inverse* of A and in analogy to what we do in the real numbers, we write B as A^{-1}. We stress again that *not all* matrices are invertible. In a short while we shall see how the entries of A determine whether or not A is invertible.

We now come to two particular, easy-looking classes of matrices.

Definition. The matrix $A = \begin{bmatrix} a & 0 \\ 0 & d \end{bmatrix}$ is called a *diagonal matrix*. The entries a and d are called its *diagonal entries*.

Definition. The matrix $A = \begin{bmatrix} a & 0 \\ 0 & a \end{bmatrix}$ is called a *scalar matrix*.

So a matrix is diagonal if its off-diagonal entries are 0. And it is a scalar matrix if, in addition, the diagonal entries are equal.

Let $A = \begin{bmatrix} a & 0 \\ 0 & a \end{bmatrix}$ be a scalar matrix. Then, if we multiply any matrix B on either side by A, we get as a result a new matrix, each of whose entries are those of B merely multiplied by the number a. We *shall write* the scalar matrix $A = \begin{bmatrix} a & 0 \\ 0 & a \end{bmatrix}$ as

$A = aI$. So, for $B = \begin{bmatrix} e & f \\ g & h \end{bmatrix}$:

$$(aI)B = \begin{bmatrix} a & 0 \\ 0 & a \end{bmatrix}\begin{bmatrix} e & f \\ g & h \end{bmatrix} = \begin{bmatrix} ae & af \\ ag & ah \end{bmatrix} = \begin{bmatrix} e & f \\ g & h \end{bmatrix}\begin{bmatrix} a & 0 \\ 0 & a \end{bmatrix} = B(aI).$$

Looking back at the definition of multiplying a matrix by a scalar, we see that multiplying by the scalar matrix aI and multiplying by the scalar a gives us the same results. More precisely, for any matrix A, $(aI)A = aA$, for any $a \in \mathbb{R}$.

If B is also a scalar matrix bI, the multiplication becomes

$$(aI)(bI) = \begin{bmatrix} a & 0 \\ 0 & a \end{bmatrix}\begin{bmatrix} b & 0 \\ 0 & b \end{bmatrix} = \begin{bmatrix} ab & 0 \\ 0 & ab \end{bmatrix} = (ab)I.$$

Thus scalar matrices multiply like real numbers! As you can easily verify, scalar matrices also add like real numbers in the sense that $aI + bI = (a + b)I$. Thus the set $\mathbb{R}$ of real numbers a together with the operations of addition and multiplication of real numbers is duplicated, in a sense, by the set $\mathbb{R}I$ of scalar matrices aI and the operations of addition and multiplication for scalar matrices. The function $f(a) = aI$ from $\mathbb{R}$ to $\mathbb{R}I$ gives a *one-to-one correspondence* between $\mathbb{R}$ and this duplicate $\mathbb{R}I$.

The addition and multiplication of diagonal matrices is slightly more complicated than that of scalar matrices, because a scalar matrix depends on only one real number, whereas a diagonal matrix depends on two. Addition and multiplication for diagonal matrices goes as follows, as you can easily check:

$$\begin{bmatrix} a & 0 \\ 0 & b \end{bmatrix} + \begin{bmatrix} c & 0 \\ 0 & d \end{bmatrix} = \begin{bmatrix} a+c & 0 \\ 0 & b+d \end{bmatrix}$$

$$\begin{bmatrix} a & 0 \\ 0 & b \end{bmatrix}\begin{bmatrix} c & 0 \\ 0 & d \end{bmatrix} = \begin{bmatrix} ac & 0 \\ 0 & bd \end{bmatrix}.$$

We come to two other classes of special matrices.

Definition. The matrix $A = \begin{bmatrix} a & b \\ c & d \end{bmatrix}$ is called *upper triangular* if $c = 0$, and is called *lower triangular* if $b = 0$.

So, in an upper triangular matrix, the lower left entry is 0, and in a lower triangular matrix the upper right entry is 0.

PROBLEMS

In the following problems, capital Latin letters denote matrices in $M_2(\mathbb{R})$.

NUMERICAL PROBLEMS

1. Prove that $0A = A0 = 0$ for every $A \in M_2(\mathbb{R})$.

2. Evaluate the following matrices.

(a) $\begin{bmatrix} 1 & 6 \\ 7 & 2 \end{bmatrix} \begin{bmatrix} -4 & 1 \\ 3 & 5 \end{bmatrix}.$

(b) $\begin{bmatrix} -1 & 4 \\ 0 & 5 \end{bmatrix} \begin{bmatrix} 6 & 2 \\ 0 & 0 \end{bmatrix} + \begin{bmatrix} -1 & 5 \\ 1 & 1 \end{bmatrix}.$

(c) $\left(\begin{bmatrix} 1 & 2 \\ 3 & 4 \end{bmatrix} \begin{bmatrix} 5 & 6 \\ 7 & 8 \end{bmatrix} \right) \begin{bmatrix} -1 & -2 \\ 3 & 4 \end{bmatrix}$ and $\begin{bmatrix} 1 & 2 \\ 3 & 4 \end{bmatrix} \left(\begin{bmatrix} 5 & 6 \\ 7 & 8 \end{bmatrix} \begin{bmatrix} -1 & -2 \\ 3 & 4 \end{bmatrix} \right).$

(d) How do the two resulting matrices compare?

(e) $\begin{bmatrix} \frac{1}{2} & \frac{2}{3} \\ \frac{1}{4} & -6 \end{bmatrix} \begin{bmatrix} 5 & 0 \\ 0 & -5 \end{bmatrix} - \begin{bmatrix} 5 & 0 \\ 0 & -5 \end{bmatrix} \begin{bmatrix} \frac{1}{2} & \frac{2}{3} \\ \frac{1}{4} & -6 \end{bmatrix}.$

(f) $\begin{bmatrix} \frac{1}{2} & \frac{1}{2} \\ \frac{1}{2} & \frac{1}{2} \end{bmatrix} \begin{bmatrix} \frac{1}{2} & \frac{1}{2} \\ \frac{1}{2} & \frac{1}{2} \end{bmatrix}.$

(g) $\begin{bmatrix} 1 & -1 \\ \pi & -\pi \end{bmatrix} \begin{bmatrix} 1 & -1 \\ \pi & -\pi \end{bmatrix}.$

(h) $\begin{bmatrix} 1 & 2 \\ 3 & 4 \end{bmatrix} \begin{bmatrix} -2 & 1 \\ \frac{3}{2} & -\frac{1}{2} \end{bmatrix}.$

(i) $\begin{bmatrix} -2 & 1 \\ \frac{3}{2} & -\frac{1}{2} \end{bmatrix} \begin{bmatrix} 1 & 2 \\ 3 & 4 \end{bmatrix}.$

(j) $\begin{bmatrix} a & b \\ c & d \end{bmatrix} \begin{bmatrix} d & -b \\ -c & a \end{bmatrix}.$

MORE THEORETICAL PROBLEMS

Easier Problems

3. If A, B, C are in $M_2(\mathbb{R})$, show that $(AB)C = A(BC)$. [This is known as the *associative law* of multiplication. In view of this result, we shall write $(AB)C$ without parentheses, as ABC.]

Make free use of the associative law in the problems that follow.

4. If A, B, C are in $M_2(\mathbb{R})$, show that $A(B + C) = AB + AC$, and that $(B + C)A = BA + CA$. (These are the *distributive laws*.)

5. If A is a diagonal matrix whose diagonal entries are both nonzero, prove that A^{-1} exists and find it explicitly.

6. If A and B are upper triangular matrices, prove that AB is also upper triangular.

7. When is an upper triangular matrix invertible?

8. If A is invertible, prove that its inverse is unique.

9. If A and B are invertible, show that AB is invertible and express $(AB)^{-1}$ in terms of A^{-1} and B^{-1}.

10. If A is invertible and $AC = 0$, prove that $C = 0$; also, show that if $DA = 0$, then $D = 0$. Then prove the more general *cancellation laws*, which state that $AB = AC$ implies $B = C$ and $BA = CA$ implies that $B = C$ if A is invertible.

Middle-Level Problems

11. If $A, B \in M_2(\mathbb{R})$, show that $(AB - BA)(AB - BA)$ is a scalar matrix.
12. If A is such that $AB = 0$ for some $B \neq 0$, show that A cannot be invertible.
13. Find two matrices A and B such that $AB = 0$ but $BA \neq 0$. What can you say about $(BA)(BA)$?
14. Prove that we cannot find two matrices A and B such that $AB - BA = aI$ where $a \neq 0$ is in $\mathbb{R}$.
15. Find all matrices A that commute with *all* matrices. [That is, find all A such that $AB = BA$ for all $B \in M_2(\mathbb{R})$.]
16. What matrices commute with $A = \begin{bmatrix} a & 0 \\ 0 & b \end{bmatrix}$ if $a \neq b$?

Harder Problems

17. Find a necessary and sufficient condition on a, b, c, d in order that $A = \begin{bmatrix} a & b \\ c & d \end{bmatrix}$ be invertible.
18. If A is not invertible, show that we can find a $B \neq 0$ such that $AB = 0$.
19. If $AB = I$, prove that $BA = I$, hence A is invertible.
20. Find all $a \in \mathbb{R}$ such that $\begin{bmatrix} -1 & 2 \\ -2 & 0 \end{bmatrix} - aI$ is *not* invertible. How many such values a are there? What polynomial equation do these values a satisfy?
21. Do Problem 20 for the general $A = \begin{bmatrix} a & b \\ c & d \end{bmatrix}$, that is, find all $u \in \mathbb{R}$ such that $A - uI$ is not invertible.
22. If $A = \begin{bmatrix} a & b \\ c & d \end{bmatrix}$ satisfies $AA = 0$, show that $a + d = 0$ and $ad - bc = 0$.
23. If $A = \begin{bmatrix} a & b \\ c & d \end{bmatrix}$ satisfies $AA = A$, show that $a + d = 0, 1,$ or 2.
24. For $A = \begin{bmatrix} 1 & 7 \\ 0 & -1 \end{bmatrix}$, show that we can find a matrix B such that B is invertible and $B^{-1}AB = \begin{bmatrix} u & 0 \\ 0 & v \end{bmatrix}$. What must the values of u and v be?
25. Show that we cannot find an invertible matrix B such that $B^{-1} \begin{bmatrix} 1 & 0 \\ 1 & 1 \end{bmatrix} B$ is a diagonal matrix.
26. If A is a matrix such that $B^{-1}AB$ is a scalar matrix for some invertible B in $M_2(\mathbb{R})$, prove that A must be a scalar matrix and that $AB = BA$.
27. Define the *transpose* of $A = \begin{bmatrix} a & b \\ c & d \end{bmatrix}$ as $A' = \begin{bmatrix} a & c \\ b & d \end{bmatrix}$ and show that the

transpose operation satisfies the properties $(AB)' = B'A'$, $(aA)' = aA'$, and $(A + B)' = A' + B'$.

28. Find the formulas for the sum and product of upper triangular matrices. Then use Problem 27 to transfer these to formulas for the sum and product of lower triangular matrices.

1.3. SOME NOTATION

This section is intended to introduce some notational devices. In reading it, the reader may very well ask: "Why go to all this trouble for the 2×2 case where everything is so explicit and can be written in full?" Indeed, if our concerns were only the 2×2 case, it certainly would be absurd to fuss about notation. However, what we do is precisely what will be needed for the general $n \times n$ case. The 2×2 matrices afford us a good arena for acquiring some skill and familiarity with this symbolism.

Our starting point is the symbol $\sum$ (the Greek letter for S, standing for *summation*) which some readers may have encountered before in the calculus. By $\sum_{r=1}^{n} a_r$, n being an integer greater or equal to 1, we shall mean the sum $\sum_{r=1}^{n} a_r = a_1 + \cdots + a_n$ of terms a_r, where r varies over the set of integers from 1 to n. Note that the subscript r over which we are summing is a "dummy" index. We could easily call it k, or α, or indeed anything else. Thus

$$\sum_{r=1}^{n} a_r = \sum_{\alpha=1}^{n} a_\alpha = \sum_{\square=1}^{n} a_\square.$$

As an example of this $\sum$ notation, $\sum_{r=1}^{5} r^2 = 1^2 + 2^2 + 3^2 + 4^2 + 5^2$. We may occasionally need a slight variant of this notation, namely the variant $\sum_{r=m}^{n} a_r$, m and n being integers with $m \leq n$, which denotes $a_m + a_{m+1} + \cdots + a_n$. So, for example,

$$\sum_{r=-2}^{3} r^2 = (-2)^2 + (-1)^2 + 0^2 + 1^2 + 2^2 + 3^2 = \sum_{\alpha=-2}^{3} \alpha^2.$$

Another, and more complicated summation symbol which will be used, although less frequently, is the *double summation* $\sum\sum$. We express a double summation in terms of single summations:

$$\sum_{r=1}^{m} \sum_{s=1}^{n} a_{rs} = \sum_{r=1}^{m} b_r, \quad \text{where} \quad b_r = \sum_{s=1}^{n} a_{rs}.$$

An example of this is

$$\sum_{r=1}^{4} \sum_{s=1}^{3} rs = \sum_{\alpha=1}^{4} \sum_{\beta=1}^{3} \alpha\beta = \sum_{\alpha=1}^{4} b_\alpha,$$

where $b_\alpha = \sum_{\beta=1}^{3} \alpha\beta = \alpha \cdot 1 + \alpha \cdot 2 + \alpha \cdot 3 = 6\alpha$ for $\alpha = 1, 2, 3,$ and 4. For instance, $b_3 = 3 \cdot 1 + 3 \cdot 2 + 3 \cdot 3 = 18$. The reader can easily verify that the values for b_1, b_2, b_3, b_4 add up to $6(1 + 2 + 3 + 4) = 60$, so that the value of the double sum $\sum_{r=1}^{4}\sum_{s=1}^{3} rs$ is 60.

We now return to the 2×2 matrices, where we want to introduce some names and a compact means of writing matrices. Consider the matrix $A = \begin{bmatrix} a & b \\ c & d \end{bmatrix}$. Since the entry a occurs in the first row and first column of A, we call a the $(1, 1)$ *entry* of A, and since b occurs in the first row and second column of A, we call it the $(1, 2)$ *entry* of A. Similarly, c is called the $(2, 1)$ and d the $(2, 2)$ entry of A. If we denote the (r, s) *entry* of A by a_{rs}—which means that the entry is in the rth row and sth column of A—we shall write A as $[a_{rs}]$. To repeat, $A = [a_{rs}]$ denotes the array

$$A = \begin{bmatrix} a_{11} & a_{12} \\ a_{21} & a_{22} \end{bmatrix}.$$

If $a_{rs} = r + s$, then the matrix $A = [a_{rs}]$ is the matrix $\begin{bmatrix} 2 & 3 \\ 3 & 4 \end{bmatrix}$ since the terms $r + s$ are arranged in the matrix A as follows:

$$A = \begin{bmatrix} 1+1 & 1+2 \\ 2+1 & 2+2 \end{bmatrix}.$$

There is a symmetry in A, in this case across the *main diagonal* of A running from its upper left corner to its lower right corner. This symmetry is due to the fact that $a_{rs} = a_{sr}$ for all values of r and s. As another example where the matrix A does not have such a condition on its entries, suppose that $a_{rs} = r + rs$ for r and s equal to 1 and 2. Then the matrix $A = [a_{rs}]$ is the matrix $\begin{bmatrix} 2 & 3 \\ 4 & 6 \end{bmatrix}$ since the terms $r + rs$ are arranged in the matrix A as follows:

$$A = \begin{bmatrix} 1+1 \cdot 1 & 1+1 \cdot 2 \\ 2+2 \cdot 1 & 2+2 \cdot 2 \end{bmatrix}.$$

There is no such symmetry in this case, due to the fact that $a_{12} = 3$, whereas $a_{21} = 4$.

The entries of a matrix can be any real numbers. *There need not be any formula expressing a_{rs} in terms of r and s.*

What do our operations in $M_2(\mathbb{R})$ look like using these notational devices? If $A = [a_{rs}]$ and $B = [b_{rs}]$, then the sum $A + B$ is $[c_{rs}]$, where $c_{rs} = a_{rs} + b_{rs}$ for each r and s. What about the product of the two matrices above? By our rule of multiplication, $AB = [a_{rs}][b_{rs}]$ is

$$\begin{bmatrix} a_{11} & a_{12} \\ a_{21} & a_{22} \end{bmatrix}\begin{bmatrix} b_{11} & b_{12} \\ b_{21} & b_{22} \end{bmatrix} = \begin{bmatrix} a_{11}b_{11} + a_{12}b_{21} & a_{11}b_{12} + a_{12}b_{22} \\ a_{21}b_{11} + a_{22}b_{21} & a_{21}b_{12} + a_{22}b_{22} \end{bmatrix} = \begin{bmatrix} c_{11} & c_{12} \\ c_{21} & c_{22} \end{bmatrix},$$

where $c_{rs} = a_{r1}b_{1s} + a_{r2}b_{2s} = \sum_{t=1}^{2} a_{rt}b_{ts}$. Thus the entry in the second row and first column is $c_{21} = \sum_{t=1}^{2} a_{2t}b_{t1} = a_{21}b_{11} + a_{22}b_{21}$.

In this new symbolism, the matrix $A = \begin{bmatrix} a_{11} & a_{12} \\ a_{21} & a_{22} \end{bmatrix}$ is a diagonal matrix if $a_{rs} = 0$ for $r \neq s$. And $\begin{bmatrix} a_{11} & a_{12} \\ a_{21} & a_{22} \end{bmatrix}$ is a scalar matrix if it is diagonal and $a_{rr} = a_{ss}$ for all r and s, $\begin{bmatrix} a_{11} & a_{12} \\ a_{21} & a_{22} \end{bmatrix}$ is upper triangular if $a_{rs} = 0$ for $r > s$, and $\begin{bmatrix} a_{11} & a_{12} \\ a_{21} & a_{22} \end{bmatrix}$ is lower triangular if $a_{rs} = 0$ for $r < s$. For instance, $\begin{bmatrix} 1 & 0 \\ 0 & 2 \end{bmatrix}$ is diagonal since $a_{12} = a_{21} = 0$, $\begin{bmatrix} 2 & 0 \\ 0 & 2 \end{bmatrix}$ is a scalar matrix since it is diagonal and $a_{11} = a_{22}$, $\begin{bmatrix} 1 & 3 \\ 0 & 2 \end{bmatrix}$ is upper triangular since $a_{21} = 0$ and $\begin{bmatrix} 1 & 0 \\ 2 & 2 \end{bmatrix}$ is lower triangular since $a_{12} = 0$.

The last piece of notation we want to introduce here is certainly more natural to most of us than the formalisms above. This is the notation of *exponent*. We want to define A^m for any 2×2 matrix A and any *positive* integer m. We do so by following the definition familiar for real numbers:

$$A^m = \underbrace{AAAAA\ldots A}_{(m - \text{times})}$$

More formally, the two equations

$$A^1 = A, \quad A^{m+1} = A^m A$$

define A^m recursively by defining it first for $m = 1$ and then for $m + 1$ after it has been defined for m, for all $m \geq 1$. We also define $A^0 = I$, since we'll need to use expressions such as $5A^0 + 6A^3 + 4A^4$.

We cannot define A^{-m} for *every* matrix A and positive integer m *unless* A is invertible. Why? Because A^{-1}, which would denote the inverse of A, makes no sense if A has no inverse, that is, if A is not invertible. However, if A is invertible, we *define* A^{-m} to be $(A^{-1})^m$ for every positive integer m.

The usual rules of exponents, namely $A^m A^n = A^{m+n}$ and $(A^m)^n = A^{mn}$, do hold for all matrices if m and n are *positive* integers, and for all invertible matrices if m and n are integers.

PROBLEMS

Make free use of the associative law $A(BC) = (AB)C$.

NUMERICAL PROBLEMS

1. Evaluate.

   (a) $\sum_{r=1}^{6} \frac{r}{r+1}$.

(b) $\sum_{t=1}^{4} \dfrac{t+1}{t+2}$.

(c) $\sum_{f=-1}^{3} \dfrac{f}{f+5}$.

(d) $\sum_{g=-1}^{3} g^2$, $\sum_{h=-1}^{3} h^3$, and $\sum_{k=-2}^{3} k^3$.

(e) $\sum_{r=1}^{5} \sum_{s=0}^{3} \dfrac{r}{s+1}$ and $\sum_{s=0}^{3} \sum_{r=1}^{5} \dfrac{r}{s+1}$.

(f) $\sum_{b=-2}^{4} \sum_{c=-1}^{2} b(c+1)$.

2. Show that $\sum_{r=1}^{n} \dfrac{1}{r(r+1)} = 1 - \dfrac{1}{n+1}$.

3. Calculate the following matrices.

(a) $\begin{bmatrix} 0 & 1 \\ -4 & 0 \end{bmatrix}^2$.

(b) $\begin{bmatrix} 1 & 2 \\ 2 & 4 \end{bmatrix}^3$.

(c) $\begin{bmatrix} 1 & 1 \\ 0 & 1 \end{bmatrix}^{-1}$ (if it exists).

(d) $\begin{bmatrix} 1 & 1 \\ 1 & 1 \end{bmatrix}^{-1}$ (if it exists).

(e) $\begin{bmatrix} 1 & 0 \\ 1 & 1 \end{bmatrix}^{-3}$ (if it exists).

(f) $\begin{bmatrix} a & -a \\ b & -b \end{bmatrix}^2$.

4. If $A = \begin{bmatrix} 1 & 2 \\ 3 & 4 \end{bmatrix}$ and $B = \begin{bmatrix} 1 & -1 \\ 1 & -1 \end{bmatrix}$, calculate $(AB)^2$ and $A^2 B^2$. Are they equal?

5. If $A = \begin{bmatrix} 1 & 7 \\ -6 & -43 \end{bmatrix}$, show that $A^2 + 42A - I = 0$.

6. Find A^n for *all* positive integers n for the following matrices.

(a) $\begin{bmatrix} 0 & \alpha \\ \beta & 0 \end{bmatrix}$.

(b) $\begin{bmatrix} 1 & 1 \\ 0 & 1 \end{bmatrix}$.

(c) $\begin{bmatrix} 1 & 1 \\ 1 & 1 \end{bmatrix}$.

(d) $\begin{bmatrix} 1 & -1 \\ 1 & -1 \end{bmatrix}$.

MORE THEORETICAL PROBLEMS

Easier Problems

7. Find A^n for all positive integers n for

    (a) $A = \begin{bmatrix} 1 & 2 \\ 2 & 4 \end{bmatrix}$.

    (b) $A = \begin{bmatrix} 0 & 1 \\ 1 & 1 \end{bmatrix}$.

8. If A and B are in $M_2(\mathbb{R})$:

    (a) Calculate $(AB - BA)^2$.

    (b) Find $(AB - BA)^n$ for all positive integers $n > 2$.

9. Let $A = \begin{bmatrix} 0 & 0 \\ 0 & 1 \end{bmatrix}$. Show that for all $B \in M_2(\mathbb{R})$, $(AB - ABA)^2 = (BA - ABA)^2 = 0$.

10. If A and B are in $M_2(\mathbb{R})$ and A is invertible, find $(ABA^{-1})^n$ for all positive integers n, expressed "nicely" in terms of A and B.

11. If A is invertible and $B, C \in M_2(\mathbb{R})$, show that $(ABA^{-1})(ACA^{-1}) = A(BC)A^{-1}$.

12. Prove the laws of exponents, namely, if A is in $M_2(\mathbb{R})$, then for all nonnegative integers m and n, $A^m A^n = A^{m+n}$ and $(A^m)^n = A^{mn}$.

Middle-Level Problems

13. Find matrices B and C such that $B^2 = \begin{bmatrix} 2 & 3 \\ 0 & 2 \end{bmatrix}$ and $C^2 = \begin{bmatrix} 1 & -1 \\ 2 & 4 \end{bmatrix}$.

14. Find a matrix $B \neq I$ such that $B^3 = I$.

15. If $A \in M_2(\mathbb{R})$ satisfies $A^2 - A + I = 0$, find A^{3n} explicitly in terms of A for every positive n.

16. If $A \in M_2(\mathbb{R})$ satisfies $A^5 - 4A^4 + 7A^2 - 14I = 0$, show that A is invertible, and express its inverse in terms of A.

17. If $A \in M_2(\mathbb{R})$ satisfies $A^2 + uA + vI = 0$, where $u, v \in \mathbb{R}$, find necessary and sufficient conditions on u and v that A be invertible.

Harder Problems

18. Let A be invertible in $M_2(\mathbb{R})$ and let $B = \begin{bmatrix} u & v \\ w & x \end{bmatrix}$. Show that if $C = ABA^{-1} = \begin{bmatrix} a & b \\ c & d \end{bmatrix}$, then $a + d = u + x$.

19. Let A and B be as in Problem 18. If $C = ABA^{-1} = \begin{bmatrix} a & b \\ c & d \end{bmatrix}$, show that $ad - bc = ux - vw$.

20. If $A \in M_2(\mathbb{R})$ is invertible, show that $A^{-1} = aA + bI$ for some real numbers a and b.

Sec. 1.4] Trace, Transpose, and Odds and Ends

21. If A is the matrix

$$\begin{bmatrix} \cos\frac{2\pi}{k} & \sin\frac{2\pi}{k} \\ -\sin\frac{2\pi}{k} & \cos\frac{2\pi}{k} \end{bmatrix},$$

where k is a positive integer, find A^m for all m. What is the matrix A^m in the special case where $m = k$?

1.4. TRACE, TRANSPOSE, AND ODDS AND ENDS

We start off with odds and ends. Heretofore, we have not singled out any result and called it a lemma or theorem. We become more formal now and designate some results as lemmas (lemmata) or theorems. This not only provides us with a handy way of referring back to a result when we need it, but it emphasizes the result and gives us a parallel path to follow when we treat matrices of any size.

Although it may complicate what is, at present, a simple and straightforward situation, we do everything using the summation notation, $\sum$, throughout. In this way, we can get some practice and facility in playing around with this important symbol. The reader might find it useful to rewrite each proof in this section without using summation notation, to see how it looks written out.

Theorem 1.4.1. If A, B, and C are in $M_2(\mathbb{R})$, then:

1. $A + B = B + A$ *Commutative Law;*
2. $(A + B) + C = A + (B + C)$ *Associative Law for Sums;*
3. $(AB)C = A(BC)$ *Associative Law for Products;*
4. $A(B + C) = AB + AC$ and
   $(B + C)A = BA + CA$ *Distributive Laws.*

Proof: The proof of Part (1) amounts to letting $A = [a_{rs}]$, $B = [b_{rs}]$ and observing that the (r, s) entry $a_{rs} + b_{rs}$ of $A + B$ equals the (r, s) entry $b_{rs} + a_{rs}$ of $B + A$ for all r, s. The proofs of Parts (2) and (4) are just as easy and are left to the reader.

We carry out the proof of Part (3) in detail. If $A = [a_{rs}]$, $B = [b_{rs}]$, and $C = [c_{rs}]$, then if $AB = [d_{rs}]$, we know that $d_{rs} = \sum_{t=1}^{2} a_{rt}b_{ts}$. Thus $(AB)C = [d_{rs}][c_{rs}] = [f_{rs}]$, where

$$f_{rs} = \sum_{u=1}^{2} d_{ru}c_{us} = \sum_{u=1}^{2}\left[\sum_{t=1}^{2} a_{rt}b_{tu}\right]c_{us} = \sum_{u=1}^{2}\sum_{t=1}^{2}(a_{rt}b_{tu})c_{us}.$$

On the other hand, $BC = [g_{rs}]$, where $g_{rs} = \sum_{u=1}^{2} b_{ru}c_{us}$. Thus $A(BC) = [a_{rs}][g_{rs}] = [h_{rs}]$, where

$$h_{rs} = \sum_{t=1}^{2} a_{rt}g_{ts} = \sum_{t=1}^{2} a_{rt}\left[\sum_{u=1}^{2} b_{tu}c_{us}\right] = \sum_{t=1}^{2}\sum_{u=1}^{2} a_{rt}(b_{tu}c_{us}).$$

Since $f_{rs} = h_{rs}$, we see that $(AB)C = [f_{rs}] = [h_{rs}] = A(BC)$. ∎

Now that you have seen in detail why the associative law $(AB)C = A(BC)$ is true, you can see that it just amounts to showing that the (r, s) entries of the matrices $(AB)C$, $A(BC)$ are the double sums

$$\sum_{u=1}^{2} \sum_{t=1}^{2} (a_{rt}b_{tu})c_{us}, \quad \sum_{t=1}^{2} \sum_{u=1}^{2} a_{rt}(b_{tu}c_{us}).$$

These double sums are equal because the order of summation does not affect the sum and *because the terms $(a_{rt}b_{tu})c_{us}$, $a_{rt}(b_{tu}c_{us})$ are equal by the associative law for numbers!*

In view of the fact that $(AB)C = A(BC)$, we can leave off the parentheses in products of three matrices—for either bracketing gives the same result—and write either product as ABC, for A, B, C in $M_2(\mathbb{R})$. In fact, we can leave off the parentheses in products of any number of matrices.

Associative operations abound in mathematics and, of course, we can *leave off the parentheses for such other associative operations as well*. We'll do this from now on without fanfare!

We know that we can add and multiply matrices. We now introduce another function on matrices—a function from $M_2(\mathbb{R})$ to $\mathbb{R}$—which plays an important role in matrix theory.

Definition. If $A = [a_{rs}]$, then the *trace* of A, denoted by $\operatorname{tr}(A)$, is

$$\operatorname{tr}(A) = \sum_{r=1}^{2} a_{rr} = a_{11} + a_{22}.$$

In short, the trace of A is simply the sum of the elements on its diagonal. What properties does this function, trace, enjoy?

Theorem 1.4.2. If $A, B \in M_2(\mathbb{R})$ and $a \in R$, then

1. $\operatorname{tr}(A + B) = \operatorname{tr}(A) + \operatorname{tr}(B)$;
2. $\operatorname{tr}(aA) = a(\operatorname{tr}(A))$;
3. $\operatorname{tr}(AB) = \operatorname{tr}(BA)$;
4. $\operatorname{tr}(aI) = 2a$.

Proof: The proofs of Parts (1), (2), and (4) are easy and direct. We leave this to you. For the sake of practice, try to show them using the summation symbol.

We prove Part (3), not because it is difficult, but to get a further acquaintance with the summation notation.

If $A = [a_{rs}]$ and $B = [b_{rs}]$, then $AB = [c_{rs}]$, where $c_{rs} = \sum_{t=1}^{2} a_{rt}b_{ts}$. Thus

$$\operatorname{tr}(AB) = \sum_{r=1}^{2} c_{rr} = \sum_{r=1}^{2} \sum_{t=1}^{2} a_{rt}b_{tr}.$$

If $BA = [d_{rs}]$, then $d_{rs} = \sum_{u=1}^{2} b_{ru}a_{us}$, whence

$$\text{tr}(BA) = \sum_{r=1}^{2} d_{rr} = \sum_{r=1}^{2}\sum_{u=1}^{2} b_{ru}a_{ur}.$$

Since the u and r in the latter sum are dummies, we may call them what we will. Replacing u by r, and r by t, we get

$$\text{tr}(BA) = \sum_{t=1}^{2}\sum_{r=1}^{2} b_{tr}a_{rt}.$$

Since our evaluation for AB was

$$\text{tr}(AB) = \sum_{r=1}^{2}\sum_{t=1}^{2} a_{rt}b_{tr},$$

we have $\text{tr}(AB) = \text{tr}(BA)$. ∎

In Problem 18 of Section 1.3 you were asked to show that if A is invertible and $B = \begin{bmatrix} u & v \\ w & x \end{bmatrix}$, then if $C = A^{-1}BA = \begin{bmatrix} a & b \\ c & d \end{bmatrix}$, we must have $u + x = a + d$. In terms of traces, this equality,

$$u + x = a + d,$$

becomes the equality

$$\text{tr}(B) = \text{tr}(A^{-1}BA).$$

We now formally state and prove this important result.

Corollary 1.4.3. If A is invertible, then $\text{tr}(B) = \text{tr}(A^{-1}BA)$.

Proof: By the theorem, $\text{tr}(A^{-1}(BA)) = \text{tr}(BA)A^{-1}) = \text{tr}(B)$. ∎

Although we know that this corollary holds in general, it is nice to see that it is true for particular matrices. If $A = \begin{bmatrix} 1 & -2 \\ 0 & 1 \end{bmatrix}$ and $B = \begin{bmatrix} -1 & 1 \\ 3 & 3 \end{bmatrix}$, then A is invertible with inverse $A^{-1} = \begin{bmatrix} 1 & 2 \\ 0 & 1 \end{bmatrix}$. Thus

$$A^{-1}BA = \begin{bmatrix} 1 & 2 \\ 0 & 1 \end{bmatrix}\begin{bmatrix} -1 & 1 \\ 3 & 3 \end{bmatrix}\begin{bmatrix} 1 & -2 \\ 0 & 1 \end{bmatrix} = \begin{bmatrix} 5 & 7 \\ 3 & 3 \end{bmatrix}\begin{bmatrix} 1 & -2 \\ 0 & 1 \end{bmatrix} = \begin{bmatrix} 5 & -3 \\ 3 & -3 \end{bmatrix},$$

hence $\text{tr}(A^{-1}BA) = 5 + (-3) = 2 = (-1) + 3 = \text{tr}(B)$.

The trace is a function (or mapping) from $M_2(\mathbb{R})$ into the reals $\mathbb{R}$. We now introduce a mapping from $M_2(\mathbb{R})$ into $M_2(\mathbb{R})$ itself which exhibits a kind of internal reverse symmetry of the system $M_2(\mathbb{R})$.

Definition. If $a = [a_{rs}] \in M_2(\mathbb{R})$, then the *transpose* of A, denoted by A', is the matrix $A' = [b_{rs}]$, where $b_{rs} = a_{sr}$ for each r and s.

The transpose of a matrix A is, from the definition, that matrix which is obtained from A by interchanging the rows and columns of A. So if A is the matrix

$$\begin{bmatrix} -3 & 2 \\ 4 & 7 \end{bmatrix},$$

then its transpose A' is the matrix

$$\begin{bmatrix} -3 & 4 \\ 2 & 7 \end{bmatrix}.$$

What are the basic properties of transpose? You can easily guess some of them by looking at examples. For instance, the transpose A'' of the transpose A' of A is A itself. Because of their importance, we now list several of these properties formally.

Theorem 1.4.4. If A and B are in $M_2(\mathbb{R})$ and a and b are in $\mathbb{R}$, then

1. $(A')' = A'' = A$;
2. $(aA + bB)' = aA' + bB'$;
3. $(AB)' = B'A'$.

Proof: Both Parts (1) and (2) are very easy and are left to the reader. We do Part (3).

Let $A = [a_{rs}]$ and $B = [b_{rs}]$, so that $AB = [c_{rs}]$, where

$$c_{rs} = \sum_{t=1}^{2} a_{rt} b_{ts}.$$

Then $(AB)' = [c_{rs}]' = [d_{rs}]$, where $d_{rs} = c_{sr}$ for each r and s.

On the other hand, if $A' = [u_{rs}]$ and $B' = [v_{rs}]$, then $u_{rs} = a_{sr}$, $v_{rs} = b_{sr}$, and $B'A' = [v_{rs}][u_{rs}] = [w_{rs}]$, where

$$w_{rs} = \sum_{t=1}^{2} v_{rt} u_{ts} = \sum_{t=1}^{2} b_{tr} a_{st} = \sum_{t=1}^{2} a_{st} b_{tr} = c_{sr} = d_{rs}.$$

Thus $(AB)' = B'A'$. ∎

It is important to note in the formula $(AB)' = B'A'$ that the order of A and B are *reversed*. Thus the transpose mapping from $M_2(\mathbb{R})$ to itself gives a kind of reverse symmetry of the multiplicative structure of $M_2(\mathbb{R})$. At the same time, $(A + B)' = A' + B'$, so that it is also a kind of symmetry of the additive structure of $M_2(\mathbb{R})$.

Again, let's verify our result for specific matrices A and B. If $A = \begin{bmatrix} 1 & 2 \\ -3 & 4 \end{bmatrix}$ and $B = \begin{bmatrix} 0 & 1 \\ 2 & -1 \end{bmatrix}$, then

$$AB = \begin{bmatrix} 1 & 2 \\ -3 & 4 \end{bmatrix}\begin{bmatrix} 0 & 1 \\ 2 & -1 \end{bmatrix} = \begin{bmatrix} 4 & -1 \\ 8 & -7 \end{bmatrix},$$

so that $(AB)' = \begin{bmatrix} 4 & 8 \\ -1 & -7 \end{bmatrix}$. On the other hand, $A' = \begin{bmatrix} 1 & -3 \\ 2 & 4 \end{bmatrix}$ and $B' = \begin{bmatrix} 0 & 2 \\ 1 & -1 \end{bmatrix}$, whence

$$B'A' = \begin{bmatrix} 0 & 2 \\ 1 & -1 \end{bmatrix}\begin{bmatrix} 1 & -3 \\ 2 & 4 \end{bmatrix} = \begin{bmatrix} 4 & 8 \\ -1 & -7 \end{bmatrix} = (AB)'.$$

We call a matrix A *symmetric* if $A' = A$, and a matrix B *skew-symmetric* or *skew* if $B' = -B$. Thus $\begin{bmatrix} 1 & \pi \\ \pi & 0 \end{bmatrix}$ is symmetric, while $\begin{bmatrix} 0 & \pi \\ -\pi & 0 \end{bmatrix}$ is skew.

PROBLEMS

Use the summation notation in your solutions for these problems.

NUMERICAL PROBLEMS

1. If $A = \begin{bmatrix} 1 & 2 \\ 3 & 4 \end{bmatrix}$ and $B = \begin{bmatrix} 0 & 1 \\ 1 & 2 \end{bmatrix}$, show by a direct calculation that $\operatorname{tr}(AB) = \operatorname{tr}(BA)$.

2. For the matrices A and B in Problem 1, calculate $(AB)'$ and $B'A'$ and show that they are equal.

3. Let $A = \begin{bmatrix} 1 & 0 \\ 0 & 3 \end{bmatrix}$, $B = \begin{bmatrix} 0 & 1 \\ 2 & 3 \end{bmatrix}$, and $C = \begin{bmatrix} 1 & 1 \\ 0 & 2 \end{bmatrix}$. Calculate $\operatorname{tr}(ABC)$ and $\operatorname{tr}(CBA)$. Are they equal?

4. Calculate $\operatorname{tr}(AB')$ if $A = \begin{bmatrix} 0 & \frac{1}{2} \\ -\frac{1}{3} & 1 \end{bmatrix}$ and $B = \begin{bmatrix} 1 & 2 \\ 3 & 0 \end{bmatrix}$.

5. Show that ABA' is symmetric if $A = \begin{bmatrix} 1 & 1 \\ 4 & 2 \end{bmatrix}$ and $B = \begin{bmatrix} 1 & -1 \\ -1 & 7 \end{bmatrix}$.

6. Show that $AB - B'A'$ is skew-symmetric where $A = \begin{bmatrix} 2 & -3 \\ 4 & 5 \end{bmatrix}$ and $B = \begin{bmatrix} 5 & \frac{1}{2} \\ 6 & 0 \end{bmatrix}$.

MORE THEORETICAL PROBLEMS

Easier Problems

7. Prove Parts (1), (2), and (4) of Theorem 1.4.1.
8. Prove Parts (1), (2), and (4) of Theorem 1.4.2.

9. Prove Parts (1) and (2) of Theorem 1.4.4.
10. If B is any matrix in $M_2(\mathbb{R})$, show that $B = C + aI$, where $\mathrm{tr}(C) = 0$ and $a \in \mathbb{R}$.
11. If A and B are symmetric, prove that
    (a) A^n is symmetric for all positive integers n.
    (b) $AB + BA$ is symmetric.
    (c) $AB - BA$ is skew-symmetric.
    (d) ABA is symmetric.
    (e) If A is invertible, then A^{-1} is symmetric.
12. If A is skew-symmetric, show that A^n is skew-symmetric if n is a positive odd integer and A^n is symmetric if n is a positive even integer.
13. If A and B are skew-symmetric, show that
    (a) $AB + BA$ is symmetric.
    (b) $AB - BA$ is skew-symmetric.
    (c) $A(AB + BA) - (AB + BA)A$ is symmetric.
    (d) ABA is skew-symmetric.
    (e) If A is invertible, then A^{-1} is skew-symmetric.
14. If A and B *commute* (i.e., $AB = BA$) show that
    (a) If A is symmetric and B skew-symmetric, then AB is skew-symmetric.
    (b) If A and B are skew-symmetric, then AB is symmetric.
    (c) If A and B are symmetric, then AB is symmetric.
15. Produce specific matrices A and B which do not commute for which Parts (a) and (b) of Problem 14 are false.
16. If $A = A'$, show that BAB' is symmetric for all matrices B.
17. If $A' = -A$, show that BAB' is skew-symmetric for all matrices B.
18. If $A \neq 0$, prove that $\mathrm{tr}(AA') > 0$.
19. Using Problem 18, prove that if A is symmetric and $A^n = 0$ for some positive integer n, then $A = 0$.
20. If $A_1, A_2, \ldots, A_n$ are symmetric and $\mathrm{tr}(A_1^2 + A_2^2 + \cdots + A_n^2) = 0$, show that $A_1 = A_2 = \cdots = A_n = 0$.
21. Show that $\mathrm{tr}(A) = \mathrm{tr}(A')$.

Middle-Level Problems

22. Given any matrix A, show that we can write A as $A = B + C$, where B is symmetric and C is skew-symmetric.
23. Show that the B and C in Problem 22 are uniquely determined by A.
24. If $AA' = I$ and $BB' = I$, show that $(AB)(AB)' = I$.
25. If $AA' = I$, prove that
    (a) ABA^{-1} is symmetric whenever B is symmetric.
    (b) ABA^{-1} is skew-symmetric whenever B is skew-symmetric.
26. If $tr(AB) = 0$ for all matrices B, prove that $A = 0$.
27. If A is any matrix, prove that AA' and $A'A$ are both symmetric.
28. If both $\mathrm{tr}(A) = 0$ and $\mathrm{tr}(A^2) = 0$, show that $A^2 = 0$.
29. If $A^2 = 0$, prove that $\mathrm{tr}(A) = 0$.

Harder Problems

30. If $\operatorname{tr}(A) = 0$, prove that there exists an invertible matrix B such that $B^{-1}AB = \begin{bmatrix} 0 & u \\ v & 0 \end{bmatrix}$ for some u, v in $\mathbb{R}$.

31. If $\operatorname{tr}(A) = 0$, show that we can find a B and C such that $A = BC - CB$.

32. If $A^2 = A$, prove that $\operatorname{tr}(A)$ is an integer. What integers can arise this way?

33. If $A' = -A$, show that $A + aI$ is invertible for all $a \neq 0$ in $\mathbb{R}$.

34. Show that if A is any matrix, then, for $a < 0$ in $\mathbb{R}$, $AA' - aI$ is invertible.

35. If $\operatorname{tr}(ABC) = \operatorname{tr}(CBA)$ for all matrices C, prove that $AB = BA$.

36. Let A be an invertible matrix. Define the mapping * from $M_2(\mathbb{R})$ to $M_2(\mathbb{R})$ by $B^* = AB'A^{-1}$ for every B in $M_2(\mathbb{R})$. What are the necessary and sufficient conditions on A in order that * satisfy the three rules:
  (1) $B^{**} = B$ [where B^{**} denotes $(B^*)^*$];
  (2) $(B + C)^* = B^* + C^*$;
  (3) $(BC)^* = C^*B^*$;
for all B, C in $M_2(\mathbb{R})$?

1.5. DETERMINANTS

So far, while the 2×2 matrices have served as a very simple model of the various definitions, concepts, and results that will be seen to hold in the more general context, one could honestly say that these special matrices do not present an oversimplified case. The pattern outlined for them is the same pattern that we shall employ, later, in general. Even the method of proof, with only a slight variation, will go over.

Now, for the first time, we come to a situation where the 2×2 case is a vastly oversimplified one—that is, in the notion of the *determinant*. Even the very definition of the determinant in the $n \times n$ matrices is of a fairly complicated nature. Moreover, the proofs of many, if not most of the results will be of a different flavor, and of a considerably different degree of difficulty. While the general case will be a rather sticky affair, for the 2×2 matrices there are no particularly nasty points.

Keeping this in mind, we proceed to define the determinant for elements of $M_2(\mathbb{R})$.

Definition. If $A = \begin{bmatrix} a & b \\ c & d \end{bmatrix}$, then the *determinant* of A, written $\det(A)$, is defined by $\det(A) = ad - bc$.

We sometimes denote the determinant of A by $\begin{vmatrix} a & b \\ c & d \end{vmatrix}$. For instance, $\det(0) = 0$, $\det I = 1$, $\det \begin{bmatrix} a & b \\ 0 & d \end{bmatrix} = ad$, $\det \begin{bmatrix} 5 & 6 \\ -1 & 2 \end{bmatrix} = 5 \cdot 2 - (-1)6 = 16$, and $\det \begin{bmatrix} 1 & 3 \\ 5 & 15 \end{bmatrix} = \begin{vmatrix} 1 & 3 \\ 5 & 15 \end{vmatrix} = 1 \cdot 15 - 3 \cdot 5 = 0$.

The first result that we prove for the determinant is a truly trivial one.

Lemma 1.5.1. $\det(A) = \det(A')$.

Proof: If $A = \begin{bmatrix} a & b \\ c & d \end{bmatrix}$, then $A' = \begin{bmatrix} a & c \\ b & d \end{bmatrix}$ and $\det(A) = ad - bc = ad - cb = \det(A')$. ∎

Trivial though it be, Lemma 1.5.1 assures us that whatever result holds for the rows of a matrix holds equally for the columns.

Now for some of the basic properties of the determinant.

Theorem 1.5.2. In changing from one matrix to another, the following properties describe how the determinant changes:

1. If we interchange two rows of a matrix A, the determinant of the resulting matrix is $-\det(A)$ (only the sign of the determinant is changed).
2. If we add any multiple of any row of a matrix A to another, the determinant of the resulting matrix is just $\det(A)$ (the determinant does not change).
3. If we multiply a given row of a matrix A by a real number u, then the determinant of the resulting matrix is $u(\det(A))$ (the determinant is multiplied by u), that is, $\det(uA) = u(\det(A))$.
4. If two rows of A are equal, then $\det(A) = 0$.

Proof: Let $A = \begin{bmatrix} a & b \\ c & d \end{bmatrix}$. If we interchange the rows of A, we get the matrix $\begin{bmatrix} c & d \\ a & b \end{bmatrix}$, so that the first property simply asserts that $\det \begin{bmatrix} c & d \\ a & b \end{bmatrix} = -\det \begin{bmatrix} a & b \\ c & d \end{bmatrix}$, which is true because $cb - ad = -(ad - bc)$. For the second assertion, suppose that the multiple (ua, ub) of the first row is added to the second row, changing A from the matrix $\begin{bmatrix} a & b \\ c & d \end{bmatrix}$ to the matrix $\begin{bmatrix} a & b \\ ua + c & ub + d \end{bmatrix}$. Then the determinant of the new matrix is $aub + ad - (bua + bc) = ad - bc$, which is the same as the determinant of the original matrix. A similar argument works if, instead, a multiple (uc, ud) of the second row is added to the first row.

The third assertion states that

$$\det \begin{bmatrix} ua & ub \\ c & d \end{bmatrix} = \det \begin{bmatrix} u & b \\ uc & ud \end{bmatrix} = u \begin{bmatrix} a & b \\ c & d \end{bmatrix},$$

which is true because $uad - ubc = aud - buc = u(ad - bc)$.

The remaining assertion simply states the obvious, namely that $\det \begin{bmatrix} a & b \\ a & b \end{bmatrix} = ab - ba$ is 0. Nevertheless, we also prove this as a consequence of Part (1) of the theorem, because this is how it is proved in the $n \times n$ case. Namely, observe that since two different rows of A are equal, when these two rows are interchanged, the following

Sec. 1.5] Determinants 23

two things happen:

1. The matrix A does not change, so neither does its determinant.
2. By Part (1), its determinant is multiplied by a factor of -1.

So $\det(A) = -\det(A)$, whence $\det(A) = 0$. ∎

Since the determinant does not change when we replace a matrix A by its transpose, that is, we replace its rows by its columns, we have

Corollary 1.5.3. If we replace the word "row" by the word "column" in Theorem 1.5.2, all the results remain valid.

Proof: The determinant of a matrix and its transpose are the same, so that row properties of determinants imply corresponding column properties and conversely. ∎

We interrelate the algebraic property of the invertibility of a matrix with properties of the determinant.

Theorem 1.5.4. A is invertible if and only if $\det(A) \neq 0$. If $A = \begin{bmatrix} a & b \\ c & d \end{bmatrix}$ is invertible, its inverse is $\dfrac{1}{\det(A)} B$, where $B = \begin{bmatrix} d & -b \\ -c & a \end{bmatrix}$.

Proof: If $A = \begin{bmatrix} a & b \\ c & d \end{bmatrix}$, then for $B = \begin{bmatrix} d & -b \\ -c & a \end{bmatrix}$, we have $AB = \begin{bmatrix} ad - bc & 0 \\ 0 & ad - bc \end{bmatrix} = (\det(A))I$. So if $\alpha = \det(A) \neq 0$, then $A(\alpha^{-1}B) = \alpha^{-1}AB = (\alpha^{-1}\det(A))I = \alpha^{-1}\alpha I = I$. Similarly, $(\alpha^{-1}B)A = I$. Hence $\alpha^{-1}B$ is the inverse of A.

If $\alpha = \det(A) = 0$, then for the B above, $AB = 0$. Thus if A were invertible, $A^{-1}(AB) = A^{-1}0 = 0$. But $A^{-1}(AB) = (A^{-1}A)B = IB = B$. So $B = 0$. Thus $d = b = c = a = 0$. But this implies that $A = 0$, and 0 is certainly not invertible. Thus if $\det(A) = 0$, then A is not invertible. ∎

The key result—and for the general case it will be the most difficult one—intertwines the determinant of the product of matrices with that of their determinants.

Theorem 1.5.5. If A and B are matrices, then $\det(AB) = \det(A)\det(B)$.

Proof: Suppose that $A = \begin{bmatrix} a & b \\ c & d \end{bmatrix}$ and $B = \begin{bmatrix} u & v \\ w & x \end{bmatrix}$. Then $\det(A) = ad - bc$ and $\det(B) = ux - wv$. Since

$$AB = \begin{bmatrix} a & b \\ c & d \end{bmatrix}\begin{bmatrix} u & v \\ w & x \end{bmatrix} = \begin{bmatrix} au + bw & av + bx \\ cu + dw & cv + dx \end{bmatrix},$$

we can compute as follows:

$$\begin{aligned}\det(AB) &= (au + bw)(cv + dx) - (av + bx)(cu + dw) \\ &= aucv + audx + bwcv + bwdx - avcu - avdw - bxcu - bxdw \\ &= adux + bcwv - adwv - bcux \\ &= (ad - bc)(ux - wv) = \det(A)\det(B).\end{aligned}$$

This proves the assertion. ∎

An immediate consequence of the theorem is

Corollary 1.5.6. If A is invertible, then $\det(A)$ is nonzero and $\det(A^{-1}) = \dfrac{1}{\det(A)}$.

Proof: Since $AA^{-1} = I$, we know that $\det(AA^{-1}) = \det(I) = 1$. But by the theorem, $\det(AA^{-1}) = \det(A)\det(A^{-1})$, in consequence of which, $\det(A)\det(A^{-1}) = 1$. Hence $\det(A)$ is nonzero and $\det(A^{-1}) = \dfrac{1}{\det(A)}$. ∎

There is yet another important consequence of Theorem 1.5.5, namely

Corollary 1.5.7. If A is invertible and B is any matrix, then $\det(A^{-1}BA) = \det(B)$.

Proof: Since $A^{-1}BA = (A^{-1}B)A$, by the theorem we have

$$\begin{aligned}\det(A^{-1}BA) &= \det((A^{-1}B)A) = \det(A^{-1}B)\det(A) \\ &= \det(A^{-1})\det(B)\det(A) \\ &= \det(A^{-1})\det(A)\det(B) \\ &= \det(B).\end{aligned}$$

(We have made another use of the theorem, and also of Corollary 1.5.6, in this chain of equalities.) ∎

Note the similarity of Corollary 1.5.7 to that of Corollary 1.4.3.

It is helpful, even when we know a general result, to see it in a specific instance. We do this for Corollary 1.5.7. Let A be the matrix $\begin{bmatrix}1 & -2 \\ 0 & 1\end{bmatrix}$ and B be the matrix $\begin{bmatrix}3 & 4 \\ 5 & 6\end{bmatrix}$. Then A is invertible, with $A^{-1} = \begin{bmatrix}1 & 2 \\ 0 & 1\end{bmatrix}$. Therefore, we have

$$\begin{aligned}A^{-1}BA &= \begin{bmatrix}1 & 2 \\ 0 & 1\end{bmatrix}\begin{bmatrix}3 & 4 \\ 5 & 6\end{bmatrix}\begin{bmatrix}1 & -2 \\ 0 & 1\end{bmatrix} \\ &= \begin{bmatrix}13 & 16 \\ 5 & 6\end{bmatrix}\begin{bmatrix}1 & -2 \\ 0 & 1\end{bmatrix} = \begin{bmatrix}13 & -10 \\ 5 & -4\end{bmatrix}.\end{aligned}$$

Thus $\det(A^{-1}BA) = 13 \cdot (-4) - (5)(-10) = -52 + 50 = -2$ and

$$\det(B) = \det\begin{bmatrix} 3 & 4 \\ 5 & 6 \end{bmatrix} = 3 \cdot 6 - 4 \cdot 5 = -2.$$

We remind you for the nth time—probably to the point of utter boredom—that although the results for the determinant of the 2×2 matrices are established easily and smoothly, this will not be the case for the analogous results in the case of $n \times n$ matrices.

PROBLEMS

NUMERICAL PROBLEMS

1. Compute the following determinants.

   (a) $\det\begin{bmatrix} 1 & 6 \\ -6 & 1 \end{bmatrix}$.

   (b) $\det\left(\begin{bmatrix} 5 & 6 \\ 7 & 8 \end{bmatrix}\begin{bmatrix} 4 & 3 \\ -3 & 4 \end{bmatrix}\right)$.

   (c) $\det\left(\begin{bmatrix} 5 & 6 \\ 7 & 8 \end{bmatrix} + \begin{bmatrix} 4 & 3 \\ -3 & 4 \end{bmatrix}\right)$.

   (d) $\det\left(\begin{bmatrix} 1 & 2 \\ 2 & 4 \end{bmatrix} + \begin{bmatrix} 5 & -6 \\ -7 & 8 \end{bmatrix}\right)$.

2. In Part (c) of Problem 1, is it true that
   $$\det\left(\begin{bmatrix} 5 & 6 \\ 7 & 8 \end{bmatrix} + \begin{bmatrix} 4 & 3 \\ -3 & 4 \end{bmatrix}\right) = \det\begin{bmatrix} 5 & 6 \\ 7 & 8 \end{bmatrix} + \det\begin{bmatrix} 4 & 3 \\ -3 & 4 \end{bmatrix}?$$

3. For the matrix $A = \begin{bmatrix} 1 & 2 \\ 3 & 4 \end{bmatrix}$, calculate $\text{tr}(A^2) - (\text{tr}(A))^2$.

4. Find all real numbers a such that $\det\left(aI - \begin{bmatrix} 4 & 5 \\ 0 & -\pi \end{bmatrix}\right) = 0$.

5. Find all real numbers a such that $\det\left(aI - \begin{bmatrix} 3 & 4 \\ 1 & 3 \end{bmatrix}\right) = 0$.

6. Find all real numbers a such that $\det\left(aI - \begin{bmatrix} 5 & 6 \\ 6 & 7 \end{bmatrix}\right) = 0$.

7. Is there a real number a such that $\det\left(aI - \begin{bmatrix} 0 & 6 \\ -6 & 0 \end{bmatrix}\right) = 0$?

MORE THEORETICAL PROBLEMS

Easier Problems

8. Complete the proof of Theorem 1.5.2.
9. If $\det(A)$ and $\det(B)$ are rational numbers, what can you say about $\det(AB)$ and about $\det(CAC^{-1})$ if C is invertible?

10. If A and B are matrices such that $AB = I$, then prove—using results about determinants—that $BA = I$.
11. If A and B are invertible, show that $\det(ABA^{-1}B^{-1}) = 1$.
12. Prove for any matrix in $M_2(\mathbb{R})$ that $2\det(A) = \operatorname{tr}(A^2) - (\operatorname{tr}(A))^2$.
13. Show that every matrix A in $M_2(\mathbb{R})$ satisfies the relation $A^2 - (\operatorname{tr}(A))A + (\det(A))I = 0$.
14. From Problem 13, find the inverse of A, if A is invertible, in terms of A.
15. Using Problem 13, show that if $C = AB - BA$, then $C^2 = aI$ for some real number a.
16. Verify the results of Problem 15 for the matrix $A = \begin{bmatrix} 1 & -6 \\ \frac{2}{3} & -5 \end{bmatrix}$ and all B.
17. Calculate $\det(xI - A)$ for any matrix A.
18. If $\det(aI - A) = 0$, show that $\det(a^n I - A^n) = 0$ for all positive integers n.
19. If $\det(A) < 0$, show that there are exactly two real numbers a such that $aI - A$ is not invertible.

Middle-Level Problems

20. Show that if a is any real number, then $\det(aI - A) = a^2 - (\operatorname{tr} A)a + \det(A)$.
21. Using determinants, show for any real numbers a and b that if $A = \begin{bmatrix} a & b \\ -b & a \end{bmatrix} \ne 0$, then A is invertible.

Harder Problems

22. If $\det(A) = 1$, then show that we can find invertible matrices B and C such that $A = BCB^{-1}C^{-1}$.
23. If $A = \begin{bmatrix} a & b \\ b & a \end{bmatrix}$ with $b \ne 0$, prove that there exists exactly two real numbers u such that $uI - A$ is not invertible.
24. If $A = \begin{bmatrix} a & b \\ -b & a \end{bmatrix}$ with $b \ne 0$, show that for all real numbers $u \ne 0$, $uI - A$ is invertible.

1.6. CRAMER'S RULE

The determinant was introduced in a rather ad hoc manner as a number associated with a given matrix. What the meaning or relationship of the determinant of a matrix to that matrix was neither clear nor motivated. Aside from providing a criterion for the invertibility of a matrix, the determinant seemed to have nothing much to do with anything.

It should be somewhat illuminating if we could see the determinant arise naturally, and with some bite, in a concrete context. This context will be in the solution of simultaneous linear equations.

Cramer's Rule

Suppose that we want to solve for x and y in the equations

$$ax + by = g$$
$$cx + dy = h. \qquad (1)$$

The method of solution is to eliminate one of x or y between these two equations. To eliminate y, we multiply the first equation by d, and the second one by b, and subtract. The outcome of all this is that $(ad - bc)x = dg - bh$, so, provided that $ad - bc \neq 0$, we obtain the solution

$$x = \frac{dg - bh}{ad - bc}.$$

Similarly, we get

$$y = \frac{ah - cg}{ad - bc}.$$

We can recognize $ad - bc$, $dg - bh$, and $ah - cg$ as the determinants of the matrices $\begin{bmatrix} a & b \\ c & d \end{bmatrix}$, $\begin{bmatrix} g & b \\ h & d \end{bmatrix}$, and $\begin{bmatrix} a & g \\ c & h \end{bmatrix}$, respectively. What are these matrices? The matrix $\begin{bmatrix} a & b \\ c & d \end{bmatrix}$ is simply the matrix of the coefficients of x and y in the system of equations (1). Furthermore, the matrix $\begin{bmatrix} g & b \\ h & d \end{bmatrix}$ is the matrix obtained from $\begin{bmatrix} a & b \\ c & d \end{bmatrix}$ by replacing the first column—the one associated with x—by the right-hand sides of (1). Similarly, $\begin{bmatrix} a & g \\ c & h \end{bmatrix}$ is obtained from $\begin{bmatrix} a & b \\ c & d \end{bmatrix}$ by replacing the second column—the one associated with y—by the right-hand sides of (1). *This is no fluke*, as we shall see in a moment. More important, when we have the determinant of an $n \times n$ matrix defined and under control, the exact analogue of this procedure for finding the solution for systems of linear equations will carry over. This method of solution is called *Cramer's rule*.

Consider the matrix $\begin{bmatrix} a & b \\ c & d \end{bmatrix}$ and its determinant. If x is any real number, then

$$x \left(\det \begin{bmatrix} a & b \\ c & d \end{bmatrix} \right) = \det \begin{bmatrix} ax & b \\ cx & d \end{bmatrix}$$

by one of the properties we showed for determinants. Furthermore, by another of the properties shown for determinants, we have

$$\det \begin{bmatrix} ax & b \\ cx & d \end{bmatrix} = \det \begin{bmatrix} ax + by & b \\ cx + dy & d \end{bmatrix}$$

for any real number y. Therefore, if x and y are such that $ax + by = g$ and $cx + dy = h$, the discussion above leads us to the equations

$$x\left(\det\begin{bmatrix} a & b \\ c & d \end{bmatrix}\right) = \det\begin{bmatrix} ax+by & b \\ cx+cy & d \end{bmatrix}$$

$$= \det\begin{bmatrix} g & b \\ h & d \end{bmatrix},$$

hence to the equation

$$x = \frac{\det\begin{bmatrix} g & b \\ h & d \end{bmatrix}}{\det\begin{bmatrix} a & b \\ c & d \end{bmatrix}}.$$

A similar argument works for y, showing

$$y = \frac{\det\begin{bmatrix} a & g \\ c & h \end{bmatrix}}{\det\begin{bmatrix} a & b \\ c & d \end{bmatrix}}.$$

From what we just did, two things are clear.

1. Matrices and their determinants have something to do with solving linear equations; and
2. The vanishing or nonvanishing of the determinant of the coefficients of a system of linear equations is a deciding criterion for carrying out the solution.

Let's try out Cramer's rule for a system of two specific linear equations. According to the recipe provided us by Cramer's rule, the solutions x and y to the system of two simultaneous equations

$$x - y = 7$$
$$3x + 4y = 5$$

are

$$x = \frac{\begin{vmatrix} 7 & -1 \\ 5 & 4 \end{vmatrix}}{\begin{vmatrix} 1 & -1 \\ 3 & 4 \end{vmatrix}} = \frac{33}{7} \quad \text{and} \quad y = \frac{\begin{vmatrix} 1 & 7 \\ 3 & 5 \end{vmatrix}}{\begin{vmatrix} 1 & -1 \\ 3 & 4 \end{vmatrix}} = -\frac{16}{7}.$$

You can verify that these values of x and y satisfy both of the given equations.

PROBLEMS

NUMERICAL PROBLEMS

1. Using determinants, solve the following equations for x and y.
   (a) $x + 7y = 11$
   $\frac{3}{2}x - 4y = 6.$
   (b) $\pi x - 3y = \pi^2$
   $x + 5y = \pi.$

MORE THEORETICAL PROBLEMS

Easier Problems

2. Show that the solution for x and y in
$$ax + by = g$$
$$cx + dy = h$$
is the same as the values x and y for which
$$\begin{bmatrix} a & b \\ c & d \end{bmatrix} \begin{bmatrix} x & 0 \\ y & 0 \end{bmatrix} = \begin{bmatrix} g & 0 \\ h & 0 \end{bmatrix}.$$

3. For the situation in Problem 2, show that if $\det \begin{bmatrix} a & b \\ c & d \end{bmatrix} \neq 0$, then
$$\begin{bmatrix} x & 0 \\ y & 0 \end{bmatrix} = \begin{bmatrix} a & b \\ c & d \end{bmatrix}^{-1} \begin{bmatrix} g & 0 \\ h & 0 \end{bmatrix}.$$
From this, computing $\begin{bmatrix} a & b \\ c & d \end{bmatrix}^{-1}$, find x and y.

1.7. MAPPINGS

If there is one notion that is central in every part of mathematics, it is that of a *mapping* or *function* from one set S to another set T. Let's recall a few things about mappings. Our main concerns will be with mappings of a given set S to itself (so where $S = T$).

By a *mapping* f from S to T—which we denote by $f: S \to T$—we mean a rule or mechanism that assigns to every element of S another, unique element of T. If $s \in S$ and $f: S \to T$, we write $t = f(s)$ as that element of T into which s is carried by f. We call $t = f(s)$ the *image* of s under f. Thus, if $S = T = \mathbb{R}$ and $f: S \to T$ is defined by $f(s) = s^2$, then $f(4) = 4^2 = 16$, $f(-\pi) = (-\pi)^2$, and so on.

When do we declare two mappings to be equal? A natural way is to define the two mappings f and g of S to T to be equal if they do the same thing to the same objects of S. More formally, $f = g$ if and only if $f(s) = g(s)$ for every $s \in S$.

Among the possible hordes of mappings from a set S to T, we single out some particularly decent types of mappings.

Definition. The mapping $f: S \to T$ is said to be *one-to-one*, written as $1-1$, if $f(s_1) = f(s_2)$ implies that $s_1 = s_2$ for all s_1, s_2 in S.

In other words, a mapping f is $1-1$ if *it takes distinct elements of S into distinct images*. If we look at $S = \mathbb{R}$ and the mapping $f: S \to S$ defined by $f(s) = s + 1$, we readily see that f is $1-1$. For if $f(s) = f(t)$, we have $s + 1 = f(s) = f(t) = t + 1$, which implies that $s = t$. On the other hand, the mapping $g: \mathbb{R} \to \mathbb{R}$ defined by $g(s) = s^2$ is *not* $1-1$ since $f(-2) = (-2)^2 = 2^2 = f(2)$, yet $-2 \neq 2$.

Another type of good mapping is defined in the

Definition. A mapping $f: S \to T$ is said to be *onto* if given any $t \in T$, there is an $s \in S$ such that $t = f(s)$.

In other words, the mapping f is onto if the images of S under f *fill out all of T*. Again, returning to the two examples of f and g above, where $S = T = \mathbb{R}$, since $f(s) = s + 1$, given the real number t, then $t = f(t - 1) = (t - 1) + 1$, so that f is onto, while $g(s) = s^2$ is *not* onto. If g were onto, then the negative number -1 could be expressed as a square $g(s) = s^2$ for some s, which is clearly impossible. Note, however, that if S is the set of all *positive* real numbers, then the mapping $g(s) = s^2$ maps S onto S, since every positive real number has a positive (and also, negative) square root.

The advantage of considering mappings of a set S into itself, rather than a mapping of one set S to a different set T, lies in the fact that *in the setup where $S = T$, we can define a product for two mappings of S into itself*. How? If f, g are mappings of S into S and $s \in S$, then $g(s)$ is again in S. As such, $g(s)$ is a candidate for action by f; that is, $f(g(s))$ is again an element of S. This prompts the

Definition. If f and g are mappings of S to itself, then the *product* or *composition* of f and g, written as fg, is the mapping defined by $(fg)(s) = f(g(s))$ for every $s \in S$.

So fg is that mapping obtained by first applying g and then, to the result of this, applying f. Let's look at a couple of examples. Suppose that $S = \mathbb{R}$ and $f: S \to S$ and $g: S \to S$ are defined by $f(s) = -4s + 3$ and $g(s) = s^2$. What is fg? Computing,

$$(fg)(s) = f(g(s)) = f(s^2) = -4s^2 + 3;$$

so, for instance, $(fg)(1) = -4(1)^2 + 3 = -1$, $f(\pi) = -4\pi^2 + 3$, and $f(0) = -4(0)^2 + 3 = 3$. While we are at it, what is gf? Computing,

$$(gf)(s) = g(f(s)) = g(-4s + 3) = (-4s + 3)^2 = 16s^2 - 24s + 9;$$

hence, for instance, $gf(1) = 16(1)^2 - 24(1) + 9 = 1$. Notice that $(gf)(1) = 1 \neq -1 = (fg)(1)$. Since fg and gf do not agree on $s = 1$, they are not equal as mappings, so that $fg \neq gf$. In other words, *the commutative law does not necessarily hold for two mappings of S to itself*. However, another basic law, which we should like to hold for the products of mappings, does indeed hold true. This is the *associative law*.

Lemma 1.7.1. (*Associative Law*). If f, g, and h are mappings of S to itself, then $f(gh) = (fg)h$.

Sec. 1.7] Mappings 31

Proof: Note first that since g and h are mappings of S to itself, then so is gh a mapping of S to itself. Therefore, $f(gh)$ is a mapping from S to itself. Similarly, $(fg)h$ is a mapping from S to itself. Hence it at least makes sense to ask if $f(gh) = (fg)h$.

To prove the result, we must merely show that $f(gh)$ and $(fg)h$ agree on every element of S. So if $s \in S$, then

$$(f(gh))(s) = f((gh)(s)) = f(g(h(s))),$$

while, on the other hand,

$$((fg)h)(s) = (fg)(h(s)) = f(g(h(s))),$$

by successive uses of the definition of the product of mappings. Since we thus see that $(f(gh))(s) = ((fg)h)(s)$ for every $s \in S$, by the definition of the equality of mappings, we have that $f(gh) = (fg)h$, the desired result. ∎

By virtue of the lemma, we can dispense with a particular bracketing for the product of three or more mappings, and we write such products simply as fgh, $fgghfr$ (product of mappings f, g, g, h, f, r), and so forth.

A very nice mapping is always present, that is, the *identity mapping* $e: S \to S$, that is, the mapping e which disturbs no element of S: $e(s) = s$ for all $s \in S$.

We leave as an exercise the proof of

Lemma 1.7.2. If e is the identity mapping on S, then $ef = fe = f$ for every mapping $f: S \to S$.

Before leaving this short discussion of mappings, there is one more fact about certain mappings that should be highlighted. Let $f: S \to S$ be *both* $1-1$ and onto. Thus, given $t \in S$, then $t = f(s)$ for some $s \in S$, since f is onto. Furthermore, because f is $1-1$, there is *one and only one* s that does the trick. So if we define $g: S \to S$ by the rule $g(t) = s$ if and only if $t = f(s)$, we do indeed get a mapping of S to itself. What properties does this g enjoy? If we compute $(gf)(s)$ for any $s \in S$, what do we get? Suppose that $t = f(s)$. Then, by the very definition of g, $s = g(t) = g(f(s)) = (gf)$. So $(gf)(s) = s = e(s)$ for every $s \in S$, so that $gf = e$. We call g the *inverse* of f and denote it by f^{-1}.

We summarize what we have just done in

Theorem 1.7.3. If f is $1-1$ and onto from S to S, then there exists a mapping $f^{-1}: S \to S$ such that $ff^{-1} = f^{-1}f = e$.

We compute f^{-1} for some sample f's. Let $S = \mathbb{R}$ and let f be the function from S to S defined by $f(s) = 6s + 5$. Then f is a $1-1$ mapping of S *onto* itself, as is readily verified. What, then, is f^{-1}? We claim that $f^{-1}(s) = \frac{1}{6}s - \frac{5}{6} = \frac{s-5}{6}$ for every s in $\mathbb{R}$.

(Verify!). *Do not confuse* $f^{-1}(s)$ *with* $(f(s))^{-1} = \frac{1}{f(s)}$. In our example, $f^{-1}(s) = \frac{s-5}{6}$, whereas $(f(s))^{-1} = \dfrac{1}{f(s)} = \dfrac{1}{6s+5}$ and these are not equal as functions on S. For instance, $f^{-1}(1) = \dfrac{1-5}{6} = \dfrac{-4}{6}$ while $(f(1))^{-1} = \dfrac{1}{6+5} = \dfrac{1}{11}$.

We do one more example. Let S be the set of all *positive* real numbers and suppose that $f: S \to S$ is defined by $f(s) = 1/s$. Clearly, f is a $1-1$ mapping of S onto itself, hence f^{-1} exists. What is f^{-1}? We leave it to the reader to prove that $f^{-1}(s) = 1/s = f(s)$ for all s, so that $f^{-1} = f$.

One last bit. To have a handy notation, we use exponents for the product of a function (several times) with itself. Thus $f^2 = ff, \ldots, f^{n-1}f$ for every positive integer n. So, by our definition, $f^n = \underbrace{fff \cdots f}_{(n \text{ times})}$. As we might expect, the usual rules of exponents hold, namely $f^n f^m = f^{n+m}$ and $(f^m)^n = f^{mn}$ for m and n positive integers. By f^0 we shall mean the identity mapping of S onto itself.

PROBLEMS

NUMERICAL PROBLEMS

1. Determine whether the following mappings are $1-1$, onto, or both.

    (a) $f: \mathbb{R} \to \mathbb{R}$, where f is defined by $f(s) = \dfrac{s}{s^2+1}$.

    (b) $f: \mathbb{R} \to \mathbb{R}$, where f is defined by $f(s) = s^2 + s + 1$.

    (c) $f: \mathbb{R} \to \mathbb{R}$, where f is defined by $f(s) = s^3$.

2. Let S be the set of integers and let f be a mapping from S to S. Then determine, for the following f's, whether they are $1-1$, onto, or both.

    (a) $f(s) = -s + 1$.

    (b) $f(s) = 6s$.

    (c) $f(s) = s$ if s is even and $f(s) = s + 2$ if s is odd.

3. If $f: \mathbb{R} \to \mathbb{R}$ is defined by $f(s) = 3s + 2$, find the formula for $f^2(s)$.

4. Is the "mapping" $f: \mathbb{R} \to \mathbb{R}$ defined by $f(s) = \dfrac{s}{1+s}$ really a mapping?

5. If S is the set of *positive* real numbers, verify that $f: S \to S$ defined by $f(s) = \dfrac{s}{1+s}$ really is a mapping. (Compare with Problem 4.)

6. In Problem 5, is f $1-1$? Is f onto?

7. If S is the set of all real numbers s such that $0 < s < 1$, verify that f defined by

    $f(s) = \dfrac{s}{1+s}$ is a mapping of S into itself. (Compare with Problem 4.)

8. In Problem 7, is f $1-1$? Is f onto?

9. In Problem 7, compute f^2 and f^3.

10. In Problem 7, find f^{-1}.

Sec. 1.7] Mappings 33

MORE THEORETICAL PROBLEMS

Easier Problems

11. If S is the set of positive integers, give an example of a mapping $f: S \to S$ such that
    (a) f is $1-1$ but not onto.
    (b) f is onto but not $1-1$.
12. If f maps S onto itself and is $1-1$, show that if $fg = fh$, then $g = h$.
13. If f and g are $1-1$ mappings of S onto itself, show that
    (a) fg is also $1-1$ and onto.
    (b) f^{-1} and g^{-1} exist.
    (c) $(fg)^{-1} = g^{-1}f^{-1}$.
14. Let $S = \mathbb{R}$ and consider the mapping $t_{a,b}: S \to S$, where $a \neq 0$ and b are real numbers, defined by $t_{a,b}(s) = as + b$. Then
    (a) Show that $t_{a,b}$ is $1-1$ and onto.
    (b) Find $t_{a,b}^{-1}$ in the form $t_{u,v}$.
    (c) If $t_{c,d}: S \to S$ where $t_{c,d}(s) = cs + d$, and $c \neq 0$, d are real numbers, find the formula for $t_{a,b}t_{c,d}$.
15. For what values of a and b does $t_{a,b}$ in Problem 14 satisfy $t_{a,b}^2 = e$?
16. If $t_{1,2}$ and $t_{a,b}$ are as in Problem 14, find all $t_{a,b}$ such that $t_{1,2}t_{a,b} = t_{a,b}t_{1,2}$.
17. Let S be the $x - y$ plane and let f be defined on S by the rule that f of any point (x, y) is the corresponding point $(0, y)$ on the y-axis. This mapping f is called *projection onto the y-axis*. Show that $f^2 = f$. Letting g be the analogous *projection onto the x-axis*, show that fg and gf both are the *zero function* that maps s to $(0, 0)$ for all $s \in S$.

Middle-Level Problems

18. Let S be a set having at least three elements. Show that we can find $1-1$ mappings of S into itself, f and g, such that $fg \neq gf$.
19. If f is a mapping of the set of positive real numbers into $\mathbb{R}$ such that $f(ab) = f(a) + f(b)$, and f is not identically 0, find
    (a) $f(1)$.
    (b) $f(a^3)$.
    (c) $f(a^n)$ for every positive integer n.
20. Show that $f(a^r) = rf(a)$ for any *positive rational* number r in Problem 19.
21. Let S be the $x - y$ plane $\mathbb{R} \times \mathbb{R} = \{(a,b) \mid a, b \in \mathbb{R}\}$ and let f be the reflection of the plane about the x-axis and g the rotation of the plane about the origin through $120°$ in the counterclockwise direction. Prove:
    (a) f and g are $1-1$ and onto.
    (b) $f^2 = g^3 = e$, the identity mapping on S.
    (c) $fg \neq gf$.
    (d) $fg = g^{-1}f$.
22. Let S be the set of all integers and define, for a, b integers and $a \neq 0$, the functions $u_{a,b}: S \to S$ by $u_{a,b}(s) = as + b$. Then
    (a) When is the function $u_{a,b}$ $1-1$?

(b) When is the function $u_{a,b}$ onto?

(c) When is $u_{a,b}$ both $1-1$ and onto? And when it is both $1-1$ and onto, describe its inverse.

23. Define $f(s)$ as $\dfrac{s^2+1}{s^2+2}$ for $s \in \mathbb{R}$. Describe the set of all images $f(s)$ as s ranges throughout the set $\mathbb{R}$.

24. Define $f: \mathbb{R} \to \mathbb{R}$ by $f(s) = s^3 + 6$. Then
    (a) Show that f is $1-1$ and onto.
    (b) Find f^{-1}.

25. Define $f: \mathbb{R} \to \mathbb{R}$ by $f(s) = s^2$. Then determine whether f is $1-1$ and onto.

26. "Prove" the law of exponents: $f^m f^n = f^{m+n}$ and $(f^m)^n = f^{mn}$, where m, n are nonnegative integers and f maps some set S to itself.

Harder Problems

27. (a) Is $(fg)^2$ necessarily equal to $f^2 g^2$ for f, g mappings of S into S? Prove or give a counterexample.
    (b) If f, g are $1-1$ mappings of S onto S, is $(fg)^2$ always equal to $f^2 g^2$? Prove or give a counterexample.

28. If f, g are $1-1$ mappings of S onto itself, show that a necessary and sufficient condition that $(fg)^n = f^n g^n$ for all $n > 1$ is that $fg = gf$.

29. If S is a *finite* set and $f: S \to S$ is $1-1$, show that f is onto.

30. If S is a *finite* set and $f: S \to S$ is onto, show that f is $1-1$.

Generalization of terminology to two sets S and T: If S and T are two sets, then $f: S \to T$ is said to be $1-1$ if $f(s) = f(s')$ only if $s = s'$ and f is said to be *onto* if, given $t \in T$, there is an $s \in S$ such that $t = f(s)$.

31. If $f: S \to T$ is $1-1$ and $fg = fh$ where $g, h: T \to S$ are two functions, then show that $g = h$.

32. If $g: S \to T$ is *onto* and p and q are mappings of T to S such that $pg = qg$, prove that $p = q$.

33. If $f: S \to T$ is $1-1$ and onto, show that we can find $g: T \to S$ such that $fg = e_s$, the identity mapping on S, and $gf = e_T$, the identity mapping on T.

34. Let $S = \mathbb{R}$ and let T be the set of all *positive* real numbers. Define $f: S \to T$ by $f(s) = 10^s$. Then
    (a) Prove that f is $1-1$ and onto.
    (b) Find the function g of Problem 33 for our f.

35. Let S be the set of all positive real numbers, and let $T = \mathbb{R}$. Define $f: S \to T$ by $f(s) = \log_2 s$. Then
    (a) Show that f is $1-1$ and onto.
    (b) Find the g of Problem 33 for f.

1.8. MATRICES AS MAPPINGS

It would be nice if we could strip matrices of their purely formal definitions and, instead, could find some way in which they would *act as transformations*, that is, mappings, on a reasonable set. A reasonable sort of *action* would be in terms of a nice kind of mapping of a set of familiar objects into itself.

For our set of familiar objects, we use the usual Cartesian plane $\mathbb{R}^2$. Traditionally, we have written the points of the plane $\mathbb{R}^2$ as ordered pairs (x, y) with $x, y \in \mathbb{R}$. For a purely technical reason, we now begin to write the points as $\begin{bmatrix} x \\ y \end{bmatrix}$. The entries x and y are referred to as the *coordinates* of $\begin{bmatrix} x \\ y \end{bmatrix}$.

We now let V denote $\mathbb{R}^2 = \left\{ \begin{bmatrix} x \\ y \end{bmatrix} \middle| x, y \in \mathbb{R} \right\}$. We refer to the elements (points) of V as *vectors* and to V as the *space of vectors* $\begin{bmatrix} x \\ y \end{bmatrix}$ with $x, y \in \mathbb{R}$ or, simply, as the *vector space of column vectors over* $\mathbb{R}$. In the subject of *vector analysis*, two operations are introduced in V, namely, the *addition* of vectors and the *multiplication* of a vector by a *scalar* (element of $\mathbb{R}$). We explain. If $\begin{bmatrix} a \\ b \end{bmatrix}$ and $\begin{bmatrix} c \\ d \end{bmatrix}$ are in V, we *define*
$\begin{bmatrix} a \\ b \end{bmatrix} + \begin{bmatrix} c \\ d \end{bmatrix} = \begin{bmatrix} a + c \\ b + d \end{bmatrix}$ and $d \begin{bmatrix} a \\ b \end{bmatrix} = \begin{bmatrix} da \\ db \end{bmatrix}$ for any d in $\mathbb{R}$. Pictorially, these two operations look as follows, where $P = \begin{bmatrix} a \\ b \end{bmatrix}$, $Q = \begin{bmatrix} c \\ d \end{bmatrix}$, and $R = \begin{bmatrix} a + c \\ b + d \end{bmatrix}$:

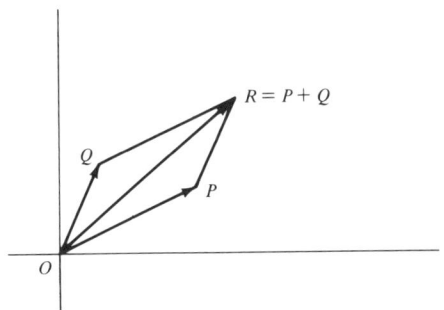

Addition: R is the sum of P and Q.

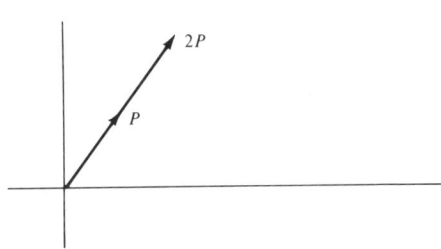

Multiplication: $2P$ is the scalar multiple $2\begin{bmatrix} a \\ b \end{bmatrix}$ of P.

So addition is performed *component-wise*, that is, by adding the coordinates (components) individually. Multiplication by a scalar a is a stretching (or shrinking) of the line by a factor of a. The stretching represented in the picture is the effect of stretching the line segment OP into the segment OQ.

With these operations, V satisfies many of the readily verified rules, such as: If $u, v,$

$w \in V$, $a, b \in \mathbb{R}$, then

$$(u + v) + w = u + (v + w),$$
$$u + v = v + u,$$
$$a(u + v) = au + av,$$
$$a(bv) = (ab)v,$$

and so on.

The structure V (the set V together with the operations of addition and multiplication by scalars $a \in \mathbb{R}$) is a particular case of the abstract notion of a *vector space*, a topic treated later in the book. The 2×2 matrices which we view in this section as mappings from this vector space V to itself are the *linear transformations* from V to V. Since the subject *linear algebra* is, more or less, simply the study of linear transformations of vector space, what we do in this section sets the stage for what is to come later in the book.

We shall let the matrices, that is, the elements of $M_2(\mathbb{R})$, act as mappings from V to itself. We accomplish this by *defining*

$$\begin{bmatrix} a & b \\ c & d \end{bmatrix} \begin{bmatrix} x \\ y \end{bmatrix} = \begin{bmatrix} ax + by \\ cx + dy \end{bmatrix}.$$

Geometrically, this can be given the interpretation of a *change of coordinates* or *coordinate transformation* on V. If $v \in V$ and $A \in M_2(\mathbb{R})$, we shall write Av as a shorthand for the specific action of A on v as described above.

Let's see two particular actions of this sort, which we can interpret geometrically. Let

$$A = \begin{bmatrix} 1 & 0 \\ 0 & -1 \end{bmatrix}.$$

Then

$$A \begin{bmatrix} x \\ y \end{bmatrix} = \begin{bmatrix} 1 & 0 \\ 0 & -1 \end{bmatrix} \begin{bmatrix} x \\ y \end{bmatrix} = \begin{bmatrix} x \\ -y \end{bmatrix}.$$

So, A carries the point $\begin{bmatrix} x \\ y \end{bmatrix}$ into its mirror image $\begin{bmatrix} x \\ -y \end{bmatrix}$ under reflection about the x-axis. So A merely is this reflection. If $B = \begin{bmatrix} 0 & -1 \\ 1 & 0 \end{bmatrix}$, then

$$B \begin{bmatrix} x \\ y \end{bmatrix} = \begin{bmatrix} 0 & -1 \\ 1 & 0 \end{bmatrix} \begin{bmatrix} x \\ y \end{bmatrix} = \begin{bmatrix} -y \\ x \end{bmatrix}.$$

So B can be interpreted as the rotation of the plane, through 90° in the counter-clockwise direction, about the origin.

From the definition of the action of a matrix on a vector, we easily can show the following basic properties of this action.

Lemma 1.8.1. If $A, B \in M_2(\mathbb{R})$, $v, w \in V$ and $a, b \in \mathbb{R}$, then:

1. $A(v + w) = Av + Aw$;
2. $A(av) = a(Av)$;
3. $(A + B)v = Av + Bv$;
4. $(aA + bB)v = aAv + bBv$.

We leave the verification of these properties as an exercise for the reader. Mappings on an abstract vector space (whatever that is) that satisfy (1) and (2) are called *linear transformations* on V, and A is said to act *linearly* on V. So, in these terms, $M_2(\mathbb{R})$ consists of linear transformations on V.

How does the composition of two matrices A and B as mapping jibe with the formal matrix product we defined for matrices? Letting A and B be the matrices

$$A = \begin{bmatrix} a & b \\ c & d \end{bmatrix} \text{ and } B = \begin{bmatrix} r & s \\ t & u \end{bmatrix},$$ then by the composition of mappings we have

$$(A \cdot B)\begin{bmatrix} x \\ y \end{bmatrix} = A\left(B\begin{bmatrix} x \\ y \end{bmatrix}\right) = A\left(\begin{bmatrix} r & s \\ t & u \end{bmatrix}\begin{bmatrix} x \\ y \end{bmatrix}\right) = A\begin{bmatrix} rx + sy \\ tx + uy \end{bmatrix}$$

$$= \begin{bmatrix} a(rx + sy) + b(tx + uy) \\ c(rx + sy) + d(tx + uy) \end{bmatrix} = \begin{bmatrix} (ar + bt)x + (as + bu)y \\ (cr + dt)x + (cs + du)y \end{bmatrix}$$

$$= \begin{bmatrix} ar + bt & as + bu \\ cr + dt & cs + du \end{bmatrix}\begin{bmatrix} x \\ y \end{bmatrix} = (AB)\begin{bmatrix} x \\ y \end{bmatrix},$$

where $A \cdot B$ is the product of A and B as mappings, and AB is the matrix product of A and B. So $A \cdot B = AB$. Thus *we see that treating matrices as mappings, or treating them as formal objects and multiplying by some weird multiplication leads to the same result for what we want the product to be*. In fact, the reason that matrix multiplication was defined as it was is precisely this fact, namely, that this formal matrix multiplication corresponds to the more natural notion of the product of mappings. Note one little by-product of all this. Since the product of mappings is associative (Lemma 1.7.1), we get, free of charge (without any messy computation), that the product of matrices satisfies the associative law.

For emphasis, we state what we just discovered as

Theorem 1.8.2. If A, B are matrices in $M_2(\mathbb{R})$ then their product as mappings, $A \cdot B$, coincides with their product as matrices in $M_2(\mathbb{R})$.

PROBLEMS

NUMERICAL PROBLEMS

1. Show that every element in V can be written in a unique way as $ae_1 + be_2$, where $a, b \in \mathbb{R}$ and $e_1 = \begin{bmatrix} 1 \\ 0 \end{bmatrix}$ and $e_2 = \begin{bmatrix} 0 \\ 1 \end{bmatrix}$ are in V.

2. Let the matrices E_{ij} be defined as follows: E_{ij} is a matrix such that the (i, j) entry is 1 and all other entries are 0.

(a) Find a formula for $E_{ij}E_{kl}$.
(b) Interpret the formula you get geometrically.
3. Find all matrices A such that $AE_{11} = E_{11}A$.
4. Find all matrices A such that $AE_{12} = E_{21}A$.
5. If A is any matrix, compute $E_{11}AE_{11}$.
6. If A is any matrix, compute $E_{12}AE_{21}$.

MORE THEORETICAL PROBLEMS

Easier Problems

7. Prove Lemma 1.8.1 in detail.
8. Give a geometric interpretation of the action of the matrix $\begin{bmatrix} -1 & 0 \\ 0 & 1 \end{bmatrix}$ on V.
9. Give a geometric interpretation of the action of the matrix $\begin{bmatrix} 1 & c \\ 0 & 1 \end{bmatrix}$ on V.
10. Give a geometric interpretation of the action of the matrix $\begin{bmatrix} \cos\theta & \sin\theta \\ -\sin\theta & \cos\theta \end{bmatrix}$ on V.
11. Under what condition on the matrix $A = \begin{bmatrix} a & b \\ c & d \end{bmatrix}$ will the new axes $x_1 = ax + by$, $y_1 = cx + dy$ intersect at 90°?
12. Show that every matrix A in $M_2(\mathbb{R})$ can be written uniquely as $A = \sum_{r=1}^{2}\sum_{s=1}^{2} a_{rs}E_{rs}$.
13. Let T be a linear transformation on V [i.e., T satisfies $T(v + w) = Tv + Tw$ and $T(av) = aT(v)$ for $v, w \in V$, $a \in \mathbb{R}$]. Show that we can realize T as some matrix $T = \begin{bmatrix} t_{11} & t_{12} \\ t_{21} & t_{22} \end{bmatrix}$. Can you see a way of determining the entries of this matrix $[t_{rs}]$?
14. Let $\varphi: V \to M_2(\mathbb{R})$ be defined by $\varphi\begin{bmatrix} x \\ y \end{bmatrix} = \begin{bmatrix} x & 0 \\ y & 0 \end{bmatrix}$. Then prove:
    (a) $\varphi(v + w) = \varphi(v) + \varphi(w)$ for $v, w \in V$.
    (b) $\varphi(av) = a\varphi(v)$ for $a \in \mathbb{R}$, $v \in V$.
    (c) If $A \in M_2(\mathbb{R})$, then $\varphi(Av) = A\varphi(v)$, where Av is the result of the action of A on v and $A\varphi(v)$ is the usual matrix product of the matrices A and $\varphi(v)$.
    (d) If $\varphi(v) = 0$, then $v = 0$.
15. If $\det(A) = 0$, show that there is a vector $v \neq 0$ in V such that $Av = 0$.

1.9. THE CAYLEY–HAMILTON THEOREM

Given a matrix A in $M_2(\mathbb{R})$, we can attach to A a polynomial, called the *characteristic polynomial* of A, the roots of which determine, to a large extent, the behavior of A as a linear transformation on V. This result will be shown later to hold quite generally for square matrices of all sizes. Interestingly enough, although this result is attributed to

Cayley, he really showed it only for the 2 × 2 and 3 × 3 matrices, where it is quite easy and straightforward. It took later mathematicians to prove it for the general case.

Consider a polynomial

$$p(x) = a_0 x^n + a x_1^{n-1} + \cdots + a_{n-1} x^1 + a_n x^0$$

with real coefficients $a_0, a_1, \ldots, a_n$. We usually write 1 in place of x^0, but it is convenient for the sake of the next definition to write it as x^0.

Definition. A matrix A is said to *satisfy* the polynomial $p(x)$ if $p(A) = 0$, where $p(A)$ denotes the matrix obtained by replacing each occurrence of x in $p(x)$ by the matrix A:

$$p(A) = a_0 A^n + a A_1^{n-1} + \cdots + a_{n-1} A^1 + a_n A^0.$$

(Here, it is convenient to adopt the convention that A^0 is the identity matrix I.)

For instance, the matrices $I = \begin{bmatrix} 1 & 0 \\ 0 & 1 \end{bmatrix}$, $A = \begin{bmatrix} 1 & 0 \\ 0 & -1 \end{bmatrix}$, and $B = \begin{bmatrix} 0 & 1 \\ -1 & 0 \end{bmatrix}$ satisfy the polynomials $p(x) = x - 1$, $p(x) = x^2 - 1$, and $p(x) = x^2 + 1$, respectively. To see the last one, we must show that $B^2 + I = 0$. But

$$B^2 = \begin{bmatrix} 0 & 1 \\ -1 & 0 \end{bmatrix} \begin{bmatrix} 0 & 1 \\ -1 & 0 \end{bmatrix} = \begin{bmatrix} -1 & 0 \\ 0 & -1 \end{bmatrix} = -I,$$

hence $B^2 + I = 0$.

We now introduce the very important polynomial associated with a given matrix.

Definition. If $A \in M_2(\mathbb{R})$, then the *characteristic polynomial* of A, denoted $P_A(x)$, is defined by

$$P_A(x) = \det(xI - A).$$

The characteristic polynomial comes up often in various sciences, such as physics and chemistry. There it is often referred to as the *secular determinant*.

For instance, going back to the matrix $B = \begin{bmatrix} 0 & 1 \\ -1 & 0 \end{bmatrix}$, we have

$$P_B(x) = \det[xI - B] = \det \begin{bmatrix} x - 0 & -1 \\ -(-1) & x - 0 \end{bmatrix} = \det \begin{bmatrix} x & -1 \\ 1 & x \end{bmatrix} = x^2 + 1.$$

Note that we showed above that B satisfies $x^2 + 1 = P_B(x)$, that is, B satisfies its characteristic polynomial. This is no accident, as we proceed to show in the

Theorem 1.9.1 (Cayley–Hamilton Theorem). A matrix satisfies its characteristic polynomial.

Proof: Since we are working with 2 × 2 matrices, the proof will be quite easy, for we can write down $P_A(x)$ explicitly. Let $A = \begin{bmatrix} a & b \\ c & d \end{bmatrix}$. Then we have

$$xI - A = \begin{bmatrix} x & 0 \\ 0 & x \end{bmatrix} - \begin{bmatrix} a & b \\ c & d \end{bmatrix} = \begin{bmatrix} x-a & -b \\ -c & x-d \end{bmatrix}.$$

Hence we have

$$P_A(x) = \det[xI - A] = \det \begin{bmatrix} x-a & -b \\ -c & x-d \end{bmatrix}$$
$$= (x-a)(x-d) - bc$$
$$= x^2 - (a+d)x + ad - bc$$
$$= x^2 - (\operatorname{tr}(A))x + (\det(A))1.$$

What we must verify is, therefore, that A satisfy the polynomial

$$x^2 - (\operatorname{tr}(A))x + (\det(A))1,$$

that is, that

$$A^2 - (\operatorname{tr}(A))A + (\det(A))I = 0.$$

This was actually given as Problem 13 of Section 1.5. However, we do it in detail here. Because

$$A^2 = \begin{bmatrix} a & b \\ c & d \end{bmatrix}^2 = \begin{bmatrix} aa+bc & ab+bd \\ ca+cd & cb+dd \end{bmatrix},$$

we have

$$A^2 - (\operatorname{tr}(A))A + (\det(A))I$$
$$= \begin{bmatrix} aa+bc & ab+bd \\ cd+ca & dd+bc \end{bmatrix} - (a+d)\begin{bmatrix} a & b \\ c & d \end{bmatrix} + (ad-bc)\begin{bmatrix} 1 & 0 \\ 0 & 1 \end{bmatrix}$$
$$= \begin{bmatrix} aa+bc-(a+d)a+ad-bc & ab+bd-(a+d)b \\ cd+ca-(a+d)c & dd+bc-(a+d)d+ad-bc \end{bmatrix} = 0.$$

Thus A satisfies the polynomial $P_A(x) = x^2 - (\operatorname{tr}(A))x + (\det(A))1$. ∎

Given a polynomial $f(x)$ with real coefficients, associated with $f(x)$ are certain "numbers," called the *roots* of $f(x)$. The number a is a root of f if $f(a) = 0$. For instance, 3 is a root of $f(x) = x^2 - 9x + 18$, since $f(3) = 3^2 - 9 \cdot 3 + 18 = 0$.

We used the vague word "numbers" above in defining a root of a polynomial. There is a very good reason for this vagueness. If we insist that the roots be real numbers, we get into a bit of a bind, for the very nice polynomial $f(x) = x^2 + 1$ has no real root, since $a^2 + 1 \neq 0$ for every real number a. On the other hand, we want our polynomials to have roots. This necessitates the enlargement of the concept of real number to some wider arena: that of *complex numbers*. We do this in the next section. When this has been accomplished, we shall not have to worry about the presence or absence of roots for the polynomials that will arise. To be precise, the powerful so-called *Fundamental Theorem of Algebra* ensures that every polynomial $f(x)$ with complex coefficients has enough complex roots $a_1, \ldots, a_n$ that it factors as a product $f(x) = (x - a_1) \cdots (x - a_n)$.

For the moment, let's act as if the polynomials we are dealing with had real roots—something that is often false. At any rate, let's see where this will lead us.

Definition. A *characteristic root* of A is a number a such that $P_A(a) = 0$.

Here, too we're still being vague as to the meaning of the word "number."

If $A = \begin{bmatrix} 1 & 6 \\ 0 & -3 \end{bmatrix}$, we readily find that $P_A(x) = \det(xI - A) = (x-1)(x+3) = x^2 + 2x - 3$, so 1 and -3 are characteristic roots of A. On the other hand, if $B = \begin{bmatrix} 0 & 1 \\ -1 & 0 \end{bmatrix}$, then, as we saw, $P_B(x) = x^2 + 1$, which has no real roots, so *for the moment* has no characteristic root.

Incidentally, what we call a characteristic root of A is often called an *eigenvalue* of A. In fact, this hybrid English–German word is the one always used by physicists.

Suppose that a is a characteristic root of A; thus $\det(aI - A) = P_A(a) = 0$. By Theorem 1.5.4, since $\det(aI - A) = 0$, $aI - A$ is *not* invertible. On the other hand, if b is a number such that $bI - A$ is not invertible, then, again by Theorem 1.5.4, $\det(bI - A) = 0$. Therefore, *the characteristic roots of A are precisely those numbers a for which $aI - A$ fails to have an inverse*. Since $P_A(x) = \det(xI - A)$ is a polynomial of degree 2, it can have at most two real roots (and, in fact, will have two real or complex roots). So $aI - A$ can fail to be invertible for *at most* two real numbers. Thus $aI - A$ is *almost always* invertible.

We summarize all of this in

Theorem 1.9.2. *The characteristic roots of A are precisely those numbers a for which the matrix $aI - A$ is not invertible. If $A \in M_2(\mathbb{R})$, then A has at most two real characteristic roots.*

In Problem 15 of Section 1.8 we asked to show that if $\det B = 0$, then there is some $v \neq 0$ in V such that $Bv = 0$. Since we need this result now, we prove it. If $B = 0$, there is nothing to prove, for *every* vector v satisfies $0v = 0$. So, suppose that $B = \begin{bmatrix} r & s \\ t & u \end{bmatrix} \neq 0$. Then some row of B does not consist entirely of zeros, say the first row. Then if $v = \begin{bmatrix} s \\ -r \end{bmatrix}$, we have

$$Bv = \begin{bmatrix} r & s \\ t & u \end{bmatrix} \begin{bmatrix} s \\ -r \end{bmatrix} = \begin{bmatrix} rs - sr \\ ts - ur \end{bmatrix} = \begin{bmatrix} 0 \\ -\det B \end{bmatrix} = \begin{bmatrix} 0 \\ 0 \end{bmatrix}.$$

Hence $Bv = 0$ and $v \neq 0$. The same argument works if it is the other row that does not consist entirely of zeros.

Suppose, now, that the real number a is a characteristic root of A. Then $aI - A$ is not invertible and $\det(aI - A) = 0$. Thus, by the above, there is a nonzero $v \in V$ such that $(aI - A)v = 0$, which is to say, $Av = (aI)v = av$. We therefore have

Theorem 1.9.3. *The real number a is a characteristic root of A if and only if for some $v \neq 0$ in V, $Av = av$.*

Definition. *An element $v \neq 0$ in V such that $Av = av$ is called a* characteristic vector *associated with a.*

Physicists often use a hybrid English–German word *eigenvector* for a characteristic vector.

Let $A = \begin{bmatrix} 5 & 12 \\ 12 & -5 \end{bmatrix}$. Then

$$P_A(x) = \det \begin{bmatrix} x-5 & -12 \\ -12 & x+5 \end{bmatrix} = (x-5)(x+5) - 12^2 = x^2 - 169,$$

whose roots are ± 13. What v are characteristic vectors associated with 13, and which with -13? If $\begin{bmatrix} 5 & 12 \\ 12 & -5 \end{bmatrix} \begin{bmatrix} x \\ y \end{bmatrix} = 13 \begin{bmatrix} x \\ y \end{bmatrix}$, then $5x + 12y = 13x$, hence $8x = 12y$, so $2x = 3y$. If we let $x = 3$, then $y = 2$. So the vector $\begin{bmatrix} 3 \\ 2 \end{bmatrix}$ is a characteristic vector of $\begin{bmatrix} 5 & 12 \\ 12 & -5 \end{bmatrix}$ associated with 13. In fact, all such vectors are of the form $\begin{bmatrix} 3s \\ 2s \end{bmatrix}$ for any $s \neq 0$. For -13, we similarly can see that $\begin{bmatrix} 2 \\ -3 \end{bmatrix}$ (and, in fact, all of its nonzero multiples $s\begin{bmatrix} 2 \\ -3 \end{bmatrix}$) is a characteristic vector associated with -13.

(Don't expect the characteristic roots always to be "nice" numbers. For instance the characteristic roots of $\begin{bmatrix} 1 & 2 \\ 3 & 4 \end{bmatrix}$ are the roots of $x^2 - 5x - 2$, which are $\dfrac{5 \pm \sqrt{33}}{2}$. Can you find the respective characteristic vectors associated with these?)

Let's return to $A = \begin{bmatrix} 5 & 12 \\ 12 & -5 \end{bmatrix}$, whose characteristic roots were 13 and -13, with corresponding characteristic vectors $\begin{bmatrix} 3 \\ 2 \end{bmatrix}$ and $\begin{bmatrix} 2 \\ -3 \end{bmatrix}$. Note the following about these characteristic vectors: If $a\begin{bmatrix} 3 \\ 2 \end{bmatrix} + b\begin{bmatrix} 2 \\ -3 \end{bmatrix} = \begin{bmatrix} 0 \\ 0 \end{bmatrix}$, then $a = b = 0$. (Prove this!) This is a rather important attribute of $\begin{bmatrix} 3 \\ 2 \end{bmatrix}$ and $\begin{bmatrix} 2 \\ -3 \end{bmatrix}$ which we treat in great detail later.

Note that if the characteristic roots of a matrix are real, we need not have two distinct ones. The easiest example is the matrix I which has only the number 1 as a characteristic root. This example illustrates that *the characteristic polynomial $P_A(x)$ need not be a polynomial of lowest degree* satisfied by A. For instance the characteristic polynomial of I is $P_I(x) = (x-1)^2$, yet I satisfies the polynomial $q(x) = x - 1$. Is there only one *monic* (leading coefficient is 1) polynomial of lowest degree satisfied by A? Yes: The difference of any two such has lower degree and is still satisfied by A, so the difference is 0.

This prompts the

Definition. The *minimal polynomial* for the matrix A is the nonzero polynomial of lowest degree satisfied by A having 1 as the coefficient of its highest term.

For 2×2 matrices, this distinction between the minimal polynomial and the characteristic polynomial of A is not very important, for if these are different, then A must be a scalar. For the general case, this is important. This is also true because of

Theorem 1.9.4 If A satisfies a polynomial $f(x)$, then every characteristic root of A is a root of $f(x)$. So every characteristic root of A is a root of the minimal polynomial of A.

Proof: Suppose that A satisfies $f(x) = a_0 x^n + a_1 x^{n-1} + \cdots + a_n x^0$. If λ is a characteristic root of A, then we can find a vector $v \neq 0$ such that $Av = \lambda v$. Thus we have

$$A^2 v = A(Av) = A(\lambda v) = \lambda(Av) = \lambda(\lambda v) = \lambda^2 v.$$

Continuing in this way, we obtain

$$A^k v = \lambda^k v \quad \text{for all } k > 0.$$

So, since $f(A) = 0$, we have

$$\begin{aligned}
0 = f(A)v &= (a_0 A^n + a_1 A^{n-1} + \cdots + a_n I)v \\
&= (a_0 A^n)v + (a_1 A^{n-1})v + \cdots + (a_n I)v \\
&= a_0 \lambda^n v + a_1 \lambda^{n-1} v + \cdots + a_n v \\
&= (a_0 \lambda^n + a_1 \lambda^{n-1} + \cdots + a_n)v.
\end{aligned}$$

Because $\mu = a_0 \lambda^n + a_1 \lambda^{n-1} + \cdots + a_{n-1} \lambda + a_n = f(\lambda)$ is a scalar and $\mu v = 0$ with $v \neq 0$, we end up with $0 = \mu = f(\lambda)$. So λ is a root of f.

Since A satisfies its minimal polynomial, every characteristic root is a root of this minimal polynomial. ∎

PROBLEMS

NUMERICAL PROBLEMS

1. Calculate the characteristic polynomial for the given matrix.

    (a) $\begin{bmatrix} 1 & -5 \\ -6 & 2 \end{bmatrix}.$

(b) $\begin{bmatrix} a & 0 \\ b & c \end{bmatrix}$.

(c) $\begin{bmatrix} \frac{1}{2} & \frac{1}{2} \\ \frac{1}{2} & \frac{1}{2} \end{bmatrix}$.

(d) $\begin{bmatrix} \frac{1}{2} & -\frac{1}{2} \\ \frac{1}{2} & -\frac{1}{2} \end{bmatrix}$.

2. In Problem 1, find the characteristic roots of the matrices given.
3. In Problem 2, find the characteristic vector for each characteristic root of each matrix given.
4. If a is *not* a characteristic root of $A = \begin{bmatrix} 1 & -5 \\ -6 & 2 \end{bmatrix}$, find the explicit form of $(aI - A)^{-1}$. What goes wrong with your calculation if a is a characteristic root?
5. In Part (c) of Problem 1 find an invertible matrix A such that

$$A \begin{bmatrix} \frac{1}{2} & \frac{1}{2} \\ \frac{1}{2} & \frac{1}{2} \end{bmatrix} A^{-1} = \begin{bmatrix} 1 & 0 \\ 0 & 0 \end{bmatrix}.$$

MORE THEORETICAL PROBLEMS

Easier Problems

6. If B is invertible and $C = BAB^{-1}$, show that $P_A(x) = P_C(x)$.
7. If A and C are as in Problem 6, and if a is a characteristic root of A with associated characteristic vector v, show that a is a characteristic root of C and exhibit its associated characteristic vector.
8. If $A^2 = 0$, find the possible forms for $P_A(x)$.
9. If $A^2 = A$, find the possible forms for $P_A(x)$.
10. Calculate $P_A(x)$ if $A = \begin{bmatrix} a & b \\ -b & a \end{bmatrix}$.
11. If A is a matrix and A' its transpose, then show that $P_A(x) = P_{A'}(x)$.
12. Using Problem 11, show that A and A' have the same characteristic roots.
13. If A is an invertible matrix, then show that the characteristic roots of A^{-1} are the inverses of the characteristic roots of A, and vice versa.
14. If a is a characteristic root of A and $f(x)$ is a polynomial with coefficients in F, show that $f(a)$ is a characteristic root of $f(A)$.

Middle-Level Problems

15. If A is invertible, express $P_{A^{-1}}(x)$ in terms of $P_A(x)$.
16. Define for A, B in $M_2(F)$ that $A \sim B$ if there exists an invertible U in $M_2(F)$ such that $B = UAU^{-1}$. (Matrices A, B such that $A \sim B$ are called *similar*.) Prove:
    (a) $A \sim A$.
    (b) If $A \sim B$, then $B \sim A$.
    (c) If $A \sim B$ and $B \sim C$, then $A \sim C$
    for all A, B, C in $M_2(F)$.

17. Show that $\begin{bmatrix} a & b \\ b & c \end{bmatrix}$ has real characteristic roots.

18. If $\det(A) < 0$, show that A has distinct, real characteristic roots.

19. If $A^2 = A$, show that $(I - A)^2 = I - A$.

20. If $A^2 = A$ is a matrix, show that every vector $v \in V$ can be written in a unique way as $v = v_1 + v_2$, where $Av_1 = v_1$ and $Av_2 = 0$. $A \neq I \neq O$

Harder Problems

21. Show that a necessary and sufficient condition that $A = \begin{bmatrix} a & b \\ c & d \end{bmatrix}$ have two distinct, real characteristic roots u and v is that for some invertible matrix C, $CAC^{-1} = \begin{bmatrix} u & 0 \\ 0 & v \end{bmatrix}$.

22. Prove that for any two matrices A and B, the characteristic polynomials for the products AB and BA are equal; that is, $P_{AB}(x) = P_{BA}(x)$.

23. Are the minimal polynomials of AB and BA always equal? If yes, prove; and if no, give a *counterexample*.

24. Find all matrices A in $M_2(\mathbb{R})$ whose characteristic roots are real and which satisfy the equation $AA' = I$.

1.10. COMPLEX NUMBERS

As we saw in looking at the matrix $B = \begin{bmatrix} 0 & 1 \\ -1 & 0 \end{bmatrix}$, whose characteristic polynomial $P_B = x^2 + 1$ has no real roots, if we restrict ourselves to working only with real numbers, we easily run into matrices that have no real characteristic roots. Yet it would be desirable if all matrices had characteristic roots—even a full complement of them. In order to reach this state of nirvana we must widen the concept of number. In doing so the set of complex numbers crops up naturally. This enlarged set of numbers play a prominent part not only in every phase of mathematics but in all sort of other fields—chemistry, physics, engineering, and so on. The analysis of electrical circuits becomes easy once one introduces the nature of complex impedance. So, aside from whatever role they will play in matrix theory, it would be highly useful to know something about them.

The right way of defining complex numbers is by formally introducing a set of symbols—constructed over the reals—and to show that with the proper definition of addition and multiplication this set of new objects behaves very well. This we shall do very soon. But to motivate the definitions we adopt a backdoor kind of approach and introduce the complex numbers via a certain set of matrices in $M_2(\mathbb{R})$.

Let $\mathscr{C}$ be the set of all matrices of the form $\begin{bmatrix} a & b \\ -b & a \end{bmatrix}$, where a and b are real numbers. If we denote the matrix $\begin{bmatrix} 0 & 1 \\ -1 & 0 \end{bmatrix}$ by J, then all the matrices in $\mathscr{C}$ look like $aI + bJ$, where a and b are any real numbers.

How do elements of C behave under the matrix operations? Since

$$J^2 = \begin{bmatrix} 0 & 1 \\ -1 & 0 \end{bmatrix}\begin{bmatrix} 0 & 1 \\ -1 & 0 \end{bmatrix} = \begin{bmatrix} -1 & 0 \\ 0 & -1 \end{bmatrix} = -I$$

we see that:

1. $(aI + bJ) + (cI + dJ) = (a + c)I + (b + d)J$, so is again in $\mathscr{C}$.
2. $(aI + bJ)(cI + dJ) = (ac - bd)I + (ad + bc)J$, so it too is again in $\mathscr{C}$.

Here we have used the fact that $J^2 = -I$.

3. $(aI + bJ)(cI + dJ) = (ac - bd)I + (ad + bc)J = (ca - bd)I + (da + cb)J = (cI + dJ)(aI + bJ)$. Thus the multiplication of elements of $\mathscr{C}$ is *commutative*.

But possibly the most important property of $\mathscr{C}$ is yet to come. If $aI + bJ \neq 0$, then one of $a \neq 0$, $b \neq 0$ holds. Hence $c = a^2 + b^2 \neq 0$. Now $(aI + bJ)(aI - bJ) = (a^2 + b^2)I$; therefore, $(aI + bJ)\left(\dfrac{a}{c}I - \dfrac{b}{c}J\right) = \dfrac{a^2 + b^2}{c}I = I$. So the inverse of $aI + bJ$ is $\dfrac{a}{c}I - \dfrac{b}{c}J$, which is also in $\mathscr{C}$. Thus we have:

4. If $aI + bJ \neq 0$ is in $\mathscr{C}$, then $(aI + bJ)^{-1}$ is also in $\mathscr{C}$.

Finally, if we "identify" the matrix aI with the real number a, we get that $\mathbb{R}$ (or something quite like it) is in $\mathscr{C}$. Thus

5. $\mathbb{R}$ is contained in $\mathscr{C}$.

So in many ways $\mathscr{C}$, which is larger than $\mathbb{R}$, behaves very much like $\mathbb{R}$ with respect to addition, multiplication, and division. But $\mathscr{C}$ has the advantage that in it $J^2 = -I$, so we have something like $\sqrt{-1}$ in $\mathscr{C}$.

With this little discussion of $\mathscr{C}$ and the behavior of its elements relative to addition and multiplication as our guide, we are ready formally to define the set of complex numbers $\mathbb{C}$. Of course, the complex numbers did not arise historically via matrices—matrices arose much later in the game than the complex numbers—but since we do have the 2×2 matrices over $\mathbb{R}$ in hand it does not hurt to motivate the complex numbers in the fashion we did above.

Definition. The complex numbers $\mathbb{C}$ is the set of all *formal symbols* $a + bi$, where a and b are any real numbers, where we define:

1. $a + bi = c + di$ if and only if $a = c$ and $b = d$.
2. $(a + bi) + (c + di) = (a + c) + (b + d)i$.
3. $(a + bi)(c + di) = (ac - bd) + (ad + bc)i$.

This last property comes from the fact that we want i^2 to be equal to -1.

If we merely write $a + 0i$ as a and $0 + bi$ as bi, then the rule of multiplication in (3) gives us $i^2 = (0 + i)(0 + i) = (0 \cdot 0 - 1 \cdot 1) + (0 \cdot 1 + 1 \cdot 0)i = -1 + 0i = -1$.

If $\alpha = a + bi$, then a is called the *real part* of α and bi the *imaginary part* of α. If $\alpha = bi$, we call α *pure imaginary*.

We assume that $i^2 = -1$ and multiply out formally to get the multiplication rule of (3). This is the best way to remember how complex numbers multiply.

Sec. 1.10]   Complex Numbers   47

As we pointed out, we *identify* $a + 0i$ with a; this implies that $\mathbb{R}$ is contained in $\mathbb{C}$.

Also, as we did for matrices, rule (3) implies that $(a + bi)(a - bi) = a^2 + b^2$, so if $a + bi \neq 0$, then

$$(a + bi)^{-1} = \frac{a}{a^2 + b^2} - \frac{b}{a^2 + b^2} i$$

and is again in $\mathbb{C}$.

If $\alpha = 2 + 3i$, we see that

$$\alpha^{-1} = \frac{2}{2^2 + 3^2} - \frac{3}{2^2 + 3^2} i = \frac{2}{13} - \frac{3}{13} i.$$

If $\beta = \frac{1}{\sqrt{2}} - \frac{1}{\sqrt{2}} i$, we leave it to the reader to show that $\beta^{-1} = \frac{1}{\sqrt{2}} + \frac{1}{\sqrt{2}} i$.

Before going any further, we document other basic rules of behavior in $\mathbb{C}$. We shall use lowercase Greek letters to denote elements of $\mathbb{C}$.

Theorem 1.10.1. $\mathbb{C}$ satisfies the following rules:

1. If $\alpha, \beta \in \mathbb{C}$, then $\alpha + \beta \in \mathbb{C}$.
2. If $\alpha, \beta \in \mathbb{C}$, then $\alpha + \beta = \beta + \alpha$.
3. For all $\alpha \in \mathbb{C}$, $\alpha + 0 = \alpha$.
4. Given $\alpha = a + bi \in \mathbb{C}$, then $-\alpha = -a - bi \in \mathbb{C}$ and $\alpha + (-\alpha) = 0$.
5. If $\alpha, \beta, \gamma \in \mathbb{C}$, then $\alpha + (\beta + \gamma) = (\alpha + \beta) + \gamma$.

These rules specify how $\mathbb{C}$ behaves relative to $+$. We now specify how it behaves relative to multiplication.

6. If $\alpha, \beta \in \mathbb{C}$, then $\alpha\beta \in \mathbb{C}$.
7. If $\alpha, \beta \in \mathbb{C}$, then $\alpha\beta = \beta\alpha$.
8. If $\alpha, \beta, \gamma \in \mathbb{C}$, then $\alpha(\beta\gamma) = (\alpha\beta)\gamma$.
9. The complex number $1 = 1 + 0i$ satisfies $\alpha 1 = \alpha$ for all $\alpha \in \mathbb{C}$.
10. Given $\alpha \neq 0 \in \mathbb{C}$, then $\alpha^{-1} \in \mathbb{C}$, where $\alpha^{-1}\alpha = 1$.

The final rule weaves together addition and multiplication.

11. If $\alpha, \beta, \gamma \in \mathbb{C}$, then $\alpha(\beta + \gamma) = \alpha\beta + \alpha\gamma$.

Proof: We leave all parts of Theorem 1.10.1, except for (8), as an exercise for the reader. How do we prove (8)? Let $\alpha = a + bi$, $\beta = c + di$, $\gamma = g + hi$. Then, by our rule for the product, $\alpha\beta = (a + bi)(c + di) = (ac - bd) + (ad + bc)i$; hence

$$\begin{aligned}(\alpha\beta)\gamma &= ((ac - bd) + (ad + bc)i)(g + hi) \\ &= (ac - bd)g - (ad + bc)h + ((ac - bd)h + (ad + bc)g)i \\ &= (a(cg - dh) - b(dg + ch)) + (a(ch + dg) + b(-dh + cg))i \\ &= (a + bi)((cg - dh) + (ch + dg)i) \\ &= (a + bi)((c + di)(g + hi)) = \alpha(\beta\gamma).\end{aligned}$$

This long chain is easy, but looks a little frightening. ∎

The properties specified for $\mathbb{C}$ by (1)–(11) define what is called a *field* in mathematics. So $\mathbb{C}$ is an example of a field. Other examples of fields are $\mathbb{R}$, $\mathbb{Q}$, the rational numbers, and $T = \{a + b\sqrt{2} \,|\, a, b \text{ rational}\}$. Notice that rules (1)–(11), while forming a long list, are really a reminder of the usual rules of arithmetic to which we are so accustomed.

$\mathbb{C}$ has the property that $\mathbb{R}$ fails to have, namely, we can find roots in $\mathbb{C}$ of any quadratic polynomials having their coefficients in $\mathbb{C}$ (something which is all too false in $\mathbb{R}$). To head toward this goal we first establish that every element in $\mathbb{C}$ has a *square root* in $\mathbb{C}$.

Theorem 1.10.2. *If α is in $\mathbb{C}$, then there exists a β in $\mathbb{C}$ such that $\alpha = \beta^2$.*

Proof: Let $\alpha = a + bi$. If $b = 0$, we leave to the reader to show that there is a complex number β such that $\beta^2 = a = \alpha$. So we may assume that $b \neq 0$. We want to find *real* numbers x and y such that $\alpha = a + bi = (x + yi)^2$. Now, by our rule of multiplication, $(x + yi)^2 = (x^2 - y^2) + 2xyi$. So we want $a + bi = (x^2 - y^2) + 2xyi$, that is, $a = x^2 - y^2$ and $b = 2xy$. Substituting $y = b/2x$ in $a = x^2 - y^2$, we come up with $a = x^2 - b^2/4x^2$, hence $4x^4 - 4ax^2 - b^2 = 0$. This is a quadratic equation in x^2, so we can solve for x^2 using the *quadratic formula*. This yields that

$$x^2 = \frac{4a \pm \sqrt{16a^2 + 16b^2}}{8} = \frac{a \pm \sqrt{a^2 + b^2}}{2}.$$

Since $b \neq 0$, $a^2 + b^2 > a^2$ and $\sqrt{a^2 + b^2}$ is therefore larger in size than a. So x^2 cannot be $\dfrac{a - \sqrt{a^2 + b^2}}{2}$ for a real value of x, since $\dfrac{a - \sqrt{a^2 + b^2}}{2} < 0$. So we are forced to conclude that

$$x = \pm\sqrt{\frac{a + \sqrt{a^2 + b^2}}{2}}$$

are the *only real solutions* for x. The formula for x makes sense since $a + \sqrt{a^2 + b^2}$ is *positive*. (Prove!) So we have our required x. What is y? Since $y = b/2x$ we get the value for y from this using the value of x obtained above. The x and y, which are real numbers, then satisfy $(x + yi)^2 = a + bi = \alpha$. Thus Theorem 1.10.2 is proved. Note that if $\alpha \neq 0$, we can find exactly two square roots of α in $\mathbb{C}$. ∎

An immediate consequence of Theorem 1.10.2 is that any quadratic equation with coefficients in $\mathbb{C}$ has two solutions in $\mathbb{C}$. This is

Theorem 1.10.3. *Given α, β, γ in $\mathbb{C}$, with $\alpha \neq 0$, we can always find a solution of $\alpha x^2 + \beta x + \gamma = 0$ for some x in $\mathbb{C}$.*

Proof: The usual quadratic formula, namely

$$x = \frac{-\beta \pm \sqrt{\beta^2 - 4\alpha\gamma}}{2\alpha},$$

holds as well here as it does for real coefficients. (Prove!) Now, by Theorem 1.10.2, since $\beta^2 - 4\alpha\gamma$ is in $\mathbb{C}$, its square root, $\sqrt{\beta^2 - 4\alpha\gamma}$, is also in $\mathbb{C}$. Thus x is in $\mathbb{C}$. Since x is a solution to $\alpha x^2 + \beta x + \gamma = 0$, the theorem is proved. ∎

In fact, Theorem 1.10.3 actually shows that $\alpha x^2 + \beta x + \gamma = 0$ has two *distinct* solutions in $\mathbb{C}$ provided that $\beta^2 - 4\alpha\gamma \neq 0$.

We have another nice operation, that of taking the *complex conjugate* of a complex number.

Definition. If $\alpha = a + bi$ is in $\mathbb{C}$, then the *complex conjugate* of α, denoted by $\bar{\alpha}$, is $\bar{\alpha} = a - bi$.

The properties governing the behavior of "conjugacy" are contained in

Lemma 1.10.4. For any complex numbers α and β we have:

1. $\overline{(\bar{\alpha})} = \bar{\bar{\alpha}} = \alpha$;
2. $\overline{(\alpha + \beta)} = \bar{\alpha} + \bar{\beta}$;
3. $\overline{(\alpha\beta)} = \bar{\alpha}\bar{\beta}$;
4. $\alpha\bar{\alpha}$ is real and nonnegative. Moreover, $\alpha\bar{\alpha} > 0$ if $\alpha \neq 0$.

Proof: The proofs of all the parts are immediate from the definitions of the sum, product, and complex conjugate. We leave Parts (1), (2), and (4) as exercises, but do prove Part (3).

Let $\alpha = a + bi$, $\beta = c + di$. Thus $\alpha\beta = (a + bi)(c + di) = (ac - bd) + (ad + bc)i$, so

$$\overline{(\alpha\beta)} = \overline{(ac - bd) + (ad + bc)i} = (ac - bd) - (ad + bc)i.$$

On the other hand, $\bar{\alpha} = a - bi$ and $\bar{\beta} = c - di$, whence

$$\bar{\alpha}\bar{\beta} = (a - bi)(c - di)$$
$$= (ac - (-b)(-d)) + (a(-d) + (-b)c)i$$
$$= (ac - bd) - (ad + bc)i = \overline{(\alpha\beta)}. \blacksquare$$

Yet another notion plays an important role in dealing with complex numbers. This is the notion of *absolute value*, which, in a sense, measures the size of a complex number.

Definition. If $\alpha = a + bi$, then the *absolute value* of α, $|\alpha|$, is defined by $|\alpha| = \sqrt{a^2 + b^2}$.

Note that $|\alpha| = \sqrt{a^2 + b^2} = \sqrt{\alpha\bar{\alpha}}$.
The properties of $|\cdot|$ are outlined in

Lemma 1.10.5. If α and β are complex numbers, then

1. $\alpha + \bar{\alpha}$ is real and $\alpha + \bar{\alpha} \leq 2|\alpha|$;

2. $|\alpha\beta| = |\alpha||\beta|$;
3. $|\alpha + \beta| \leq |\alpha| + |\beta|$ (triangle inequality).

Proof: (1) Let $\alpha = a + bi$. Then $\alpha + \bar{\alpha} = 2a \leq 2\sqrt{a^2 + b^2} = 2|\alpha|$.
(2) $|\alpha\beta| = \sqrt{\alpha\beta\bar{\alpha}\bar{\beta}} = \sqrt{\alpha\bar{\alpha}\beta\bar{\beta}} = \sqrt{\alpha\bar{\alpha}}\sqrt{\beta\bar{\beta}} = |\alpha||\beta|$ since $\alpha\bar{\alpha}$ is real and nonnegative.
(3) $|\alpha + \beta|^2 = (\alpha + \beta)\overline{(\alpha + \beta)} = \alpha\bar{\alpha} + \alpha\bar{\beta} + \bar{\alpha}\beta + \beta\bar{\beta}$
$= |\alpha|^2 + (\alpha\bar{\beta}) + \overline{(\alpha\bar{\beta})} + |\beta|^2 \leq |\alpha|^2 + 2|\alpha\bar{\beta}| + |\beta|^2$
$= |\alpha|^2 + 2|\alpha||\beta| + |\beta|^2 = (|\alpha| + |\beta|)^2,$

since $(\alpha\bar{\beta}) + \overline{(\alpha\bar{\beta})} \leq 2|\alpha\bar{\beta}| = 2|\alpha||\beta|$. Taking square roots, we obtain $|\alpha + \beta| \leq |\alpha| + |\beta|$. ∎

Why the name "triangle inequality" for the inequality in Part (3)? We can identify the complex number $\alpha = a + bi$ with the point (a, b) in the Euclidean plane:

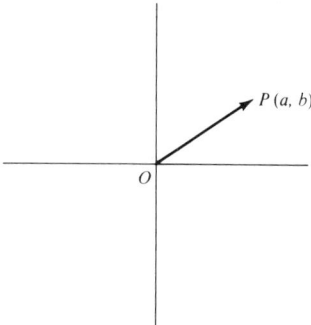

Then $|\alpha| = \sqrt{a^2 + b^2}$ is the length of the line segment OP. If we recall the diagram for the addition of complex numbers as points of the plane:

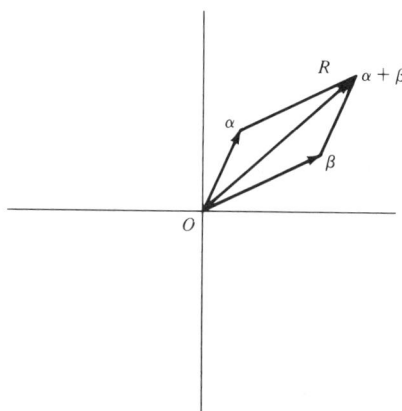

then the point $R, (a + c, b + d)$, is that point associated with $\alpha + \beta$, where $\alpha = a + bi$, $\beta = c + di$.

The triangle inequality then merely asserts that the length of OR is at most that of OP plus that of PR, in other words, in a triangle the length of any side is at most the sum of the lengths of the other two sides.

Again turning to the geometric interpretation of complex numbers, if $\alpha = a + bi$, it corresponds to the point P in

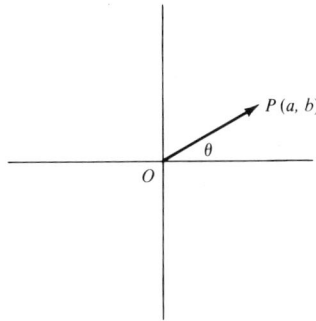

Thus $r = \sqrt{a^2 + b^2}$ is the length of OP, and the angle θ—called the *argument* of α—is determined by $\sin(\theta) = b/r$ if we restrict θ to be in the range $-\pi$ to π. Note that $a = r\cos(\theta)$ and $b = r\sin(\theta)$; thus $\alpha = a + bi = r\cos(\theta) + r\sin(\theta)i = r(\cos(\theta) + i\sin(\theta))$. This is called the *polar form* of the complex number α.

Before leaving the subject, we make a remark related to the result of Theorem 1.10.3. There it was shown that any quadratic polynomial having complex coefficients has a complex root. This is but a special case of a beautiful and very important result due to Gauss which is known as the *Fundamental Theorem of Algebra*. This theorem asserts that any polynomial, of whatever positive degree, with complex coefficients always has a complex root. We shall need this result when we treat the $n \times n$ matrices.

PROBLEMS

NUMERICAL PROBLEMS

1. Carry out the following.
   (a) $(6 - 5i)(7 + \frac{3}{2}i)$.
   (b) $(16 + 4i)(\frac{1}{2} - \frac{2}{3}i)^{-1}$.
   (c) $(4 - 3i)(16 + 12i)$.
   (d) $(4 + 7i)\overline{(8 - 6i)}$.
2. Find $\sqrt{i}$ in the form $a + bi$.
3. Find $\sqrt{6 + 7i}$.
4. Solve the equation $x^2 - ix + 5 = 0$.
5. What is $|\cos(\theta) + i\sin(\theta)|$?
6. Show that $\sqrt{\cos(\theta) + i\sin(\theta)} = \cos\left(\dfrac{\theta}{2}\right) + i\sin\left(\dfrac{\theta}{2}\right)$.
7. Use the result of Problem 6 to find $\sqrt{\dfrac{1}{2} + \dfrac{\sqrt{3}}{2}i}$.

MORE THEORETICAL PROBLEMS

Easier Problems

8. Show that $T = \{a + b\sqrt{2} \mid a, b \text{ rational numbers}\}$ is a field.
9. Prove all the various parts of Theorem 1.10.1.
10. Verify that the quadratic formula gives you the roots of $\alpha x^2 + \beta x + \gamma$ even when α, β, and γ are complex numbers.
11. Prove Parts (1), (2), and (4) of Lemma 1.10.4. Show that $\alpha + \bar{\alpha}$ is twice the real part of α, and $\alpha - \bar{\alpha}$ is twice the imaginary part of α.
12. Let $\alpha \neq 0$ be a complex number. Suppose that β is a complex number such that $\alpha\beta$ is real. Show that $\beta = u\bar{\alpha}$ for some *real* number u.
13. If $\alpha = r[\cos(\theta) + i\sin(\theta)]$ and $\beta = s[\cos(\varphi) + i\sin(\varphi)]$ are two complex numbers in polar form, show that $\alpha\beta = rs[\cos(\theta + \varphi) + i\sin(\theta + \varphi)]$. (Thus the argument of the product of two complex numbers is the sum of their arguments.)
14. Show that $T = \{a + bi \mid a, b \text{ rational numbers}\}$ is a field.

Middle-Level Problems

15. If α is a root of the polynomial $p(x) = a_0 x^n + a_1 x^{n-1} + \cdots + a_n$, where the a_j are all *real*, show that $\bar{\alpha}$ is also a root of $p(x)$.
16. From the result of Problem 13, prove De Moivre's Theorem, namely that $[\cos(\theta) + i\sin(\theta)]^n = \cos(n\theta) + i\sin(n\theta)$ for every integer $n \geq 0$.
17. Prove De Moivre's Theorem even if $n < 0$.
18. Express α^{-1}, where $\alpha = r[\cos(\theta) + i\sin(\theta)] \neq 0$, in polar form.
19. Show that $\left(\cos\left(\frac{2\pi}{k}\right) + \sin\left(\frac{2\pi}{k}\right)\right)^k = 1$ and $\left(\cos\left(\frac{2\pi}{k}\right) + \sin\left(\frac{2\pi}{k}\right)\right)^m \neq 1$ for every $0 < m < k$.

Harder Problems

20. Given a complex number $a \neq 0$ show that for every integer $n > 0$ we can find n distinct complex numbers $b_1, \ldots, b_n$ such that $a = b_r^n$. (So complex numbers have nth roots.)
21. Find the necessary and sufficient conditions on $\alpha, \beta \in \mathbb{C}$ so that $|\alpha + \beta| = |\alpha| + |\beta|$.

1.11. $M_2(\mathbb{C})$

We have discussed $M_2(\mathbb{R})$, the 2×2 matrices over the real numbers, $\mathbb{R}$. Now that we have the complex numbers, $\mathbb{C}$, at our disposal we can enlarge our domain of discourse from $M_2(\mathbb{R})$ to $M_2(\mathbb{C})$. What is the advantage of passing from $\mathbb{R}$ to $\mathbb{C}$? For one thing, since the characteristic polynomial of a matrix in $M_2(\mathbb{R})$ [and in $M_2(\mathbb{C})$] is quadratic, by Theorem 1.10.3, we know that this quadratic polynomial has roots in $\mathbb{C}$. Therefore, *every matrix in $M_2(\mathbb{C})$ will have characteristic roots, and these will be in $\mathbb{C}$.*

How do we carry out this passage from $M_2(\mathbb{R})$ to $M_2(\mathbb{C})$? In the most obvious way! *Define equality, addition, and multiplication of matrices in $M_2(\mathbb{C})$, that is, matrices having complex numbers as their entries, exactly as we did for matrices having real* entries. Everything we have done, so far, carries over in its entirety. We leave this verification to the reader. We will have some of these verifications as problems.

We shall let $M_2(\mathbb{C})$ act on something—namely the set of 2-tuples having complex coordinates. As before, let $W = \left\{ \begin{bmatrix} \xi \\ \vartheta \end{bmatrix} \middle| \xi, \vartheta \in \mathbb{C} \right\}$. We let $M_2(\mathbb{C})$ act on W as follows:

If $A = \begin{bmatrix} \alpha & \beta \\ \gamma & \delta \end{bmatrix} \in M_2(\mathbb{C})$ and if $\begin{bmatrix} \xi \\ \vartheta \end{bmatrix} \in W$, then $Aw = \begin{bmatrix} \alpha & \beta \\ \gamma & \delta \end{bmatrix} \begin{bmatrix} \xi \\ \vartheta \end{bmatrix} = \begin{bmatrix} \alpha\xi + \beta\vartheta \\ \gamma\xi + \delta\vartheta \end{bmatrix}$.

Because of the nice properties of $\mathbb{C}$ relative to addition, multiplication, and division, what we did for $M_2(\mathbb{R})$ carries over *verbatim* to $M_2(\mathbb{C})$.

PROBLEMS

NUMERICAL PROBLEMS

1. Multiply

   (a) $\begin{bmatrix} 1 & i \\ -i & 1 \end{bmatrix} \begin{bmatrix} 1 & -i \\ i & 1 \end{bmatrix}$.

   (b) $\begin{bmatrix} 3+i & 0 \\ \frac{1}{2} - \frac{1}{3}i & 1 \end{bmatrix} \begin{bmatrix} 3-i & 0 \\ 6+7i & 1 \end{bmatrix}$.

   (c) $\begin{bmatrix} i & 1 \\ 1 & i \end{bmatrix} \begin{bmatrix} 4 & 5 \\ 6 & 3-2i \end{bmatrix} \begin{bmatrix} i & 1 \\ 1 & i \end{bmatrix}^{-1}$.

2. Calculate the determinant of the matrices in Problem 1 directly and also by using the fact that $\det(AB) = \det(A)\det(B)$.

3. Find the characteristic roots of

   (a) $\begin{bmatrix} 1 & i \\ -i & 1 \end{bmatrix}$.

   (b) $\begin{bmatrix} 5 & 6 \\ -6 & 5 \end{bmatrix}$.

   (c) $\begin{bmatrix} \frac{(4-i)-2}{2} & 4 \\ \frac{(i+3)}{3} & 0 \end{bmatrix}$.

4. In Problem 3, find the characteristic vectors for the matrices for each characteristic root.

5. Find all matrices A in $M_2(\mathbb{C})$ such that

   $$A \begin{bmatrix} 1 & 2 \\ -2 & 1 \end{bmatrix} = \begin{bmatrix} 1 & 2 \\ -2 & 1 \end{bmatrix} A.$$

6. Find all matrices A in $M_2(\mathbb{C})$ such that $AB = BA$ for all $B \in M_2(\mathbb{C})$.

7. Find all α such that for every invertible A in $M_2(\mathbb{C})$,

   $$\det\left(A \begin{bmatrix} 2+i & \alpha \\ 1-i & 1+i \end{bmatrix} A^{-1} \right) = 0.$$

MORE THEORETICAL PROBLEMS

Easier Problems

8. For what values of $\beta \in \mathbb{C}$ is $\begin{bmatrix} u & \beta \\ -\beta & u \end{bmatrix}$ invertible for all *real* $u \neq 0$?

9. Prove that in $M_2(\mathbb{C})$:
   (a) $\det(AB) = \det(A)\det(B)$
   (b) $\operatorname{tr}(AB) = \operatorname{tr}(BA)$
   (c) $\operatorname{tr}(\alpha A + B) = \alpha(\operatorname{tr}(A)) + \operatorname{tr}(B)$
   (d) $(AB)' = B'A'$

   for all $A, B \in M_2(\mathbb{C})$ and all $\alpha \in \mathbb{C}$.

10. Define the mapping $\bar{}$ in $M_2(\mathbb{C})$ by $\overline{\begin{bmatrix} \alpha & \beta \\ \gamma & \delta \end{bmatrix}} = \begin{bmatrix} \bar{\alpha} & \bar{\beta} \\ \bar{\gamma} & \bar{\delta} \end{bmatrix}$. Prove:
    (a) $\bar{\bar{A}} = \overline{(\bar{A})} = A$
    (b) $\overline{(A + B)} = \bar{A} + \bar{B}$
    (c) $\overline{(\alpha A)} = \bar{\alpha}\bar{A}$
    (d) $\overline{(AB)} = \bar{A}\bar{B}$

    for all $A, B \in M_2(\mathbb{C})$ and all $\alpha \in \mathbb{C}$.

11. Show that $\overline{(AB)'} = \overline{B'A'}$.

12. If $A \in M_2(\mathbb{C})$ is invertible, show that $\overline{A^{-1}} = \bar{A}^{-1}$.

13. If $A \in M_2(\mathbb{C})$ define A^* by $A^* = \bar{A}'$. Prove:
    (a) $A^{**} = (A^*)^* = A$
    (b) $(\alpha A + B)^* = \bar{\alpha}A^* + B^*$
    (c) $(AB)^* = B^*A^*$

    for all $A, B \in M_2(\mathbb{C})$ and $\alpha \in \mathbb{C}$.

14. Show that $AA^* = 0$ if and only if $A = 0$.

15. Show that $(A + A^*)^* = A + A^*$, $(A - A^*)^* = -(A - A^*)$ and $(AA^*)^* = AA^*$ for $A \in M_2(\mathbb{C})$.

16. If $B \in M_2(\mathbb{C})$, show that $B = A + C$, where $A^* = A$, $C^* = -C$. Moreover, show that A and C are uniquely determined by B.

17. If $A^* = \pm A$ and $A^2 = 0$, show that $A = 0$.

Middle-Level Problems

18. If $A^* = A$, prove that the characteristic roots of A are real.

19. If $A^* = -A$, prove that the characteristic roots of A are pure imaginaries.

20. If $A^*A = I$ and α is a characteristic root of A, what can you say about $|\alpha|$?

21. If $A^*A = I$ and $B^*B = I$, what is a simple form for $(AB)^{-1}$?

Harder Problems

22. Prove that for $A \in M_2(\mathbb{C})$ the characteristic roots of A^*A are *real* and *nonnegative*.

23. If $A^* = -A$, show that:
    (a) $I - A$ and $I + A$ are invertible.
    (b) If $B = (I - A)(I + A)^{-1}$, then $B^*B = I$.

Sec. 1.12] Inner Products 55

24. If $AB = 0$, is $A^*B^* = 0$? Either prove or give a counterexample.
25. If $A \in M_2(\mathbb{C})$, show that $\text{tr}(AA^*) > 0$.
26. If $A(AA^* - A^*A) = (AA^* - A^*A)A$, show that $AA^* = A^*A$.
27. If $A^* = A$, show that we can find a B such that $B^*B = I$ and $BAB^{-1} = \begin{bmatrix} \alpha & 0 \\ 0 & \beta \end{bmatrix}$.
What are α and β in relation to A?

1.12. INNER PRODUCTS

For those readers who have been exposed to vector analysis, the concept of *dot product* is not a new one. Given $W = \left\{ \begin{bmatrix} \xi \\ \vartheta \end{bmatrix} \middle| \xi, \vartheta \in \mathbb{C} \right\}$ if $v, w \in W$ and $v = \begin{bmatrix} \alpha \\ \beta \end{bmatrix}$ and $w = \begin{bmatrix} \gamma \\ \delta \end{bmatrix}$ we define the *inner product* of v and w.

Definition. If $v = \begin{bmatrix} \alpha \\ \beta \end{bmatrix}$ and $w = \begin{bmatrix} \gamma \\ \delta \end{bmatrix}$ are in W, then their *inner product*, denoted by (v, w), is defined by $(v, w) = \alpha\bar{\gamma} + \beta\bar{\delta}$.

The reason we use the complex conjugate, rather than defining (v, w) as $\alpha\gamma + \beta\delta$, is because we want (v, v) to be nonzero if $v \neq 0$. For instance, if $v = \begin{bmatrix} 1 \\ i \end{bmatrix}$, then, if we defined the inner product without complex conjugates, (v, v) *would* be $(1)(1) + (i)(i) = 1 - 1 = 0$, yet $v \neq 0$. You might very well ask why we want to avoid such a possibility. The answer is that we want (v, v) to give us the length (actually, the square of the length) of v and we want a nonzero element to have positive length.

EXAMPLE

Let $v = w = \begin{bmatrix} 1 + i \\ 2i \end{bmatrix}$. Then $(v, w) = (1 + i)(1 - i) + (2i)(-2i) = 2 + 4 = 6$.

We summarize some of the basic behavior of $(\cdot, \cdot)$ in

Lemma 1.12.1. If $u, v, w \in W$ and $\sigma \in \mathbb{C}$, then

1. $(v + w, u) = (v, u) + (w, u)$.
2. $(u, v + w) = (u, v) + (u, w)$.
3. $(v, w) = \overline{(w, v)}$.
4. $(\sigma v, w) = \sigma(v, w) = (v, \bar{\sigma}w)$.
5. $(v, v) \geq 0$ is real and $(v, v) = 0$ if and only if $v = 0$.

Proof: The proof is straightforward, following directly from the definition of the inner product. We do verify Parts (3), (4), and (5).

To see Part (3), let $v = \begin{bmatrix} \alpha \\ \beta \end{bmatrix}$ and let $w = \begin{bmatrix} \gamma \\ \delta \end{bmatrix}$. Then, by definition, $(v, w) = \alpha\bar{\gamma} + \beta\bar{\delta}$

while $(w, v) = \gamma\bar{\alpha} + \delta\bar{\beta}$. Noticing that

$$\gamma\bar{\alpha} + \delta\bar{\beta} = \bar{\bar{\gamma}}\bar{\alpha} + \bar{\bar{\delta}}\bar{\beta} = \overline{\alpha\bar{\gamma} + \beta\bar{\delta}},$$

we get that $(v, w) = \overline{(w, v)}$.

To see Part (4), note that $(\sigma v, w) = (\sigma\alpha)\bar{\gamma} + (\sigma\beta)\bar{\delta} = \sigma(\alpha\bar{\gamma} + \beta\bar{\delta}) = \sigma(v, w)$ because $\sigma v = \sigma\begin{bmatrix}\alpha\\\beta\end{bmatrix} = \begin{bmatrix}\sigma\alpha\\\sigma\beta\end{bmatrix}$.

Finally, if $v = \begin{bmatrix}\alpha\\\beta\end{bmatrix} \neq 0$, then at least one of α or β is not 0. Thus $(v, v) = \alpha\bar{\alpha} + \beta\bar{\beta} = |\alpha|^2 + |\beta|^2 \neq 0$, and is, in fact, positive. This proves Part (5). ∎

We call v *orthogonal* to w if $(v, w) = 0$. For example, the vector $v = \begin{bmatrix}2\\3\end{bmatrix}$ is orthogonal to $w = \begin{bmatrix}-3\\2\end{bmatrix}$:

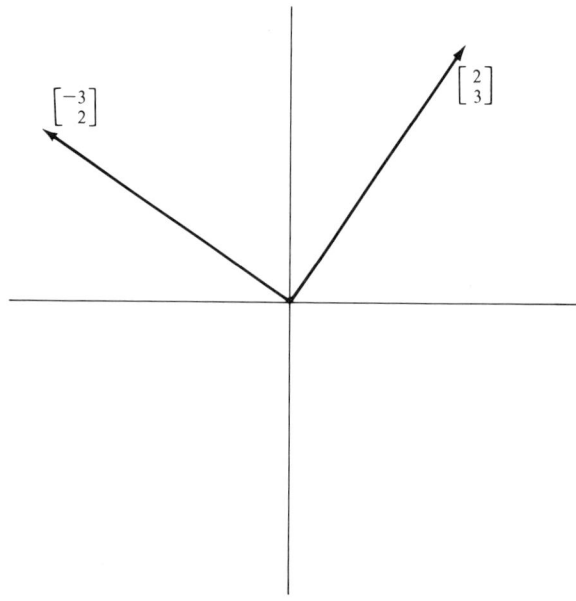

Note that if v is orthogonal to w, then w is orthogonal to v because v is orthogonal to w if and only if $(v, w) = 0$. But then, by Part (3) of Lemma 1.12.1, $(w, v) = \overline{(v, w)} = \bar{0} = 0$. So w is orthogonal to v.

Clearly, $v = 0$ is orthogonal to every element of W. Is it the only vector (element of W) with this property? Suppose that $(v, w) = 0$ for all w in W; then certainly $(v, v) = 0$. By Part (5) of Lemma 1.12.1 we obtain that $v = 0$. So

Lemma 1.12.2

1. If v is orthogonal to w, then w is orthogonal to v.
2. If v is orthogonal to all of W, then $v = 0$.

Sec. 1.12]   **Inner Products**

3. If $v^\perp = \{w \in W \,|\, (v, w) = 0\}$, then:
   (a) $w_1, w_2 \in v^\perp$ implies that $w_1 + w_2 \in v^\perp$.
   (b) $\alpha w \in v^\perp$ for all $\alpha \in \mathbb{C}$ and $w \in v^\perp$.

We leave the proof of Part (3) to the reader.

A natural question is to investigate the relationship between acting on a vector by a matrix and the inner product. We do this now.

Definition. If $A \in M_2(\mathbb{C})$, we let $A^* = (\bar{A})'$.

Theorem 1.12.3. If $A \in M_2(\mathbb{C})$ and $v, w \in W$, then $(Av, w) = (v, A^*w)$.

Proof: We verify this result by computing (Av, w) and (v, A^*w) and comparing the results.

Let $A = \begin{bmatrix} \alpha & \beta \\ \gamma & \delta \end{bmatrix}$ and $v = \begin{bmatrix} \varepsilon \\ \phi \end{bmatrix}, w = \begin{bmatrix} \zeta \\ \eta \end{bmatrix}$. Then $Av = \begin{bmatrix} \alpha & \beta \\ \gamma & \delta \end{bmatrix}\begin{bmatrix} \varepsilon \\ \phi \end{bmatrix} = \begin{bmatrix} \alpha\varepsilon + \beta\phi \\ \gamma\varepsilon + \delta\phi \end{bmatrix}$,

hence $(Av, w) = (\alpha\varepsilon + \beta\phi)\bar{\zeta} + (\gamma\varepsilon + \delta\phi)\bar{\eta} = \alpha\varepsilon\bar{\zeta} + \beta\phi\bar{\zeta} + \gamma\varepsilon\bar{\eta} + \delta\phi\bar{\eta}$. On the other hand, $A^* = \begin{bmatrix} \bar{\alpha} & \bar{\gamma} \\ \bar{\beta} & \bar{\delta} \end{bmatrix}$, so

$$A^*w = \begin{bmatrix} \bar{\alpha} & \bar{\gamma} \\ \bar{\beta} & \bar{\delta} \end{bmatrix}\begin{bmatrix} \zeta \\ \eta \end{bmatrix} = \begin{bmatrix} \bar{\alpha}\zeta + \bar{\gamma}\eta \\ \bar{\beta}\zeta + \bar{\delta}\eta \end{bmatrix},$$

whence

$$(v, A^*w) = \varepsilon\overline{(\bar{\alpha}\zeta + \bar{\gamma}\eta)} + \phi\overline{(\bar{\beta}\zeta + \bar{\delta}\eta)} = \varepsilon(\overline{\bar{\alpha}\zeta} + \overline{\bar{\gamma}\eta}) + \phi(\overline{\bar{\beta}\zeta} + \overline{\bar{\delta}\eta})$$
$$= \varepsilon(\alpha\bar{\zeta} + \gamma\bar{\eta}) + \phi(\beta\bar{\zeta} + \delta\bar{\eta}) = \alpha\varepsilon\bar{\zeta} + \gamma\varepsilon\bar{\eta} + \beta\phi\bar{\zeta} + \delta\phi\bar{\eta},$$

which we see is exactly what we obtained for (Av, w). Thus Theorem 1.12.3 is proved. ∎

Note a few consequences of the theorem. These are things that were done directly—in the problems in the preceding section—by a direct computation. For the tactical reason that the proof we are about to give works in general, we derive these consequences more abstractly from Theorem 1.12.3.

Theorem 1.12.4. If $A, B \in M_2(\mathbb{C})$ and $\alpha \in \mathbb{C}$, then
1. $A^{**} = (A^*)^* = A$;
2. $(A + B)^* = (A^* + B^*)$;
3. $(\alpha A)^* = \bar{\alpha} A^*$;
4. $(AB)^* = B^*A^*$.

Proof: The proofs of all the parts will depend on something we worked out earlier, namely, if $(v, w) = 0$ for all $w \in W$, then $v = 0$.

We start with Part (1). If $v, w \in W$, then

$$(Av, w) = (v, A^*w) = \overline{(A^*w, v)} = \overline{(w, A^{**}v)} = (A^{**}v, w),$$

using Lemma 1.12.1 and the definition of A^*. Thus

$$(Av - A^{**}v, w) = (Av, w) - (A^{**}v, w) = 0,$$

hence $((A - A^{**})v, w) = 0$ for all $w \in W$. Therefore, $(A - A^{**})v = 0$ for all $v \in W$. This implies that $A^{**} = A$.

For Part (2) notice that

$$(Av, w) + (Bv, w) = ((A + B)v, w) = (v, (A + B)^*w).$$

However,

$$(Av, w) + (Bv, w) = (v, A^*w) + (v, B^*w) = (v, (A^* + B^*)w).$$

Comparing these evaluations leads us to $(A + B)^* = A^* + B^*$.

Part (3) we leave to the reader.

Finally, for Part (4), we have

$$(ABv, w) = (A(Bv), w) = (Bv, A^*w) = (v, B^*(A^*w)) = (v, B^*A^*w);$$

however, $(ABv, w) = (v, (AB)^*w)$. Comparing the results yields that $(AB)^* = B^*A^*$. ∎

Another interesting question about the interaction of multiplication by a matrix and the inner product is: For what A is (v, w) *preserved*, that is, for what A is $(Av, Aw) = (v, w)$? Clearly, if $(Av, Aw) = (v, w)$, then $(v, A^*Aw) = (v, w)$, hence $A^*Aw = w$ for all $w \in W$. This forces $A^*A = I$. On the other hand, if $A^*A = I$, then for all $v, w \in W$, $(A^*Av, w) = (Iv, w) = (v, w)$. But $(A^*Av, w) = (Av, A^{**}w) = (Av, Aw)$, by Theorem 1.12.4. Thus $(Av, Aw) = (v, w)$.

We have proved

Theorem 1.12.5. For $A \in M_2(\mathbb{C})$, $(Av, Aw) = (v, w)$ for all $v, w \in W$ if and only if $A^*A = I$.

Definition. A matrix $A \in M_2(\mathbb{C})$ is called *unitary* if $A^*A = I$.

Unitary matrices are very nice in that they preserve orthogonality and also (v, w).

Definition. If v is in W, then the *length* of v, denoted by $\|v\|$, is defined by $\|v\| = \sqrt{(v, v)}$.

So if $v = \begin{bmatrix} \xi \\ \vartheta \end{bmatrix} \in W$, then $\|v\| = \sqrt{\xi\bar{\xi} + \vartheta\bar{\vartheta}}$. Note that if ξ and ϑ are *real*, then $\|v\| = \sqrt{\xi^2 + \vartheta^2}$, the usual length of the line segment join the origin to the point (ξ, ϑ) in the plane.

Sec. 1.12] Inner Products 59

We know that any unitary matrix preserves length, for if A is unitary, then $\|Av\|$ equals $\sqrt{(Av, Av)} = \sqrt{(v, v)} = \|v\|$. What about the other way around? Suppose that $(Av, Av) = (v, v)$ for all $v \in W$; is A unitary?

The answer, as we see in the next theorem, is "yes."

Theorem 1.12.6. If $A \in M_2(\mathbb{C})$ is such that $(Aw, Aw) = (w, w)$ for all $w \in W$, then A is unitary [hence $(Av, Aw) = (v, w)$ for all $v, w \in W$].

Proof: Let $v, w \in W$. Then

$$(A(v + w), A(v + w)) = (Av + Aw, Av + Aw)$$
$$= (Av, Av) + (Aw, Aw) + (Av, Aw) + (Aw, Av).$$

Since $(Av, Av) = (v, v)$ and $(Aw, Aw) = (w, w)$ we get

$$(Av, Aw) + (Aw, Av) = (v, w) + (w, v). \tag{1}$$

Since (1) is true for all $v, w \in W$ and since $iw \in W$, if we replace w by iw in (1), we get

$$(Av, iAw) + (iAw, Av) = (v, iw) + (iw, v), \tag{2}$$

and so

$$\bar{i}(Av, Aw) + i(Aw, Av) = \bar{i}(v, w) + i(w, v). \tag{3}$$

However, $\bar{i} = -i$; thus (3) becomes

$$i(-(Av, Aw) + (Aw, Av)) = i(-(v, w) + (w, v)), \tag{4}$$

which gives us

$$(Aw, Av) - (Av, Aw) = (w, v) - (v, w). \tag{5}$$

If we add the result in (1) to that in (5), we end up with $2(Aw, Av) = 2(w, v)$, hence $(Aw, Av) = (w, v)$. Therefore, A is unitary. ∎

Thus to express that a matrix in $M_2(\mathbb{C})$ is unitary, we can say it succinctly as: *A preserves length*.

Let's see how easy and noncomputational the proof of the following fact becomes: *If $A^* = A$, then the characteristic roots of A are all real*. To prove it by computing it directly can be very messy. Let's see how it goes using inner products.

Let $\alpha \in \mathbb{C}$ be a characteristic root of $A \in M_2(\mathbb{C})$, where $A^* = A$. Thus $Av = \alpha v$ for some $v \neq 0 \in V$. Therefore,

$$\alpha(v, v) = (\alpha v, v) = (Av, v) = (v, A^*v) = (v, Av) = (v, \alpha v) = \bar{\alpha}(v, v).$$

Since $(v, v) > 0$ we get $\alpha = \bar{\alpha}$, hence α is real. We have proved

Theorem 1.12.7. If $A^* = A$, then the characteristic roots of A are real.

Using a similar technique we prove

Theorem 1.12.8. If $A \in M_2(\mathbb{C})$ is unitary and if α is a characteristic root of A, then $|\alpha| = 1$.

Proof: Let α be a characteristic root of A and $v \neq 0$ a corresponding characteristic vector. Thus $Av = \alpha v$. Hence

$$\alpha(v, v) = (Av, v) = (v, A^*v) = (v, A^{-1}v).$$

But if $Av = \alpha v$, then $A^{-1}(Av) = \alpha A^{-1}v$, that is, $A^{-1}v = \frac{1}{\alpha}v$. Therefore, returning to our calculation above, $\alpha(v,v) = (v, A^{-1}v) = \left(v, \frac{1}{\alpha}v\right) = \frac{1}{\bar{\alpha}}(v,v)$. Thus $\alpha\bar{\alpha}(v,v) = (v,v)$ and since $(v,v) \neq 0$, we get $\alpha\bar{\alpha} = 1$. Thus $|\alpha| = \sqrt{\alpha\bar{\alpha}} = \sqrt{1} = 1$ ∎

Before closing this section, we should introduce the names for some of these things.

1. A^* is called the *Hermitian adjoint*.
2. If $A^* = A$, then A is called *Hermitian*.
3. If $A^* = -A$, then A is called *skew-Hermitian*.
4. If $A^*A = AA^*$, then A is called *normal*.

The word "Hermitian" comes from the name of the French mathematician Hermite. We use these terms freely in the problems.

PROBLEMS

NUMERICAL PROBLEMS

1. Determine the values for a (if any) such that the vectors $\begin{bmatrix} 1+\alpha \\ 1-\alpha \end{bmatrix}$ and $\begin{bmatrix} 1-\alpha \\ 3 \end{bmatrix}$ are orthogonal.

2. Describe the set of vectors $\begin{bmatrix} \alpha \\ \beta \end{bmatrix}$ orthogonal to $\begin{bmatrix} 3 \\ 4 \end{bmatrix}$.

3. Describe the set of vectors $\begin{bmatrix} \alpha \\ \beta \end{bmatrix}$ orthogonal to $\begin{bmatrix} 1 \\ 4 \end{bmatrix}$.

4. Describe the set of vectors $\begin{bmatrix} \alpha \\ \beta \end{bmatrix}$ which are orthogonal to $\begin{bmatrix} 3 \\ 4 \end{bmatrix}$ as well as to $\begin{bmatrix} 1 \\ 4 \end{bmatrix}$.

MORE THEORETICAL PROBLEMS

Easier Problems

5. Given the matrix $A = \begin{bmatrix} \alpha & \beta \\ \gamma & \delta \end{bmatrix} \in M_2(\mathbb{C})$, if $v = \begin{bmatrix} \xi \\ \vartheta \end{bmatrix} \in W$, find the formula for (Av, v) in terms of $\alpha, \beta, \gamma, \delta, \xi$, and ϑ.

Sec. 1.12]   Inner Products   61

6. If A is unitary and has *real* entries, show that $\det(A) = \pm 1$.
7. If A is unitary and has real entries, show that if $\det(A) = 1$, then
$$A = \begin{bmatrix} \cos(\theta) & -\sin(\theta) \\ \sin(\theta) & \cos(\theta) \end{bmatrix} \quad \text{for some } \theta.$$
8. If A and B are unitary, show that AB is unitary.

Middle-Level Problems

9. If α is a characteristic root of A, show that α^n is a characteristic root of A^n for all $n \geq 1$.
10. If $(Av, v) = 0$ for all v with real coordinates, where $A \in M_2(\mathbb{C})$, determine the form of A if $A \neq 0$.
11. Show that if A is as in Problem 10, then $A \neq BB^*$ for all $B \in M_2(\mathbb{C})$.
12. If $A \in M_2(\mathbb{R})$ is such that for all $v = \begin{bmatrix} \xi \\ \vartheta \end{bmatrix}$, with ξ and ϑ real, $(Av, Av) = (v, v)$, is A^*A necessarily equal to I? Either prove or give a counterexample.
13. Prove Parts (1) and (2) of Lemma 1.12.1.
14. Prove Part (3) of Lemma 1.12.2.
15. If $B \in M_2(\mathbb{C})$, show that $B = C + D$, where C is Hermitian and D is skew-Hermitian.
16. If A is both Hermitian and unitary, what possible values can the characteristic roots of A have?
17. If A is Hermitian and $A^n = I$ for some $n \geq 1$, show that $A^2 = I$.
18. If A is normal and if for some $v \neq 0$ in W, $Av = \alpha v$, prove that $A^*v = \bar{\alpha}v$.
19. If $\text{tr}(AA^*) = 0$, show that $A = 0$.
20. Show that $\det(A^*) = \overline{\det(A)}$.
21. If $A^* = -A$ and $A^n = I$ for some $n > 0$, show that $A^4 = I$.
22. If A is skew-Hermitian, show that its characteristic roots are pure imaginaries, making use of the inner product on W.
23. If A is normal and invertible, show that $B = A^*A^{-1}$ is unitary.
24. Is the result of Problem 23 valid if A is not normal? Either prove or give a counterexample.
25. Define, for $v, w \in W$, $\langle v, w \rangle = (Av, w)$, where $A \in M_2(\mathbb{C})$. What conditions must A satisfy in order that $\langle \cdot, \cdot \rangle$ satisfies that five properties in Lemma 1.12.1?
26. Prove that if $A \in M_2(\mathbb{C})$ and if $p(x) = \det(xI - A)$, then $p(A) = 0$. [This is the Cayley–Hamilton Theorem for $M_2(\mathbb{C})$.]
27. If U is unitary, show that UBU^{-1} is Hermitian if B is Hermitian and skew-Hermitian if B is skew-Hermitian.
28. If A and B are in $M_2(\mathbb{R})$ and $B = CAC^{-1}$, where $C \in M_2(\mathbb{C})$, show that we can find a $D \in M_2(\mathbb{R})$ such that $B = DAD^{-1}$.
29. If α is a characteristic root of A and if $p(x)$ is a polynomial with complex coefficients, show that $p(\alpha)$ is a characteristic root of $p(A)$.
30. If $p(A) = 0$ for some polynomial $p(x)$ with complex coefficients, show that if α is a characteristic root of A, then $p(\alpha) = 0$.

CHAPTER 2

Systems of Linear Equations

2.1. INTRODUCTION

If you should happen to glance at a variety of books on matrix theory or linear algebra you would find that many of them, if not most, start things rolling by considering systems of linear equations and their solutions. A great number of these books go even further and state that this subject—systems of linear equations—is the most important part and central focus of linear algebra. We certainly do not subscribe to this point of view. We can think of many topics in linear algebra which play an even more important role in mathematics and allied fields. To name a few such (and these will come up in the course of the book): the theory of characteristic roots, the Cayley–Hamilton Theorem, the diagonalization of Hermitian matrices, vector spaces and linear transformations on them, inner product spaces, and so on. These are areas of linear algebra which find everyday use in a host of theoretical and applied problems.

Be that as it may, we do not want to minimize the importance of the study of systems of linear equations. Not only does this topic stand firmly on its own feet, but the results and techniques we shall develop in this chapter will crop up often as tools in the subsequent material of this book.

Many of the things that we do with linear equations lend themselves readily to an algorithmic description. In fact, it seems that a recipe for attack is available for almost all the results we shall obtain here. Wherever possible, we shall stress this algorithmic approach and summarize what was done into a "method for ...."

In starting with 2×2 matrices and their properties, one is able to get to the heart of linear algebra easily and quickly, and to get an idea of how, and in what direction, things will run in general. The results on systems of linear equations will provide us with a rich set of techniques to study a large cross section of ideas and results in matrix theory and linear algebra.

Sec. 2.1] Introduction 63

We start things off with some examples.

1. The system of equations

$$2x + 3y = 5$$
$$4x + 5y = 9$$

can be solved by using the first equation to express y in terms of x as

$$y = (\tfrac{1}{3})(5 - 2x),$$

then substituting this expression for y into the second equation and solving for x, then solving for y:

$$4x + 5(\tfrac{1}{3})(5 - 2x) = 9$$
$$12x + 25 - 10x = 27$$
$$x = 1$$
$$y = (\tfrac{1}{3})(5 - 2) = 1.$$

For this system of equations, the solution $x = 1$, $y = 1$ is the only solution. Geometrically, these equations are represented by lines in the plane which are not parallel and so intersect in a single point. The point of intersection is the solution $\begin{bmatrix} x \\ y \end{bmatrix} = \begin{bmatrix} 1 \\ 1 \end{bmatrix}$ to the system of equations.

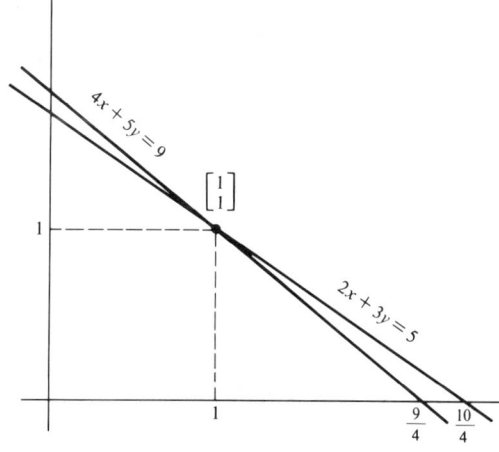

2. The system of equations

$$2x + 3y = 5$$
$$4x + 6y = 10$$

is easy to solve because any solution to the first equation is also a solution to the second. Geometrically, these equations are represented by lines, and it turns out that these lines are really the same line. The points on this line are the solutions $\begin{bmatrix} x \\ y \end{bmatrix}$ to the system of equations.

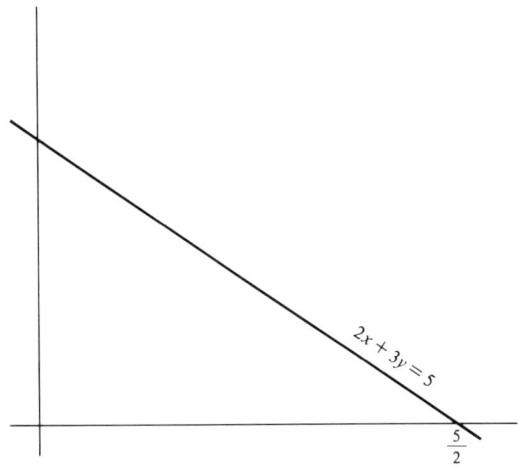

If we take any value x, we can solve for y and we get the corresponding solution

$$x \text{ (any value)}$$
$$y = (\tfrac{1}{3})(5 - 2x).$$

Because we give the variable x any value we want and then give y the corresponding value $(\tfrac{1}{3})(5 - 2x)$, which depends on the value we give to x, we call x an *independent variable* and y a *dependent variable*. However, if instead we had chosen any value for y and set $x = (\tfrac{1}{2})(5 - 3y)$, we would have called y the independent variable any x the dependent variable.

3. The system of equations

$$2x + 3y = 5$$
$$4x + 6y = 20$$

has no solution, since if x and y satisfy the first equation, they cannot satisfy the second. Geometrically, these equations represent parallel lines that do not intersect, as illustrated in the diagram on page 65.

The examples above illustrate the principle that

the set of solutions $\begin{bmatrix} x \\ y \end{bmatrix} \in \mathbb{R}^{(2)}$ *to a system of linear equations in two variables*

x *and* y *is either a point, a line, the empty set, or all of* $\mathbb{R}$.

This principle even takes care of cases where all coefficients are 0. For example, the set of solutions of the system consisting of the one equation $0x + 0y = 2$ is the empty set,

Sec. 2.1] Introduction 65

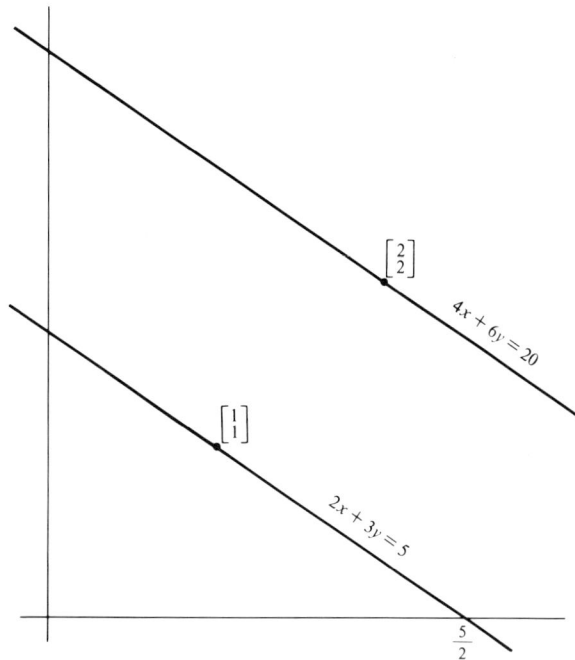

and the set of solutions of the system consisting of the one equation $0x + 0y = 0$ is all of $\mathbb{R}^{(2)}$. A similar principle holds for any number of variables. If the number of variables is three, for instance this principle is that

the set of solutions $\begin{bmatrix} x \\ y \\ z \end{bmatrix}$ *to a system of linear equations in three variables x, y, z*

is one of the following: a point, a line, a plane, the empty set, or all of $\mathbb{R}^{(3)}$.

The reason for this is quite simple. If there is only one equation $ax + by + cz = d$, then if a, b, and c are 0, the set of solutions $\begin{bmatrix} x \\ y \\ z \end{bmatrix}$ is $\mathbb{R}^{(3)}$ if $d = 0$ and the empty set if $d \neq 0$. If, on the other hand, the equation is *nonzero* (meaning that at least one coefficient on the left-hand side is nonzero), the set of solutions is a plane. Suppose that we now add another nonzero equation $fx + gy + hz = e$, ending up with the system of equations

$$ax + by + cz = d$$
$$fx + gy + hz = e.$$

The set of solutions to the second equation is also a plane, so the set of solutions to the system of two equations is the intersection of two planes. This will be a line or a plane or the empty set. Each time we add another nonzero equation, the new solution set is the intersection of the old solution set with the plane of solutions to the new equation.

If the old solution set was a line, then the intersection of that line with this plane is either a point, a line, or the empty set. In the case of a system

$$ax + by + cz = d$$
$$fx + gy + hz = k$$
$$rx + sy + tz = u$$

of three nonzero equations, each equation is represented by a plane. We leave it as an exercise for the reader to show that if two of these planes are parallel, then the set of solutions is a line, a plane, or the empty set, and each of these possibilities really can happen. Assuming, on the other hand, that no two of these planes are parallel, we find that their intersection is a point, a line, or the empty set:

Intersection is a point:

Intersection is a line:

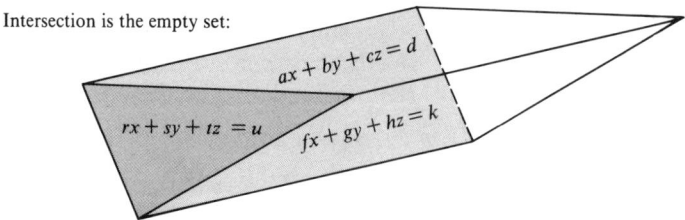

Intersection is the empty set:

EXAMPLE

Suppose that we have a supply of 6000 units of S and 8000 units of T, materials used in manufacturing products P, Q, R. If each unit of P uses 2 units of S and 0

Sec. 2.1]   Introduction

units of T, each unit of Q uses 3 units of S and 4 units of T, and each unit of R uses 1 unit of S and 4 units of T, how many units p, q, and r of P, Q, and R should we make if we want to use up the entire supply?

To answer this, we write down the system of linear equations

$$2p + 3q + 1r = 6000$$
$$0p + 4q + 4r = 8000$$

which represents the given information. Why do we call these equations *linear*? Because the variables p, q, r occur in these equations only to the *first* power.

The rectangular 2×3 array of numbers $\begin{bmatrix} 2 & 3 & 1 \\ 0 & 4 & 4 \end{bmatrix}$ is called the *matrix of coefficients* (or coefficient matrix) of this system of equations.

Since there are as many variables as equations (in fact, there is an extra *independent* variable r), and since the *coefficient matrix* $\begin{bmatrix} 2 & 3 & 1 \\ 0 & 4 & 4 \end{bmatrix}$ has non-zero entries 2 and 4 on the diagonal going down from the top left-hand corner of the matrix with only 0 below it, we can solve this system of equations. How do we do this? We use a process of *back substitution*.

What do we mean by this? We work backward. From the second equation, $4q + 4r = 8000$ we get that $q = 2000 - r$; this expresses q in terms of r. We now substitute this expression for q in the first equation to obtain $2p + 3(2000 - r) + 1r = 6000$ or $2p = 2r$, hence $p = r$. Thus we have also expressed p in terms of r. Assigning any value to r gives us values for p and q which are solutions to our system of equations. For instance, if we use $r = 500$, we get the solution $p = 500$ and $q = 2000 - 500 = 1500$. Since p and q are determined by r we call r an *independent variable* of this system. The set of solutions $\begin{bmatrix} r \\ 2000 - r \\ r \end{bmatrix}$ is a line, which is just the intersection of the two planes defined by the two equations in the system.

We let a matrix A—even if only rectangular instead of square—operate on column vectors v using a rule similar to what we used for 2×2 matrices. If A is an $m \times n$ matrix, that is, has m row and n columns, it operates on column vectors v with

n entries. Namely, we "multiply" the first row of A by v to get the first entry of Av, the second row of A by v to get the second entry of Av, and so on. What does it mean to multiply a row of A times the column vector v? Just multiply the first entries together, then the second entries, and so on. When this has been done, add them up. Of course, the number of columns of A must equal the number of entries of v for it to make sense when we "multiply" a row of A times v in this fashion. And, of course, the vector Av we end up with has m (the number of rows of A) entries, instead of the number n of entries of v. So when we let the 2×3 matrix $\begin{bmatrix} 2 & 3 & 1 \\ 0 & 4 & 4 \end{bmatrix}$ operate on the vector $\begin{bmatrix} p \\ q \\ r \end{bmatrix}$ of three entries, we perform the product

$$2p + 3q + 1r = 6000$$

to get the first entry and the product

$$0p + 4q + 4r = 8000$$

to get the second. So we get a vector $\begin{bmatrix} 6000 \\ 8000 \end{bmatrix}$ with two entries. In this way we can represent the system of equations

$$2p + 3q + 1r = 6000$$
$$0p + 4q + 4r = 8000$$

by a corresponding *matrix equation*

$$\begin{bmatrix} 2 & 3 & 1 \\ 0 & 4 & 4 \end{bmatrix} \begin{bmatrix} p \\ q \\ r \end{bmatrix} = \begin{bmatrix} 6000 \\ 8000 \end{bmatrix}.$$

Now we can represent the information given in the system of equation in the *matrix* $\begin{bmatrix} 2 & 3 & 1 \\ 0 & 4 & 4 \end{bmatrix}$ and *column vector* $\begin{bmatrix} 6000 \\ 8000 \end{bmatrix}$. We then can represent the information found after solving the system as the column vector $\begin{bmatrix} p \\ q \\ r \end{bmatrix} = \begin{bmatrix} r \\ 2000 - r \\ r \end{bmatrix}$. We equate solving the system of equations

$$2p + 3q + 1r = 6000$$
$$0p + 4q + 4r = 8000$$

Sec. 2.1]   Introduction   69

for p, q, r with solving the matrix equation

$$\begin{bmatrix} 2 & 3 & 1 \\ 0 & 4 & 4 \end{bmatrix} \begin{bmatrix} p \\ q \\ r \end{bmatrix} = \begin{bmatrix} 6000 \\ 8000 \end{bmatrix}$$

for $\begin{bmatrix} p \\ q \\ r \end{bmatrix}$.

We can use systems of equations and their corresponding matrix equations interchangeably, since each contains all the entries needed to write down the other. Let's use subscripts to keep track of equations and variables in the most general *system*

$$\begin{aligned} a_{11}x_1 + \cdots + a_{1n}x_n &= y_1 \\ \vdots \qquad \qquad \vdots \qquad &\quad \vdots \\ a_{m1}x_1 + \cdots + a_{mn}x_n &= y_m \end{aligned}$$

of m linear equations in n variables. The $m \times n$ array or *matrix* $\begin{bmatrix} a_{11} & \cdots & a_{1n} \\ \vdots & & \vdots \\ a_{m1} & \cdots & a_{mn} \end{bmatrix}$
is called the *coefficient matrix* of the system of equations. As in our example, we equate the system of equations with the corresponding *matrix equation*

$$\begin{bmatrix} a_{11} & \cdots & a_{1n} \\ \vdots & & \vdots \\ a_{m1} & \cdots & a_{mn} \end{bmatrix} \begin{bmatrix} x_1 \\ \vdots \\ x_n \end{bmatrix} = \begin{bmatrix} y_1 \\ \vdots \\ y_m \end{bmatrix}.$$

We can abbreviate this equation as $Ax = y$, where $A = \begin{bmatrix} a_{11} & \cdots & a_{1n} \\ \vdots & & \vdots \\ a_{m1} & \cdots & a_{mn} \end{bmatrix}$ and x
and y are the column vectors $x = \begin{bmatrix} x_1 \\ \vdots \\ x_n \end{bmatrix}$ and $y = \begin{bmatrix} y_1 \\ \vdots \\ y_m \end{bmatrix}$. The *main diagonal of A*
holds the *diagonal entries* $a_{11}, \ldots, a_{dd}$, where $d = m$ if $m < n$ and $d = n$ otherwise.
For example, the diagonal entries of $\begin{bmatrix} 1 & 2 & 3 \\ 5 & 6 & 7 \\ 9 & 8 & 7 \\ 5 & 4 & 3 \end{bmatrix}$ and $\begin{bmatrix} 1 & 2 & 3 \\ 5 & 6 & 7 \\ 9 & 8 & 7 \end{bmatrix}$ are 1, 6, and 7,
whereas those of $\begin{bmatrix} 1 & 2 & 3 \\ 5 & 6 & 7 \end{bmatrix}$ are 1 and 6.

We say that the matrix A is *upper triangular* if the entries a_{rs} with $r > s$ (those below the main diagonal) are all 0. So $\begin{bmatrix} 2 & 3 & 1 \\ 0 & 4 & 4 \\ 0 & 0 & 7 \\ 0 & 0 & 0 \end{bmatrix}$ and $\begin{bmatrix} 2 & 3 & 1 \\ 0 & 4 & 4 \\ 0 & 0 & 7 \end{bmatrix}$ are upper triangular and $\begin{bmatrix} 2 & 3 & 1 \\ 0 & 4 & 4 \\ 0 & 0 & 7 \\ 0 & 0 & 3 \end{bmatrix}$ and $\begin{bmatrix} 2 & 3 & 1 \\ 5 & 4 & 4 \\ 0 & 0 & 7 \end{bmatrix}$ are not.

Just as for the system of two equations in three variables represented by the matrix equation $\begin{bmatrix} 2 & 3 & 1 \\ 0 & 4 & 4 \end{bmatrix} \begin{bmatrix} p \\ q \\ r \end{bmatrix} = \begin{bmatrix} 6000 \\ 8000 \end{bmatrix}$, we can solve the matrix equation $\begin{bmatrix} 2 & 3 & 1 \\ 0 & 4 & 4 \\ 0 & 0 & 7 \end{bmatrix} \begin{bmatrix} p \\ q \\ r \end{bmatrix} = \begin{bmatrix} 6000 \\ 8000 \\ 7000 \end{bmatrix}$ by back substitution, getting $r = 1000$ from the last equation, then $a = 1000$ from the next-to-last equation, and $p = 1000$ from the first equation, that is, $\begin{bmatrix} p \\ q \\ r \end{bmatrix} = \begin{bmatrix} 1000 \\ 1000 \\ 1000 \end{bmatrix}$. In this case, however, the number of equations equals the number of variables, and $\begin{bmatrix} p \\ q \\ r \end{bmatrix} = \begin{bmatrix} 1000 \\ 1000 \\ 1000 \end{bmatrix}$ is the *only* solution.

When the coefficient matrix of a matrix equation is an upper triangular matrix having at least as many columns as rows and the entries on the main diagonal are all nonzero, it is very easy to solve the equation.

EXAMPLE

To solve $\begin{bmatrix} 1 & 0 & 2 & 4 & 5 \\ 0 & 2 & 2 & 4 & 4 \\ 0 & 0 & 1 & 3 & 4 \end{bmatrix} \begin{bmatrix} a \\ b \\ c \\ d \\ e \end{bmatrix} = \begin{bmatrix} 8 \\ 8 \\ 2 \end{bmatrix}$, let's first write down the corresponding system of linear equations

$$1a + 0b + 2c + 4d + 5e = 8$$
$$2b + 2c + 4d + 4e = 8$$
$$1c + 3d + 4e = 2.$$

To solve this system, let's rewrite it with the last $5 - 3 = 2$ variables on the right-hand side:

$$1a + 0b + 2c = 8 - 4d - 5e$$

Sec. 2.1] Introduction

$$2b + 2c = 8 - 4d - 4e$$
$$1c = 2 - 3d - 4e.$$

We solve these equations starting with the last equation and then working backward, getting

$$c = 2 - 3d - 4e$$
$$2b + 2c = 8 - 4d - 4e$$
$$b = 4 - 2d - 2e - c$$
$$= 4 - 2d - 2e - 2 + 3d + 4e$$
$$= 2 + d + 2e$$
$$1a + 0b + 2c = 8 - 4d - 5e$$
$$a = 8 - 4d - 5e - 2c$$
$$= 8 - 4 + 6d + 8e - 4d - 5e,$$
$$= 4 + 2d + 3e,$$

that is, getting

$$\begin{bmatrix} a \\ b \\ c \\ d \\ e \end{bmatrix} = \begin{bmatrix} 4 + 2d + 3e \\ 2 + d + 2e \\ 2 - 3d - 4e \\ d \\ e \end{bmatrix}.$$

We now give the general result.

Theorem 2.1.1. Suppose that $m \leq n$ and the coefficient matrix of a system of m linear equations in n variables is upper triangular with only nonzero entries on the main diagonal. Then we can always solve the system by back substitution. If $m < n$, there are $n - m$ independent variables in the solution. On the other hand, if $m = n$ there is exactly one solution.

Proof: Instead of using the matrix equation, we shall play with the problem as a problem on linear equations. Since we have assumed that the coefficient matrix is upper triangular with nonzero entries on the main diagonal, the system of linear equations that we are dealing with looks like

$$\begin{aligned} a_{11}x_1 + a_{12}x_2 + a_{13}x_3 + \cdots + a_{1,m}x_m + \cdots + a_{1,n}x_n &= y_1 \\ a_{22}x_2 + a_{23}x_3 + \cdots + a_{2,m}x_m + \cdots + a_{2,n}x_n &= y_2 \\ a_{33}x_3 + \cdots + a_{3,m}x_m + \cdots + a_{3,n}x_n &= y_3 \\ &\vdots \\ a_{m-1,m-1}x_{m-1} + a_{m-1,m}x_m + \cdots + a_{m-1,n}x_n &= y_{m-1} \\ a_{m,m}x_m + \cdots + a_{m,n}x_n &= y_m, \end{aligned} \quad (1)$$

where $a_{11} \neq 0, a_{22} \neq 0, \ldots, a_{m,m} \neq 0$, and $y_1, \ldots, y_m$ are given numbers.

In all the equations above, move the terms involving $x_{m+1}, \ldots, x_n$ to the right-hand side. The result is a system of m linear equations

$$
\begin{aligned}
a_{11}x_1 + a_{12}x_2 + a_{13}x_3 + \cdots + a_{1,m}x_m &= y_1 - (a_{1,m+1}x_{m+1} + \cdots + a_{1,n}x_n) \\
a_{22}x_2 + a_{23}x_3 + \cdots + a_{2,m}x_m &= y_2 - (a_{2,m+1}x_{m+1} + \cdots + a_{2,n}x_n) \\
a_{33}x_3 + \cdots + a_{3,m}x_m &= y_3 - (a_{3,m+1}x_{m+1} + \cdots + a_{3,n}x_n) \\
&\vdots \\
a_{m-1,m-1}x_{m-1} + a_{m-1,m}x_m &= y_{m-1} - (a_{m-1,m+1}x_{m+1} + \cdots + a_{m-1,n}x_n) \\
a_{mm}x_m &= y_m - (a_{m,m+1}x_{m+1} + \cdots + a_{m,n}x_n).
\end{aligned}
\tag{2}
$$

The solution of the systems (1) and (2) are the same. To solve system (2) is not hard. Start from the last equation

$$a_{mm}x_m = y_m - (a_{m,m+1}x_{m+1} + \cdots + a_{m,n}x_n).$$

Since $a_{mm} \neq 0$, we can solve for x_m merely by dividing by a_{mm}. Notice that x_m is expressed in terms of $x_{m+1}, \ldots, x_n$. Now feed this value of x_m into the second-to-last equation,

$$a_{m-1,m-1}x_{m-1} + a_{m-1,m}x_m = y_{m-1} - (a_{m-1,m+1}x_{m+1} + \cdots + a_{m-1,n}x_n).$$

Since $a_{m-1,m-1} \neq 0$ we can solve for x_{m-1} in terms of $x_m, x_{m+1}, \ldots, x_n$. Since we now have x_m expressed in terms of $x_{m+1}, \ldots, x_n$, we have x_{m-1} also expressed in such terms.

Continue the process. Feed the values of x_{m-1} and x_m just found into the third-to-last equation. Because $a_{m-2,m-2} \neq 0$ we can solve for x_{m-2} in terms of $x_{m-1}, x_m, x_{m+1}, \ldots, x_n$. But since x_m, x_{m-1} are expressible in terms of $x_{m+1}, \ldots, x_n$, we get that x_{m-2} is also so expressible.

Repeat the procedure, climbing up through the equations to reach x_1. In this way we get that each of $x_1, \ldots, x_m$ is expressible in terms of $x_{m+1}, \ldots, x_n$, and we are free to assign values to $x_{m+1}, \ldots, x_n$ at will. Each such assignment of values for $x_{m+1}, \ldots, x_n$ leads to a solution of the systems (1) and (2).

Because $x_{m+1}, \ldots, x_n$ can take on any values independently we call them the *independent variables*. So if $m < n$, we get what we claimed, namely that the solution to system (1) are given by $n - m$ independent variables whose values we can assign arbitrarily.

We leave the verification to the reader that if $m = n$ there is one and only one solution to the system (1). ∎

Because of the importance of the method outlined in the proof of this theorem, we summarize it here

Method to solve a system of linear equations by back substitution if the coefficient matrix is upper triangular with nonzero diagonal entries

1. If $m < n$, let the $n - m$ extra variable $x_{m+1}, \ldots, x_n$ have any values, and move them to the right-hand side of the equations.
2. Find the value for the last dependent variable x_m from the last equation.
3. Eliminate that variable by substituting its value back in place of it in all the preceding equations.

Sec. 2.1] Introduction

4. Repeat (2), (3), and (4) for the new last remaining dependent variable using the new last remaining equation to get its value and eliminate it and the equation. Continue until the values for all dependent variables have been found.

PROBLEMS

NUMERICAL PROBLEMS

1. Represent each of the following systems of equations by a matrix equation.
   (a) $1u + 4k + 6h = 6$
       $0u + 3k + 5h = 5.$
   (b) $3s + 5t + 8h = \pi$
       $10t + 3r + 5w = \frac{1}{2}.$
   (c) $2a + 3b + 6c = 6$
       $0a + 4b + 5c = 5$
       $1a + 4b + 6c = 6$
       $3b + 5c = 5.$
   (d) $2a + 3b + 6c = 3$
       $4b + 5c = 5$
       $1d + 4a + 6e = 2$
       $6b + 5e = 5.$
   (e) $1u + 4k + 6h = 3$
       $2u + 3h + 4x = 8.$

2. Find all solutions for each of the following triangular systems of equations.
   (a) $2x + 3w = 6$
       $4w = 5.$
   (b) $2a + 3b + 6c = 6$
       $4b + 5c = 5$
       $6c = 6.$
   (c) $4a + 3b + 6c = 6$
       $4b + 5c = 5$
       $6c = 6$
   (d) $2a + 3b + 6c = 6$
       $4b + 5c = 5$
       $6c = 6.$
       $4d + 6e = 2.$

3. Describe the system of equations corresponding to the following matrix equations, and for each say how many independent variables must appear in its general solution.

   (a) $\begin{bmatrix} 1 & 2 & 1 & 1 & 5 \\ 0 & 2 & 1 & 2 & 1 \\ 0 & 0 & 1 & 3 & 4 \end{bmatrix} \begin{bmatrix} a \\ b \\ c \\ d \\ e \end{bmatrix} = \begin{bmatrix} 4 \\ 4 \\ 4 \end{bmatrix}.$

(b) $\begin{bmatrix} 1 & 1 & 5 \\ 0 & 1 & 1 \\ 0 & 1 & 4 \end{bmatrix} \begin{bmatrix} w \\ q \\ d \end{bmatrix} = \begin{bmatrix} 554 \\ 1084 \\ 54 \end{bmatrix}.$

(c) $\begin{bmatrix} 2 & 1 & 1 & 5 \\ 2 & 1 & 2 & 1 \\ 0 & 1 & 3 & 4 \end{bmatrix} \begin{bmatrix} a \\ s \\ f \\ g \end{bmatrix} = \begin{bmatrix} 44 \\ 41 \\ 14 \end{bmatrix}.$

4. Solve the system of equations in Part (a) of Problem 3 by back substitution.
5. Find a nontrivial solution to the system

$$x - y - z - w = 0$$
$$-2y - 4z - 5w = 0.$$

MORE THEORETICAL PROBLEMS

Easier Problems

6. Show that the system of equations

$$ax + by + cz = u$$
$$0x + fy + gz = v$$
$$0x + hy + kz = w$$

has a unique solution for every u, v, w if and only if $a(fk - gh)$ is nonzero.

7. Give a formula for the solution x, y, z to the system of equations in Problem 6 which works as long as $a(fk - gh)$ is nonzero.

8. Prove that if $m = n$ in Theorem 2.1.1, then there is one and only one solution to the system (1).

2.2. EQUIVALENT SYSTEMS

Suppose that $Ax = y$ represents a system of m linear equations in n variables. We ask:

1. Are there solutions x to the equation $Ax = y$?
2. If so, how do we find x using the entries of A and y?

If $m \leq n$ and A is upper triangular with only nonzero entries on the main diagonal, we know by Theorem 2.1.1 that the system can be solved by back substitution. Otherwise, we find the answers to these questions by *eliminating variables* and replacing the given system by an *equivalent system* which we can use to get the answers. Let's look at an example.

EXAMPLE

To solve the system

$$2p + 3q + r = 6000$$
$$-4p - 2q + 2r = -4000$$

represented by the matrix equation $\begin{bmatrix} 2 & 3 & 1 \\ -4 & -2 & 2 \end{bmatrix} \begin{bmatrix} p \\ q \\ r \end{bmatrix} = \begin{bmatrix} 6000 \\ -4000 \end{bmatrix}$, we use

the first equation to modify the second equation by adding two times $(2p+3q+r)$ to its left-hand side and two times 6000 to its right-hand side, getting

$$0p + 4q + 4r = 8000.$$

So we can replace our original system by the system

$$2p + 3q + r = 6000$$
$$0p + 4q + 4r = 8000$$

obtained from it by adding 2 times the first equation to the second equation. Since the second system is obtained from the first by adding 2 times the first equation to the second, and since the first system could be obtained from the second by subtracting 2 times the first equation from the second, *the two systems have the same solutions*. The second system, represented by the matrix equation

$\begin{bmatrix} 2 & 3 & 1 \\ 0 & 4 & 4 \end{bmatrix} \begin{bmatrix} p \\ q \\ r \end{bmatrix} = \begin{bmatrix} 6000 \\ 8000 \end{bmatrix}$, has an upper triangular coefficient matrix, so it can

be solved by back substitution, by Theorem 2.1.1. In fact, we already did this in our first example, getting the solution $\begin{bmatrix} r \\ 2000 - r \\ r \end{bmatrix}$. Checking, we see that

$\begin{bmatrix} 2 & 3 & 1 \\ -4 & -2 & 2 \end{bmatrix} \begin{bmatrix} r \\ 2000 - r \\ r \end{bmatrix} = \begin{bmatrix} 6000 \\ -4000 \end{bmatrix}$, as expected.

Of course, if we modify a system of equations by multiplying all terms of one of its equations by a nonzero value or by interchanging two of the equations, the solutions again will be the same for the original system and the modified system.

EXAMPLE

The system

$$0p + 4q + 4r = 8000$$
$$2p + 3q + r = 6000$$

has the same solutions as the system

$$2p + 3q + r = 6000$$
$$0p + 4q + 4r = 8000$$

obtained from it by interchanging rows 1 and 2. This system, in turn has the same solutions as the system

$$2p + 3q + r = 6000$$
$$0p + q + r = 2000$$

obtained from it by multiplying row 2 by $\frac{1}{4}$. So, since the solutions to the last system are $\begin{bmatrix} r \\ 2000 - r \\ r \end{bmatrix}$, where $r \in F$, these are the solutions to the first system as well.

These operations come up often enough that we give them a name in the

Definition. The following three kinds of operations are called *elementary operations*:

1. The operation of adding u times equation s to equation r (where $r \neq s$);
2. The operation of interchanging equations r and s (where $r \neq s$);
3. The operation of multiplying equation r by u (where $u \neq 0$).

Each elementary operation can be reversed by a corresponding *inverse* operation. The inverse operations are

1'. The operation of adding $-u$ times equation s to equation r (where $r \neq s$);
2'. The operation of interchanging equations s and r (where $r \neq s$);
3'. The operation of by multiplying equation r by $1/u$ (where $u \neq 0$).

EXAMPLE

To solve the system

$$2p + 3q + r = 6000$$
$$-4p - 2q + 2r = -4000,$$

we first got the system

$$2p + 3q + r = 6000$$
$$0p + 4q + 4r = 8000$$

by adding 2 times the first equation to the second equation. If we now were to apply the inverse operation of adding -2 times the first equation to the second equation, we would get back the system,

$$2p + 3q + r = 6000$$
$$-4p - 2q + 2r = -4000,$$

with which we started.

Since each elementary operation can be reversed in this fashion,

The solutions to a system obtained from a given system by any elementary operation are the same as the solutions to the original system.

To release the full power of this observation, we make the

Definition. Two systems of equations are *equivalent* if the second can be obtained from the first by applying elementary operations successively a finite number of times.

The relation of equivalence satisfies the following properties:

1. *Any system of equations is equivalent to itself.*
2. *If a system of equations represented by $Ax = y$ is equivalent to a system of equations represented by $Bx = z$, then the system of equations represented by $Bx = z$ is equivalent to the system of equations represented by $Ax = y$.*
3. *If the system of equations represented by $Ax = y$ is equivalent to the system of equations represented by $Bx = z$ and if the system of equations represented by $Bx = z$ is equivalent to the system of equations represented by $Cx = w$, then the system of equations represented by $Ax = y$ is equivalent to $Cx = w$.*

Why? Multiplying each equation by 1 of the system of equations represented by the equation $Ax = y$, we find that there is no change and we end up with the system of equations that we started with. This proves (1). For (2), suppose that the system of equations represented by the equation $Ax = y$ is equivalent to the system of equations represented by the equation $Bx = z$. Then the system of equations represented by the equation $Bx = z$ is obtained from the system of equations represented by the equation $Ax = y$ by performing, successively, a finite number of elementary operations. Let's now *reverse* each of these elementary operations and apply the reverse operations successively in *reverse order* starting with the system of equations represented by the equation $Bx = z$. Then we get back the system of equations that we started with. Since the reverse operations are also elementary operations, it follows that the system of equations represented by the equation $Bx = z$ is equivalent to the system of equations represented by the equation $Ax = y$, which proves (2). For (3), suppose that starting with the system of equations represented by the equation $Ax = y$, we apply elementary operations successively a finite number of times, ending up with the system of equations represented by the equation $Bx = z$. And then, starting with the equation $Bx = z$, we apply elementary operations successively a finite number of times, ending up with the system of equations represented by the equation $Cx = w$. Then, starting with the system of equations represented by the equation $Ax = y$, we can apply all of these elementary operations in succession, first applying those used to get to the system represented by $Bx = z$ and then those used to get to $Cx = w$. So if the system of equations represented by $Ax = y$ is equivalent to the system of equations represented by $Bx = z$ and if the system of equations represented by $Bx = z$ is equivalent to the system of equations represented by $Cx = w$, then the system of equations represented by $Ax = y$ is equivalent to $Cx = w$, which proves (3).

These properties of equivalence become very important when we take into account

Theorem 2.2.1. If two matrix equations $Ax = y$ and $Bx = z$ are equivalent, then they have the same solutions x.

Proof: We already know that if a matrix equation is obtained from another matrix equation by a single elementary operation, the new equation has the same solutions as the old. So, using this time and time again, the new equation still has the same solutions as the old. ∎

PROBLEMS

NUMERICAL PROBLEMS

1. For each matrix A listed below, given the system of equations represented by $Ax = 0$, find an equivalent system of equations represented by $Bx = 0$, where B is upper triangular, and solve (find all solutions of) the latter.

   (a) $A = \begin{bmatrix} 1 & 4 & 6 \\ 1 & 7 & 11 \end{bmatrix}$.

   (b) $A = \begin{bmatrix} 2 & 4 & 6 \\ 5 & 5 & 61 \end{bmatrix}$.

   (c) $A = \begin{bmatrix} 1 & 2 & 3 \\ 4 & 5 & 6 \\ 7 & 8 & 9 \end{bmatrix}$.

2. For each matrix A and vector y listed below, given the system of equations represented by $Ax = y$, find an equivalent system of equations represented by $Bx = z$, where B is upper triangular, and solve the latter.

   (a) $A = \begin{bmatrix} 1 & 6 \\ 1 & 11 \end{bmatrix}$ and $y = \begin{bmatrix} 6 \\ 11 \end{bmatrix}$.

   (b) $A = \begin{bmatrix} 1 & 4 & 6 \\ 1 & 7 & 11 \end{bmatrix}$ and $y = \begin{bmatrix} 6 \\ 11 \end{bmatrix}$.

   (c) $A = \begin{bmatrix} 1 & 2 & 3 \\ 4 & 5 & 6 \\ 7 & 9 & 9 \end{bmatrix}$ and $y = \begin{bmatrix} 16 \\ 1 \\ 0 \end{bmatrix}$.

3. Show that the system of equations

$$x - y + z - w = 0$$
$$2x + 3y + 4z + 5w = 0$$

$$x - y + 2z - 2w = 0$$
$$5x + 5y + 9z + 9w = 0$$

has a nontrivial solution, by finding a system of three equations having the same solutions.

MORE THEORETICAL PROBLEMS

Easier Problems

4. Show by three examples that the set of solutions x to an equation $Ax = y$ with upper triangular coefficient matrix can be any one of the following: empty (i.e., no solutions), infinite, or consisting of exactly one solution.

Middle-Level Problems

5. Let A be an upper triangular $m \times n$ matrix with real entries and let $y = \begin{bmatrix} y_1 \\ \vdots \\ y_m \end{bmatrix}$.

Show that the set of solutions x to the equations $Ax = y$ must be one of the following: empty, infinite, or consisting of exactly one solution.

2.3. ELEMENTARY ROW OPERATIONS. ECHELON MATRICES

Elementary operations on systems of equations can be represented very efficiently by operations on matrices.

EXAMPLE

To solve the system

$$2p + 3q + r = 6000$$
$$-4p - 2q + 2r = -4000$$

we first got the system

$$2p + 3q + r = 6000$$
$$0p + 4q + 4r = 8000$$

by adding 2 times the first equation to the second equation. Instead, we can represent the system

$$2p + 3q + r = 6000$$
$$-4p - 2q + 2r = -4000$$

by the matrix $\begin{bmatrix} 2 & 3 & 1 \\ -4 & -2 & 2 \end{bmatrix}$ and vector $\begin{bmatrix} 6000 \\ -4000 \end{bmatrix}$, which we combine into the *augmented matrix* $\begin{bmatrix} 2 & 3 & 1 & 6000 \\ -4 & -2 & 2 & -4000 \end{bmatrix}$. We then add 2 times the first row to the second, getting $\begin{bmatrix} 2 & 3 & 1 & 6000 \\ 0 & 4 & 4 & 8000 \end{bmatrix}$, which represents the system

$$2p + 3q + r = 6000$$
$$0p + 4q + 4r = 8000.$$

Since the coefficient matrix is now upper triangular, we can solve the system by back substitution.

Taking this point of view, we can concentrate on how the operations affect the matrix representing the system. When the system of equations is *homogeneous*, that is, when the corresponding matrix equation is of the form $Ax = 0$, we do not even bother to augment the coefficient matrix A, so we then just concentrate on how the operations affect the coefficient matrix.

EXAMPLE

The homogeneous equations

$$\begin{bmatrix} 2 & 3 & 1 \\ -4 & -2 & 2 \end{bmatrix} \begin{bmatrix} p \\ q \\ r \end{bmatrix} = \begin{bmatrix} 0 \\ 0 \end{bmatrix} \quad \text{and} \quad \begin{bmatrix} 2 & 3 & 1 \\ 0 & 4 & 4 \end{bmatrix} \begin{bmatrix} p \\ q \\ r \end{bmatrix} = \begin{bmatrix} 0 \\ 0 \end{bmatrix}$$

have the same solutions since adding 2 times row 1 of $\begin{bmatrix} 2 & 3 & 1 \\ -4 & -2 & 2 \end{bmatrix}$ to row 2 gives us the new coefficient matrix $\begin{bmatrix} 2 & 3 & 1 \\ 0 & 4 & 4 \end{bmatrix}$. Solving the latter, we get the solution $\begin{bmatrix} p \\ q \\ r \end{bmatrix} = \begin{bmatrix} r \\ -r \\ r \end{bmatrix}$, which has $3 - 2 = 1$ independent variables.

These operations on matrices are just as important as are the corresponding operations on systems of equations. We now give them names.

Definition. The following three kinds of operations on $m \times n$ matrices are called *elementary row operations*:

1. The operation of adding u times row s to row r (where $r \neq s$), which we denote Add $(r, s; u)$.
2. The operation of interchanging rows r and s (where $r \neq s$), which we denote Interchange (r, s).
3. The operation of multiplying row r by u (where $u \neq 0$), which we denote Multiply $(r; u)$.

Sec. 2.3] Elementary Row Operations. Echelon Matrices

In our notation for the elementary row operations, we have arranged the subscripts first, followed by a semicolon, followed by a scalar in (1) and (3). Let's try out this notation by looking at what we get when we apply the operations Add$(r, s; u)$, Interchange(r, s), Multiply$(r; u)$ to the $n \times n$ identity matrix:

1. If we apply the operation Add$(r, s; u)$ to the $n \times n$ identity matrix I, we get the matrix which has 1's on the diagonal, (r, s) entry u and all other entries 0. For example, if $n = 3$, the operation Add$(2, 3; u)$ changes the identity matrix to the matrix $\begin{bmatrix} 1 & 0 & 0 \\ 0 & 1 & u \\ 0 & 0 & 1 \end{bmatrix}$ whose $(2, 3)$ entry is u.

2. If we apply the operation Interchange(r, s) to the $n \times n$ identity matrix I, it interchanges rows r and s. So we get the matrix which has 0 as (r, r) entry, 0 as (s, s) entry, 1 as (r, s) entry, 1 as (s, r) entry; all other diagonal entries 1; and all other entries 0. For example, Interchange$(2, 3)$ changes the 3×3 identity matrix to the matrix $\begin{bmatrix} 1 & 0 & 0 \\ 0 & 0 & 1 \\ 0 & 1 & 0 \end{bmatrix}$.

3. If we apply the operation Multiply$(r; u)$ to the $n \times n$ identity matrix I, we get the matrix which has (r, r) entry u, all other diagonal entries 1 and all other entries 0. For example, the operation Multiply$(2; u)$ changes the 3×3 identity matrix to the matrix $\begin{bmatrix} 1 & 0 & 0 \\ 0 & u & 0 \\ 0 & 0 & 1 \end{bmatrix}$ whose $(2, 2)$ entry is u.

Matrices such as $\begin{bmatrix} 1 & 0 & 0 \\ 0 & 1 & u \\ 0 & 0 & 1 \end{bmatrix}$, $\begin{bmatrix} 1 & 0 & 0 \\ 0 & 0 & 1 \\ 0 & 1 & 0 \end{bmatrix}$, $\begin{bmatrix} 1 & 0 & 0 \\ 0 & u & 0 \\ 0 & 0 & 1 \end{bmatrix}$ obtained by applying elementary row operations to the identity matrix are called *elementary matrices*. These matrices turn out to be very useful.

Each of the three elementary row operations on matrices can be reversed by a corresponding *inverse* operation, in the same way that we reversed the elementary operations on equations. These inverse operations are

1'. The operation of adding $-u$ times row s to row r (where $r \neq s$), which we denote Add$(r, s; -u)$.
2'. The operation of interchanging rows s and r (where $r \neq s$), which we denote Interchange(s, r).
3'. The operation of multiplying row r by $1/u$ (where $u \neq 0$), which we denote Multiply$(r; 1/u)$.

Of course, Interchange(r, s) and Interchange(s, r) are equal and the net effect of carrying out Interchange(r, s) twice is not to change the matrix at all. So the inverse of the operation Interchange(r, s) is just Interchange(r, s) itself.

EXAMPLE

When we apply the elementary row operations Add $(2, 3; 5)$, Interchange $(2, 3)$, Multiply $(3; 7)$ to the matrix $\begin{bmatrix} 1 & 3 & 4 \\ 3 & 2 & 8 \\ 2 & 3 & 1 \end{bmatrix}$, we get the matrices

$$\begin{bmatrix} 1 & 3 & 4 \\ 3+5\cdot 2 & 2+5\cdot 3 & 8+5\cdot 1 \\ 2 & 3 & 1 \end{bmatrix}, \begin{bmatrix} 1 & 3 & 4 \\ 2 & 3 & 1 \\ 3 & 2 & 8 \end{bmatrix}, \begin{bmatrix} 1 & 3 & 4 \\ 3 & 2 & 8 \\ 7\cdot 2 & 7\cdot 3 & 7\cdot 1 \end{bmatrix},$$

respectively. Then when we apply the inverse operation Add $(2, 3; -5)$ to

$$\begin{bmatrix} 1 & 3 & 4 \\ 3+5\cdot 2 & 2+5\cdot 3 & 8+5\cdot 1 \\ 2 & 3 & 1 \end{bmatrix},$$

we get back the matrix

$$\begin{bmatrix} 1 & 3 & 4 \\ 3+5\cdot 2+(-5)\cdot 2 & 2+5\cdot 3+(-5)\cdot 3 & 8+5\cdot 1+(-5)\cdot 1 \\ 2 & 3 & 1 \end{bmatrix} = \begin{bmatrix} 1 & 3 & 4 \\ 3 & 2 & 8 \\ 2 & 3 & 1 \end{bmatrix}.$$

Similarly, we get back the matrix $\begin{bmatrix} 1 & 3 & 4 \\ 3 & 2 & 8 \\ 2 & 3 & 1 \end{bmatrix}$ when we apply the inverse operations Interchange $(3, 2)$, Multiply $(3; \frac{1}{7})$ to $\begin{bmatrix} 1 & 3 & 4 \\ 2 & 3 & 1 \\ 3 & 2 & 8 \end{bmatrix}, \begin{bmatrix} 1 & 3 & 4 \\ 3 & 2 & 8 \\ 7\cdot 2 & 7\cdot 3 & 7\cdot 1 \end{bmatrix}$:

$$\text{Interchange }(3, 2) \begin{bmatrix} 1 & 3 & 4 \\ 2 & 3 & 1 \\ 3 & 2 & 8 \end{bmatrix} = \begin{bmatrix} 1 & 3 & 4 \\ 3 & 2 & 8 \\ 2 & 3 & 1 \end{bmatrix}$$

$$\text{Multiply }(3; \tfrac{1}{7}) \begin{bmatrix} 1 & 3 & 4 \\ 3 & 2 & 8 \\ 7\cdot 2 & 7\cdot 3 & 7\cdot 1 \end{bmatrix} = \begin{bmatrix} 1 & 3 & 4 \\ 3 & 2 & 8 \\ (\tfrac{1}{7})7\cdot 2 & (\tfrac{1}{7})7\cdot 3 & (\tfrac{1}{7})7\cdot 1 \end{bmatrix}$$

$$= \begin{bmatrix} 1 & 3 & 4 \\ 3 & 2 & 8 \\ 2 & 3 & 1 \end{bmatrix}.$$

The counterpart for matrices of our definition of equivalence for systems of equations is

Definition. Two matrices are *row equivalent* if the second can be obtained from the first by successively applying elementary row operations a finite number of times.

Sec. 2.3]   Elementary Row Operations. Echelon Matrices   83

The relation of row equivalence of matrices satisfies the following properties, which you can verify by following in the footsteps that we made in verifying the properties equivalence of systems of equations:

1. *Any matrix A is row equivalent to itself.*
2. *If A is row equivalent to B, then B is row equivalent to A.*
3. *If A is row equivalent to B and if B is row equivalent to C, then A is row equivalent to C.*

We now get the following counterpart of Theorem 2.2.1 for matrices. We leave it as an exercise for the reader to prove it, along the same lines as we proved Theorem 2.2.1.

Theorem 2.3.1.   If two matrices A and B are row equivalent, then the homogeneous equations $Ax = 0$ and $Bx = 0$ have the same solutions x.

This theorem provides us with an efficient way of solving the homogeneous equation $Ax = 0$. How? The coefficient matrix A is row equivalent to an echelon matrix B, defined below, and the equation $Bx = 0$ can be solved by a variation of the method of back substitution described in Section 2.1.

Definition.   An $m \times n$ matrix B is an *echelon matrix* if

1. The *leading entry* (first nonzero entry) of each nonzero row is 1; and
2. For any two consecutive rows, either the second of them is 0 or both rows are nonzero and the leading entry of the second of them is to the right of the leading entry of the first of them.

EXAMPLE

The matrices

$$\begin{bmatrix} 1 & 2 & 3 & 4 \\ 0 & 0 & 1 & 3 \\ 0 & 0 & 0 & 1 \\ 0 & 0 & 0 & 0 \end{bmatrix}, \begin{bmatrix} 1 & 4 & 0 & 0 & 0 \\ 0 & 0 & 0 & 0 & 1 \\ 0 & 0 & 0 & 0 & 0 \\ 0 & 0 & 0 & 0 & 0 \end{bmatrix}$$

are echelon matrices. The matrices

$$\begin{bmatrix} 1 & 2 & 3 & 4 \\ 0 & 0 & 1 & 3 \\ 0 & 0 & 1 & 1 \\ 0 & 0 & 0 & 1 \end{bmatrix}, \begin{bmatrix} 1 & 4 & 0 & 0 & 0 \\ 0 & 0 & 0 & 1 & 0 \\ 0 & 0 & 0 & 0 & 0 \\ 0 & 0 & 0 & 0 & 1 \end{bmatrix},$$

on the other hand, are not.

Though the leading entries of their nonzero rows are 1, each has two consecutive rows, where the leading entry 1 of the second does not occur to the right of a leading entry 1 of the first. For $\begin{bmatrix} 1 & 2 & 3 & 4 \\ 0 & 0 & 1 & 3 \\ 0 & 0 & 1 & 1 \\ 0 & 0 & 0 & 1 \end{bmatrix}$, these are the second and third rows; and for $\begin{bmatrix} 1 & 4 & 0 & 0 & 0 \\ 0 & 0 & 0 & 1 & 0 \\ 0 & 0 & 0 & 0 & 0 \\ 0 & 0 & 0 & 0 & 1 \end{bmatrix}$, the third and fourth.

An $m \times n$ matrix A can be row reduced to an echelon matrix as follows.

Method to row reduce an $m = n$ matrix A to an echelon matrix

1. Reduce A to a matrix B with 0's in the $(r, 1)$ position for $r = 2, \ldots, n$ as follows:
   (a) If the first column of A is zero, let $B = A$.
   (b) If the first column of A is not zero, then
      (i) If the a_{11} entry is 0, look for the first nonzero entry a_{r1} and interchange rows 1 and r.
      (ii) Multiply row 1 by a_{11}^{-1} and add $-a_{r1}$ times row 1 to row r for all $r > 1$.
2. If done, then stop. Otherwise, pass to the matrix C obtained from B by ignoring the first row and column of B.
3. Replace n by $n - 1$ and A by the $(n - 1) \times (n - 1)$ matrix C. Then repeat this entire process for the new n and A.

We state this result for future reference.

Theorem 2.3 ?. Any $m \times n$ matrix is row equivalent to an echelon matrix.

EXAMPLE

To solve $\begin{bmatrix} 2 & 3 & 1 \\ -2 & -3 & 3 \end{bmatrix} \begin{bmatrix} a \\ b \\ c \end{bmatrix} = 0$, we reduce $\begin{bmatrix} 2 & 3 & 1 \\ -2 & -3 & 3 \end{bmatrix}$ to an echelon matrix by a variation of the algorithm above. Starting with $\begin{bmatrix} 2 & 3 & 1 \\ -2 & -3 & 3 \end{bmatrix}$, we add the first row to the second to get $\begin{bmatrix} 2 & 3 & 1 \\ 0 & 0 & 4 \end{bmatrix}$. We continue on to multiply the first row by $\frac{1}{2}$, then the second row by $\frac{1}{4}$, getting the echelon matrix $\begin{bmatrix} 1 & \frac{3}{2} & \frac{1}{2} \\ 0 & 0 & 1 \end{bmatrix}$. So the solutions to $\begin{bmatrix} 2 & 3 & 1 \\ -2 & 1 & 3 \end{bmatrix} \begin{bmatrix} a \\ b \\ c \end{bmatrix} = 0$ are the solutions

Sec. 2.4] Solving Systems of Linear Equations

to the homogeneous equation $\begin{bmatrix} 1 & \frac{3}{2} & \frac{1}{2} \\ 0 & 0 & 1 \end{bmatrix} \begin{bmatrix} a \\ b \\ c \end{bmatrix} = 0$, which can be found by back substitution to be $\begin{bmatrix} a \\ b \\ c \end{bmatrix} = \begin{bmatrix} -(\frac{3}{2})b \\ b \\ 0 \end{bmatrix}$.

PROBLEMS

NUMERICAL PROBLEMS

1. Reduce the following matrices to echelon matrices using elementary row operations.

   (a) $\begin{bmatrix} 1 & 2 & 3 & 4 \\ 1 & 2 & 1 & 3 \\ 1 & 3 & 1 & 1 \end{bmatrix}$.

   (b) $\begin{bmatrix} 1 & 2 & 3 & 4 \\ 2 & 2 & 2 & 2 \\ 4 & 3 & 2 & 1 \end{bmatrix}$.

   (c) $\begin{bmatrix} 1 & 2 & 3 & 1 & 2 & 3 \\ 2 & 3 & 2 & 3 & 2 & 3 \\ 3 & 5 & 5 & 4 & 4 & 6 \end{bmatrix}$.

   (d) $\begin{bmatrix} 3 & 4 \\ 2 & 2 \\ 2 & 1 \end{bmatrix}$.

2. Find some nonzero solution to $Ax = 0$ (or show that none exists) for each matrix A listed in Problem 1.

MORE THEORETICAL PROBLEMS

Easier Problems

3. Prove the following properties of row equivalence.
   (1) Any matrix A is row equivalent to itself.
   (2) If A is row equivalent to B, then B is row equivalent to A.
   (3) If A is row equivalent to B and if B is row equivalent to C, then A is row equivalent to C.

2.4. SOLVING SYSTEMS OF LINEAR EQUATIONS

Let A be an $m \times n$ matrix and suppose that we have computed an echelon matrix B row equivalent to A, by the reduction algorithm given above. Then the solutions to the homogeneous equation $Ax = 0$ are the same as the solutions to the homogeneous equation $Bx = 0$ and the latter can be gotten by back substitution. Since B is an echelon matrix, the number of independent variables in the general solution to $Bx = 0$ by back substitution is $n - m'$, where m' is the number of nonzero rows of B. Why? Let's look at an example before we give the answer in general. The solution to

$$\begin{bmatrix} 1 & 2 & 3 & 4 \\ 0 & 0 & 1 & 3 \\ 0 & 0 & 0 & 0 \end{bmatrix} \begin{bmatrix} a \\ b \\ c \\ d \end{bmatrix} = \begin{bmatrix} 0 \\ 0 \\ 0 \end{bmatrix} \text{ is } \begin{bmatrix} a \\ b \\ c \\ d \end{bmatrix} = \begin{bmatrix} -2b + 9d - 4d \\ b \\ -3d \\ d \end{bmatrix},$$

which has $2 = 4 - 2$ independent variables b and d. Note that b and d correspond to columns 2 and 4, which are the columns that do not contain leading entries 1 of nonzero rows. This gives the answer for our example. In the general case, the answer is the same. The independent variables correspond to those columns that do not contain a leading entry of 1. So there are $n - m'$ of them, where m' is the number of nonzero rows of B. This proves

Theorem 2.4.1. Let A be an $m \times n$ matrix and suppose that A is row equivalent to an echelon matrix B. Then the general solution x to the homogeneous equation $Ax = 0$ obtained by solving the homogeneous equation $Bx = 0$ by back substitution contains $n - m'$ independent variables, where m' is the number of nonzero rows of B.

A useful special case of Theorem 2.4.1 is

Corollary 2.4.2. Let A be an $m \times n$ matrix and suppose that A is row equivalent to an echelon matrix B. Then 0 is the only solution to the homogeneous equation $Ax = 0$ if and only if B has n nonzero rows.

This corollary, in turn, leads to the

Corollary 2.4.3. Let A be an $m \times n$ matrix with $m < n$. Then the homogeneous equation $Ax = 0$ has a nonzero solution.

Proof: Take B to be an echelon matrix which is row equivalent to A. Since $m < n$ and m is the number of rows, B cannot have n nonzero rows. So $Ax = 0$ has a nonzero solution by Corollary 2.4.2. ∎

From the discussion above we see that the method of back substitution to solve the matrix equation $Bx = z$ when the coefficient matrix B is an $m \times n$ echelon matrix is just the following slight refinement of the method of back substitution that we gave in a special case in Section 2.1. Of course, *this method works only when $z_r = 0$ whenever row r of B is 0*, since otherwise $Bx = z$ has no solution x.

Method to solve a system of linear equations by back substitution if the coefficient matrix is an echelon matrix

1. Let the $n - m$ extra variables $x_{s_1}, \ldots, x_{s_{n-m}}$ corresponding to the columns $s_1, \ldots, s_{n-m}$ which do not contain leading entries 1 of nonzero rows have any values, and move them to the right-hand side of the equations.

Sec. 2.4]   Solving Systems of Linear Equations

2. Find the value for the last remaining dependent variable from the last remaining equation and discard that equation.
3. Eliminate that variable by substituting its value back in place of it in all of the preceding equations.
4. Repeat (2), (3), and (4) for the next last remaining dependent variable using the next last remaining equation to get its value and eliminate it and the equation. Continue until the values for all dependent variables have been found.

EXAMPLE

To solve the equation $\begin{bmatrix} 1 & 2 & 3 & 4 \\ 0 & 0 & 1 & 3 \\ 0 & 0 & 0 & 0 \end{bmatrix} \begin{bmatrix} a \\ b \\ c \\ d \end{bmatrix} = \begin{bmatrix} 8 \\ 4 \\ 0 \end{bmatrix}$ by back substitution, we look

for the columns that do not contain leading entries 1 of nonzero rows, namely columns 2 and 4. The corresponding *extra* variables are b and d, so we move them to the right-hand side, getting the system of equations

$$1a + 3c = 8 - 2b - 4d$$
$$0a + 1c = 4 - 3d$$
$$0a + 0a = 0.$$

We now solve this system by back substitution, getting $c = 4 - 3d$ and then $a = 8 - 2b - 4d - 3(4 - 3d) = -4 - 2b + 5d$ or

$$\begin{bmatrix} a \\ b \\ c \\ d \end{bmatrix} = \begin{bmatrix} -4 - 2b + 5d \\ b \\ 4 - 3d \\ d \end{bmatrix}.$$

How can we use this method to solve a *nonhomogeneous* matrix equation $Ax = y$ for x? We find an equivalent matrix equation $Bx = z$, where B is an echelon matrix. This is done by using the method of Section 2.3 to row reduce A to an echelon matrix B with the variation that each row operation during the reduction process is applied both to the coefficient matrix and to the column vector, ultimately changing A to B and y to z.

In practice, this can be done by forming the *augmented matrix* $[A, y]$ and reducing it to an echelon matrix $[B, z]$. Then the equations $Ax = y$ and $Bx = z$ are equivalent and the solutions to $Ax = y$ are found as the solutions to $Bx = z$ using the method of back substitution. So we have

Method to solve $Ax = y$ for x

1. Form the augmented matrix $[A, y]$ and reduce it to an echelon matrix $[B, z]$.
2. Find x as the general solution to $Bx = z$ by back substitution.

EXAMPLE

To find the solutions to $\begin{bmatrix} 1 & 2 & 3 & 4 \\ 1 & 2 & 4 & 7 \\ 2 & 4 & 6 & 8 \end{bmatrix} \begin{bmatrix} a \\ b \\ c \\ d \end{bmatrix} = \begin{bmatrix} 8 \\ 12 \\ 16 \end{bmatrix}$, reduce the augmented matrix $\begin{bmatrix} 1 & 2 & 3 & 4 & 8 \\ 1 & 2 & 4 & 7 & 12 \\ 2 & 4 & 6 & 8 & 16 \end{bmatrix}$ to the echelon matrix $\begin{bmatrix} 1 & 2 & 3 & 4 & 8 \\ 0 & 0 & 1 & 3 & 4 \\ 0 & 0 & 0 & 0 & 0 \end{bmatrix}$. Since the third entry of the vector $\begin{bmatrix} 8 \\ 4 \\ 0 \end{bmatrix}$ on the right-hand side is 0, we can solve the equation $\begin{bmatrix} 1 & 2 & 3 & 4 \\ 0 & 0 & 1 & 3 \\ 0 & 0 & 0 & 0 \end{bmatrix} \begin{bmatrix} a \\ b \\ c \\ d \end{bmatrix} = \begin{bmatrix} 8 \\ 4 \\ 0 \end{bmatrix}$ by back substitution. We did this in the example above, getting $\begin{bmatrix} a \\ b \\ c \\ d \end{bmatrix} \begin{bmatrix} -4 - 2b + 5d \\ b \\ 4 - 3d \\ d \end{bmatrix}$.

How do we know when $Ax = y$ has a solution? Because $Bx = z$ has a solution if and only if $z_r = 0$ whenever row r of B consists of 0's, we have

Method to determine whether $Ax = y$ has a solution

1. Form the augmented matrix [A, y] and reduce it to an echelon matrix [B, z].
2. Then $Ax = y$ has a solution if and only if $z_r = 0$ whenever row r of B consists of 0's.

When $Ax = y$ does have a solution, the discussion at the beginning of this section shows that Theorem 2.4.1 has the following counterpart for nonhomogeneous equations.

Theorem 2.4.4. Let A be an $m \times n$ matrix and suppose that the equation $Ax = y$ is equivalent to the equation $Bx = z$, where B is an echelon matrix. If $Ax = y$ has a solution, then the general solution x to the equation $Ax = y$ obtained by solving the equation $Bx = z$ by back substitution contains $n - m'$ independent variables where m' is the number of nonzero rows of B.

The number of nonzero rows of an $m \times n$ echelon matrix B is called its *rank*, denoted rank (B). Our theorem gives the formula

$n = $ rank $(B) + $ number of independent variables in the solutions of $Bx = 0$.

We discuss this formula in more depth in Section 4.4.

PROBLEMS

NUMERICAL PROBLEMS

1. Show that each of the following equations has a nonzero solution.

(a) $\begin{bmatrix} 1 & 2 & 3 & 4 \\ 1 & 2 & 1 & 3 \\ 1 & 3 & 1 & 1 \end{bmatrix} x = 0.$

(b) $\begin{bmatrix} 5 & 3 & 4 \\ 1 & 1 & 2 \\ 6 & 4 & 6 \end{bmatrix} x = 0.$

(c) $\begin{bmatrix} 2 & 5 & 5 & 4 & 2 & 2 & 2 & 3 & 4 \\ 1 & 2 & 4 & 5 & 8 & 7 & 7 & 6 & 3 \\ 1 & 1 & 1 & 1 & 3 & 2 & 2 & 3 & 4 & 4 \end{bmatrix} x = 0.$

2. Solve the equation $\begin{bmatrix} 1 & 4 & 3 & 2 \\ 0 & 0 & 1 & 3 \\ 0 & 0 & 0 & 0 \end{bmatrix} x = \begin{bmatrix} 3 \\ 3 \\ 0 \end{bmatrix}$ using the method of back substitution.

3. Solve the equation $\begin{bmatrix} 1 & 2 & 3 & 2 \\ 2 & 4 & 1 & 3 \\ 3 & 6 & 4 & 5 \end{bmatrix} x = \begin{bmatrix} 4 \\ 7 \\ 11 \end{bmatrix}$ by forming the augmented matrix to find an equivalent equation whose coefficient matrix is an echelon matrix and solving the latter by the method of back substitution.

4. Determine whether the equation $\begin{bmatrix} 1 & 2 & 3 & 2 \\ 2 & 4 & 6 & 2 \\ 3 & 6 & 9 & 3 \end{bmatrix} x = \begin{bmatrix} 4 \\ 7 \\ 1 \end{bmatrix}$ has a solution.

MORE THEORETICAL PROBLEMS

Easier Problems

5. Represent the system of equations

$$1u + 4k + 6h = 6$$
$$0u + 3k + 5h = 5$$

by a matrix equation and show that the set of all solutions is the set of vectors $v + w$, where v is one solution and $w = \begin{bmatrix} u \\ k \\ h \end{bmatrix}$ is any solution to the homogeneous system of equations

$$1u + 4k + 6h = 0$$
$$0u + 3k + 5h = 0.$$

CHAPTER 3

The $n \times n$ Matrices

With the experience of the 2×2 matrices behind us the transition to the $n \times n$ case will be rather straightforward and should not present us with many large difficulties. In fact, for a large part, the development of the $n \times n$ matrices will closely follow the line used in the 2×2 case. So we have some inkling of what results to expect and how to go about establishing these results. All this will stand us in good stead now.

3.1. THE OPENING

What better place to start than the beginning? So we start things off with our cast of players. If $\mathbb{R}$ is the set of real numbers and $\mathbb{C}$ that of complex numbers, by $M_n(\mathbb{R})$ and $M_n(\mathbb{C})$ we shall mean the set of all square arrays, with n rows and n columns, of real or complex numbers, respectively.

We could speak about $m \times n$ matrices, even when $m \neq n$, but we shall restrict our attention for now to $n \times n$ matrices. We call them *square matrices*.

What are they? An $n \times n$ matrix is a square array

$$\begin{bmatrix} a_{11} & a_{12} & \cdots & a_{1n} \\ a_{21} & a_{22} & \cdots & a_{2n} \\ \vdots & \vdots & & \vdots \\ a_{n1} & a_{n2} & \cdots & a_{nn} \end{bmatrix},$$

where all the a_{rs} are in $\mathbb{C}$. We use $M_n(\mathbb{C})$ to denote the set of all $n \times n$ matrices over $\mathbb{C}$, that is, the set of all such square arrays. Similarly, $M_n(\mathbb{R})$ is the set of such arrays where all the a_{rs} are restricted to being in $\mathbb{R}$. Since $\mathbb{R} \subset \mathbb{C}$ we have $M_n(\mathbb{R}) \subset M_n(\mathbb{C})$.

We don't want to keep specifying whether we are working over $\mathbb{R}$ or $\mathbb{C}$ at any given time. So we shall merely speak about matrices. For most things it will not be important if these matrices are in $M_n(\mathbb{R})$ or $M_n(\mathbb{C})$. If we simply say that we are working with $M_n(F)$, where F is understood to be either $\mathbb{R}$ or $\mathbb{C}$, then what we say applies in both cases. When the results do depend on the number system we are using, it will be so pointed out.

If A is the matrix above, we call the a_{rs} the *entries* of A, and a_{rs} is called the (r, s) *entry*, meaning that it is that entry in row r and column s.

In analogy with what we did for the 2×2 matrices we want to be able to express when two matrices are equal, how to add them, and how to multiply them. We shall use the shorthand $A = (a_{rs})$ to denote the matrix A above. [Mathematicians usually—out of habit—write (a_{ij}), but to avoid confusion with the complex number i, where $i^2 = -1$, we prefer to use other subscripts.] So (a_{rs}) will be that matrix whose (r, s) entry is a_{rs}. In all that follows n is a fixed integer with $n \geq 2$ and all matrices are in $M_n(F)$.

Definition. The two matrices $A = (a_{rs})$ and $B = (b_{rs})$ are defined to be *equal*, $A = B$, if and only if $a_{uv} = b_{uv}$ for all u and v.

So, for instance, $\begin{bmatrix} x & y & 1 \\ 0 & 5 & z \\ 0 & 1 & 1 \end{bmatrix} = \begin{bmatrix} 1 & 3 & 1 \\ 0 & 5 & \pi \\ 0 & 1 & 1 \end{bmatrix}$ if and only if $x = 1$, $y = 3$, and $z = \pi$. Note that for *no* choice of x, y, z can $\begin{bmatrix} x & y & 1 \\ 0 & 5 & 3 \\ 0 & 1 & 1 \end{bmatrix}$ equal $\begin{bmatrix} 1 & 3 & 1 \\ 0 & 5 & \pi \\ 0 & 0 & 1 \end{bmatrix}$, for these differ in the $(3, 2)$ entry.

Now that we have equality of two matrices defined we pass on to the next notion, that of the *addition* of matrices.

Definition. Given the matrices $A = (a_{rs})$ and $B = (b_{rs})$, then $A + B = (c_{rs})$, where $c_{uv} = a_{uv} + b_{uv}$ for all u and v.

By this definition it is clear that the sum of matrices is again a matrix, that is, if $A, B \in M_n(\mathbb{C})$, then $A + B \in M_n(\mathbb{C})$, and if $A, B \in M_n(\mathbb{R})$, then $A + B \in M_n(\mathbb{R})$. We add matrices by adding their corresponding entries.

As before, the matrix all of whose entries are 0—which we shall write merely as 0—has the property that $A + 0 = A$ for all matrices A. Similarly, if $A = (a_{rs})$, then the matrix (b_{rs}), where $b_{rs} = -a_{rs}$, is written as $-A$ and has the property that $A + (-A) = 0$. Note that addition of matrices satisfies the *commutative law*, namely, that $A + B = B + A$. For example,

$$\begin{bmatrix} 2 & 0 & 0 \\ 1 & 3 & 1 \\ 0 & 0 & 9 \end{bmatrix} + \begin{bmatrix} 2 & 0 & 0 \\ 2 & 3 & 4 \\ 0 & 0 & 8 \end{bmatrix} = \begin{bmatrix} 4 & 0 & 0 \\ 3 & 6 & 5 \\ 0 & 0 & 17 \end{bmatrix} = \begin{bmatrix} 2 & 0 & 0 \\ 2 & 3 & 4 \\ 0 & 0 & 8 \end{bmatrix} + \begin{bmatrix} 2 & 0 & 0 \\ 1 & 3 & 1 \\ 0 & 0 & 9 \end{bmatrix}.$$

If $A = \begin{bmatrix} 1 & i & 0 & 2 \\ 3+i & 0 & \frac{1}{2} & -1 \\ i & 0 & 0 & -3i \\ 1 & 1 & 1 & 0 \end{bmatrix}$ and $B = \begin{bmatrix} 0 & -i & 1 & 3 \\ i & i & i & -i \\ 1 & -i & 1 & -1 \\ 0 & i & 0 & -i \end{bmatrix}$ are two 4×4

matrices, then $A + B$ is the matrix

$$\begin{bmatrix} 1 & 0 & 1 & 5 \\ 3 + 2i & i & \frac{1}{2} + i & -1 - i \\ 1 + i & -1 & 1 & -1 - 3i \\ 1 & 1 + i & 1 & -i \end{bmatrix}.$$

We single out some nice-looking matrices, as we did in the 2×2 case.

Definition. The matrix $A = (a_{rs})$ is a *diagonal matrix* if $a_{uv} = 0$ for $u \ne v$.

So a diagonal matrix is one in which the entries off the main diagonal are all 0. For example, the matrices $\begin{bmatrix} 2 & 0 & 0 \\ 0 & 3 & 0 \\ 0 & 0 & 9 \end{bmatrix}$ and $\begin{bmatrix} 99 & 0 \\ 0 & 99 \end{bmatrix}$ are diagonal matrices, whereas the matrix $A = \begin{bmatrix} 0 & 0 & 1 \\ 0 & 1 & 0 \\ 1 & 0 & 0 \end{bmatrix}$ is not, since a_{13} and a_{31} are not 0.

Among the diagonal matrices there are some that are nicer than others. They are the so-called scalar matrices.

Definition. A diagonal matrix is called a *scalar matrix* if all the entries on the main diagonal are equal.

Thus a scalar matrix looks like

$$\begin{bmatrix} a & 0 & 0 & \cdots & 0 \\ 0 & a & 0 & \cdots & 0 \\ 0 & 0 & a & \cdots & 0 \\ \vdots & & & \cdots & \vdots \\ 0 & 0 & 0 & \cdots & a \end{bmatrix},$$

where a is a number. We shall write this matrix as aI, where I is the matrix

$$I = \begin{bmatrix} 1 & 0 & 0 & \cdots & 0 \\ 0 & 1 & 0 & \cdots & 0 \\ 0 & 0 & 1 & \cdots & 0 \\ \vdots & & & \cdots & \vdots \\ 0 & \cdots & \cdots & \cdots & 1 \end{bmatrix}.$$

Here, in writing I, wherever we wrote no entry that entry is understood to be 0. So the matrix aI is

$$aI = \begin{bmatrix} a & 0 & 0 & \cdots & 0 \\ 0 & a & 0 & \cdots & 0 \\ 0 & 0 & a & \cdots & 0 \\ \vdots & & & \cdots & \vdots \\ 0 & \cdots & \cdots & \cdots & a \end{bmatrix}.$$

Notice that $aI + bI = (a + b)I$; for instance,

$$1I + \pi I = \begin{bmatrix} 1 & 0 \\ 0 & 1 \end{bmatrix} + \begin{bmatrix} \pi & 0 \\ 0 & \pi \end{bmatrix} = \begin{bmatrix} 1 + \pi & 0 \\ 0 & 1 + \pi \end{bmatrix} = (1 + \pi)I.$$

So the scalar matrices aI behave like the scalars (numbers) a themselves—real or complex as the case may be—under addition.

Definition. We define *multiplication by a scalar* as follows. If $A = (a_{rs})$ and b is a number, then $bA = (c_{rs})$, where for each u and v, $c_{uv} = ba_{uv}$.

So, for example, $(-5)\begin{bmatrix} 1 & 2 & 3 \\ 0 & 1 & 6 \\ 1 & -1 & 4 \end{bmatrix} = \begin{bmatrix} -5 & -10 & -15 \\ 0 & -5 & -30 \\ -5 & 5 & -20 \end{bmatrix}.$

The result of multiplying A by the scalar -1 is the matrix $(-1)A = -A$ which we introduced earlier.

In a moment we shall see that this type of multiplication is merely a special case of the product defined for any matrices.

We now come to the trickiest of the operations, that of the *product* of two matrices. With what we saw for the 2×2 matrices, the definition we are about to make should not be as mysterious or confusing as it might otherwise be.

Definition. Given the matrices $A = (a_{rs})$ and $B = (b_{rs})$, then $AB = (c_{rs})$, where $c_{uw} = \sum_{\varphi=1}^{n} a_{u\varphi}b_{\varphi w}$.

Note that the subscript φ over which the sum above is taken is a dummy index. We could equally well call it anything. So

$$c_{uw} = \sum_{\varphi=1}^{n} a_{u\varphi}b_{\varphi w} = \sum_{\square=1}^{n} a_{u\square}b_{\square w} = \sum_{v=1}^{n} a_{uv}b_{vw}.$$

To highlight the dummy variables while you get used to them, we'll use Greek letters for them in this section and part of the next. After that, we'll just use Latin letters, to keep the notation simple.

Note also that if $A, B \in M_n(\mathbb{C})$, then $AB \in M_n(\mathbb{C})$, and if $A, B \in M_n(\mathbb{R})$, then $AB \in M_n(\mathbb{R})$. We express this by saying that $M_n(\mathbb{C})$ [and $M_n(\mathbb{R})$] is *closed* with respect to products.

Let's compute the product of some matrices.

If $A = \begin{bmatrix} 1 & 2 & 3 \\ 3 & 2 & 1 \\ 0 & -1 & 0 \end{bmatrix}$ and $B = \begin{bmatrix} i & 0 & 0 \\ 0 & -i & 0 \\ 0 & 0 & 1 \end{bmatrix}$, then

$$AB = \begin{bmatrix} 1(i) + 2(0) + 3(0) & 1(0) + 2(-i) + 3(0) & 1(0) + 2(0) + 3(1) \\ 3(i) + 2(0) + 1(0) & 3(0) + 2(-i) + 1(0) & 3(0) + 2(0) + 1(1) \\ 0(i) + (-1)0 + 0(0) & 0(0) + (-1)(-i) + 0(0) & 0(0) + (-1)(0) + 0(1) \end{bmatrix}$$

$$= \begin{bmatrix} i & -2i & 3 \\ 3i & -2i & 1 \\ 0 & i & 0 \end{bmatrix}.$$

As a way of remembering how to form the product AB:

The (s, t) entry is the "dot product" of the row s of A by column t of B.

In the case above, for example, the $(2, 3)$ entry is the dot product $3(0) + 2(0) + 1(1)$ of row $2 = (3, 2, 1)$ of A and $(0, 0, 1)$, which is column 3 written horizontally.

Note that the matrix I introduced above satisfies $AI = IA = A$ for every matrix A. So I acts in matrices as the number 1 does in numbers. Note also that $(aI)A = aA$, so multiplication by a scalar is merely a special case of matrix multiplication. (Prove!) For this reason, we call I the *identity matrix*.

Similarly, the matrix $0I$, which we denote simply by 0, satisfies $A0 = 0A = 0$ for all A. We call $0 = 0I$ the *zero matrix*.

As we already saw in the 2×2 matrices, matrices do not, in general, satisfy the *commutative law* of multiplication; that is, AB need not equal BA. Remember also that in matrices it is possible that $AB = 0$ but $A \neq 0$ and $B \neq 0$.

In the $n \times n$ matrices there are n^2 particular matrices that play a key role. These are called the *matrix units*, E_{rs}, which are defined as follows: E_{rs} is the matrix whose (r, s) entry is 1 and all of whose other entries are 0. So, for instance, in $M_3(\mathbb{C})$,

$$E_{32} = \begin{bmatrix} 0 & 0 & 0 \\ 0 & 0 & 0 \\ 0 & 1 & 0 \end{bmatrix} \quad \text{and} \quad E_{22} = \begin{bmatrix} 0 & 0 & 0 \\ 0 & 1 & 0 \\ 0 & 0 & 0 \end{bmatrix}.$$

One of the important properties of the E_{rs} is that every matrix can be expressed as a *combination* of them with scalar coefficients, that is, $A = (a_{rs}) = \sum_\rho \sum_\sigma a_{\rho\sigma} E_{\rho\sigma}$.

For example,

$$\begin{bmatrix} 2 & 0 & 0 \\ 0 & 5 & 0 \\ 0 & 8 & 0 \end{bmatrix} = 2\begin{bmatrix} 1 & 0 & 0 \\ 0 & 0 & 0 \\ 0 & 0 & 0 \end{bmatrix} + 5\begin{bmatrix} 0 & 0 & 0 \\ 0 & 1 & 0 \\ 0 & 0 & 0 \end{bmatrix} + 8\begin{bmatrix} 0 & 0 & 0 \\ 0 & 0 & 0 \\ 0 & 1 & 0 \end{bmatrix} = 2E_{11} + 5E_{22} + 8E_{32}.$$

Another very important thing about the E_{rs}'s is how they multiply. The equations

$$E_{rs}E_{uv} = 0 \quad \text{if } s \neq u$$

$$E_{rs}E_{sv} = E_{rv}$$

hold for all r, s, u, v. Instances of this are $E_{31}E_{22} = \begin{bmatrix} 0 & 0 & 0 \\ 0 & 0 & 0 \\ 1 & 0 & 0 \end{bmatrix}\begin{bmatrix} 0 & 0 & 0 \\ 0 & 1 & 0 \\ 0 & 0 & 0 \end{bmatrix} = 0$ and

$$E_{32}E_{22} = \begin{bmatrix} 0 & 0 & 0 \\ 0 & 0 & 0 \\ 0 & 1 & 0 \end{bmatrix}\begin{bmatrix} 0 & 0 & 0 \\ 0 & 1 & 0 \\ 0 & 0 & 0 \end{bmatrix} = \begin{bmatrix} 0 & 0 & 0 \\ 0 & 0 & 0 \\ 0 & 1 & 0 \end{bmatrix} = E_{32}.$$

We leave these important things for the reader to verify (they occur in the problem set to follow).

We compute—anticipating the arithmetic properties that hold for matrices—the product of some combination of the E_{rs}'s. Let $A = 2E_{11} + 3E_{12} + E_{33}$ and $B = E_{12} - E_{31}$. Then AB is

$$(2E_{11} + 3E_{12} + E_{33})(E_{12} - E_{31})$$
$$= 2E_{11}E_{12} + 3E_{12}E_{12} + E_{33}E_{12} - 2E_{11}E_{31} - 3E_{12}E_{31} - E_{33}E_{31}$$
$$= 2E_{12} - E_{31}$$

by the rules of multiplying the matrix units. As matrices, we have

$$A = \begin{bmatrix} 2 & 3 & 0 \\ 0 & 0 & 0 \\ 0 & 0 & 1 \end{bmatrix}, \quad B = \begin{bmatrix} 0 & 1 & 0 \\ 0 & 0 & 0 \\ -1 & 0 & 0 \end{bmatrix}, \quad AB = \begin{bmatrix} 0 & 2 & 0 \\ 0 & 0 & 0 \\ -1 & 0 & 0 \end{bmatrix} = 2E_{12} - E_{31}.$$

Not surprisingly, the answer we get this way for AB is exactly the same as if we computed AB by our definition of the product of the matrices A and B.

In the 2×2 matrices we found that the introduction of exponents was a handy and useful device for representing the powers of matrix. Here, too, it will be equally useful. As before, we define for an $n \times n$ matrix A, the *power* A^m for positive integer values m to be $A^m = \underbrace{AAA \cdots A}_{(m \text{ times})}$. We define A^0 to be I. For negative integers m there is also a problem, for it cannot be defined in a sensible way for all matrices. We would want A^{-1} to have the meaning: that matrix such that $AA^{-1} = I$. However, for some matrices A, no such A^{-1} exists!

Definition. The matrix A is said to be *invertible* if there exists a matrix B such that $AB = BA = I$.

If A is invertible, there is only one matrix B such that $AB = BA = I$ (prove!), and we denote this matrix B as A^{-1}.

Even for 2×2 matrices we saw that A^{-1} need not exist for every matrix A. As we proceed, we shall find the necessary and sufficient criterion that a given matrix have an *inverse* A^{-1}, that is, that it be invertible.

Finally, for nonnegative integers r and s the usual rules of exponents hold, namely $A^r A^s = A^{r+s}$ and $(A^r)^s = A^{rs}$. (Verify!) If A is invertible, we define A^{-m}, for m a positive integer, by $A^{-m} = (A^{-1})^m$. If A is invertible, then $A^r A^s = A^{r+s}$ and $(A^r)^s = A^{rs}$ hold for *all* integers r and s (if we understand A^0 to be I). (Verify!)

We close this section with an omnibus theorem about addition, multiplication, and how they interact. Its proof we leave as a massive, although generally easy, exercise for the reader, since our proof of the 2×2 version of it is a road map for the general proof.

Theorem 3.1.1. For matrices the following hold:

1. $A + B = B + A$ — *Commutativity;*
2. $(A + B) + C = A + (B + C)$ — *Associativity of Sums;*
3. $A + 0 = A$.
4. $A + (-A) = 0$.
5. $A0 = 0A = 0$;
6. $AI = IA = A$;
7. $(AB)C = A(BC)$ — *Associativity of Products;*
8. $A(B + C) = AB + AC$; and
   $(B + C)A = BA + CA$ — *Distributivity.*

PROBLEMS

NUMERICAL PROBLEMS

1. Calculate.

   (a) $\begin{bmatrix} 1 & -6 & 0 & \pi \\ \pi & i & 0 & i \\ 0 & 0 & 1 & 0 \\ 1 & 0 & i & i \end{bmatrix} \begin{bmatrix} 1 & 2 & 3 & 4 \\ 0 & 1 & 2 & 3 \\ 0 & 0 & 1 & 2 \\ 0 & 0 & 0 & 1 \end{bmatrix}$.

   (b) $\begin{bmatrix} \frac{1}{2} & \frac{1}{3} & \frac{1}{4} \\ -\frac{1}{2} & -\frac{1}{3} & -\frac{1}{4} \\ 1 & 1 & 1 \end{bmatrix}^2$.

   (c) $\begin{bmatrix} 0 & 1 & 0 & 0 \\ 0 & 0 & 1 & 0 \\ 0 & 0 & 0 & 1 \\ 1 & 0 & 0 & 0 \end{bmatrix}^4$.

   (d) $\begin{bmatrix} 1 & -1 & 2 \\ 3 & 1 & 4 \\ 5 & 6 & 0 \end{bmatrix} \left(\begin{bmatrix} 1 & 2 & 0 \\ -10 & -9 & 8 \\ 7 & 6 & 4 \end{bmatrix} + \begin{bmatrix} 3 & 0 & 2 \\ 10 & 8 & -9 \\ -7 & -6 & 4 \end{bmatrix} \right)$.

   (e) $\begin{bmatrix} \frac{1}{3} & \frac{1}{3} & \frac{1}{3} \\ \frac{1}{3} & \frac{1}{3} & \frac{1}{3} \\ \frac{1}{3} & \frac{1}{3} & \frac{1}{3} \end{bmatrix} \begin{bmatrix} \frac{2}{3} & -\frac{1}{3} & -\frac{1}{3} \\ -\frac{1}{3} & \frac{2}{3} & -\frac{1}{3} \\ -\frac{1}{3} & -\frac{1}{3} & \frac{2}{3} \end{bmatrix}$.

(f) $\begin{bmatrix} 0 & 0 & i \\ 0 & i & 0 \\ i & 0 & 0 \end{bmatrix} \begin{bmatrix} 0 & 0 & 3 \\ 0 & 3 & 0 \\ 3 & 0 & 0 \end{bmatrix} - \begin{bmatrix} 0 & 0 & 3 \\ 0 & 3 & 0 \\ 3 & 0 & 0 \end{bmatrix} \begin{bmatrix} 0 & 0 & i \\ 0 & i & 0 \\ i & 0 & 0 \end{bmatrix}.$

2. Find all matrices M such that $M \begin{bmatrix} 1 & 2 & 3 \\ 0 & 2 & 3 \\ 0 & 0 & 3 \end{bmatrix} = \begin{bmatrix} 1 & 2 & 3 \\ 0 & 2 & 3 \\ 0 & 0 & 3 \end{bmatrix} M.$

3. (a) Find a matrix $A \neq 0$ such that $A \begin{bmatrix} 1 & 1 & 1 \\ 3 & 3 & 3 \\ 0 & 5 & 6 \end{bmatrix} = 0.$

   (b) Find the form of all matrices A such that $A \begin{bmatrix} 1 & 1 & 1 \\ 3 & 3 & 3 \\ 0 & 5 & 6 \end{bmatrix} = 0.$

   (c) Find the form of all matrices A such that $\begin{bmatrix} 1 & 1 & 1 \\ 3 & 3 & 3 \\ 0 & 5 & 6 \end{bmatrix} A = 0.$

4. Show that $A = \begin{bmatrix} 0 & 1 & 0 \\ 0 & 0 & 1 \\ -a_3 & -a_2 & -a_1 \end{bmatrix}$ satisfies $A^3 + a_1 A^2 + a_2 A + a_3 I = 0.$

MORE THEORETICAL PROBLEMS

Easier Problems

5. Prove Theorem 3.1.1 in its entirety.
6. Show that $(aI)A = aA.$
7. Prove that
   (a) $(0I)(bI) = 0I$ and $(1I)(bI) = bI.$
   (b) $(aI) + (bI) = (a + b)I.$
   (c) $(aI)(bI) = (ab)I.$
8. Show that $(aI)A = A(aI)$ for all matrices $A.$
9. If D is a diagonal 3×3 matrix all of whose diagonal elements are distinct, show that if $AD = DA$, then A must be a diagonal matrix.
10. Prove the product rule for the matrix units, namely, $E_{rs}E_{uv} = 0$ if $s \neq u$ and $E_{rs}E_{sv} = E_{rv}.$
11. Show that $\sum_{\sigma=1}^{n} E_{\sigma\sigma} = I.$

Middle-Level Problems

12. Using the fact that $A = (a_{rs}) = \sum_{\rho=1}^{n} \sum_{\sigma=1}^{n} a_{\rho\sigma} E_{\rho\sigma}$ and $B = (b_{rs}) = \sum_{\rho=1}^{n} \sum_{\sigma=1}^{n} b_{\rho\sigma} E_{\rho\sigma}$, reconstruct what the product AB must be, by exploiting the multiplication rules for the E_{rs}'s.
13. Prove the associative law $A(BC) = (AB)C$ using the representation of A, B, C as

combinations of the matrix units. (This should give you a lot of practice with playing with sums as $\sum$'s.)

14. If $AB = 0$, is $BA = 0$? Either prove or produce a counterexample.
15. If M is a matrix such that $MA = AM$ for all matrices A, show that M must be a scalar matrix.
16. If A is invertible and $AB = 0$, show that $B = 0$.
17. If A is invertible, show that its inverse is unique.
18. If A and B are invertible, show that AB is also invertible and express $(AB)^{-1}$ in terms of A^{-1} and B^{-1}.
19. If $A^m = 0$ and $AB = B$, prove that $B = 0$.
20. If $A^m = 0$, prove that $aI - A$ is invertible for all scalars $a \neq 0$.

Harder Problems

21. If $A^2 = A$, prove that $(AB - ABA)^2 = 0$ for all matrices B.
22. If $AB = I$, prove that $BA = I$.
23. If $C = BAB^{-1}$, where A is as in Problem 4 and B is invertible, prove that $C^3 + a_1 C^2 + a_2 C + a_3 I = 0$.
24. Prove that $(BAB^{-1})(BVB^{-1}) = B(AV)B^{-1}$ for all matrices A, V, and invertible B.
25. What does Problem 24 tell you about $(BAB^{-1})^k$ for $k \geq 1$ an integer?
26. Let B, C be invertible matrices. Define $\phi_B: M_n(\mathbb{C}) \to M_n(\mathbb{C})$ by $\phi_B(A) = BAB^{-1}$ for all $A \in M_n(\mathbb{C})$. Similarly, define $\phi_C(A) = CAC^{-1}$ for all $A \in M_n(\mathbb{C})$. Prove that $\phi_B \phi_C = \phi_{BC}$.
27. Given a fixed matrix A define $d: M_n(\mathbb{C}) \to M_n(\mathbb{C})$ by $d(X) = AX - XA$ for all $X \in M_n(\mathbb{C})$. Then:
    (a) Prove that $d(XY) = d(X)Y + Xd(Y)$ and $d(X + Y) = d(X) + d(Y)$ for all X, $Y \in M_n(\mathbb{C})$. (What does this remind you of?)
    (b) Determine $d(I)$, where I is the identity.
    (c) Determine $d(aI)$.
    (d) Determine $d(A^3)$.

3.2. MATRICES AS OPERATORS

For the 2×2 matrices we found that we could define an action on the set of all 2-tuples, which was rather nice and which gave some sense to the product of matrices as we had defined them. In Chapter 2 we did a similar thing in expressing a system of linear equations as $Av = w$, where A is an $m \times n$ coefficient matrix and v an element of $F^{(n)}$ (see below). We now do the same for the $n \times n$ matrices.

Let F be either the set of real or complex numbers. By $F^{(n)}$ we shall mean the set of n-tuples $\begin{bmatrix} a_1 \\ a_2 \\ \vdots \\ a_n \end{bmatrix}$, where $a_1, a_2, \ldots, a_n$ are all in F.

In $F^{(n)}$ we introduce an algebraic structure by defining two operations—addition and multiplication by a scalar—and it is on this structure that we shall allow the $n \times n$ matrices over F to act.

But first we need a way of recognizing when two *vectors*—that is, elements of $F^{(n)}$—are equal.

Definition. If $v = \begin{bmatrix} a_1 \\ \vdots \\ a_n \end{bmatrix}$ and $w = \begin{bmatrix} b_1 \\ \vdots \\ b_n \end{bmatrix}$ are in $F^{(n)}$, then we *define* $v = w$ if and only if $a_j = b_j$ for every $j = 1, 2, \ldots, n$.

We call a_r the rth *coordinate* or *component of* v. So two vectors are equal only when their corresponding coordinates are equal.

Now that we are able to tell when two vectors are equal, we are ready to introduce the two operations, mentioned above, in $F^{(n)}$.

Definition. If $v = \begin{bmatrix} a_1 \\ \vdots \\ a_n \end{bmatrix}$ and $w = \begin{bmatrix} b_1 \\ \vdots \\ b_n \end{bmatrix}$ are in $F^{(n)}$, we *define* $v + w = \begin{bmatrix} a_1 + b_1 \\ \vdots \\ a_n + b_n \end{bmatrix}$.

For example, if $v = \begin{bmatrix} 1 \\ 2 \\ 3 \end{bmatrix}$ and $w = \begin{bmatrix} \pi \\ 2.3 \\ 3 \end{bmatrix}$, then $v + w = \begin{bmatrix} 1 + \pi \\ 2 + 2.3 \\ 3 + 3 \end{bmatrix} = \begin{bmatrix} 1 + \pi \\ 4.3 \\ 6 \end{bmatrix}$.

So to add two vectors, just add their corresponding coordinates. Note that the vector $\begin{bmatrix} 0 \\ \vdots \\ 0 \end{bmatrix}$, all of whose coordinates are 0, has the key property that $v + \begin{bmatrix} 0 \\ \vdots \\ 0 \end{bmatrix} = v$ for every $v \in F^{(n)}$. We denote $\begin{bmatrix} 0 \\ \vdots \\ 0 \end{bmatrix}$ simply as 0. Note also that the vector $u = \begin{bmatrix} -a_1 \\ \vdots \\ -a_n \end{bmatrix}$ has the property that $u + v = 0$, where $v = \begin{bmatrix} a_1 \\ \vdots \\ a_n \end{bmatrix}$. We write u as $-v$.

We now come to the second operation, that of multiplying a vector in $F^{(n)}$ by an element of F. We call this *multiplication by a scalar*.

Definition. If $v = \begin{bmatrix} a_1 \\ \vdots \\ a_n \end{bmatrix} \in F^{(n)}$ and $t \in F$, then we define $tv = \begin{bmatrix} ta_1 \\ \vdots \\ ta_n \end{bmatrix}$.

For instance, if $w = \begin{bmatrix} \pi \\ 2.3 \\ 3 \end{bmatrix}$, then $4w = \begin{bmatrix} 4\pi \\ 9.2 \\ 12 \end{bmatrix}$.

So to multiply a vector v by a scalar t, just multiply each component of v by t. Note that $(-1)v$ is the same as $-v$ defined above.

The basic rules governing the behavior of addition and multiplication by a scalar are contained in

Theorem 3.2.1. If u, v, w are in $F^{(n)}$ and a, b are in F, then av and $v + w$ are in $F^{(n)}$ and

1. $v + w = w + v$;
2. $u + (v + w) = (u + v) + w$;
3. $v + 0 = v$ and $v + (-v) = 0$;
4. $a(v + w) = av + aw$;
5. $(a + b)v = av + bv$;
6. $1v = v$;
7. $0v = 0$, $a0 = 0$;
8. $a(bv) = (ab)v$.

Proof: The proofs of all these parts of the theorem are easy and direct. We pick two sample ones, Parts (4) and (8), to show how such a proof should run.

Suppose that $v = \begin{bmatrix} s_1 \\ \vdots \\ s_n \end{bmatrix}$ and $w = \begin{bmatrix} t_1 \\ \vdots \\ t_n \end{bmatrix}$ are in $F^{(n)}$ and $a \in F$. Then $v + w = \begin{bmatrix} s_1 + t_1 \\ \vdots \\ s_n + t_n \end{bmatrix}$, hence

$$a(v + w) = \begin{bmatrix} a(s_1 + t_1) \\ \vdots \\ a(s_n + t_n) \end{bmatrix} = \begin{bmatrix} as_1 + at_1 \\ \vdots \\ as_n + at_n \end{bmatrix} = \begin{bmatrix} as_1 \\ \vdots \\ as_n \end{bmatrix} + \begin{bmatrix} at_1 \\ \vdots \\ at_n \end{bmatrix} = a\begin{bmatrix} s_1 \\ \vdots \\ s_n \end{bmatrix} + a\begin{bmatrix} t_1 \\ \vdots \\ t_n \end{bmatrix}$$

$$= av + aw.$$

Here we have used the basic definition of addition and multiplication by a scalar several times. This proves Part (4).

Now to demonstrate Part (8). If $v = \begin{bmatrix} s_1 \\ \vdots \\ s_n \end{bmatrix}$, then $bv = \begin{bmatrix} bs_1 \\ \vdots \\ bs_n \end{bmatrix}$, so

$$a(bv) = \begin{bmatrix} a(bs_1) \\ \vdots \\ a(bs_n) \end{bmatrix} = \begin{bmatrix} (ab)s_1 \\ \vdots \\ (ab)s_n \end{bmatrix} = (ab)\begin{bmatrix} s_1 \\ \vdots \\ s_n \end{bmatrix} = (ab)v.$$

This shows Part (8) to be proved.

The proofs of all the other parts are along the lines used for Parts (4) and (8). ∎

Since we have two zeros running around—that of F and that of $F^{(n)}$—we might get some confusion by using the same symbol for them. But there really is no danger of this because $0v = 0$ for any $v \in F^{(n)}$, where the zero on the left is that of F and the one on the right that of $F^{(n)}$, and $a0 = 0$ for any $a \in F$, where the zero on the left is that of $F^{(n)}$ and the one on the right that of $F^{(n)}$ also. To avoid notational extravaganzas, we use the symbol 0 everywhere for both.

The properties described in Theorem 3.2.1 make of $F^{(n)}$ what is called in mathematics a *vector space over F*. Abstract vector spaces will play a prominent role in the last half of the book.

If $n = 3$ and $F = \mathbb{R}$, then $F^{(3)}$ is the usual 3-dimensional Euclidean space that we run into in geometry. So it is natural to call $F^{(n)}$ the *n-dimensional space* over F.

The space $F^{(n)}$ gives us a natural domain on which to define an action by matrices in $M_n(F)$, which we do now in the

Definition. Suppose that $A(a_{rs})$ is in $M_n(F)$ and $v = (v_s) = \begin{bmatrix} v_1 \\ \vdots \\ v_n \end{bmatrix}$ is in $F^{(n)}$. Then we define $Av = (w_r) = \begin{bmatrix} w_1 \\ \vdots \\ w_n \end{bmatrix}$, where for each $r = 1, 2, \ldots, n$, $w_r = \sum_{\sigma=1}^{n} a_{r\sigma} v_\sigma$.

Let's try this definition out for a few matrices and vectors. If

$$A = \begin{bmatrix} 1 & 2 & -1 \\ \pi & 1 & 1 \\ 1 & \pi & 3 \end{bmatrix}$$

and

$$v = \begin{bmatrix} 0 \\ 1 \\ 0 \end{bmatrix},$$

then $Av = \begin{bmatrix} 1 & 2 & -1 \\ \pi & 1 & 1 \\ 1 & \pi & 3 \end{bmatrix} \begin{bmatrix} 0 \\ 1 \\ 0 \end{bmatrix} = \begin{bmatrix} 2 \\ 1 \\ \pi \end{bmatrix}$. For example, the first entry of Av is $1 \cdot 0 + 2 \cdot 1 + (-1) \cdot 0 = 2$. Perhaps a little more interesting is the example

$$A = \begin{bmatrix} 1 & -1 & -1 & 1 \\ 1 & 0 & 0 & 0 \\ 0 & 1 & 0 & 0 \\ -5 & -5 & -5 & 5 \end{bmatrix}$$

and

$$v = \begin{bmatrix} 0 \\ 0 \\ \pi + 1 \\ \pi + 1 \end{bmatrix},$$

for we notice that $Av = 0$. So it is possible for a nonzero matrix to knock-out a nonzero vector.

How do matrices in $M_n(F)$ behave as mappings on $F^{(n)}$? Given $A = (a_{rs})$ and $B = (b_{rs})$, then $A + B = (a_{rs} + b_{rs})$, so for any vector $v = \begin{bmatrix} v_1 \\ \vdots \\ v_n \end{bmatrix}$,

$$(A + B)\begin{bmatrix} v_1 \\ \vdots \\ v_n \end{bmatrix} = \begin{bmatrix} t_1 \\ \vdots \\ t_n \end{bmatrix},$$

where $t_r = \sum_{\sigma=1}^{n} (a_{r\sigma} + b_{r\sigma})v_\sigma = \sum_{\sigma=1}^{n} a_{r\sigma}v_\sigma + \sum_{\sigma=1}^{n} b_{r\sigma}v_\sigma = A\begin{bmatrix} v_1 \\ \vdots \\ v_n \end{bmatrix} + B\begin{bmatrix} v_1 \\ \vdots \\ v_n \end{bmatrix}$. In short, $(A + B)v = Av + Bv$. A similar argument shows that $(aA)v = a(Av) = A(av)$. Also, $A(v + w) = Av + Aw$. These are easy to verify. (Do so!) These last two behaviors of A on vectors (on scalar products and on sums) are very important. We use them to make the

Definition. A mapping $T: F^{(n)} \to F^{(n)}$ is a *linear transformation on $F^{(n)}$* if:

1. $T(v + w) = T(v) + T(w)$
2. $T(av) = aT(v)$

for every $a \in F$ and for any $v, w \in F^{(n)}$.

EXAMPLE

For $n = 3$, the mapping $T: F^{(n)} \to F^{(n)}$ defined by $T\begin{bmatrix} a \\ b \\ c \end{bmatrix} = \begin{bmatrix} 3c - b \\ 3b - a \\ a \end{bmatrix}$ is a *linear transformation on $F^{(n)}$* since

$$T\left(\begin{bmatrix} a \\ b \\ c \end{bmatrix} + \begin{bmatrix} g \\ h \\ k \end{bmatrix}\right) = T\begin{bmatrix} a + g \\ b + h \\ c + k \end{bmatrix} = \begin{bmatrix} 3(c + k) - (b + h) \\ 3(b + h) - (a + g) \\ a + g \end{bmatrix}$$

$$= \begin{bmatrix} 3c - b \\ 3b - a \\ a \end{bmatrix} + \begin{bmatrix} 3k - h \\ 3h - g \\ g \end{bmatrix} = T\begin{bmatrix} a \\ b \\ c \end{bmatrix} + T\begin{bmatrix} g \\ h \\ k \end{bmatrix}$$

and

$$T\left(d\begin{bmatrix} a \\ b \\ c \end{bmatrix}\right) = T\left(\begin{bmatrix} da \\ db \\ dc \end{bmatrix}\right) = \begin{bmatrix} 3dc - db \\ 3db - da \\ da \end{bmatrix} = d\begin{bmatrix} 3c - b \\ 3b - a \\ a \end{bmatrix} = dT\begin{bmatrix} a \\ b \\ c \end{bmatrix}$$

for all $\begin{bmatrix} a \\ b \\ c \end{bmatrix}, \begin{bmatrix} g \\ h \\ k \end{bmatrix}$ in $F^{(n)}$ and all d in F.

Sec. 3.2] Matrices as Operators 103

By this definition, every $n \times n$ matrix A defines by its action $Av (v \in F^{(n)})$ a linear transformation on $F^{(n)}$. The opposite is also true, namely, a linear transformation on $F^{(n)}$ is given by a matrix, as in the following example.

EXAMPLE

Let $n = 3$ and define $T: F^{(n)} \to F^{(n)}$ by $T\begin{bmatrix} a \\ b \\ c \end{bmatrix} = \begin{bmatrix} 3c - b \\ 3b - a \\ a \end{bmatrix}$ as in the example

above. Then $T\begin{bmatrix} 1 \\ 0 \\ 0 \end{bmatrix} = \begin{bmatrix} 0 \\ -1 \\ 1 \end{bmatrix}$, $T\begin{bmatrix} 0 \\ 1 \\ 0 \end{bmatrix} = \begin{bmatrix} -1 \\ 3 \\ 0 \end{bmatrix}$, $T\begin{bmatrix} 0 \\ 0 \\ 1 \end{bmatrix} = \begin{bmatrix} 3 \\ 0 \\ 0 \end{bmatrix}$. But then it is clear

that the matrix $\begin{bmatrix} 0 & -1 & 3 \\ -1 & 3 & 0 \\ 1 & 0 & 0 \end{bmatrix}$ has the same effect:

$$\begin{bmatrix} 0 & -1 & 3 \\ -1 & 3 & 0 \\ 1 & 0 & 0 \end{bmatrix} \begin{bmatrix} 1 \\ 0 \\ 0 \end{bmatrix} = \begin{bmatrix} 0 \\ -1 \\ 1 \end{bmatrix},$$

$$\begin{bmatrix} 0 & -1 & 3 \\ -1 & 3 & 0 \\ 1 & 0 & 0 \end{bmatrix} \begin{bmatrix} 0 \\ 1 \\ 0 \end{bmatrix} = \begin{bmatrix} -1 \\ 3 \\ 0 \end{bmatrix},$$

$$\begin{bmatrix} 0 & -1 & 3 \\ -1 & 3 & 0 \\ 1 & 0 & 0 \end{bmatrix} \begin{bmatrix} 0 \\ 0 \\ 1 \end{bmatrix} = \begin{bmatrix} 3 \\ 0 \\ 0 \end{bmatrix}.$$

Trying it with the general vector, we get

$$\begin{bmatrix} 0 & -1 & 3 \\ -1 & 3 & 0 \\ 1 & 0 & 0 \end{bmatrix} \begin{bmatrix} a \\ b \\ c \end{bmatrix} = \begin{bmatrix} 3c - b \\ 3b - a \\ a \end{bmatrix}.$$

You may be convinced from this example that every linear transformation is given by a matrix. This is not surprising, since we can model the general proof after the example in

Theorem 3.2.2. Let T be a linear transformation on $F^{(n)}$. Then $T = (t_{rs})$ as mappings for some matrix (t_{rs}) in $M_n(F)$.

Proof: Consider the vectors $e_1 = \begin{bmatrix} 1 \\ 0 \\ \vdots \\ 0 \end{bmatrix}$, $e_2 = \begin{bmatrix} 0 \\ 1 \\ \vdots \\ 0 \end{bmatrix}, \ldots, e_n = \begin{bmatrix} 0 \\ 0 \\ \vdots \\ 1 \end{bmatrix}$, that is, those

vectors e_r whose rth component is 1 and all of whose other components are 0. Given any vector $v \in F^{(n)}$, if $v = \begin{bmatrix} a_1 \\ a_2 \\ \vdots \\ a_n \end{bmatrix}$, we see that $v = a_1 e_1 + a_2 e_2 + \cdots + a_n e_n$. Thus

$$T(v) = T(a_1 e_1 + a_2 e_2 + \cdots + a_n e_n) = a_1 T(e_1) + a_2 T(e_2) + \cdots + a_n T(e_n).$$

Therefore, if we knew what T did to each of $e_1, \ldots, e_n$, we would know what T did to any vector. Now Te_s as an element of $F^{(n)}$ is representable as $Te_s = t_{1s} e_1 + t_{2s} e_2 + \cdots + t_{ns} e_n$ for each $s = 1, 2, \ldots, n$. Consider the action of the matrix (t_{rs}) on $F^{(n)}$. As noted above, to know how (t_{rs}) behaves on $F^{(n)}$, it is enough to know what it does to $e_1, \ldots, e_n$. Computing yields

$$(t_{rs})e_1 = \begin{bmatrix} t_{11} & t_{12} & \cdots & t_{1n} \\ t_{21} & t_{22} & \cdots & t_{2n} \\ \vdots & \vdots & & \vdots \\ t_{n1} & t_{n2} & \cdots & t_{nn} \end{bmatrix} \begin{bmatrix} 1 \\ 0 \\ \vdots \\ 0 \end{bmatrix} = \begin{bmatrix} t_{11} \\ t_{21} \\ \vdots \\ t_{n1} \end{bmatrix}$$

$$= t_{11} e_1 + t_{21} e_2 + \cdots + t_{n1} e_n = Te_1.$$

Similarly, we get that $(t_{rs})e_u = Te_u$ for every $u = 1, 2, \ldots, n$. Thus T and (t_{rs}) agree on each of $e_1, \ldots, e_n$, from which it follows that T and (t_{rs}) agree on all elements of $F^{(n)}$. Hence $T = (t_{rs})$. ∎

Definition. Let T be a linear transformation on $F^{(n)}$. Then the matrix (t_{rs}) in $M_n(F)$ such that $T = (t_{rs})$ as mappings of $F^{(n)}$, that is, the matrix (t_{rs}) such that

$$Te_s = t_{1s} e_1 + t_{2s} e_2 + \cdots + t_{ns} e_n \text{ for } 1 \leq s \leq n$$

is called the *matrix* of T.

This is a constructive definition in the sense that to find the matrix of T it tells us to use as column s of this matrix the vector Te_s.

EXAMPLE

If T is the mapping on $F^{(4)}$ defined by $T \begin{bmatrix} a_1 \\ a_2 \\ a_3 \\ a_4 \end{bmatrix} = \begin{bmatrix} a_2 \\ 2a_3 \\ 3a_4 \\ 0 \end{bmatrix}$, then

$$Te_1 = T\begin{bmatrix} 1 \\ 0 \\ 0 \\ 0 \end{bmatrix} = \begin{bmatrix} 0 \\ 0 \\ 0 \\ 0 \end{bmatrix}, \quad Te_2 = T\begin{bmatrix} 0 \\ 1 \\ 0 \\ 0 \end{bmatrix} = \begin{bmatrix} 1 \\ 0 \\ 0 \\ 0 \end{bmatrix},$$

$$Te_3 = T\begin{bmatrix}0\\0\\1\\0\end{bmatrix} = \begin{bmatrix}0\\2\\0\\0\end{bmatrix}, \quad Te_4 = T\begin{bmatrix}0\\0\\0\\1\end{bmatrix} = \begin{bmatrix}0\\0\\3\\0\end{bmatrix}.$$

Therefore, the matrix of T is $\begin{bmatrix}0&1&0&0\\0&0&2&0\\0&0&0&3\\0&0&0&0\end{bmatrix}$. Checking out

$$\begin{bmatrix}0&1&0&0\\0&0&2&0\\0&0&0&3\\0&0&0&0\end{bmatrix}\begin{bmatrix}a_1\\a_2\\a_3\\a_4\end{bmatrix} = \begin{bmatrix}a_2\\2a_3\\3a_4\\0\end{bmatrix},$$

we see that, indeed, the matrix $T = \begin{bmatrix}0&1&0&0\\0&0&2&0\\0&0&0&3\\0&0&0&0\end{bmatrix}$ gives T.

We should like to see how the product of two matrices behaves in terms of mappings. Since matrices A and B in $M_n(F)$ define mappings of $F^{(n)}$ into itself, it is fairly natural to ask how $A \circ B$—the product (i.e., composite) of A and B as mappings—relates to AB, the product of A and B as matrices. Let $A = (a_{rs})$ and $B = (b_{rs})$; so if $x = \begin{bmatrix}x_1\\ \vdots \\ x_n\end{bmatrix} \in F^{(n)}$ then $Bx = \begin{bmatrix}y_1\\ \vdots \\ y_n\end{bmatrix}$, where $y_s = \sum_{t=1}^{n} b_{st}x_t$ for $s = 1, 2, \ldots, n$. Thus

$$(A \circ B)(x) = A(Bx) = A\begin{bmatrix}y_1\\ \vdots \\ y_n\end{bmatrix} = \begin{bmatrix}z_1\\ \vdots \\ z_n\end{bmatrix},$$

where

$$z_r = \sum_{s=1}^{n} a_{rs}y_s = \sum_{s=1}^{n} a_{rs}\left(\sum_{t=1}^{n} b_{st}x_t\right) = \sum_{s=1}^{n}\sum_{t=1}^{n} a_{rs}b_{st}x_t$$
$$= \sum_{t=1}^{n}\sum_{s=1}^{n} a_{rs}b_{st}x_t = \sum_{t=1}^{n}\left(\sum_{s=1}^{n} a_{rs}b_{st}\right)x_t.$$

This implies that z_r, which was defined as the rth entry of $(A \circ B)(x)$, is also the rth entry of $(AB)x$, since AB is the matrix $\left(\sum_{s=1}^{n} a_{rs}b_{st}\right)$. Thus $(A \circ B)(x)$ and $(AB)x$ have the same rth entry for all r, which implies that $(A \circ B)(x) = (AB)x$. Since this is true for all x, we can conclude that $A \circ B = AB$. In other words, *the composition of A and B as* *mappings agrees with their product as matrices.* In fact, this is the basic reason for which the product of matrices is defined in the way it is—which at first glance seems contrived.

The result above merits being singled out, and we do so in

Theorem 3.2.3. *The mapping defined by AB on $F^{(n)}$ coincides with the mapping $A \circ B$—the composition of the mapping A and B—defined by $(A \circ B)v = A(Bv)$ for every $v \in F^{(n)}$.*

Although we know it is true in general, let's check out Theorem 3.2.3 in a specific instance. Let $A = \begin{bmatrix} 1 & 2 & 4 \\ -2 & 0 & 1 \\ 3 & -1 & 0 \end{bmatrix}$ and $B = \begin{bmatrix} 0 & 1 & 0 \\ 1 & 1 & 1 \\ 0 & 0 & 1 \end{bmatrix}$; then the *matrix product* AB of A and B is $\begin{bmatrix} 2 & 3 & 6 \\ 0 & -2 & 1 \\ -1 & 2 & -1 \end{bmatrix}$. So if $v = \begin{bmatrix} x_1 \\ x_2 \\ x_3 \end{bmatrix} \in F^{(3)}$, then

$$(AB)v = \begin{bmatrix} 2 & 3 & 6 \\ 0 & -2 & 1 \\ -1 & 2 & -1 \end{bmatrix} \begin{bmatrix} x_1 \\ x_2 \\ x_3 \end{bmatrix} = \begin{bmatrix} 2x_1 + 3x_2 + 6x_3 \\ -2x_2 + x_3 \\ -x_1 + 2x_2 - x_3 \end{bmatrix}.$$

On the other hand,

$$Bv = \begin{bmatrix} 0 & 1 & 0 \\ 1 & 1 & 1 \\ 0 & 0 & 1 \end{bmatrix} \begin{bmatrix} x_1 \\ x_2 \\ x_3 \end{bmatrix} = \begin{bmatrix} x_2 \\ x_1 + x_2 + x_3 \\ x_3 \end{bmatrix};$$

thus

$$(A \circ B)v = A(Bv) = \begin{bmatrix} 1 & 2 & 4 \\ -2 & 0 & 1 \\ 3 & -1 & 0 \end{bmatrix} \begin{bmatrix} x_2 \\ x_1 + x_2 + x_3 \\ x_3 \end{bmatrix}$$

$$= \begin{bmatrix} x_2 + 2(x_1 + x_2 + x_3) + 4x_3 \\ -2x_2 + x_3 \\ 3x_2 - (x_1 + x_2 + x_3) \end{bmatrix} = \begin{bmatrix} 2x_1 + 3x_2 + 6x_3 \\ -2x_2 + x_3 \\ -x_1 + 2x_2 - x_3 \end{bmatrix}$$

$$= (AB)v.$$

So in this specific example we see that $A \circ B = AB$, as the theorem tells us that it should be.

PROBLEMS

In the following problems, $A \circ B$ will denote the product of A and B as mappings. Also, if T is a linear transformation, we shall mean by $m(T)$ the matrix of T.

NUMERICAL PROBLEMS

1. Compute the vector $-\begin{bmatrix} 1 \\ 2 \\ -3 \end{bmatrix}$.

2. Compute the vector $\begin{bmatrix} 4 \\ -2 \\ -3 \end{bmatrix} - \begin{bmatrix} 6 \\ -4 \\ 5 \end{bmatrix}$.

3. Compute the vector $3\begin{bmatrix} 4 \\ -2 \\ -3 \end{bmatrix}$.

4. Compute the vector $8\begin{bmatrix} 4 \\ -2 \\ -3 \end{bmatrix} + (-4)\left(\begin{bmatrix} 4 \\ -2 \\ -3 \end{bmatrix} - \begin{bmatrix} 6 \\ -4 \\ 5 \end{bmatrix}\right)$.

5. Compute the vector $\begin{bmatrix} 3 & 2 & 1 \\ -2 & 3 & 1 \\ -4 & -3 & 1 \end{bmatrix}\begin{bmatrix} 6 \\ -4 \\ 5 \end{bmatrix}$.

6. Compute the vectors

$$\begin{bmatrix} 3 & 2 & 5 \\ -2 & 2 & 3 \\ -1 & -3 & 1 \end{bmatrix}\begin{bmatrix} 3 \\ -2 \\ -1 \end{bmatrix}, \begin{bmatrix} 3 & 2 & 5 \\ -2 & 2 & 3 \\ -1 & -3 & 1 \end{bmatrix}\begin{bmatrix} 3 \\ 2 \\ 1 \end{bmatrix}, \begin{bmatrix} 3 & 2 & 5 \\ -2 & 2 & 3 \\ -1 & -3 & 1 \end{bmatrix}\begin{bmatrix} 3 \\ 1 \\ 1 \end{bmatrix}.$$

7. Compute the matrix $\begin{bmatrix} 3 & 2 & 5 \\ -2 & 2 & 3 \\ -1 & -3 & 1 \end{bmatrix}\begin{bmatrix} 3 & 3 & 3 \\ -2 & 2 & 1 \\ -1 & -1 & 1 \end{bmatrix}$. How does the result relate to the three vectors found in Problem 6?

8. What is the matrix of the linear transformation that maps the vector $\begin{bmatrix} a \\ b \\ c \end{bmatrix}$ to the vector $\begin{bmatrix} u \\ v \\ w \end{bmatrix}$ such that $\begin{bmatrix} 0 & 0 & 0 \\ a & b & c \\ 0 & 0 & 0 \end{bmatrix}\begin{bmatrix} 3 & 3 & 3 \\ -2 & 2 & 1 \\ -1 & -1 & 1 \end{bmatrix} = \begin{bmatrix} 0 & 0 & 0 \\ u & v & w \\ 0 & 0 & 0 \end{bmatrix}$?

9. What is the matrix of the linear transformation that maps the vector $\begin{bmatrix} a \\ b \\ c \end{bmatrix}$ to the vector $\begin{bmatrix} u \\ v \\ w \end{bmatrix}$ such that $\begin{bmatrix} a & b & c \\ 0 & 0 & 0 \\ 0 & 0 & 0 \end{bmatrix}\begin{bmatrix} 3 & 3 & 3 \\ -2 & 2 & 1 \\ -1 & -1 & 1 \end{bmatrix} = \begin{bmatrix} u & v & w \\ 0 & 0 & 0 \\ 0 & 0 & 0 \end{bmatrix}$?

10. Let S be the linear transformation whose matrix is $\begin{bmatrix} 1 & 2 & 2 \\ 3 & 3 & 3 \\ 4 & 4 & 4 \end{bmatrix}$ and let T be the linear transformation whose matrix is $\begin{bmatrix} 1 & 3 & 4 \\ 2 & 3 & 4 \\ 2 & 3 & 4 \end{bmatrix}$. Compute the matrix of the linear transformation $S \circ T$ and verify directly that the vectors $(S \circ T)(v)$ and $(ST)(v)$ are the same for all v in $F^{(3)}$.

MORE THEORETICAL PROBLEMS

Easier Problems

11. Show that the identity mapping T on $F^{(n)}$ is a linear transformation of $F^{(n)}$ whose matrix is the unit matrix I.

12. If A is an invertible matrix and A^{-1} is its inverse, show that $A \circ A^{-1}$ is the identity mapping.

13. Verify that the T's given are linear transformations on the appropriate $F^{(n)}$ and for each find $m(T)$, the matrix of T.

(a) $T \begin{bmatrix} a_1 \\ a_2 \\ a_3 \\ a_4 \end{bmatrix} = \begin{bmatrix} a_2 \\ a_1 \\ a_4 \\ a_3 \end{bmatrix}$ for all $\begin{bmatrix} a_1 \\ a_2 \\ a_3 \\ a_4 \end{bmatrix}$ in $F^{(4)}$.

(b) $T \begin{bmatrix} a_1 \\ a_2 \\ a_3 \\ a_4 \\ a_5 \end{bmatrix} = \begin{bmatrix} 0 \\ a_1 \\ \frac{1}{2} a_2 \\ a_3 \\ \frac{1}{4} a_4 \end{bmatrix}$ on $F^{(5)}$.

(c) $T \begin{bmatrix} a_1 \\ a_2 \\ a_3 \\ a_4 \end{bmatrix} = \begin{bmatrix} ia_4 \\ -a_1 \\ 3a_2 \\ -ia_3 \end{bmatrix}$ on $F^{(4)}$ ($F = \mathbb{C}$).

(d) $T \begin{bmatrix} a_1 \\ a_2 \\ a_3 \end{bmatrix} = \begin{bmatrix} -a_3 + a_2 - a_1 \\ a_1 \\ a_2 \end{bmatrix}$ on $F^{(3)}$.

14. Let A be the T in Part (a) and B the T in Part (c) of Problem 13. Find the matrix $m(A \circ B)$ and verify directly the assertion of Theorem 3.2.3 that $m(A \circ B) = m(A)m(B)$.

15. If T on $F^{(4)}$ is defined by $T \begin{bmatrix} a_1 \\ a_2 \\ a_3 \\ a_4 \end{bmatrix} = \begin{bmatrix} a_4 \\ a_1 \\ a_2 \\ a_3 \end{bmatrix}$ for all $\begin{bmatrix} a_1 \\ a_2 \\ a_3 \\ a_4 \end{bmatrix}$ in $F^{(4)}$, show that $T^4 =$

$T \circ T \circ T \circ T$ is the identity mapping on $F^{(4)}$, find $m(T)$, and show by a direct calculation that $m(T)^4 = I$.

16. If $v \in F^{(n)}$ and $A \in M_n(F)$ is invertible, show that $A^{-1}(Av) = v$.
17. If A in $M_n(F)$ is invertible and $Av = 0$ for $v \in F^{(n)}$, show that $v = 0$.
18. If two columns of the matrix A are identical, show that $Av = 0$ for some $v \neq 0$, hence A cannot be invertible.
19. If A in $M_3(F)$ is not invertible, show that you can find a $v \neq 0$ in $F^{(3)}$ such that $Av = 0$.
20. If $A \in M_n(F)$ is such that the vectors $w_1, \ldots, w_n$, where w_s is column s of A, satisfy a relation of the form
$$a_1 w_1 + a_2 w_2 + \cdots + a_n w_n = 0,$$
where the $a_r \in F$ and not all of them are 0, then show that $Av = 0$ for some $v \neq 0$ in $F^{(n)}$ by describing the coordinates of such a v.

21. Let T be defined by $T\begin{bmatrix} x_1 \\ x_2 \\ x_3 \end{bmatrix} = \begin{bmatrix} x_2 \\ x_3 \\ -cx_1 - bx_2 - ax_3 \end{bmatrix}$, where $a, b, c \in F$, for all $\begin{bmatrix} x_1 \\ x_2 \\ x_3 \end{bmatrix}$ in $F^{(3)}$. Show that T is a linear transformation on $F^{(3)}$ and find $A = m(T)$. Then show that $A^3 + aA^2 + bA + cI = 0$.

22. Let $v_1 = \begin{bmatrix} 1 \\ 0 \\ \vdots \\ 0 \end{bmatrix}, v_2 = \begin{bmatrix} 0 \\ 1 \\ \vdots \\ 0 \end{bmatrix}, \ldots, v_n = \begin{bmatrix} 0 \\ 0 \\ \vdots \\ 1 \end{bmatrix}$ be in $F^{(n)}$. Suppose that T is defined on $F^{(n)}$ by $T(a_1 v_1 + a_2 v_2 + \cdots + a_n v_n) = a_1 v_2 + a_2 v_3 + \cdots + a_{n-1} v_n$ for all $a_1, \ldots, a_n$ in F. Show that T is a linear transformation on $F^{(n)}$, find $A = m(T)$, and show that $A^n = 0$.

3.3. TRACE

In discussing the 2×2 matrices, one of the concepts introduced there was the trace of a matrix. For the matrix $\begin{bmatrix} a & b \\ c & d \end{bmatrix}$ we defined its trace as $a + d$, the sum of its diagonal elements. This notion was a useful one for the 2×2 matrices. One would hope its extension to the $n \times n$ case would be equally useful.

Given the matrix $A = (a_{rs})$ in $M_n(F)$, we do the obvious and make the

Definition. If $A = (a_{rs}) \in M_n(F)$, then the *trace* of A, written $\text{tr}(A)$, is defined as $\text{tr}(a) = \sum_{t=1}^{n} a_{tt}$.

In words, $\text{tr}(A)$ *is the sum of the diagonal entries of* A.

Fortunately, all the theorems we proved for the trace in $M_2(F)$ go over smoothly and exactly to $M_n(F)$. Note that $\operatorname{tr}(I) = n$.

Theorem 3.3.1. If $A, B \in M_n(F)$ and $a \in F$, then

1. $\operatorname{tr}(A + B) = \operatorname{tr}(A) + \operatorname{tr}(B)$;
2. $\operatorname{tr}(aA) = a \operatorname{tr}(A)$;
3. $\operatorname{tr}(AB) = \operatorname{tr}(BA)$.

Proof: Suppose that $A = (a_{rs})$ and $B = (b_{rs})$. Then $\operatorname{tr}(A) = \sum_{t=1}^{n} a_{tt}$ and $\operatorname{tr}(B) = \sum_{t=1}^{n} b_{tt}$. By the way we add matrices $A + B = (c_{rs})$, where $c_{rs} = a_{rs} + b_{rs}$; thus

$$\operatorname{tr}(A + B) = \sum_{t=1}^{n} c_{tt} = \sum_{t=1}^{n} (a_{tt} + b_{tt}) = \sum_{t=1}^{n} a_{tt} + \sum_{t=1}^{n} b_{tt} = \operatorname{tr}(A) + \operatorname{tr}(B).$$

Similarly, $\operatorname{tr}(aA) = a \operatorname{tr}(A)$, whose proof we leave to the reader.

If $AB = (d_{rs})$ and $BA = (g_{rs})$, we know that $d_{rt} = \sum_{s=1}^{n} a_{rs} b_{st}$ and $g_{rt} = \sum_{s=1}^{n} b_{rs} a_{st}$ every r and t. (Remember, the index of summation, s for d_{rt} and s for g_{rt}, is a dummy index. We can designate it however we like.) Thus

$$\operatorname{tr}(AB) = \sum_{r=1}^{n} d_{rr} = \sum_{r=1}^{n} \sum_{s=1}^{n} a_{rs} b_{sr}.$$

Similarly,

$$\operatorname{tr}(BA) = \sum_{r=1}^{n} g_{rr} = \sum_{r=1}^{n} \sum_{s=1}^{n} b_{rs} a_{sr}.$$

Since the r and s are dummy indices, if we interchange the letters r by s in the first expression, we get

$$\operatorname{tr}(BA) = \sum_{r=1}^{n} \sum_{s=1}^{n} b_{rs} a_{sr} = \sum_{s=1}^{n} \sum_{r=1}^{n} a_{sr} b_{rs} = \operatorname{tr}(AB). \qquad \blacksquare$$

If you have trouble playing around with these summations $\sum$, try the argument given for 3×3 matrices. It should clarify what is going on.

As always, it is nice to see that the general theorem holds for specific cases. So let us illustrate that $\operatorname{tr}(AB) = \operatorname{tr}(BA)$ with the example

$$A = \begin{bmatrix} 1 & 2 & 0 & -1 \\ 0 & i & 0 & i \\ i & 1 & 1 & i \\ 1+i & 0 & 0 & 1 \end{bmatrix} \quad \text{and} \quad B = \begin{bmatrix} 0 & 1 & 0 & 0 \\ 1 & 1+i & 1 & i \\ 1 & 0 & 2 & 3 \\ 0 & 1 & 4 & -5 \end{bmatrix}.$$

Then, to know $\operatorname{tr}(AB)$ we don't need to know all the entries of AB, just its diagonal entries. So

$$AB = \begin{bmatrix} 2 & & \\ & i(1+i)+i & * \\ & & 1+2+4i \\ * & & -5 \end{bmatrix},$$

where we are not interested in the *; so

$$\operatorname{tr}(AB) = 2 + i(i+1) + 1 + 1 + 2 + 4i - 5 = -1 + 6i.$$

We also have $BA = \begin{bmatrix} 0 & & \\ & 2+(1+i)i+1 & * \\ & & 2 \\ * & & i+4i-5 \end{bmatrix}$, whence

$$\begin{aligned}\operatorname{tr}(BA) &= 0 + 2 + (1+i)i + 1 + 2 + i + 4i - 5 \\ &= 2 + i(1+i) + i + 1 + 2 + 4i - 5 \\ &= -1 + 6i = \operatorname{tr}(AB).\end{aligned}$$

The third part of Theorem 3.3.1 has an equivalent statement, namely

Corollary 3.3.2. For $A, B \in M_n(F)$, $\operatorname{tr}(AB - BA) = 0$.

Proof: By Parts (1) and (2) of the theorem, $\operatorname{tr}(AB - BA) = \operatorname{tr}(AB) + \operatorname{tr}(-BA) = \operatorname{tr}(AB) - \operatorname{tr}(BA)$, which is 0 by Part (3) of the theorem. ∎

This corollary itself has a corollary, namely

Corollary 3.3.3. If $A, B \in M_n(F)$ and A is invertible, then $\operatorname{tr}(B) = \operatorname{tr}(ABA^{-1})$.

Proof: Let $C = BA^{-1}$; then $\operatorname{tr}(ABA^{-1}) = \operatorname{tr}(AC) = \operatorname{tr}(CA) = \operatorname{tr}(BA^{-1}A) = \operatorname{tr}(BI) = \operatorname{tr}(B)$. ∎

If E_{rs} is a matrix unit, that is, a matrix whose (r, s) entry is 1 and all of whose other entries are 0, then it is clear that $\operatorname{tr}(E_{rs}) = 0$ if $r \neq s$ and $\operatorname{tr}(E_{rr}) = 1$ for all r and s. We remind you of the basic multiplication rule for these matrix units E_{rs}, namely, $E_{rs}E_{uv} = 0$ if $s \neq u$ and $E_{rs}E_{sv} = E_{rv}$.

Suppose that f is a function from $M_n(F)$ to F which satisfies the four basic properties of trace, that is,

1. $f(I) = n$
2. $f(A + B) = f(A) + f(B)$
3. $f(aA) = af(A)$
4. $f(AB) = f(BA)$

for all $A, B \in M_n(F)$ and all $a \in F$.

Thus $f(E_{rs}E_{sr}) = f(E_{sr}E_{rs})$ by (4); since $E_{rs}E_{sr} = E_{rr}$ and $E_{sr}E_{rs} = E_{ss}$, we get that $f(E_{rr}) = f(E_{ss})$ for all r and s. However, $I = E_{11} + E_{22} + \cdots + E_{nn}$; hence by Part (2) we have $n = f(I) = f(E_{11} + E_{22} + \cdots + E_{nn}) = nf(E_{rr})$ since all $f(E_{rr})$ are equal. This tells us that $f(E_{rr}) = 1$ for every r. That is, $f(E_{rr}) = \text{tr}(E_{rr})$.

On the other hand, if $r \neq s$, then $E_{ts}E_{rt} = 0$. So $0 = f(E_{ts}E_{rt}) = f(E_{rt}E_{ts}) = f(E_{rs})$, so $f(E_{rs}) = 0$ if $r \neq s$. Hence $f(E_{rs}) = \text{tr}(E_{rs})$ if $r \neq s$. Thus f and trace agree on all matrix units, that is, $\text{tr}(E_{rs}) = f(E_{rs})$ for all r and s.

Given the matrix $A = (a_{rs})$ in $M_n(F)$, then $A = \sum_{r=1}^{n} \sum_{s=1}^{n} a_{rs} E_{rs}$. Hence, by applying (2) and (3), we obtain that

$$f(A) = f\left(\sum_{r=1}^{n} \sum_{s=1}^{n} a_{rs} E_{rs}\right) = \sum_{r=1}^{n} \sum_{s=1}^{n} a_{rs} f(E_{rs}) = \sum_{r=1}^{n} \sum_{s=1}^{n} a_{rs} \text{tr}(E_{rs})$$

$$= \text{tr}\left(\sum_{r=1}^{n} \sum_{s=1}^{n} a_{rs} E_{rs}\right) = \text{tr}(A).$$

So f and trace agree on all matrices A, that is, $f(A) = \text{tr}(A)$ for all $A \in M_n(F)$. We have proved

Theorem 3.3.4. If f is a mapping from $M_n(F)$ to F such that

1. $f(I) = n$
2. $f(A + B) = f(A) + f(B)$
3. $f(aA) = af(A)$
4. $f(AB) = f(BA)$

for all $A, B \in M_n(F)$ and all $a \in F$, then $f(A) = \text{tr}(A)$ for all $A \in M_n(F)$.

What the theorem points out is that trace is the *unique* function from $M_n(F)$ to F which satisfies the trace-like properties (1)–(4).

We change subjects for a moment. Let's consider a sample product. The matrix units E_{rs} are nicely behaved matrices. We want to see what multiplication by an E_{rs} does. Before doing it in general, we do it for the 3×3 matrices. If

$$A = \begin{bmatrix} 1 & 2 & 3 \\ 4 & 5 & 6 \\ 7 & 8 & 9 \end{bmatrix} \quad \text{and} \quad E_{23} = \begin{bmatrix} 0 & 0 & 0 \\ 0 & 0 & 1 \\ 0 & 0 & 0 \end{bmatrix},$$

then

$$AE_{23} = \begin{bmatrix} 1 & 2 & 3 \\ 4 & 5 & 6 \\ 7 & 8 & 9 \end{bmatrix} \begin{bmatrix} 0 & 0 & 0 \\ 0 & 0 & 1 \\ 0 & 0 & 0 \end{bmatrix} = \begin{bmatrix} 0 & 0 & 2 \\ 0 & 0 & 5 \\ 0 & 0 & 8 \end{bmatrix}$$

that is, the matrix whose third column is the second column of A, and all of whose other columns are columns of zeros. The calculation did not depend on the particular entries of A (1 2 3 etc.) and the conclusions we reached holds equally well for any

3 × 3 matrix A. Multiplication by E_{23} picked out the second column of A and shifted it to the third column, making everything else 0.

Let's try it from the other side, that is,

$$E_{23}A = \begin{bmatrix} 0 & 0 & 0 \\ 0 & 0 & 1 \\ 0 & 0 & 0 \end{bmatrix}\begin{bmatrix} 1 & 2 & 3 \\ 4 & 5 & 6 \\ 7 & 8 & 9 \end{bmatrix} = \begin{bmatrix} 0 & 0 & 0 \\ 7 & 8 & 9 \\ 0 & 0 & 0 \end{bmatrix}.$$

So multiplication by E_{23} *from the left* picked out the third row of A and shifted it to the second row, leaving everything else as 0.

What we did does not depend on the fact that we were working with 3 × 3 matrices, nor with E_{23}. We state: If $A \in M_n(F)$, then multiplying A by E_{rs} *from the right* shifts the rth *column* of A to become the sth *column* and leaves every other entry as 0. Multiplying A by E_{uv} *from the left* shifts the vth *row* to the uth *row* and has every other entry 0.

Let's see what happens when we multiply by E_{uv} from the left and E_{rs} from the right. We do it with the example of the 3 × 3 matrix above, with $E_{uv} = E_{11}$ and $E_{rs} = E_{23}$. Thus

$$E_{11}AE_{23} = \begin{bmatrix} 1 & 0 & 0 \\ 0 & 0 & 0 \\ 0 & 0 & 0 \end{bmatrix}\begin{bmatrix} 1 & 2 & 3 \\ 4 & 5 & 6 \\ 7 & 8 & 9 \end{bmatrix}\begin{bmatrix} 0 & 0 & 0 \\ 0 & 0 & 1 \\ 0 & 0 & 0 \end{bmatrix} = \begin{bmatrix} 1 & 0 & 0 \\ 0 & 0 & 0 \\ 0 & 0 & 0 \end{bmatrix}\begin{bmatrix} 0 & 0 & 2 \\ 0 & 0 & 5 \\ 0 & 0 & 8 \end{bmatrix}$$

$$= \begin{bmatrix} 0 & 0 & 2 \\ 0 & 0 & 0 \\ 0 & 0 & 0 \end{bmatrix} = 2\begin{bmatrix} 0 & 0 & 1 \\ 0 & 0 & 0 \\ 0 & 0 & 0 \end{bmatrix} = 2E_{13}.$$

Noting that 2 is the $(1,2)$-entry of A, we could state this as: $E_{11}AE_{23} = a_{12}E_{13}$.

What we did for the particular 3 × 3 matrix holds in general for any matrix in any $M_n(F)$. Waving our hands a little, we state: If $A = (a_{rs}) \in M_n(F)$, then $E_{uv}AE_{jk} = a_{vj}E_{uk}$. So we squeeze A down, by this operation, to a matrix with only one nonzero entry a_{vj} occurring in the (u,k) place.

Suppose now that A is a matrix in $M_n(F)$ such that $\text{tr}(AX) = 0$ *for all* matrices X in $M_n(F)$. So $\text{tr}(YAX) = \text{tr}(AXY) = 0$, since $XY \in M_n(F)$ for all X and Y. In particular, $\text{tr}(E_{1u}AE_{v1}) = 0$. But

$$E_{1u}AE_{v1} = \begin{bmatrix} a_{uv} & 0 & \cdots & 0 \\ 0 & 0 & \cdots & 0 \\ \vdots & \vdots & & \vdots \\ 0 & 0 & \cdots & 0 \end{bmatrix},$$

hence

$$0 = \text{tr}(E_{1u}AE_{v1}) = \text{tr}\begin{bmatrix} a_{uv} & 0 & \cdots & 0 \\ 0 & 0 & \cdots & 0 \\ \vdots & \vdots & & \vdots \\ 0 & 0 & \cdots & 0 \end{bmatrix} = a_{uv}.$$

This holds for every u and every v; hence all entries of A are zero. In short, $A = 0$. We have proved

Theorem 3.3.5. *If $A \in M_n(F)$ is such that $\operatorname{tr}(AX) = 0$ for all X in $M_n(F)$, then $A = 0$.*

Although Theorem 3.3.5 is not central to our development of matrix theory and represents a slight detour from our main path, it is an amusing result. In fact, when one goes deeply into matrix theory, it is even an important result. For us it gives practice in multiplying by the matrix units from the left and the right—a technique that can often stand us in good stead. In the harder exercises we shall lay out—in a series of problems—how one proves a very important fact about the set of all $n \times n$ matrices.

PROBLEMS

NUMERICAL PROBLEMS

1. Find $\operatorname{tr}(A)$ for

   (a) $A = \begin{bmatrix} 1 & 2 & 3 \\ 4 & 5 & 6 \\ 7 & 8 & 9 \end{bmatrix} \begin{bmatrix} 7 & 8 & 9 \\ 4 & 5 & 6 \\ 1 & 2 & 3 \end{bmatrix}$.

   (b) $A = \begin{bmatrix} 1 & -1 & 0 \\ 0 & 0 & 1 \\ 2 & -3 & 0 \end{bmatrix} \left(\begin{bmatrix} 1 & .5 & .5 \\ 1 & 1 & 1 \\ 3 & 0 & -2 \end{bmatrix} + \begin{bmatrix} 0 & \pi & 0 \\ \pi & 0 & 0 \\ 0 & 0 & \pi \end{bmatrix} \right)$.

   (c) $A = \begin{bmatrix} 1 & 0 & 3 \\ 0 & 1 & 0 \\ 0 & 0 & 1 \end{bmatrix} \begin{bmatrix} -2 & .3 & 0 \\ 0 & 5 & 1 \\ 0 & 0 & -3 \end{bmatrix} \begin{bmatrix} 1 & 0 & -3 \\ 0 & 1 & 0 \\ 0 & 0 & 1 \end{bmatrix}$.

   (d) $A = \begin{bmatrix} 1 & 0 & 0 \\ 0 & 2 & 0 \\ 0 & 0 & 3 \end{bmatrix} \begin{bmatrix} -1 & 0 & -2 \\ -3 & 0 & -4 \\ 5 & 0 & 6 \end{bmatrix} \begin{bmatrix} 1 & 0 & 0 \\ 0 & .5 & 0 \\ 0 & 0 & .3 \end{bmatrix}$.

   (e) $A = \begin{bmatrix} .25 & .25 & .25 & .25 \\ .25 & .25 & .25 & .25 \\ .25 & .25 & .25 & .25 \\ .25 & .25 & .25 & .25 \end{bmatrix}^2$.

2. By a direct computation find all 3×3 matrices A such that $\operatorname{tr}(AB) = 0$, where
   $$B = \begin{bmatrix} 1 & 0 & 0 \\ 0 & 2 & 0 \\ 0 & 0 & 3 \end{bmatrix}.$$

3. Verify by a direct calculation of the products that $\operatorname{tr}(AB) = \operatorname{tr}(BA)$, where
   $$A = \begin{bmatrix} 0 & 1 & 0 \\ .5 & 0 & 1 \\ 3 & 0 & -1 \end{bmatrix} \text{ and } B = \begin{bmatrix} .2 & .3 & .4 \\ 0 & 1 & 0 \\ 0 & 6 & 5 \end{bmatrix}.$$

4. For what values of a is $\operatorname{tr}(A) = 0$, where $A = \begin{bmatrix} 0 & 1 & 0 \\ a & 0 & 1 \\ 0 & 0 & -a \end{bmatrix}$?

5. For what values of a is $\operatorname{tr}(A) = 0$, where
$$A = \begin{bmatrix} 0 & 1 & 0 \\ a & 0 & 1 \\ 0 & 0 & -a \end{bmatrix} \begin{bmatrix} a & 1 & 0 \\ 1 & -a & 0 \\ 0 & 0 & -a \end{bmatrix}?$$

MORE THEORETICAL PROBLEMS

Easier Problems

6. Show that $\operatorname{tr}(ABC) = \operatorname{tr}(CAB)$ for all matrices C.
7. Show that $\operatorname{tr}((AB)^m) = \operatorname{tr}((BA)^m)$ for all positive integers m.
8. If $A \in M_2(F)$ and $A^2 = 0$, show that $\operatorname{tr}(A) = 0$.
9. If $A \in M_2(F)$ and $\operatorname{tr}(A) = 0 = \operatorname{tr}(A^2)$, show that $A^2 = 0$.

Middle-Level Problems

10. If $A = \begin{bmatrix} a_1 & b_1 & c_1 \\ 0 & a_2 & b_2 \\ 0 & 0 & a_3 \end{bmatrix}$ is such that $\operatorname{tr}(A) = \operatorname{tr}(A^2) = \operatorname{tr}(A^3) = 0$, prove that $a_1 = a_2 = a_3 = 0$ and that $A^3 = 0$.

11. If $A = \begin{bmatrix} a_1 & & & * \\ & a_2 & & \\ & & \ddots & \\ 0 & & & a_n \end{bmatrix}$, where every entry below the main diagonal is 0 and where the entries above the main diagonal are arbitrary elements of F, show that if $A^k = 0$ for some k, then $\operatorname{tr}(A) = 0$.

12. If $A \in M_n(F)$, show that $A = aI + B$ for some $a \in F$, $B \in M_n(F)$, and $\operatorname{tr}(B) = 0$.

Harder Problems

13. If f is a mapping from $M_n(F)$ to F satisfying
   (a) $f(A + B) = f(A) + f(B)$
   (b) $f(aA) = af(A)$
   (c) $f(AB) = f(BA)$
   for all $A, B \in M_n(F)$ and all $a \in F$, show that for all A, $f(A) = \mu \operatorname{tr}(A)$ for some fixed μ in F.

14. How would you describe the μ in Problem 13?

15. If $\operatorname{tr}(ABC) = \operatorname{tr}(CBA)$ for given matrices A and B in $M_n(F)$ and for all $C \in M_n(F)$, prove that $AB = BA$.

In the problems that follow, W is a subset of $M_n(F)$, $W \neq (0)$, with the following three properties:

(a) $A, B \in W$ implies that $A + B \in W$.
(b) $A \in W$, $X \in M_n(F)$ implies that $AX \in W$.
(c) $A \in W$, $Y \in M_n(F)$ implies that $YA \in W$.

16. Show that if $A \neq 0$ is in W, then some matrix unit E_{st} is in W. (*Hint:* Play with the matrix units and A.)

17. Using the result of Problem 16, show that if $A \neq 0$ is in W, then *all* matrix units E_{jk} are in W.

18. From Problem 17 show that *all* matrices in $M_n(F)$ are in W, that is, $W = M_n(F)$. [The result in Problem 18 is a famous theorem in algebra: It is usually stated as: $M_n(F)$ *is a simple ring.*]

19. Give an example of a subset $V \neq (0)$ and $V \neq M_n(F)$ that satisfies properties (1) and (2) in the definition of W.

20. Give an example of a subset $U \neq (0)$ and $U \neq M_n(F)$ that satisfies properties (1) and (3) in the definition of W.

3.4. TRANSPOSE AND HERMITIAN ADJOINT

In working with the 2×2 matrices with real entries we introduced the concept of the transpose of a matrix. If you recall, the transpose of a matrix was the matrix obtained by interchanging the rows with the columns of the given matrix. Later, after we had discussed the complex numbers and were working with matrices whose entries were complex numbers, we talked about the Hermitian adjoint of a matrix, A. It was like the transpose, but with a twist. The Hermitian ajoint A^* of a matrix A was obtained by first taking the transpose of A and then applying the operation of complex conjugation to all the entries of this transposed matrix. Of couse, for a matrix with real entries its transpose and its Hermitian adjoint are the same.

For these operations on the matrices we found that certain rules pertained. We shall find the exact analogue of these rules, here, for the $n \times n$ case. But first we make the formal definitions that we need.

Definition. Given $A = (a_{rs}) \in M_n(F)$, then the *transpose* of A, denoted by A', is the matrix $A' = (b_{rs})$, where for each r and s, $b_{rs} = a_{sr}$.

So, for example, if $A = \begin{bmatrix} 1 & i & 2 \\ 3 & 7 & 0 \\ 1+i & 0 & 0 \end{bmatrix}$, then $A' = \begin{bmatrix} 1 & 3 & 1+i \\ i & 7 & 0 \\ 2 & 0 & 0 \end{bmatrix}$.

We immediately pass on to the Hermitian adjoint.

Definition. If $A = (a_{rs}) \in M_n(\mathbb{C})$, then the *Hermitian adjoint* A^* of A is defined by $A^* = (c_{rs})$, where $c_{rs} = \bar{a}_{sr}$, the $\bar{}$ indicating the complex conjugate of a_{sr}.

Sec. 3.4] Transpose and Hermitian Adjoint

Thus, for the example we used above, $A = \begin{bmatrix} 1 & i & 2 \\ 3 & 7 & 0 \\ 1+i & 0 & 0 \end{bmatrix}$, we have

$$A^* = \begin{bmatrix} \bar{1} & \bar{3} & \overline{1+i} \\ \bar{i} & \bar{7} & \bar{0} \\ \bar{2} & \bar{0} & \bar{0} \end{bmatrix} = \begin{bmatrix} 1 & 3 & 1-i \\ -i & 7 & 0 \\ 2 & 0 & 0 \end{bmatrix}.$$

When we discuss matrices with complex entries we shall seldom—possibly never—use the transpose; instead, we shall always use the Hermitian adjoint. Accordingly, we shall prove the basic properties only for the Hermitian adjoint, leaving the corresponding results for transpose as exercises for the reader.

Theorem 3.4.1. If $A, B \in M_n(\mathbb{C})$ and $a \in \mathbb{C}$, then

1. $(A^*)^* = A^{**} = A$;
2. $(A + B)^* = A^* + B^*$;
3. $(aA)^* = \bar{a}A^*$;
4. $(AB)^* = B^*A^*$.

Proof: Before getting down to the details of the proof, note that the rules for $*$ and $'$ are the same, with the exception of (3), where for transpose $(aA') = aA'$. Note the important property of $*$ in (4), which says that on taking the Hermitian adjoint we reverse the order of the matrices involved.

Now to the proof itself. If $A = (a_{rs})$, then $A^* = (b_{rs})$, where $b_{rs} = \bar{a}_{sr}$. Thus $(A^*)^* = (c_{rs})$, where $c_{rs} = \bar{b}_{sr} = \bar{\bar{a}}_{rs} = a_{rs}$. Thus $(A^*)^* = A$. This proves (1). To get (2), if $A = (a_{rs})$ and $B = (b_{rs})$, then $A^* = (c_{rs})$, where $c_{rs} = \bar{a}_{sr}$, and $B^* = (d_{rs})$, where $d_{rs} = \bar{b}_{sr}$. Since $A + B = (u_{rs})$, where $u_{rs} = a_{rs} + b_{rs}$, we have that

$$(A + B)^* = (v_{rs}),$$

where

$$v_{rs} = \bar{u}_{sr} = \overline{a_{sr} + b_{sr}} = \bar{a}_{sr} + \bar{b}_{sr}.$$

On the other hand,

$$A^* + B^* = (c_{rs}) + (d_{rs}) = (c_{rs} + d_{rs}),$$

and

$$c_{rs} + d_{rs} = \bar{a}_{sr} + \bar{b}_{sr}.$$

Therefore, we see, on comparing these evaluations, that indeed

$$(A + B)^* = A^* + B^*.$$

To prove (3) is even easier. With the notation used above, $aA = (aa_{rs})$ and $(aA)^* = (w_{rs})$, where $w_{rs} = \overline{(aa_{sr})} = \bar{a}\bar{a}_{sr}$, whence

$$(aA)^* = (\bar{a}\bar{a}_{sr}) = \bar{a}A^*.$$

Finally, we come to the hardest of the four. Again with the notation above, $AB = (a_{rs})(b_{rs}) = (t_{rs})$, where $t_{rs} = \sum_{k=1}^{n} a_{rk}b_{ks}$. Thus

$$(AB)^* = (f_{rs}), \text{ where } f_{rs} = \bar{t}_{rs} = \sum_{k=1}^{n} \bar{a}_{sk}\bar{b}_{kr}.$$

On the other hand,

$$B^*A^* = (d_{rs})(c_{rs}) = (f_{rs}),$$

since $\sum_{k=1}^{n} d_{rk}c_{ks} = \sum_{k=1}^{n} \bar{b}_{kr}\bar{a}_{sk} = \sum_{k=1}^{n} \bar{a}_{sk}\bar{b}_{kr} = a_{rs}$. Thus $B^*A^* = (AB)^*$, as claimed. ∎

Let's see (4) in a particular example. If

$$A = \begin{bmatrix} i & 1+i & 3 \\ 1-i & 4 & i \\ 3 & i & 6+3i \end{bmatrix} \quad \text{and} \quad B = \begin{bmatrix} 1 & -i & .5i \\ 0 & 0 & 1+i \\ 5+6i & 0 & 1 \end{bmatrix},$$

then

$$AB = \begin{bmatrix} i + 3(5+6i) & -i^2 & .5i^2 + (1+i)^2 + 3 \\ 1 - i + i(5+6i) & (1-i)(-i) & .5i(1-i) + 4(1+i) + i \\ 3 + (6+3i)(5+6i) - 3i & -3i & 1.5i - i(1+i) + (6+3i) \end{bmatrix}$$

$$= \begin{bmatrix} 15 + 19i & 1 & 2.5 + 2i \\ -5 + 4i & -1 - i & 4.5 + 5.5i \\ 15 + 51i & -3i & 7 + 3.5i \end{bmatrix}.$$

(Check!). So

$$(AB)^* = \begin{bmatrix} 15 - 19i & -5 - 4i & 15 - 51i \\ 1 & -1 + i & 3i \\ 2.5 - 2i & 4.5 - 5.5i & 7 - 3.5i \end{bmatrix}.$$

We now compute B^*A^*. Since

$$A^* = \begin{bmatrix} -i & 1+i & 3 \\ 1-i & 4 & +i \\ 3 & -i & 6-3i \end{bmatrix} \quad \text{and} \quad B^* = \begin{bmatrix} 1 & 0 & 5-6i \\ i & 0 & 0 \\ -.5i & i-i & 1 \end{bmatrix}$$

the product B^*A^* is

$$\begin{bmatrix} -i + 3(5-6i) & 1(1+i) - i(5-6i) & 3 + (5-6i)(6-3i) \\ (-i)i & i(1+i) & 3i \\ (-.5i)(-i) + (1-i)^2 + 3 & -.5i(1+i) + 4(1-i) - i & -1.5i + (1-i) + (6-3i) \end{bmatrix},$$

which simplifies to

$$\begin{bmatrix} 15 - 19i & -5 - 4i & 15 - 51i \\ 1 & -1+i & 3i \\ 2.5 - 2i & 4.5 - 5.5i & 7 - 3.5i \end{bmatrix} = (AB)^*.$$

Thus we see that $(AB)^* = B^*A^*$, as it should be.

As we mentioned earlier, for the transpose we have

$$(A')' = A, (A + B)' = A' + B', (aA)' = aA', (AB)' = B'A'.$$

We should like to interrelate the trace and Hermitian adjoint. Let $A = (a_{rs}) \in M_n(\mathbb{C})$. Then $A^* = (b_{rs})$, where $b_{rs} = \bar{a}_{sr}$. Thus $AA^* = (a_{rs})(b_{rs}) = (c_{rs})$, where

$$c_{rs} = \sum_{k=1}^{n} a_{rk} b_{ks} = \sum_{k=1}^{n} a_{rk} \bar{a}_{sk}.$$

In particular, the diagonal entries, c_{rr}, are given by $c_{rr} = \sum_{k=1}^{n} a_{rk} \bar{a}_{rk} = \sum_{k=1}^{n} |a_{rk}|^2$. Thus $\operatorname{tr}(AA^*) = \sum_{r=1}^{n} \left(\sum_{k=1}^{n} |a_{rk}|^2 \right)$. Suppose that $\operatorname{tr}(AA^*) = 0$; then $\sum_{r=1}^{n} \left(\sum_{k=1}^{n} |a_{rk}|^2 \right) = 0$. Since $|a_{rk}|^2 \geq 0$, the only way this sum can be 0 is if each $a_{rk} = 0$ for every r and k. But this merely tells us that $A = 0$. We have proved

Theorem 3.4.2. If A is in $M_n(\mathbb{C})$, then $\operatorname{tr}(AA^*) = 0$ if and only if $A = 0$. In fact, $\operatorname{tr}(AA^*) > 0$ if $A \neq 0$.

This result is false if we use transpose instead of Hermitian adjoint. For example, if $A = \begin{bmatrix} 1 & i \\ -2i & 2 \end{bmatrix}$, then

$$AA' = \begin{bmatrix} 1 & i \\ -2i & 2 \end{bmatrix} \begin{bmatrix} 1 & -2i \\ i & 2 \end{bmatrix} = \begin{bmatrix} 1 + i^2 & -2i + 2i \\ -2i + 2i & 4i^2 + 4 \end{bmatrix} = \begin{bmatrix} 0 & 0 \\ 0 & 0 \end{bmatrix},$$

yet $A \neq 0$. Of course, if all the entries of A are real, then $A^* = A'$, so in this instance we do have that $\operatorname{tr}(AA') = 0$ forces $A = 0$.

To illustrate what is going on in the proof of Theorem 3.4.2, let us look at it for a matrix of small size, namely a 3×3 matrix. Let $A = \begin{bmatrix} a_{11} & a_{12} & a_{13} \\ a_{21} & a_{22} & a_{23} \\ a_{31} & a_{32} & a_{33} \end{bmatrix}$; then

$A^* = \begin{bmatrix} \bar{a}_{11} & \bar{a}_{21} & \bar{a}_{31} \\ \bar{a}_{12} & \bar{a}_{22} & \bar{a}_{32} \\ \bar{a}_{13} & \bar{a}_{23} & \bar{a}_{33} \end{bmatrix}$. Since we want to look at $\operatorname{tr}(AA^*)$, our only concern will be with the diagonal entries of AA^*. These are

$$a_{11}\bar{a}_{11} + a_{12}\bar{a}_{12} + a_{13}\bar{a}_{13} = |a_{11}|^2 + |a_{12}|^2 + |a_{13}|^2,$$
$$a_{21}\bar{a}_{21} + a_{22}\bar{a}_{22} + a_{23}\bar{a}_{23} = |a_{21}|^2 + |a_{22}|^2 + |a_{23}|^2,$$

and finally,

$$a_{31}\bar{a}_{31} + a_{32}\bar{a}_{32} + a_{33}\bar{a}_{33} = |a_{31}|^2 + |a_{32}|^2 + |a_{33}|^2.$$

Therefore,

$$\operatorname{tr}(AA^*) = |a_{11}|^2 + |a_{12}|^2 + |a_{13}|^2 + |a_{21}|^2 + |a_{22}|^2 + |a_{23}|^2 + |a_{31}|^2 + |a_{32}|^2 + |a_{33}|^2$$

and we see that if $A \neq 0$, then some entry is not 0; hence $\operatorname{tr}(AA^*)$ is positive. Also, $\operatorname{tr}(AA^*) = 0$ forces each a_{jk} to be 0; hence $A = 0$.

Associated with the Hermitian adjoint are two particular classes of matrices.

Definition. $A \in M_n(\mathbb{C})$ is a *Hermitian* matrix if $A = A^*$, and A is a *skew-Hermitian* matrix if $A^* = -A$.

EXAMPLES

1. The matrix $\begin{bmatrix} -1 & 1+i & 3 \\ 1-i & 4 & +i \\ 3 & -i & 2 \end{bmatrix}$ is Hermitian.

2. The matrix $\begin{bmatrix} -i & 1+i & 3 \\ -1+i & 4i & +i \\ -3 & +i & 0 \end{bmatrix}$ is skew-Hermitian.

3. The matrix $\begin{bmatrix} -1 & 1 & 3 \\ 1 & 4 & 3 \\ 3 & 3 & 2 \end{bmatrix}$ is Hermitian with real entries.

In case $A \in M_n(\mathbb{R})$, A is Hermitian, as in (3) above, we just call it a (real) *symmetric matrix*. Thus a real matrix A is symmetric matrix if and only if it equals its transpose A'.

Given $A \in M_n(\mathbb{C})$, then $A = \dfrac{A + A^*}{2} + \dfrac{A - A^*}{2}$; however,

$$(A + A^*)^* = A^* + A^{**} = A^* + A$$

and

$$(A - A^*)^* = A^* - A^{**} = A^* - A = -(A - A^*).$$

This says that $A + A^*$ is Hermitian and $A - A^*$ is skew-Hermitian. So *A is the sum of the Hermitian and skew-Hermitian matrices* $\dfrac{A + A^*}{2}$ *and* $\dfrac{A - A^*}{2}$. Furthermore, if $A = B + C$, where B is Hermitian and C is skew-Hermitian, then $A^* = (B + C)^* = B^* + C^* = B - C$. Thus $B = \dfrac{A + A^*}{2}$ and $C = \dfrac{A - A^*}{2}$. (Prove!) So there is *only one way* in which we can decompose A as a sum of a Hermitian and a skew-Hermitian matrix. We summarize this in

Lemma 3.4.3. Given $A \in M_n(\mathbb{C})$, then $A = \dfrac{A + A^*}{2} + \dfrac{A - A^*}{2}$ is a decomposition of A as a sum of a Hermitian and a skew-Hermitian matrix. Moreover, this decomposition is unique in that if $A = B + C$, where B is Hermitian and C is skew-Hermitian, then $B = \dfrac{A + A^*}{2} + \dfrac{A - A^*}{2}$.

It is easy to write down a great number of Hermitian matrices. As we saw above, $A + A^*$ is Hermitian for any A. Similarly, since $(AA^*)^* = (A^*)^*A^* = AA^*$, AA^* is Hermitian. A similar computation shows that XAX^* is Hermitian, for any matrix X, if A is Hermitian. Finally, note that if A is skew-Hermitian, then A^2 is Hermitian for $(A^2)^* = (A^*)^2 = (-A)^2 = A^2$.

For skew-Hermitian matrices we also have easy means of producing them. For instance, for any matrix A the matrix $A - A^*$ is skew-Hermitian, and for any matrix X the matrix XAX^* is skew-Hermitian if A is skew-Hermitian.

In general, the product of two Hermitian matrices is *not* Hermitian. If $A = A^*$ and $B = B^*$ and $(AB)^* = AB$, we get, since $(AB)^* = B^*A^* = BA$, that AB is Hermitian only if A and B commute.

There are many combinations of Hermitian and skew-Hermitian matrices which end up as Hermitian or skew-Hermitian. Some of these will appear in the problems.

PROBLEMS

NUMERICAL PROBLEMS

1. For the given matrices A write down A', A^*, and $A' - A^*$.

   (a) $A = \begin{bmatrix} i & 2 & \frac{1+i}{1-i} \\ (1+i)^2 & 3 & 5 \\ \frac{1}{2}i & \frac{1}{3}i & \frac{1}{4}i \end{bmatrix}$.

(b) $A = \begin{bmatrix} 1 & i & -i & 0 \\ -i & \pi & 0 & 1+i \\ i & 0 & \pi^2 & 6 \\ 0 & 1-i & 0 & \pi^3 \end{bmatrix}.$

(c) $A = \begin{bmatrix} 1 & i & -i & 0 \\ i & \pi & 0 & 1+i \\ -i & 0 & \pi^2 & 6 \\ 0 & 1+i & 6 & \pi^3 \end{bmatrix}.$

(d) $A = \begin{bmatrix} 1 & i & 2i \\ 1+i & 1-i & 0 \\ 0 & 1 & 1 \end{bmatrix}^2.$

2. For the given matrices A and B write down A^*, B^* and verify that $(AB)^* = B^*A^*$.

(a) $A = \begin{bmatrix} 1 & 2 & 1 \\ -1 & 6 & i \\ i & 6 & -1 \end{bmatrix},\ B = \begin{bmatrix} 1 & 0 & 1+i \\ 0 & 2-i & 0 \\ 3i & 1 & 1 \end{bmatrix}.$

(b) $A = \begin{bmatrix} 1 & i & i^2 & i^3 \\ -i & 0 & 1 & 2 \\ i^2 & 1 & i & 0 \\ -i^3 & 2 & 0 & i \end{bmatrix},\ B = \begin{bmatrix} 1 & 1 & 1 & 1 \\ 1 & 1 & 1 & 1 \\ 1 & 1 & 1 & 1 \\ 1 & 1 & 1 & 1 \end{bmatrix}.$

(c) $A = \begin{bmatrix} 1 & i & -1 \\ 0 & 2 & i \\ 3-2i & 0 & 1 \end{bmatrix},\ B = \begin{bmatrix} 1 & 0 & 3-2i \\ i & 2 & 0 \\ -1 & i & 1 \end{bmatrix}.$

(d) $A = \begin{bmatrix} 1 & i & i^2 \\ -i & 2 & 1+i \\ i^2 & 1-i & 3 \end{bmatrix},\ B = \begin{bmatrix} 1 & 0 & 0 \\ 0 & 1 & 0 \\ 0 & 0 & 1 \end{bmatrix}.$

3. For what values of a and b are the following matrices Hermitian?

(a) $A = \begin{bmatrix} 1 & \frac{1+i}{i} & \frac{1-i}{1+i} \\ a & a & 0 \\ b & 0 & b \end{bmatrix}.$

(b) $A = \begin{bmatrix} 1+i & 0 \\ b & a \end{bmatrix} \begin{bmatrix} 1-i & 0 \\ a & b \end{bmatrix}.$

(c) $A = \begin{bmatrix} a & 0 & b \\ 0 & a & a \\ i & 1 & a \end{bmatrix}.$

(d) $A = \begin{bmatrix} a & 0 & b \\ 0 & a & a \\ i & 1 & a \end{bmatrix}^2.$

Sec. 3.4] Transpose and Hermitian Adjoint 123

4. By a direct calculation show that if $A = \begin{bmatrix} 1 & i & 0 \\ i & 0 & i \\ 0 & i & 0 \end{bmatrix}$ and B is Hermitian, then AB is Hermitian only if $AB = BA$.

5. Compute $\operatorname{tr}(AA^*)$ for

   (a) $A = \begin{bmatrix} 1 & \frac{1+i}{1-i} & 2 \\ -i & 0 & 0 \\ 0 & 1 & 2 \end{bmatrix}$.

   (b) $A = \begin{bmatrix} i & 0 & 1 & 0 \\ 0 & i & 2+i & 0 \\ 1 & 2-i & i & \pi \\ 0 & 0 & \pi & i \end{bmatrix}$.

   (c) $A = \begin{bmatrix} i & 2 & 3 & 4 \\ 2i & 0 & 0 & 0 \\ 3i & 0 & 0 & 0 \\ 4i & 0 & 0 & 0 \end{bmatrix}$.

MORE THEORETICAL PROBLEMS

Easier Problems

6. Show that if $'$ is the transpose, then
   (a) $(A')' = A$
   (b) $(A + B) = A' + B'$
   (c) $(aA)' = aA'$
   (d) $(AB)' = B'A'$
   for all $A, B \in M_n(F)$ and all $a \in F$.

7. Show that the following matrices are Hermitian.
   (a) $AB + BA$, where $A^* = A$, $B^* = B$.
   (b) $AB - BA$, where $A^* = A$, $B^* = -B$.
   (c) $AB + BA$, where $A^* = -A$, $B^* = -B$.
   (d) A^{2n}, where $A^* = -A$.

8. Show that the following matrices are skew-Hermitian.
   (a) $AB - BA$, where $A^* = A$, $B^* = B$.
   (b) $AB + BA$, where $A^* = A$, $B^* = -B$.
   (c) $AB - BA$, where $A^* = -A$, $B^* = -B$.
   (d) $A^2B - BA^2$, where $A^* = -A$, $B^* = B$.

9. If X is any matrix in $M_n(\mathbb{C})$, show that XAX^* is Hermitian if A is Hermitian, and XAX^* is skew-Hermitian if A is skew-Hermitian.

10. If $A_1, \ldots, A_n \in M_n(\mathbb{C})$, show that $(A_1 A_2 \ldots A_n)^* = A_n^* A_{n-1}^* \ldots A_1^*$.

Middle-Level Problems

11. If $A_1, \ldots, A_n \in M_n(\mathbb{C})$ and $\operatorname{tr}(A_1 A_1^* + A_2 A_2^* + \cdots + A_n A_n^*) = 0$, prove that $A_1 = A_2 = \cdots = A_n = 0$.
12. If $A^* = A$, show that if $A^m = 0$, then $A = 0$.
13. If A is invertible, then $(A^*)^{-1} = (A^{-1})^*$.

Harder Problems

14. If $A^* = -A$ and $AB = B$, show that $B = 0$. [*Hint:* Consider $B^*B = (AB)^*(AB)$.]
15. Generalize the result of Problem 14 to show that if $A^* = -A$ and $AB = aB$, where $a \neq 0$ is a real number, then $B = 0$.
16. If $A^* = A$ and $AB = (ai)B$, where $a \neq 0$ is real, show that $B = 0$.
17. If $A^* = A$ and $AB = (a + bi)B$, where a, b are real and $B \neq 0$, show that $b = 0$. (As we shall see later, this implies that the characteristic roots of a Hermitian matrix are real.)
18. Given any matrix $A \in M_n(\mathbb{C})$ show that if for some $B \neq 0$, $AA^*B = aB$, $a \in \mathbb{R}$, then $a \geq 0$. (A consequence of this is that the characteristic roots of AA^* are not only real but are nonnegative.)
19. Give an example, in $M_3(\mathbb{C})$ of a matrix A such that $A^* = -A$, yet A is invertible.
20. Suppose that $A \in M_n(\mathbb{C})$ is such that $AB = BA$ for all $B = B^*$ in $M_n(\mathbb{C})$. Show that $A = I$ for some $a \in \mathbb{C}$, that is, that A is a scalar matrix. (*Hint:* For D Hermitian, the matrix iD is skew-Hermitian.)

3.5. INNER PRODUCT SPACES

In Section 1.12 the topic under discussion was the inner product on the space W of 2-tuples whose components were in $\mathbb{C}$. The inner product of $v = \begin{bmatrix} x_1 \\ x_2 \end{bmatrix}$ and $w = \begin{bmatrix} y_1 \\ y_2 \end{bmatrix}$, where v and w are in W, was defined as $(v, w) = x_1 \bar{y}_1 + x_2 \bar{y}_2$. We then saw some properties obtained by W relative to its inner product.

What we shall do for $\mathbb{C}^{(n)}$, the set of n-tuples over $\mathbb{C}$, will be the exact analogue of what we said above.

Definition. If $v = \begin{bmatrix} x_1 \\ \vdots \\ x_n \end{bmatrix}$ and $w = \begin{bmatrix} y_1 \\ \vdots \\ y_n \end{bmatrix}$ are in $\mathbb{C}^{(n)}$, then their *inner product*, denoted by (v, w), is defined by $(v, w) = \sum_{j=1}^{n} x_j \bar{y}_j$.

For instance, the inner product of $\begin{bmatrix} 41 \\ e \\ 12 \end{bmatrix}$ and $\begin{bmatrix} 1 \\ f \\ 2 \end{bmatrix}$ is $41 + e\bar{f} + 24$.

Everything we did before, in Chapter 1, carries over immediately to our present context. So you can be a little briefer in the discussion and the proofs. If you are perplexed by anything that comes up, go back to Section 1.12 to see how it was done there. We state, without proof, the first of these results, leaving the proof to the reader.

Lemma 3.5.1. If u, v, w are in $\mathbb{C}^{(n)}$ and $a \in \mathbb{C}$, then

1. $(v + w, u) = (v, u) + (w, u)$;
2. $(u, v + w) = (u, v) + (u, w)$;
3. $(v, w) = \overline{(w, v)}$;
4. $(av, w) = a(v, w) = (v, \bar{a}w)$;
5. $(v, v) \geq 0$ is real and $(v, v) = 0$ if and only if $v = 0$.

Note that property (4) says we can pull a scalar a out of the symbol $(\cdot, \cdot)$ if it occurs as a multiplier of the first entry of $(\cdot, \cdot)$, but we can only pull it out as $\bar{a}$ if it occurs as a multiplier of the second one. That is, if $a \in \mathbb{C}$, then $(v, aw) = \bar{a}(v, w)$.

Definition. If $v, w \in \mathbb{C}^{(n)}$, then v is said to be *orthogonal* to w if $(v, w) = 0$.

Notice that if v is orthogonal to w, then w is orthogonal to v, since $(w, v) = \overline{(v, w)} = \bar{0} = 0$. Also, if $(v, w) = 0$ for *all* $w \in \mathbb{C}^{(n)}$, then, in particular, $(v, v) = 0$, and so, by Part (5) of Lemma 3.5.1, $v = 0$.

Definition. $v^\perp = \{w \in \mathbb{C}^{(n)} \mid (v, w) = 0\}$.

EXAMPLE

If $v = \begin{bmatrix} 2 \\ i \\ 3 \end{bmatrix}$, then $v^\perp = \left\{ \begin{bmatrix} x \\ y \\ z \end{bmatrix} \middle| 2\bar{x} + i\bar{y} + 3\bar{z} = 0 \right\}$. If w_1, w_2 are in $v^\perp$, then $(v, w_1 + w_2) = (v, w_1) + (v, w_2) = 0 + 0 = 0$, hence $w_1 + w_2$ is again in $v^\perp$. Also, if $a \in \mathbb{C}$ and w is in $v^\perp$, then $(v, aw) = \bar{a}(v, w) = \bar{a}0 = 0$, so aw is in $v^\perp$.

We summarize what was just done in

Lemma 3.5.2

1. If v is orthogonal to w, then w is orthogonal to v.
2. If v is orthogonal to all elements of $\mathbb{C}^{(n)}$, then $v = 0$.
3. If $v^\perp = \{w \in \mathbb{C}^{(n)} \mid (v, w) = 0\}$, then:
    (a) w_1, w_2 in $v^\perp$ implies that $w_1 + w_2$ is in $v^\perp$.
    (b) a in $\mathbb{C}$, w in $v^\perp$ implies that aw is in $v^\perp$.

In Chapter 1 we interrelated the inner product and the Hermitian adjoint. We should like to do so here as well. All the proofs given will be the exact duplicates of those in Section 1.12. So we shall give relatively few proofs and just wave our hands with a "look back in Chapter 1 for the method of proof." However, since the next result involves the summation symbol and many readers do not feel comfortable with it, we do the proof in detail.

Theorem 3.5.3. If $A \in M_n(\mathbb{C})$ and $v, w \in \mathbb{C}^{(n)}$, then $(Av, w) = (v, A^*w)$, where A^* is the Hermitian adjoint of A.

Proof: Let $v = \begin{bmatrix} x_1 \\ \vdots \\ x_n \end{bmatrix}$ and $w = \begin{bmatrix} y_1 \\ \vdots \\ y_n \end{bmatrix}$ be in $\mathbb{C}^{(n)}$ and $A = (a_{rs})$ in $M_n(\mathbb{C})$. Thus

$Av = (a_{rs}) \begin{bmatrix} x_1 \\ \vdots \\ x_n \end{bmatrix} = \begin{bmatrix} z_1 \\ \vdots \\ z_n \end{bmatrix}$, where for each r, $z_r = \sum_{s=1}^{n} a_{rs} x_s$. Therefore,

$$(Av, w) = \sum_{r=1}^{n} z_r \bar{y}_r = \sum_{r=1}^{n} \sum_{s=1}^{n} (a_{rs} x_s) \bar{y}_r = \sum_{r=1}^{n} \sum_{s=1}^{n} a_{rs} x_s \bar{y}_r.$$

On the other hand, $A^* = (b_{rs})$, where $b_{rs} = \bar{a}_{sr}$. So $A^*w = (b_{rs}) \begin{bmatrix} y_1 \\ \vdots \\ y_n \end{bmatrix} = \begin{bmatrix} t_1 \\ \vdots \\ t_n \end{bmatrix}$,

where for each r, $t_r = \sum_{s=1}^{n} b_{rs} y_s = \sum_{s=1}^{n} \bar{a}_{sr} y_s$. Therefore,

$$(v, A^*w) = \sum_{r=1}^{n} x_r \bar{t}_r = \sum_{r=1}^{n} \sum_{s=1}^{n} x_r (\overline{\bar{a}_{sr} y_s}) = \sum_{r=1}^{n} \sum_{s=1}^{n} x_r a_{sr} \bar{y}_s.$$

If we call r by the letter s and s by the letter r (don't forget, these are just dummy indices) we have

$$(v, A^*w) = \sum_{r=1}^{n} \sum_{s=1}^{n} x_s a_{rs} \bar{y}_r = \sum_{r=1}^{n} \sum_{s=1}^{n} a_{rs} x_s \bar{y}_r = (Av, w),$$

from above. This finishes the proof. ■

From here on—in this section—every result and every proof parrots virtually verbatim those in Section 1.12. We leave the proofs as exercises. Again, if you get stuck trying to do the exercises, look back at the relevant part in Chapter 1. But before doing this, see if you can come up with proofs of your own.

Theorem 3.5.4. If $A, B \in M_n(\mathbb{C})$ and $a \in \mathbb{C}$, then

1. $(A^*)^* = A^{**} = A$;
2. $(A + B)^* = A^* + B^*$;
3. $(aA)^* = \bar{a}A^*$;
4. $(AB)^* = B^*A^*$.

Although we did this in Section 3.4 by direct computations, see if you can carry out the proofs using inner products and the important fact that $(Av, w) = (v, A^*w)$. [You will also need that $v = 0$ if $(v, w) = 0$ for all w.]

We go down the list of theorems.

Theorem 3.5.5. An element $A \in M_n(\mathbb{C})$ is such that $(Av, Aw) = (v, w)$ for all v, w in $\mathbb{C}^{(n)}$ if and only if $A^*A = I$.

Definition. A matrix A in $M_n(\mathbb{C})$ is called *unitary* if $A^*A = I$.

For instance, the matrix $A = (1/\sqrt{2})\begin{bmatrix} 1 & i \\ i & 1 \end{bmatrix}$ is unitary.

So a unitary matrix is one that does not disturb the inner product of two vectors in $\mathbb{C}^{(n)}$. Equivalently, a unitary matrix is an invertible matrix whose inverse is A^*. In the case of the unitary matrix $A = (1/\sqrt{2})\begin{bmatrix} 1 & i \\ i & 1 \end{bmatrix}$, its inverse is

$$A^* = (1/\sqrt{2})\begin{bmatrix} 1 & -i \\ -i & 1 \end{bmatrix}.$$

Definition. If $v \in \mathbb{C}^{(n)}$, then the *length* of v, denoted by $\|v\|$, is defined by $\|v\| = \sqrt{(v, v)}$.

For example, the vector $\begin{bmatrix} 1 \\ i \end{bmatrix}$ has length 2 and both columns of the matrix $(1/\sqrt{2})\begin{bmatrix} 1 & -i \\ -i & 1 \end{bmatrix}$ have length 1.

Theorem 3.5.6. If $A \in M_n(\mathbb{C})$ is such that $(Aw, Aw) = (w, w)$ for all $w \in \mathbb{C}^{(n)}$, then A is unitary [hence $(Av, Aw) = (v, w)$ for all v and w in $\mathbb{C}^{(n)}$].

If $(Aw, Aw) = (w, w)$, then $\|Aw\| = \|w\|$. So unitary matrices preserve length and any matrix that preserves length must be unitary.

At the end of Section 3.4, the following occurred: If $A = A^*$ and for some $B \neq D$, $AB = aB$, where $a \in \mathbb{C}$, then a is real. What this really says is that the characteristic roots of A are real. Do you recall from Chapter 1 what a characteristic root of A is? We define it anew.

Definition. If $A \in M_n(\mathbb{C})$, then $a \in \mathbb{C}$ is a *characteristic root* of A if for some $v \neq 0$ in $\mathbb{C}^{(n)}$, $Av = av$. Such a vector v is then called a *characteristic vector* associated with a.

The same proof as that of Theorem 1.12.7 gives us the very important

Theorem 3.5.7. If A is Hermitian, that is, if $A^* = A$, then the characteristic roots of A are real.

It is hard to exaggerate the importance of this theorem, not only in mathematics, but also in physics, chemistry, and statistics.

Finally, the exact analogue of Theorem 1.12.8 is

Theorem 3.5.8. If $A \in M_n(\mathbb{C})$ is unitary, then $|a| = 1$ for any characteristic root of A.

PROBLEMS

NUMERICAL PROBLEMS

1. Calculate the inner product (v, w) for the given v and w.

(a) $v = \begin{bmatrix} 1 \\ -3 \\ 2 \end{bmatrix}$, $w = \begin{bmatrix} 0 \\ 1 \\ 5 \end{bmatrix}$.

(b) $v = \begin{bmatrix} i \\ 1+i \\ 3i \\ 0 \end{bmatrix}$, $w = \begin{bmatrix} 1-i \\ 1-i \\ 2 \\ 7 \end{bmatrix}$.

(c) $v = \begin{bmatrix} 1 \\ 0 \\ \vdots \\ 0 \end{bmatrix}$, $w = \begin{bmatrix} 0 \\ 1 \\ \vdots \\ 0 \end{bmatrix}$.

(d) $v = \begin{bmatrix} \frac{i}{2} \\ \pi^2 \\ 0 \\ -\frac{1}{3} \end{bmatrix}$, $w = \begin{bmatrix} \pi \\ \pi^2 \\ \pi^3 \\ i \end{bmatrix}$.

2. For the v and w in Problem 1, calculate (w, v) and show in each that $(w, v) = \overline{(v, w)}$.

3. Find the length, $\sqrt{(v, v)}$, for the given v.

(a) $v = \begin{bmatrix} i \\ 2 \\ -\frac{1}{3} + i \\ 4i \end{bmatrix}$.

(b) $v = \begin{bmatrix} 1 \\ 2 \\ 3 \\ 4 \\ 5 \end{bmatrix}$.

(c) $v = \begin{bmatrix} i \\ 2i \\ 3i \\ 4i \\ 5i \end{bmatrix}$.

(d) $v = \begin{bmatrix} a \\ 0 \\ \vdots \\ 0 \end{bmatrix}$.

4. For the given matrices A and vectors v and w, calculate (Av, w), (v, A^*w) and verify that $(Av, w) = (v, A^*w)$.

(a) $A = \begin{bmatrix} 1 & i & 0 \\ 3-i & 2i & 7 \\ 6 & 5 & 0 \end{bmatrix}$, $v = \begin{bmatrix} 1 \\ 2 \\ 3 \end{bmatrix}$, $w = \begin{bmatrix} i \\ i^3 \\ 2 \end{bmatrix}$.

(b) $A = \begin{bmatrix} 1 & 2 & 3 & 4 \\ 2i & 0 & 1 & 2 \\ 3i & 0 & 0 & 1 \\ 4i & 1 & 2 & 0 \end{bmatrix}$, $v = \begin{bmatrix} i \\ -17i \\ 6 \\ 1 \end{bmatrix}$, $w = \begin{bmatrix} 7+i \\ 4 \\ 0 \\ 3i \end{bmatrix}$.

(c) $A = \begin{bmatrix} 0 & i & 0 & 0 \\ 0 & 0 & -i & 0 \\ 0 & 0 & 0 & 1 \\ 3i & 0 & 0 & 0 \end{bmatrix}$, $v = \begin{bmatrix} 1 \\ 2 \\ 3 \\ 4 \end{bmatrix}$, $w = \begin{bmatrix} 4i \\ -3i \\ 2i \\ i \end{bmatrix}$.

5. For the given v, find the form of all elements w in $v^\perp$.

(a) $v = \begin{bmatrix} 1 \\ -1 \\ 2 \\ 3 \end{bmatrix}$.

(b) $v = \begin{bmatrix} 1 \\ -i \\ 2 \\ 3i \end{bmatrix}$.

(c) $v = \begin{bmatrix} 1 \\ 0 \\ \vdots \\ 0 \end{bmatrix}$.

(d) $v = \begin{bmatrix} 0 \\ i \\ 0 \\ \vdots \\ 0 \end{bmatrix}$.

6. For the vectors $v = \begin{bmatrix} 1 \\ i \\ 2 \\ 0 \end{bmatrix}$, $w = \begin{bmatrix} 7-i \\ 4 \\ 3i \\ 2-i \end{bmatrix}$, recalling that the length of a vector u, $\|u\|$, is $\sqrt{(u,u)}$, show that $\|v + w\| \le \|v\| + \|w\|$.

7. Find all matrices A such that for $v = \begin{bmatrix} 1 \\ 2 \\ 3 \end{bmatrix}$ and $w = \begin{bmatrix} 3i \\ 2i \\ i \end{bmatrix}$, $(Av, Aw) = (v, w)$.

8. If $v, w \in \mathbb{C}^{(3)}$, compute (Av, w) and (v, A^*w), where $A = \begin{bmatrix} 0 & i & 0 \\ 2i & 0 & 0 \\ 0 & 3 & 3i \end{bmatrix}$, and show that $(Av, w) = (v, A^*w)$.

9. Prove that the following matrices are unitary by calculating A^*A.

(a) $A = \begin{bmatrix} 0 & 1 & 0 & 0 \\ 1 & 0 & 0 & 0 \\ 0 & 0 & 0 & 1 \\ 0 & 0 & 1 & 0 \end{bmatrix}$.

(b) $A = \begin{bmatrix} 1-i & 0 \\ 0 & 2+i \end{bmatrix} \begin{bmatrix} 1+i & 0 \\ 0 & 2-i \end{bmatrix}^{-1}$.

(c) $A = \begin{bmatrix} 0 & 0 & 0 & i \\ 0 & 0 & i & 0 \\ 0 & i & 0 & 0 \\ i & 0 & 0 & 0 \end{bmatrix}$.

MORE THEORETICAL PROBLEMS

Easier Problems

10. Write out a complete proof of Theorem 3.5.4, following in the footsteps of what we did in Chapter 1.

11. Repeat Problem 10 for Theorem 3.5.5.

12. Repeat Problem 10 for Theorems 3.5.6, 3.5.7, and 3.5.8.

13. If $A \in M_n(\mathbb{R})$ is such that $(Av, Av) = (v, v)$ for all $v \in \mathbb{R}^{(n)}$, is A^*A necessarily equal to I? Either prove or produce a counterexample.

14. If A is both Hermitian and unitary, what possible values can the characteristic roots of A have?

15. If A is Hermitian and B is unitary, prove that BAB^{-1} is Hermitian.

16. If A is skew-Hermitian, show that the characteristic roots of A are pure imaginary.

17. If $A \in M_n(\mathbb{C})$ and $a \in \mathbb{C}$ is a characteristic root of A, prove that $aI - A$ is *not* invertible.

18. If $A \in M_n(\mathbb{C})$ and $a \in \mathbb{C}$ is a characteristic root of A, prove that for some matrix $B \neq 0$ in $M_n(\mathbb{C})$, $AB = aB$.

Middle-Level Problems

19. Let A be a real symmetric matrix. Prove that if a is a characteristic root of A, then a is real, and that there exists a vector $v \neq 0$ *having all its components real* such that $Av = av$.

20. Call a matrix *normal* if $AA^* = A^*A$. If A is normal and invertible, show that $B = A^*A^{-1}$ is unitary.

21. Is the result in Problem 20 correct if we drop the assumption that A is normal? Either prove or produce a counterexample.

22. If A is skew-Hermitian, show that both $(I - A)$ and $(I + A)$ are invertible and that $C = (I - A)(I + A)^{-1}$ is unitary.

23. If A and B are unitary, prove that AB is unitary and A^{-1} is also unitary.

24. Show that the matrix $A = \begin{bmatrix} 0 & 0 & 1 & 0 \\ 1 & 0 & 0 & 0 \\ 0 & 1 & 0 & 0 \\ 0 & 0 & 0 & 1 \end{bmatrix}$ is unitary. Can you find the characteristic roots of A?

25. Let the entries of $A = \begin{bmatrix} a & b & c \\ b & d & e \\ c & e & f \end{bmatrix}$ be real. Suppose that $g \in \mathbb{C}$ and there exists $v = \begin{bmatrix} x \\ y \\ z \end{bmatrix} \in M_3(\mathbb{C})$ such that $v \neq 0$ and $Av = gv$. Show by direct computations that g is real and we can pick v so that its components are real.

26. Let $A \in M_n(\mathbb{C})$. Define a new function $\langle \cdot, \cdot \rangle$ by $\langle v, w \rangle = (Av, w)$ for all $v, w \in \mathbb{C}^{(n)}$. What conditions must A satisfy in order that $\langle \cdot, \cdot \rangle$ satisfies all the properties of an inner product as contained in the statement of Lemma 3.5.1?

27. If $\|v\| = \sqrt{(v,v)}$, prove the *triangle inequality* $\|v + w\| \leq \|v\| + \|w\|$.

3.6. BASES OF $F^{(n)}$

In studying $F^{(n)}$, where F is the real, $\mathbb{R}$, or the complexes, $\mathbb{C}$, the set of elements

$$e_1 = \begin{bmatrix} 1 \\ 0 \\ \vdots \\ 0 \end{bmatrix}, \; e_2 = \begin{bmatrix} 0 \\ 1 \\ \vdots \\ 0 \end{bmatrix}, \ldots, e_n = \begin{bmatrix} 0 \\ 0 \\ \vdots \\ 1 \end{bmatrix},$$

where e_j is that element whose jth component is 1 and all of whose other components are 0, played a special role. Their most important property was that given any $v \in F^{(n)}$, then

$$v = a_1 e_1 + a_2 e_2 + \cdots + a_n e_n,$$

where $a_1, a_2, \ldots, a_n$ are in F. Moreover, these a_r are unique in the sense that if

$$v = a_1 e_1 + a_2 e_2 + \cdots + a_n e_n = b_1 e_1 + b_2 e_2 + \cdots + b_n e_n,$$

where the b_j are in F, then

$$a_1 = b_1, a_2 = b_2, \ldots, a_n = b_n.$$

These properties are of great importance and we will abstract from them the notions of *linear independence*, *span*, and *basis*.

If $v_1, \ldots, v_r$ are elements of $F^{(n)}$ and $a_1, \ldots, a_r$ are elements of F, a vector $a_1 v_1 + \cdots + a_r v_r$ is called a *linear combination* of $v_1, \ldots, v_r$ over F. For example, any vector $v = \begin{bmatrix} a \\ b \\ c \end{bmatrix}$ of $F^{(3)}$ is a linear combination $ae_1 + be_2 + ce_3$ of e_1, e_2, e_3 over F, and $F^{(3)}$ is the set of all such linear combinations. The only linear combination $ae_1 + be_2 + ce_3$ of e_1, e_2, e_3 which is 0 is the *trivial* linear combination $0e_1 + 0e_2 + 0e_3$, so that e_1, e_2, e_3 are linearly independent in the sense of

Definition. The elements $v_1, v_2, \ldots, v_r$ in $F^{(n)}$ are said to be *linearly independent over F* if $a_1 v_1 + a_2 v_2 + \cdots + a_r v_r = 0$ only if each of $a_1, a_2, \ldots, a_r$ is 0.

If $v_1, v_2, \ldots, v_r \in F^{(n)}$ are linearly independent over F, suppose that $v = a_1 v_1 + \cdots + a_r v_r = b_1 v_1 + \cdots + b_r v_r$, where the a_j's and b_j's are in F. Then $(a_1 - b_1) v_1 + \cdots + (a_r - b_r) v_r = 0$, which since the elements $v_1, \ldots, v_r$ are linearly independent over F, forces $a_1 - b_1 = 0, \ldots, a_r - b_r = 0$, that is, $a_1 = b_1, a_2 = b_2, \ldots, a_r = b_r$. So the a_j *are unique*. This could serve as an alternative definition of linear independence.

EXAMPLE

If $F = \mathbb{R}$ and $z_1 = \begin{bmatrix} 1 \\ 2 \\ 3 \end{bmatrix}$, $z_2 = \begin{bmatrix} -1 \\ 3 \\ -3 \end{bmatrix}$, $z_3 = \begin{bmatrix} 0 \\ 1 \\ 1 \end{bmatrix}$ in $F^{(3)}$, are they linearly independent over F? We must see if we can find a_1, a_2, a_3, not all 0, such that $a_1 z_1 + a_2 z_2 + a_3 z_3 = 0$. Since

$$a_1 z_1 + a_2 z_2 + a_3 z_3 = a_1 \begin{bmatrix} 1 \\ 2 \\ 3 \end{bmatrix} + a_2 \begin{bmatrix} -1 \\ 3 \\ -3 \end{bmatrix} + a_3 \begin{bmatrix} 0 \\ 1 \\ 1 \end{bmatrix} = \begin{bmatrix} a_1 - a_2 \\ 2a_1 + 3a_2 + a_3 \\ 3a_1 - 3a_2 + a_3 \end{bmatrix},$$

for this to be 0 we must satisfy:

$$a_1 - a_2 = 0$$
$$2a_1 + 3a_2 + a_3 = 0$$
$$3a_1 - 3a_2 + a_3 = 0.$$

You can check that the only solution to these three equations is $a_1 = 0, a_2 = 0$, and $a_3 = 0$. Thus z_1, z_2, z_3 are linearly independent over $F = \mathbb{R}$.

Can we always determine easily whether vectors $v_1, \ldots, v_r$ are linearly independent? Write $v_1 = \begin{bmatrix} a_{11} \\ \vdots \\ a_{n1} \end{bmatrix}$, $v_2 = \begin{bmatrix} a_{12} \\ \vdots \\ a_{n2} \end{bmatrix}$, $\ldots$, $v_r = \begin{bmatrix} a_{1r} \\ \vdots \\ a_{nr} \end{bmatrix}$, so that the linear combination $x_1 v_1 + x_2 v_2 + \cdots + x_r v_r$ becomes

$$x_1 \begin{bmatrix} a_{11} \\ \vdots \\ a_{n1} \end{bmatrix} + x_2 \begin{bmatrix} a_{12} \\ \vdots \\ a_{n2} \end{bmatrix} + \cdots + x_r \begin{bmatrix} a_{1r} \\ \vdots \\ a_{nr} \end{bmatrix} = \begin{bmatrix} x_1 a_{11} + x_2 a_{12} + \cdots + x_r a_{1r} \\ \vdots \\ x_1 a_{n1} + x_2 a_{n2} + \cdots + x_r a_{nr} \end{bmatrix}$$

$$= \begin{bmatrix} a_{11} & \cdots & a_{1r} \\ \vdots & & \vdots \\ a_{n1} & \cdots & a_{nr} \end{bmatrix} \begin{bmatrix} x_1 \\ \vdots \\ x_r \end{bmatrix}.$$

Then the v_j are the columns of the matrix $\begin{bmatrix} a_{11} & \cdots & a_{1r} \\ \vdots & & \vdots \\ a_{n1} & \cdots & a_{nr} \end{bmatrix}$ and they are linearly independent if and only if $\begin{bmatrix} 0 \\ \vdots \\ 0 \end{bmatrix}$ is the *only* solution to the equation

$$\begin{bmatrix} a_{11} & \cdots & a_{1r} \\ \vdots & & \vdots \\ a_{n1} & \cdots & a_{nr} \end{bmatrix} \begin{bmatrix} x_1 \\ \vdots \\ x_r \end{bmatrix} = \begin{bmatrix} 0 \\ \vdots \\ 0 \end{bmatrix}.$$

Sec. 3.6] Bases of $F^{(n)}$ 133

This, in turn, is equivalent to the condition that an echelon matrix row equivalent to
$\begin{bmatrix} a_{11} & \cdots & a_{1r} \\ \vdots & & \vdots \\ a_{n1} & \cdots & a_{nr} \end{bmatrix}$ has r nonzero rows, by Corollary 2.4.2. So we have

Method to determine if $v_1, \ldots, v_r$ in $F^{(n)}$ are linearly independent

1. Form the matrix $A = [v_1, \ldots, v_r]$ whose columns are the v_i and row reduce A to an echelon matrix B.
2. Then the vectors $v_1, \ldots, v_r$ are linearly independent if and only if B has r nonzero rows.

EXAMPLE

Let's now use *this* method to determine whether the vectors

$$\begin{bmatrix} 1 \\ 2 \\ 3 \end{bmatrix}, \begin{bmatrix} -1 \\ 3 \\ -3 \end{bmatrix}, \begin{bmatrix} 0 \\ 1 \\ 1 \end{bmatrix}$$

considered in the preceding example are linearly independent. We apply elementary row operations to the matrix $\begin{bmatrix} 1 & -1 & 0 \\ 2 & 3 & 1 \\ 3 & -3 & 1 \end{bmatrix}$, getting

$$\begin{bmatrix} 1 & -1 & 0 \\ 0 & 5 & 1 \\ 3 & -3 & 1 \end{bmatrix}, \begin{bmatrix} 1 & -1 & 0 \\ 0 & 5 & 1 \\ 0 & 0 & 1 \end{bmatrix}, \begin{bmatrix} 1 & -1 & 0 \\ 0 & 1 & \frac{1}{5} \\ 0 & 0 & 1 \end{bmatrix}.$$

Since the last matrix we get is an echelon matrix with three nonzero rows, the three given vectors are linearly independent.

If the elements $v_1, v_2, \ldots, v_r$ are not linearly independent over F, they are called *linearly dependent over F*.

EXAMPLE

Let's also use our method to determine whether the vectors

$$\begin{bmatrix} 1 \\ 1 \\ 1 \end{bmatrix}, \begin{bmatrix} -i \\ 1-i \\ 0 \end{bmatrix}, \begin{bmatrix} 0 \\ i \\ -1 \end{bmatrix}$$

are linearly independent over $\mathbb{C}$. We apply elementary row operations to the matrix $\begin{bmatrix} 1 & -i & 0 \\ 1 & 1-i & i \\ 0 & & -1 \end{bmatrix}$, getting

$$\begin{bmatrix} 1 & -i & 1 \\ 0 & 1 & i \\ 1 & 0 & -1 \end{bmatrix}, \begin{bmatrix} 1 & -i & 0 \\ 0 & 1 & i \\ 0 & i & -1 \end{bmatrix}, \begin{bmatrix} 1 & -i & 0 \\ 0 & 1 & i \\ 0 & 0 & 0 \end{bmatrix}.$$

Since the last matrix we get is an echelon matrix with only two nonzero rows, the three given vectors are linearly dependent over $\mathbb{C}$. If we so wish, we can easily get a nonzero linear combination by solving the equation $\begin{bmatrix} 1 & -i & 0 \\ 0 & 1 & i \\ 0 & 0 & 0 \end{bmatrix} \begin{bmatrix} a \\ b \\ c \end{bmatrix} = 0$, getting a nonzero solution $\begin{bmatrix} a \\ b \\ c \end{bmatrix} = \begin{bmatrix} 1 \\ -i \\ 1 \end{bmatrix}$. So we can use the values $1, -i, 1$ to express 0 as a nontrivial linear combination of $\begin{bmatrix} 1 \\ 1 \\ 1 \end{bmatrix}, \begin{bmatrix} -i \\ 1-i \\ 0 \end{bmatrix}, \begin{bmatrix} 0 \\ i \\ -1 \end{bmatrix}$:

$$0 = 1\begin{bmatrix} 1 \\ 1 \\ 1 \end{bmatrix} - i\begin{bmatrix} -i \\ 1-i \\ 0 \end{bmatrix} + 1\begin{bmatrix} 0 \\ i \\ -1 \end{bmatrix}.$$

Since any vector $v = \begin{bmatrix} a \\ b \\ c \end{bmatrix}$ of $F^{(3)}$ is a linear combination $ae_1 + be_2 + ce_3$ of e_1, e_2, e_3 over F and $F^{(3)}$ is the set of all such linear combinations, $\langle e_1, e_2, e_3 \rangle = F^{(3)}$ and e_1, e_2, e_3 span $F^{(3)}$ in the sense of the

Definition. The set of all linear combinations of $v_1, \ldots, v_r$ over F is denoted $\langle v_1, \ldots, v_r \rangle$ and is called the *span* of $v_1, \ldots, v_r$. If V is the set $\langle v_1, \ldots, v_r \rangle$, we say that *the vectors* $v_1, \ldots, v_r$ *span* V.

How do we know whether a vector v is in the span of vectors $v_1 = \begin{bmatrix} a_{11} \\ \vdots \\ a_{n1} \end{bmatrix}$, $v_2 = \begin{bmatrix} a_{12} \\ \vdots \\ a_{n2} \end{bmatrix}, \ldots, v_r = \begin{bmatrix} a_{1r} \\ \vdots \\ a_{nr} \end{bmatrix}$? The linear combination $x_1 v_1 + x_2 v_2 + \cdots + x_r v_r$ is

$$\begin{bmatrix} a_{11} & \cdots & a_{1r} \\ \vdots & & \vdots \\ a_{n1} & \cdots & a_{nr} \end{bmatrix} \begin{bmatrix} x_1 \\ \vdots \\ x_r \end{bmatrix},$$

Sec. 3.6]   Bases of $F^{(n)}$ 135

as we've just seen. So, expressing v as a linear combination of $\begin{bmatrix} a_{11} \\ \vdots \\ a_{n1} \end{bmatrix}, \ldots, \begin{bmatrix} a_{1r} \\ \vdots \\ a_{nr} \end{bmatrix}$ is the same as finding a solution $\begin{bmatrix} x_1 \\ \vdots \\ x_r \end{bmatrix}$ to the equation $\begin{bmatrix} a_{11} & \cdots & a_{1r} \\ \vdots & & \vdots \\ a_{n1} & \cdots & a_{nr} \end{bmatrix} \begin{bmatrix} x_1 \\ \vdots \\ x_r \end{bmatrix} = v$.

So, by Section 2.4, we have a straightforward way of settling whether or not v is a linear combination of $v_1, \ldots, v_r$.

Method to express $v \in F^{(n)}$ as a linear combination of $v_1, \ldots, v_r \in F^{(n)}$

1. Form the matrix $A = [v_1, \ldots, v_r]$ whose columns are the v_i.
2. Form the augmented matrix $[A, v]$ and row reduce it to an echelon matrix $[B, w]$.
3. Then v is expressed as a linear combination $v = Ax$ of the v_i if and only if x is a solution to $Bx = w$.

EXAMPLE

Let's use this method to try to express the vectors $\begin{bmatrix} 1 \\ 2 \\ 6 \end{bmatrix}$ and $\begin{bmatrix} 1 \\ 2 \\ 3 \end{bmatrix}$ as linear combinations of the vectors $\begin{bmatrix} 1 \\ 1 \\ 2 \end{bmatrix}, \begin{bmatrix} 2 \\ 3 \\ 5 \end{bmatrix}, \begin{bmatrix} 3 \\ 5 \\ 8 \end{bmatrix}, \begin{bmatrix} 4 \\ 5 \\ 9 \end{bmatrix}$. We first form the matrix $A = \begin{bmatrix} 1 & 2 & 3 & 4 \\ 1 & 3 & 5 & 5 \\ 2 & 5 & 8 & 9 \end{bmatrix}$, which we will use for both vectors.

To try to express $\begin{bmatrix} 1 \\ 2 \\ 6 \end{bmatrix}$ as a linear combination of $\begin{bmatrix} 1 \\ 1 \\ 2 \end{bmatrix}, \begin{bmatrix} 2 \\ 3 \\ 5 \end{bmatrix}, \begin{bmatrix} 3 \\ 5 \\ 8 \end{bmatrix}, \begin{bmatrix} 4 \\ 5 \\ 9 \end{bmatrix}$, we form the augmented matrix $\begin{bmatrix} A, \begin{matrix} 1 \\ 2 \\ 6 \end{matrix} \end{bmatrix} = \begin{bmatrix} 1 & 2 & 3 & 4 & 1 \\ 1 & 3 & 5 & 5 & 2 \\ 2 & 5 & 8 & 9 & 6 \end{bmatrix}$ and row reduce it to the echelon matrix $\begin{bmatrix} 1 & 2 & 3 & 4 & 1 \\ 0 & 1 & 2 & 1 & 1 \\ 0 & 0 & 0 & 0 & 1 \end{bmatrix} = \begin{bmatrix} B, \begin{matrix} 1 \\ 1 \\ 1 \end{matrix} \end{bmatrix}$.

Since

$$\begin{bmatrix} 1 & 2 & 3 & 4 \\ 0 & 1 & 2 & 1 \\ 0 & 0 & 0 & 0 \end{bmatrix} x = \begin{bmatrix} 1 \\ 1 \\ 1 \end{bmatrix}$$

has no solution x, no linear combination $\begin{bmatrix} 1 & 2 & 3 & 4 \\ 1 & 3 & 5 & 5 \\ 2 & 5 & 8 & 9 \end{bmatrix} x$ of

$$\begin{bmatrix} 1 \\ 1 \\ 2 \end{bmatrix}, \begin{bmatrix} 2 \\ 3 \\ 5 \end{bmatrix}, \begin{bmatrix} 3 \\ 5 \\ 8 \end{bmatrix}, \begin{bmatrix} 4 \\ 5 \\ 9 \end{bmatrix}$$

equals $\begin{bmatrix} 1 \\ 2 \\ 6 \end{bmatrix}$.

To try to express $\begin{bmatrix} 1 \\ 2 \\ 3 \end{bmatrix}$ as a linear combination of $\begin{bmatrix} 1 \\ 1 \\ 2 \end{bmatrix}, \begin{bmatrix} 2 \\ 3 \\ 5 \end{bmatrix}, \begin{bmatrix} 3 \\ 5 \\ 8 \end{bmatrix}, \begin{bmatrix} 4 \\ 5 \\ 9 \end{bmatrix}$, we

form the augmented matrix $\left[A, \begin{matrix} 1 \\ 2 \\ 3 \end{matrix} \right] = \begin{bmatrix} 1 & 2 & 3 & 4 & 1 \\ 1 & 3 & 5 & 5 & 2 \\ 2 & 5 & 8 & 9 & 3 \end{bmatrix}$ and row reduce it to the

echelon matrix $\begin{bmatrix} 1 & 2 & 3 & 4 & 1 \\ 0 & 1 & 2 & 1 & 1 \\ 0 & 0 & 0 & 0 & 0 \end{bmatrix} = \left[B, \begin{matrix} 1 \\ 1 \\ 0 \end{matrix} \right]$, Since $\begin{bmatrix} 1 & 2 & 3 & 4 \\ 0 & 1 & 2 & 1 \\ 0 & 0 & 0 & 0 \end{bmatrix} x = \begin{bmatrix} 1 \\ 1 \\ 0 \end{bmatrix}$ has

the solution $x = \begin{bmatrix} -1 \\ 1 \\ 0 \\ 0 \end{bmatrix}$, the linear combination

$$\begin{bmatrix} 1 & 2 & 3 & 4 \\ 1 & 3 & 5 & 5 \\ 2 & 5 & 8 & 9 \end{bmatrix} \begin{bmatrix} -1 \\ 1 \\ 0 \\ 0 \end{bmatrix} = -1 \begin{bmatrix} 1 \\ 1 \\ 2 \end{bmatrix} + 1 \begin{bmatrix} 2 \\ 3 \\ 5 \end{bmatrix} + 0 \begin{bmatrix} 3 \\ 5 \\ 8 \end{bmatrix} + 0 \begin{bmatrix} 4 \\ 5 \\ 9 \end{bmatrix}$$

equals $\begin{bmatrix} 1 \\ 2 \\ 3 \end{bmatrix}$.

We could handle both cases *at once* by reducing the *doubly* augmented matrix $\begin{bmatrix} 1 & 2 & 3 & 4 & 1 & 1 \\ 1 & 3 & 5 & 5 & 2 & 2 \\ 2 & 5 & 8 & 9 & 6 & 3 \end{bmatrix}$ to the echelon matrix $\begin{bmatrix} 1 & 2 & 3 & 4 & 1 & 1 \\ 0 & 1 & 2 & 1 & 1 & 1 \\ 0 & 0 & 0 & 0 & 1 & 0 \end{bmatrix}$, which

shows at once that we must look for the solutions to $\begin{bmatrix} 1 & 2 & 3 & 4 \\ 0 & 1 & 2 & 1 \\ 0 & 0 & 0 & 0 \end{bmatrix} x = \begin{bmatrix} 1 \\ 1 \\ 1 \end{bmatrix}$

for the case $\begin{bmatrix} 1 \\ 2 \\ 6 \end{bmatrix}$ (there clearly are none) and to $\begin{bmatrix} 1 & 2 & 3 & 4 \\ 0 & 1 & 2 & 1 \\ 0 & 0 & 0 & 0 \end{bmatrix} x = \begin{bmatrix} 1 \\ 1 \\ 0 \end{bmatrix}$ for the case $\begin{bmatrix} 1 \\ 2 \\ 3 \end{bmatrix}$.

Now that we have defined linear independence and span, we can define the notion of basis of $F^{(n)}$ over F.

Definition. The elements $v_1, \ldots, v_r$ in $F^{(n)}$ form a *basis* of $F^{(n)}$ over F if:

$v_1, \ldots, v_r$ are linearly independent over F;

$v_1, \ldots, v_r$ span $F^{(n)}$.

From what we did above, we see that if $v_1, v_2, \ldots, v_r$ is a basis of $F^{(n)}$ over F and if $v \in F^{(n)}$, then $v = a_1 v_1 + a_2 v_2 + \cdots + a_r v_r$, where the a_j are in F, *in one and only one way*. In fact, we could use this as the definition of a basis of $F^{(n)}$ over F.

Theorem 3.6.1. *If a finite set of vectors spans $F^{(n)}$, then it has a subset $v_1, \ldots, v_r$ (which is a basis for $F^{(n)}$).*

Proof: Take $v_1, \ldots, v_r$ to be a subset with r as small as possible which spans $F^{(n)}$. To show that $v_1, \ldots, v_r$ is a basis for $F^{(n)}$, it suffices to show that $v_1, \ldots, v_r$ are linearly independent. Suppose, to the contrary, that there is a linear combination $x_1 v_1 + \cdots + x_r v_r = 0$ with $x_t \neq 0$. Then $v_1, \ldots, v_{t-1}, v_{t+1}, \ldots, v_r$ span $F^{(n)}$ since v_t can be expressed in terms of them. But this contradicts our choice of $v_1, \ldots, v_r$ as a spanning set with as few elements as possible. So there can be no such linear combination; that is, $v_1, \ldots, v_r$ are linearly independent. ∎

Of course, the columns $e_1, e_2, \ldots, e_n$ of the $n \times n$ identity matrix form a basis of $F^{(n)}$ over F, which we call the *canonical basis* of $F^{(n)}$ over F.

Why, in our definition of basis for $F^{(n)}$, did we not simply use n instead of r? Are there bases $v_1, \ldots, v_r$ for $F^{(n)}$ where r is not equal to n? The answer is "no"! Why, then, do we not use n instead of r? We will *prove* that r equals n. We then know that r and n are equal without having to *assume* so. This has many and great consequences.

Theorem 3.6.2. *Let $v_1, \ldots, v_r$ be a linear independent set of vectors in $F^{(n)}$ and suppose that the vectors $v_1, \ldots, v_r$ are contained in $\langle w_1, \ldots, w_s \rangle$. Then $r \leq s$.*

Proof: We can show instead that if $r > s$, then we can find a linear dependence $x_1 v_1 + x_2 v_2 + \cdots + x_r v_r = 0$ (not all of the $x_1, \ldots, x_r$ are 0). To do this, write

$$v_1 = a_{11} w_1 + a_{21} w_2 + \cdots + a_{s1} w_s$$
$$\vdots \qquad \qquad \vdots$$
$$v_r = a_{1r} w_1 + a_{2r} w_2 + \cdots + a_{sr} w_s.$$

Then the equation $0 = x_1v_1 + x_2v_2 + \cdots + x_rv_r$ becomes

$$0 = x_1a_{11}w_1 + x_1a_{21}w_2 + \cdots + x_1a_{s1}w_s +$$
$$\vdots \qquad \qquad \vdots$$
$$x_ra_{1r}w_1 + x_ra_{2r}w_2 + \cdots + x_ra_{sr}w_s$$
$$= (a_{11}x_1 + a_{12}x_2 + \cdots + a_{1r}x_r)w_1 +$$
$$\vdots \qquad \qquad \vdots$$
$$(a_{s1}x_1 + a_{s2}x_2 + \cdots + a_{sr}x_r)w_s.$$

Since $r > s$, the system of equations

$$a_{11}x_1 + a_{12}x_2 + \cdots + a_{1r}x_r = 0$$
$$\vdots \qquad \qquad \vdots$$
$$a_{s1}x_1 + a_{s2}x_2 + \cdots + a_{sr}x_r = 0$$

has a nonzero solution, by Corollary 2.4.3, so that $x_1v_1 + x_2v_2 + \cdots + x_rv_r = 0$. So they are linearly dependent over F. ∎

The following corollaries are straightforward; you may do them as easy exercises.

Corollary 3.6.3. Let $v_1,\ldots,v_r$ and $w_1,\ldots,w_s$ be two linear independent sets of vectors in $F^{(n)}$ such that the spans $\langle v_1,\ldots,v_r \rangle$, $\langle w_1,\ldots,w_s \rangle$ are equal. Then $r = s$.

Corollary 3.6.4. Any basis for $F^{(n)}$ has n elements.

Corollary 3.6.5. Any set of n linearly independent elements of $F^{(n)}$ is a basis of $F^{(n)}$.

Proof: If $v_1,\ldots,v_n$ are linearly independent elements of $F^{(n)}$ and if v is an arbitrary element of $F^{(n)}$, we must show that v is a linear combination of $v_1,\ldots,v_n$. Suppose not. Then the vectors $v_1,\ldots,v_n, v$ are linearly dependent by Theorem 3.6.2. Letting $x_1v_1 + \cdots + x_nv_n + x_{n+1}v = 0$ where not all of the x_s are 0, we must have $x_{n+1} \neq 0$. Why? If $x_{n+1} = 0$, then $x_1v_1 + \cdots + x_nv_n = 0$, which by the linear independence of $v_1,\ldots,v_n$, forces that $x_1 = x_2 = \cdots = x_n = 0$. In other words, all the coefficients are 0, which is not the case. But then, since $x_{n+1} \neq 0$:

$$v = -\frac{1}{x_{n+1}}(x_1v_1 + \cdots + x_nv_n)$$
$$= \left(-\frac{x_1}{x_{n+1}}\right)v_1 + \cdots + \left(-\frac{x_n}{x_{n+1}}\right)v_n.$$

Corollary 3.6.6. Any set of n elements that spans $F^{(n)}$ is a basis of $F^{(n)}$.

Proof: By Theorem 3.6.1, any set of n elements that spans $F^{(n)}$ contains a basis for $F^{(n)}$ which, by Corollary 3.6.4, has n elements. So the set itself is a basis. ∎

PROBLEMS

NUMERICAL PROBLEMS

1. Determine if the following elements are linearly independent or linearly dependent.

   (a) $\begin{bmatrix} 1 \\ 1 \\ 0 \end{bmatrix}, \begin{bmatrix} 1 \\ 0 \\ 1 \end{bmatrix}, \begin{bmatrix} 0 \\ 1 \\ 1 \end{bmatrix}$ in $\mathbb{C}^{(3)}$.

   (b) $\begin{bmatrix} 1 \\ i \\ 0 \end{bmatrix}, \begin{bmatrix} i \\ 0 \\ 1 \end{bmatrix}, \begin{bmatrix} 0 \\ i \\ 1 \end{bmatrix}, \begin{bmatrix} i \\ i \\ i \end{bmatrix}$ in $\mathbb{C}^{(3)}$.

   (c) $\begin{bmatrix} 1 \\ 2 \end{bmatrix}, \begin{bmatrix} 2 \\ 1 \end{bmatrix}$ in $\mathbb{R}^{(2)}$.

   (d) $\begin{bmatrix} 1 \\ 2 \\ 3 \\ 4 \end{bmatrix}, \begin{bmatrix} i \\ -i \\ 1 \\ 4 \end{bmatrix}, \begin{bmatrix} 2i \\ 4i \\ 6i \\ 8i \end{bmatrix}$ in $\mathbb{C}^{(4)}$.

   (e) $\begin{bmatrix} 0 \\ 0 \\ 0 \\ 0 \end{bmatrix}, \begin{bmatrix} 1 \\ 1 \\ 0 \\ 0 \end{bmatrix}, \begin{bmatrix} 0 \\ 1 \\ 0 \\ 0 \end{bmatrix}$ in $\mathbb{R}^{(4)}$.

   (f) $\begin{bmatrix} 1 \\ 2 \\ 0 \end{bmatrix}, \begin{bmatrix} i \\ 0 \\ 1+i \end{bmatrix}, \begin{bmatrix} 2+i \\ 4 \\ 1+i \end{bmatrix}$ in $\mathbb{C}^{(3)}$.

2. Which of sets of the elements in Problem 1 form a basis of their respective $F^{(n)}$'s?

3. In $F^{(3)}$ verify that Ae_1, Ae_2, and Ae_3 form a basis over F, where

$$A = \begin{bmatrix} 1 & 1 & 3 \\ -1 & 0 & 0 \\ 0 & 0 & 2 \end{bmatrix}.$$

4. For what values of a in F do the elements Ae_1, Ae_2, Ae_3 *not* form a basis of $F^{(3)}$ over F if $A = \begin{bmatrix} 1 & 2 & 3 \\ -1 & 0 & 0 \\ 1 & 0 & a \end{bmatrix}$?

5. Let $A = \begin{bmatrix} 1 & 0 & 2 \\ 4 & -5 & 6 \\ 2 & 0 & 4 \end{bmatrix}$. Show that Ae_1, Ae_2, and Ae_3 do *not* form a basis of $F^{(3)}$ over F.

6. In $F^{(2)}$ show that any three elements must be linearly dependent over F.
7. In $F^{(3)}$ show that any four elements are linearly dependent over F.
8. If $\begin{bmatrix} 1 & 0 & a \\ 2 & 1 & b \\ 3 & 2 & c \end{bmatrix}$ is not invertible, show that the elements

$$\begin{bmatrix} 1 \\ 2 \\ 3 \end{bmatrix}, \begin{bmatrix} 0 \\ 1 \\ 2 \end{bmatrix}, \begin{bmatrix} a \\ b \\ c \end{bmatrix}$$

must be linearly dependent.

9. Show that if the matrix in Problem 8 is invertible, then the vectors

$$\begin{bmatrix} 1 \\ 2 \\ 3 \end{bmatrix}, \begin{bmatrix} 0 \\ 1 \\ 2 \end{bmatrix}, \begin{bmatrix} a \\ b \\ c \end{bmatrix}$$

are linearly independent.

MORE THEORETICAL PROBLEMS

Easier Problems

10. Prove Corollaries 3.6.3 and 3.6.4.

3.7. CHANGE OF BASIS OF $F^{(n)}$

Starting with the canonical basis $e_1, \ldots, e_n$ as one basis for $F^{(n)}$ over F, let's also take $w_1, \ldots, w_n$ to be another basis of $F^{(n)}$ over F and consider the matrix C whose columns are $w_1, \ldots, w_n$.

Definition. We call C the *matrix of the change of basis* from the basis $e_1, \ldots, e_n$ to the basis $w_1, \ldots, w_n$.

EXAMPLE

Recall from an example in Section 3.6 with $F = \mathbb{R}$ and

$$z_1 = \begin{bmatrix} 1 \\ 2 \\ 3 \end{bmatrix}, z_2 = \begin{bmatrix} -1 \\ 3 \\ -3 \end{bmatrix}, z_3 = \begin{bmatrix} 0 \\ 1 \\ 1 \end{bmatrix}$$

that z_1, z_2, z_3 are linearly independent over $\mathbb{R}$. Thus z_1, z_2, z_3 forms a basis for

$R^{(3)}$. (Prove!) The corresponding matrix of the change of basis is

$$C = \begin{bmatrix} 1 & -1 & 0 \\ 2 & 3 & 1 \\ 3 & -3 & 1 \end{bmatrix}.$$

What properties does the matrix C of the change of basis enjoy? Can it be any old matrix? No! As we now show, C must be an *invertible* matrix.

Theorem 3.7.1. *If C is the matrix of the change of basis from $e_1, \ldots, e_n$ to the basis $w_1, \ldots, w_n$, then C is invertible.*

Proof: By the definition of C, $Ce_1 = w_1, \ldots, Ce_n = w_n$. Define the mapping $S: F^{(n)} \to F^{(n)}$ as follows: Given $v \in F^{(n)}$, express v as $v = b_1 w_1 + \cdots + b_n w_n$ with the $b_j \in F$ and define S by $Sv = b_1 e_1 + \cdots + b_n e_n$. Note that $Sw_r = e_r$. We leave it to the reader to show that S is also a linear transformation on $F^{(n)}$, and so is a matrix (s_{rs}). Since $(SC)e_r = S(Ce_r) = Sw_r = e_r$ we see that SC is the identity mapping as far as the elements $e_1, \ldots, e_n$ are concerned. But this implies that SC is the identity mapping on $F^{(n)}$, since $e_1, \ldots, e_n$ form a basis of $F^{(n)}$ over F. (Prove!) But then we have that $(s_{rs})(c_{rs}) = I$. Similarly, we have $(c_{rs})(s_{rs})$, since $CSw_r = Ce_r = w_r$ and the w_r also form a basis. Since $SC = CS = I$, C is invertible with inverse S. ∎

So given any basis $w_1, \ldots, w_n$ of $F^{(n)}$ over F, the matrix C whose columns are $w_1, \ldots, w_n$ is an invertible matrix. On the other hand, if B is an invertible matrix in $M_n(F)$ we claim that its columns $z_1 = Be_1, \ldots, z_n = Be_n$ form a basis of $F^{(n)}$ over F. Why? To begin with, since B is invertible, we know that given $v \in F^{(n)}$, then $v = (BB^{-1})v = B(B^{-1}v)$, so $v = Bw$, where $w = B^{-1}v$ is in $F^{(n)}$. Furthermore, $w = a_1 e_1 + \cdots + a_n e_n$, where the a_r are in F and are unique. Thus

$$v = Bw = B(a_1 e_1 + \cdots + a_n e_n) = a_1 Be_1 + \cdots + a_n Be_n = a_1 z_1 + \cdots + a_n z_n.$$

So every element in $F^{(n)}$ is realizable in the form $a_1 z_1 + \cdots + a_n z_n$, the second requisite condition defining a basis.

To prove that $z_1, \ldots, z_n$ is indeed a basis of $F^{(n)}$ over F, we must verify that $b_1 z_1 + \cdots + b_n z_n = 0$ forces $b_1 = b_2 = \cdots = b_n = 0$. But if $b_1 z_1 + \cdots + b_n z_n = 0$, then $0 = A(b_1 z_1 + \cdots + b_n z_n)$, where $A = B^{-1}$. Since $Be_r = z_r$, we have that $Az_r = e_r$ for every r. Therefore,

$$0 = A(b_1 z_1 + \cdots + b_n z_n) = b_1 Az_1 + \cdots + b_n Az_n = b_1 e_1 + \cdots + b_n e_n.$$

Because $e_1, \ldots, e_n$ is a basis of $F^{(n)}$ over F we have that each $b_j = 0$. So $z_1, \ldots, z_n$ are linearly independent over F. So we see that $z_1 = Be_1, \ldots, z_n = Be_n$ form a basis of $F^{(n)}$ over F.

Combining what we just did with Theorem 3.7.1, we get a description of *all possible* bases of $F^{(n)}$ over F, namely,

Theorem 3.7.2. *Given any invertible matrix A in $M_n(F)$, then $w_1 = Ae_1, \ldots, w_n = Ae_n$*

form a basis of $F^{(n)}$ over F. Moreover, given any basis $z_1, \ldots, z_n$ of $F^{(n)}$ over F, then for some invertible matrix A in $M_n(F)$, $z_r = Ae_r$ for $1 \leq r \leq n$.

PROBLEMS

NUMERICAL PROBLEMS

1. Show that $\begin{bmatrix} 1 \\ 1 \\ 0 \\ 1 \end{bmatrix}, \begin{bmatrix} 1 \\ 0 \\ 1 \\ 1 \end{bmatrix}, \begin{bmatrix} 0 \\ 0 \\ 1 \\ 1 \end{bmatrix}, \begin{bmatrix} 0 \\ 1 \\ 0 \\ 2 \end{bmatrix}$ form a basis of $F^{(4)}$ over F.

2. Find the matrix of the change of basis from the basis $e_1, \ldots, e_n$ to that of Problem 1.

3. Show that we can make $\begin{bmatrix} 0 \\ 0 \\ 1 \\ 1 \end{bmatrix}$ the first column of some matrix A in $M_4(F)$ such that A is invertible.

4. Show that we can make $\begin{bmatrix} 1 \\ 1 \\ 0 \\ 1 \end{bmatrix}, \begin{bmatrix} 0 \\ 1 \\ 0 \\ 1 \end{bmatrix}$ the first and second columns of some matrix A in $M_4(F)$ such that A is invertible.

5. Show that we can make

$$\begin{bmatrix} 1 \\ 2 \\ 3 \\ 4 \end{bmatrix}, \begin{bmatrix} 0 \\ 0 \\ 1 \\ 1 \end{bmatrix}, \begin{bmatrix} 0 \\ 0 \\ 0 \\ 1 \end{bmatrix}$$

the first, second, and fourth columns of some matrix A in $M_4(F)$ such that A is invertible.

6. Determine for which real numbers a, b, c, d the four vectors

$$a\begin{bmatrix} 1 \\ 1 \\ 0 \\ 1 \end{bmatrix} + b\begin{bmatrix} 1 \\ 0 \\ 1 \\ 1 \end{bmatrix} + c\begin{bmatrix} 0 \\ 0 \\ 1 \\ 1 \end{bmatrix} + d\begin{bmatrix} 0 \\ 1 \\ 1 \\ 1 \end{bmatrix}, \begin{bmatrix} 1 \\ 0 \\ 0 \\ 2 \end{bmatrix}, \begin{bmatrix} 0 \\ 0 \\ 1 \\ 1 \end{bmatrix}, \begin{bmatrix} 0 \\ 1 \\ 0 \\ 2 \end{bmatrix}$$

are linearly independent.

7. Determine for which real numbers a, b, c, d the four vectors

$$a\begin{bmatrix}1\\1\\0\\1\end{bmatrix} + b\begin{bmatrix}1\\0\\1\\1\end{bmatrix}, \quad c\begin{bmatrix}0\\0\\1\\1\end{bmatrix} + d\begin{bmatrix}0\\1\\0\\2\end{bmatrix}, \quad \begin{bmatrix}1\\0\\1\\1\end{bmatrix}, \quad \begin{bmatrix}0\\1\\0\\2\end{bmatrix}$$

are linearly independent.

8. Compute the matrices $A = \begin{bmatrix}0&1&0&0\\c&0&0&d\\1&0&0&0\\0&0&1&0\end{bmatrix}^2$ and $B = \begin{bmatrix}0&1&0&0\\c&0&0&d\\1&0&0&0\\0&0&1&0\end{bmatrix}^3$ and determine the values of c and d for which their columns are bases of $F^{(4)}$.

9. For what values of c and d are the vectors

$$\begin{bmatrix}0&1&0&0\\c&0&0&d\\1&0&0&0\\0&0&1&0\end{bmatrix}^2\begin{bmatrix}1\\1\\0\\1\end{bmatrix}, \quad \begin{bmatrix}0&1&0&0\\c&0&0&d\\1&0&0&0\\0&0&1&0\end{bmatrix}^2\begin{bmatrix}1\\0\\1\\2\end{bmatrix},$$

$$\begin{bmatrix}0&1&0&0\\c&0&0&d\\1&0&0&0\\0&0&1&0\end{bmatrix}^2\begin{bmatrix}0\\0\\1\\1\end{bmatrix}, \quad \begin{bmatrix}0&1&0&0\\c&0&0&d\\1&0&0&0\\0&0&1&0\end{bmatrix}^2\begin{bmatrix}0\\1\\0\\2\end{bmatrix}$$

linearly independent?

10. Show by direct methods that the columns of the matrix $\begin{bmatrix}1&3&4&e\\3&1&4&d\\1&0&0&0\\0&1&0&0\end{bmatrix}$ are a basis for $F^{(4)}$ if and only if the columns of the matrix $\begin{bmatrix}1&3&1&0\\3&1&0&1\\4&4&0&0\\e&d&0&0\end{bmatrix}$ are a basis for $F^{(10)}$.

MORE THEORETICAL PROBLEMS

Easier Problems

11. Find the matrix of the change of basis from the basis $e_1, \ldots, e_n$ to the basis $w_1 = e_2$, $w_2 = e_3, \ldots, w_{n-1} = e_n, w_n = e_1$.

12. If $w \neq 0$ is in $F^{(3)}$, show that we can find elements w_2 and w_3 in $F^{(3)}$ so that w, w_2, w_3 form a basis of $F^{(3)}$ over F.

13. If $A = (a_{jk})$ is in $M_4(F)$, show that A is invertible if and only if

$$\begin{bmatrix} a_{11} \\ a_{21} \\ a_{31} \\ a_{41} \end{bmatrix}, \begin{bmatrix} a_{12} \\ a_{22} \\ a_{32} \\ a_{42} \end{bmatrix}, \begin{bmatrix} a_{13} \\ a_{23} \\ a_{33} \\ a_{43} \end{bmatrix}, \begin{bmatrix} a_{14} \\ a_{24} \\ a_{34} \\ a_{44} \end{bmatrix}$$

are linearly independent over F.

14. Given $w \neq 0$ in $F^{(4)}$, show that we can make w the first column of some matrix A in $M_4(F)$ such that A is invertible.

15. If A in Problem 14 is found, what can you say about the columns of A?

16. Do Problem 14 for the explicit vector $w = \begin{bmatrix} i \\ 0 \\ -i \\ 5+i \end{bmatrix}$ in $\mathbb{C}^{(4)}$.

17. Find a basis of $\mathbb{C}^{(4)}$ over $\mathbb{C}$ in which the vector $w = \begin{bmatrix} i \\ 0 \\ -i \\ 5+i \end{bmatrix}$ is the first basis element.

Middle-Level Problems

18. Given $w \neq 0$ in $F^{(n)}$, show that you can find elements $w_2, \ldots, w_n$ in $F^{(n)}$ such that $w, w_2, \ldots, w_n$ form a basis of $F^{(n)}$ over F.

19. Generalize Problem 18 as follows: Given $v_1, \ldots, v_m$ in $F^{(n)}$ which are linearly independent over F, and $m < n$, show that you can find elements $w_{m+1}, \ldots, w_n$ in $F^{(n)}$ such that $v_1, \ldots, v_m, w_{m+1}, \ldots, w_n$ form a basis of $F^{(n)}$ over F.

20. A matrix P in $M_n(F)$ is called a *permutation matrix* if its nonzero entries consist of exactly one 1 in each row and in each column. What does P do to the canonical basis of $F^{(n)}$?

21. Using the results of Problem 20, show that a permutation matrix is invertible. What is its inverse?

22. In $F^{(4)}$ if w_1, w_2, w_3 are nonzero elements, show that there is some element $v \in F^{(4)}$ such that v cannot be expressed as $v = a_1 w_1 + a_2 w_2 + a_3 w_3$ for any choice of a_1, a_2, a_3 in F.

23. In $F^{(4)}$ show that any five elements are linearly dependent over F.

Harder Problems

24. Show that if $w_1, \ldots, w_m$ are in $F^{(n)}$, where $m < n$, then there is an element v in $F^{(n)}$ which cannot be expressed as $v = a_1 w_1 + \cdots + a_m w_m$ for any $a_1, \ldots, a_m$ in F.

25. If $w_1, \ldots, w_n$ in $F^{(n)}$ are such that *every* $v \in F^{(n)}$ can be expressed as

$$v = a_1 w_1 + \cdots + a_n w_n,$$

where $a_1, \ldots, a_n \in F$, show that $w_1, \ldots, w_n$ is a basis for $F^{(n)}$.

26. If P and Q are permutation matrices in $M_n(F)$, prove that PQ is also a permutation matrix.

27. How many permutation matrices are there in $M_n(F)$?

28. If P is a permutation matrix, prove that P' is actually equal to P^{-1}.

3.8. INVERTIBLE MATRICES

In Section 3.7, we showed that an $n \times n$ matrix A is invertible if and only if its columns form a basis for $F^{(n)}$. This gives us a description of all possible bases of $F^{(n)}$ over F in terms of invertible matrices. But given an $n \times n$ matrix A, how do we know whether it is invertible? We give some equivalent conditions in

Theorem 3.8.1. Let A be an $n \times n$ matrix and let $v_1, \ldots, v_n$ denote the columns of A. Then the following conditions are equivalent:

1. The only solution to $Ax = 0$ is $x = 0$.
2. A is $1-1$;
3. The columns of A are linearly independent;
4. The columns of A form a basis;
5. A is onto;
6. A is invertible.

Proof: The strategy of the proof is to show first that (1) implies (2), (2) implies (3), (3) implies (4), (4) implies (5), and (5) implies (1). From this it will follow that the first five conditions are all equivalent. These conditions certainly imply that A is invertible, condition (6). Conversely, if A is invertible, then A is certainly $1-1$, so that A satisfies (2). But then it follows that all six conditions must be equivalent.

So it suffices to show that the first five conditions are equivalent, which we now do.

(1) *implies* (2): If the only solution to $Ax = 0$ is $x = 0$, and if $Au = Av$, then $A(u - v) = Au - Av = 0$ implies that $u - v = 0$, that is, $u = v$. So A is $1-1$ and hence (1) implies (2).

(2) *implies* (3): The columns of A are $Ae_1, Ae_2, \ldots, Ae_n$. So if $x_1 Ae_1 + \cdots + x_n Ae_n = 0$, then $A \begin{bmatrix} x_1 \\ \vdots \\ x_n \end{bmatrix} = 0$, which implies that $\begin{bmatrix} x_1 \\ \vdots \\ x_n \end{bmatrix} = 0$ because A is $1-1$. So the columns $Ae_1, \ldots, Ae_n$ of A are linearly independent and (2) implies (3).

(3) *implies* (4): If the columns of A are linearly independent, they are a basis for $F^{(n)}$ by Corollary 3.6.5. So (3) implies (4).

(4) *implies* (5): If the columns of A form a basis for $F^{(n)}$, any $y \in F^{(n)}$ can be expressed as a linear combination $x_1 Ae_1 + \cdots + x_n Ae_n$ of the columns of A, so $y = \begin{bmatrix} x_1 \\ \vdots \\ x_n \end{bmatrix}$. So (4) implies (5).

(5) *implies* (1): If A is onto, its columns $Ae_1, \ldots, Ae_n$ span $F^{(n)}$ and so are a basis for $F^{(n)}$ by Corollary 3.6.6. ∎

We now get

Corollary 3.8.2. If $A, B \in M_n(F)$ and AB is invertible, then A and B are invertible.

Proof: Since AB is invertible, AB is $1-1$ and onto, by Theorem 3.8.1. Since AB is $1-1$, B is also $1-1$ (Prove!), so B is invertible by Theorem 3.8.1. Since AB is onto, A is also onto (Prove!), so A is invertible by Theorem 3.8.1. ∎

Since an $n \times n$ matrix A is invertible if and only if its columns are linearly independent, by Theorem 3.8.1, we can use the method of Section 3.6 for determining whether the columns of A are linearly independent as a

Method to determine whether an $n \times n$ matrix is invertible

1. Row reduce A to an echelon matrix B.
2. Then A is invertible if and only if all diagonal entries of B are 1.

EXAMPLE

To see whether $A = \begin{bmatrix} 1 & 1 & 1 \\ 1 & 2 & 1 \\ 1 & 2 & 2 \end{bmatrix}$ is invertible, we row reduce A to an echelon matrix $\begin{bmatrix} 1 & 1 & 1 \\ 0 & 1 & 0 \\ 0 & 0 & 1 \end{bmatrix}$. Since its diagonal entries all equal 1, A is invertible.

Let A be an invertible matrix with columns $v_1, \ldots, v_n$. We know that the $v_1, \ldots, v_n$ form a basis, so we can find scalars c_{rs} such that $e_s = \sum_{r=1}^{n} c_{rs} v_r$ for all s. Then

$$e_s = \sum_{r=1}^{n} c_{rs} v_r = \sum_{r=1}^{n} c_{rs} A e_r = A\left(\sum_{r=1}^{n} c_{rs} e_r\right)$$

and $A^{-1} e_s = \sum_{r=1}^{n} c_{rs} e_r$. This means that column s of A^{-1} is $\sum_{r=1}^{n} c_{rs} e_r$; that is, c_{rs} is the (r, s) entry of A^{-1}. Is this useful? Yes! It enables us to find the entries of A^{-1} by expressing the e_s as linear combinations of $v_1, \ldots, v_n$.

To be specific, let's use our method from the preceding section to express the vector e_s as linear combinations of $v_1, \ldots, v_n$. Following our method with $A = [v_1, \ldots, v_n]$, we form the augmented matrix $[A, e_s]$ and row reduce it to an echelon matrix $[B, z_s]$. Then B is an upper triangular matrix with diagonal entries all equal to 1. So we can further row reduce $[B, z_s]$ until B becomes the identity matrix and we get a matrix $[I, w_s]$. Then $Ax = e_s$ if and only if $Ix = w_s$, that is, if and only if $x = w_s$. So w_s is the solution to the equation $Ax = e_s$ for each s. It follows that w_s is $A^{-1} e_s$; that is, w_s is column s of A^{-1} for each s. So we have computed the columns of A^{-1} and we get A^{-1} as $[w_1, \ldots, w_n]$.

In practice, we compute all of the w_s at once. Instead of forming the *singly* augmented matrices $[A, e_s]$ and reducing each of them to $[I, w_s]$, we form the *n-fold*

Sec. 3.8]   Invertible Matrices   **147**

augmented matrix $[A, e_1, \ldots, e_n] = [A, I]$ and row reduce it to the matrix $[I, w_1, \ldots, w_n] = [I, A^{-1}]$. This gives the

Method to compute the inverse of an invertible $n \times n$ matrix A

1. Form the $n \times 2n$ matrix $[A, I]$ whose columns are the columns v_i of A followed by the columns e_k of I.
2. Row reduce $[A, I]$ to the echelon matrix $[I, Z]$. (If you find that this is not possible because the first n entries of some row become 0, then A is not invertible.)
3. Then Z is the matrix A^{-1}.

EXAMPLE

Suppose that we wish to determine whether the matrix $A = \begin{bmatrix} 1 & 2 & 3 \\ 1 & 3 & 5 \\ 2 & 5 & 9 \end{bmatrix}$ is invertible. And if so, we want to compute A^{-1}. To try to compute the inverse of A, we form the matrix $[A, I] = \begin{bmatrix} 1 & 2 & 3 & 1 & 0 & 0 \\ 1 & 3 & 5 & 0 & 1 & 0 \\ 2 & 5 & 9 & 0 & 0 & 1 \end{bmatrix}$ and row reduce it to the

matrix $\begin{bmatrix} 1 & 2 & 3 & 1 & 0 & 0 \\ 0 & 1 & 2 & -1 & 1 & 0 \\ 0 & 1 & 3 & -2 & 0 & 1 \end{bmatrix}$, then to the echelon matrix

$$\begin{bmatrix} 1 & 2 & 3 & 1 & 0 & 0 \\ 0 & 1 & 2 & -1 & 1 & 0 \\ 0 & 0 & 1 & -1 & -1 & 1 \end{bmatrix}.$$

Since the diagonal elements are all 1, the matrix is invertible and we continue until it is row reduced to $\begin{bmatrix} 1 & 2 & 0 & 4 & 3 & -3 \\ 0 & 1 & 0 & 1 & 3 & -2 \\ 0 & 0 & 1 & -1 & -1 & 1 \end{bmatrix}$ and finally to

$\begin{bmatrix} 1 & 0 & 0 & 2 & -3 & 1 \\ 0 & 1 & 0 & 1 & 3 & -2 \\ 0 & 0 & 1 & -1 & -1 & 1 \end{bmatrix} = [I, A^{-1}]$.

PROBLEMS

NUMERICAL PROBLEMS

1. Determine those values for a for which the matrix $\begin{bmatrix} 1 & a & 0 \\ 2 & 1 & 0 \\ 1 & 3 & 1 \end{bmatrix}$ is not invertible.

2. Determine those values for a for which the matrix $\begin{bmatrix} 1 & a & 0 \\ 3 & 4 & 1 \\ 1 & 3 & 1 \end{bmatrix}$ is not invertible.

3. Determine those values for a for which the matrix $\begin{bmatrix} 2 & a & 1 \\ 3 & 4 & 1 \\ 1 & 3 & 1 \end{bmatrix}$ is not invertible.

4. Compute the inverse of $\begin{bmatrix} 2 & 3 & 1 \\ 3 & 4 & 1 \\ 1 & 3 & 1 \end{bmatrix}$ by row reduction of $[A, I]$.

MORE THEORETICAL PROBLEMS

Easier Problems

5. Let A and B be $n \times n$ matrices such that $AB = I$ and $Ax = 0$ only if $x = 0$. Without using the results of Sections 3.6 through 3.8, show that $A(BA - I) = 0$ and use this to show that $AB = BA$ and A is invertible.

6. Compute the inverse of the matrix $\begin{bmatrix} 1 & a & 0 \\ 1 & 1 & 1 \\ 0 & 0 & 1 \end{bmatrix}$ for all a for which it exists.

7. Show that if an $n \times n$ matrix A is invertible and B is row equivalent to A, then B is invertible.

8. Show that an $n \times n$ matrix A is row equivalent to I if and only if A is invertible.

Middle-Level Problems

9. Show that if A is row equivalent to B, then AC is row equivalent to BC for any $n \times n$ matrices A, B, C.

10. Using Problems 8 and 9, show that two $n \times n$ matrices A and B are row equivalent if and only if $B = UA$ for some invertible $n \times n$ matrix U.

3.9. MATRICES AND BASES

Given $F^{(n)}$, it is not aware of what particular basis one uses for it. As far as it is concerned, one basis is just as good as any other. If T is a linear transformation on $F^{(n)}$, all $F^{(n)}$ cares about is that T satisfies $T(v + w) = T(v) + T(w)$ and $T(av) = aT(v)$ for all $v, w \in F^{(n)}$ and all $a \in F$. How we represent T—as a matrix or otherwise—is our business and not that of $F^{(n)}$. For *us*, the canonical basis is often best because its components are so simple—but $F^{(n)}$ could not care less.

Given a linear transformation T on $F^{(n)}$ we found, by using the canonical basis, that T is a matrix. However, the canonical basis is not holy. *We could talk in an analogous way about the matrix of T in any basis.*

To do this, let $v_1, \ldots, v_n$ be a basis of $F^{(n)}$ over F and let T be a linear transformation on $F^{(n)}$. If we knew $Tv_1, \ldots, Tv_n$, we would know how T acts on any element v in $F^{(n)}$. Why? Since $v \in F^{(n)}$ and $v_1, \ldots, v_n$ is a basis of $F^{(n)}$ over F, we know that $v = a_1 v_1 + \cdots + a_n v_n$, where the $a_1, \ldots, a_n$ are in F and are unique. So $Tv = T(a_1 v_1 + \cdots + a_n v_n) = a_1 T v_1 + \cdots + a_n T v_n$. So knowing each of $Tv_1, \ldots, Tv_n$ allows us to know *everything* about how T acts on arbitrary elements of $F^{(n)}$.

Because Tv_s is in $F^{(n)}$ for every $s = 1, \ldots, n$ and since $v_1, \ldots, v_n$ is a basis of $F^{(n)}$ over F, Tv_s is realizable in a unique way as $Tv_s = b_{1s} v_1 + b_{2s} v_2 + \cdots + b_{ns} v_n$. If we have the n^2 elements b_{rs} in hand, then we know T exactly how it acts on $F^{(n)}$.

Definition. We call the matrix (b_{rs}) such that $Tv_s = b_{1s} v_1 + b_{2s} v_2 + \cdots + b_{ns} v_n$ for all s the *matrix of T in the basis $v_1, \ldots, v_n$*.

In summation notation, the condition on the entries b_{rs} of the matrix of T in the basis $v_1, \ldots, v_n$ is

$$Tv_s = \sum_{r=1}^{n} b_{rs} v_r.$$

When we showed that any linear transformation over $F^{(n)}$ was a matrix, what we really showed was that its matrix in the canonical basis $e_1, \ldots, e_n$, namely, the matrix having the vectors $Te_1, \ldots, Te_n$ as its columns, also mapped e_r to Te_r for all r.

Before going on, let's look at an example. Let T be the matrix $\begin{bmatrix} 1 & 2 & 3 \\ -1 & 4 & 7 \\ 5 & 0 & 1 \end{bmatrix}$ and regard T as a linear transformation of $F^{(3)}$. Then the matrix of T (as linear transformation) in the canonical basis is $\begin{bmatrix} 1 & 2 & 3 \\ -1 & 4 & 7 \\ 5 & 0 & 1 \end{bmatrix}$. The vectors

$$v_1 = \begin{bmatrix} 0 \\ 1 \\ 1 \end{bmatrix}, \quad v_2 = \begin{bmatrix} 1 \\ 0 \\ 1 \end{bmatrix}, \quad v_3 = \begin{bmatrix} 1 \\ 1 \\ 0 \end{bmatrix}$$

can readily be shown to be a basis of $F^{(3)}$ over F. (Do it!) How do we go about finding the matrix of the T above in the basis v_1, v_2, v_3? To find the expression for $Tv_1 = \begin{bmatrix} 1 & 2 & 3 \\ -1 & 4 & 7 \\ 5 & 0 & 1 \end{bmatrix} \begin{bmatrix} 0 \\ 1 \\ 1 \end{bmatrix} = \begin{bmatrix} 5 \\ 11 \\ 1 \end{bmatrix}$ in terms of v_1, v_2, v_3, we follow the method in Section 3.6 for expressing $\begin{bmatrix} 5 \\ 11 \\ 1 \end{bmatrix}$ as a linear combination of v_1, v_2, v_3. We form the augmented matrix $[v_1, v_2, v_3, v] = \begin{bmatrix} 0 & 1 & 1 & 5 \\ 1 & 0 & 1 & 11 \\ 1 & 1 & 0 & 1 \end{bmatrix}$ and row reduce it, getting

$\begin{bmatrix} 1 & 0 & 1 & 11 \\ 0 & 1 & 1 & 5 \\ 0 & 0 & -2 & -15 \end{bmatrix}$. Solving $\begin{bmatrix} 1 & 0 & 1 \\ 0 & 1 & 1 \\ 0 & 0 & -2 \end{bmatrix} x = \begin{bmatrix} 11 \\ 5 \\ -15 \end{bmatrix}$ for x by back substitution,

we get $x = \begin{bmatrix} \frac{7}{2} \\ -\frac{5}{2} \\ \frac{16}{2} \end{bmatrix}$. So $x = \begin{bmatrix} \frac{7}{2} \\ -\frac{5}{2} \\ \frac{15}{2} \end{bmatrix}$ is also a solution to

$$[v_1, v_2, v_3] x = \begin{bmatrix} 0 & 1 & 1 \\ 1 & 0 & 1 \\ 1 & 1 & 0 \end{bmatrix} x = \begin{bmatrix} 5 \\ 11 \\ 1 \end{bmatrix}$$

and we get

$$Tv_1 = \begin{bmatrix} 5 \\ 11 \\ 1 \end{bmatrix} = [v_1, v_2, v_3] \begin{bmatrix} \frac{7}{2} \\ -\frac{5}{2} \\ \frac{15}{2} \end{bmatrix} = \tfrac{7}{2} v_1 - \tfrac{5}{2} v_2 + \tfrac{15}{2} v_3.$$

By the recipe given above for determining the matrix of T in the basis v_1, v_2, v_3, the first column of the matrix of T in the basis v_1, v_2, v_3 is $\begin{bmatrix} \frac{7}{2} \\ -\frac{5}{2} \\ \frac{15}{2} \end{bmatrix}$. Similarly,

$$Tv_2 = \begin{bmatrix} 1 & 2 & 3 \\ -1 & 4 & 7 \\ 5 & 0 & 1 \end{bmatrix} \begin{bmatrix} 1 \\ 0 \\ 1 \end{bmatrix} = \begin{bmatrix} 4 \\ 6 \\ 6 \end{bmatrix} = [v_1, v_2, v_3] \begin{bmatrix} 4 \\ 2 \\ 2 \end{bmatrix} = 4v_1 + 2v_2 + 2v_3. \text{ (Verify!)}$$

So the second column of the matrix of T in the basis v_1, v_2, v_3 is $\begin{bmatrix} 4 \\ 2 \\ 2 \end{bmatrix}$. Finally,

$$Tv_3 = \begin{bmatrix} 1 & 2 & 3 \\ -1 & 4 & 7 \\ 5 & 0 & 1 \end{bmatrix} \begin{bmatrix} 1 \\ 1 \\ 0 \end{bmatrix} = \begin{bmatrix} 3 \\ -3 \\ 5 \end{bmatrix} = [v_1, v_2, v_3] \begin{bmatrix} \frac{5}{2} \\ \frac{5}{2} \\ \frac{5}{2} \end{bmatrix}$$
$$= \tfrac{5}{2} v_1 + \tfrac{5}{2} v_2 + \tfrac{1}{2} v_3. \text{ (Verify!)}$$

so the third column of the matrix of T in the basis v_1, v_2, v_3 is $\begin{bmatrix} \frac{5}{2} \\ \frac{5}{2} \\ \frac{1}{2} \end{bmatrix}$. Therefore,

the matrix of T in the basis v_1, v_2, v_3 is $\begin{bmatrix} \frac{7}{2} & 4 & \frac{5}{2} \\ -\frac{5}{2} & 2 & \frac{5}{2} \\ \frac{15}{2} & 2 & \frac{1}{2} \end{bmatrix}$. So *the two matrices*

Sec. 3.9]   Matrices and Bases   151

$$\begin{bmatrix} 1 & 2 & 3 \\ -1 & 4 & 7 \\ 5 & 0 & 1 \end{bmatrix}, \begin{bmatrix} \frac{7}{2} & 4 & \frac{5}{2} \\ -\frac{5}{2} & 2 & \frac{5}{2} \\ \frac{15}{2} & 2 & \frac{1}{2} \end{bmatrix}$$

represent the same linear transformation T, but in different bases: the first in the canonical basis, the second in the basis

$$v_1 = \begin{bmatrix} 0 \\ 1 \\ 1 \end{bmatrix}, \quad v_2 = \begin{bmatrix} 1 \\ 0 \\ 1 \end{bmatrix}, \quad v_3 = \begin{bmatrix} 1 \\ 1 \\ 0 \end{bmatrix}.$$

Forget about linear transformations for a moment; given the two matrices

$$\begin{bmatrix} 1 & 2 & 3 \\ -1 & 4 & 7 \\ 5 & 0 & 1 \end{bmatrix}, \begin{bmatrix} \frac{7}{2} & 4 & \frac{5}{2} \\ -\frac{5}{2} & 2 & \frac{5}{2} \\ \frac{15}{2} & 2 & \frac{1}{2} \end{bmatrix},$$

and viewing them just as matrices, how on earth are they related? *Because we know that they came from the same linear transformation—in different bases—common sense that they should be related. But how?* That is precisely what we are going to find out in the next theorem.

Theorem 3.9.1. *Regard the $n \times n$ matrix T as a linear transformation of $F^{(n)}$. Then the matrix A of T in a basis $v_1, \ldots, v_n$ of $F^{(n)}$ is $A = C^{-1}TC$, where C is the matrix whose columns are $v_1, \ldots, v_n$.*

Proof: Letting $A = (a_{rs})$, we have $T(v_s) = \sum_{r=1}^{n} a_{rs} v_r$ for $s = 1, \ldots, n$. Therefore,

$$C^{-1}TC(e_s) = C^{-1}T(v_s) = C^{-1}\left(\sum_{r=1}^{n} a_{rs} v_r\right)$$

$$= \sum_{r=1}^{n} a_{rs} C^{-1}(v_r) = \sum_{r=1}^{n} a_{rs} e_r.$$

Thus $C^{-1}TC(e_s) = \sum_{r=1}^{n} a_{rs} e_r$ for all s and $C^{-1}TC = A$. ∎

We first illustrate the theorem with a simple example. Let $v_1 = \begin{bmatrix} 0 \\ 1 \end{bmatrix}, v_2 = \begin{bmatrix} -1 \\ 0 \end{bmatrix}$. Then C is $C = \begin{bmatrix} 0 & -1 \\ 1 & 0 \end{bmatrix}$. Notice that $C^{-1} = \begin{bmatrix} 0 & 1 \\ -1 & 0 \end{bmatrix}$. Suppose that

$$T = \begin{bmatrix} 3 & -1 \\ 1 & 5 \end{bmatrix}.$$

Since $Tv_1 = Te_2 = \begin{bmatrix} -1 \\ 5 \end{bmatrix} = 5v_1 + 1v_2$ and $T(v_2) = T(-e_2)\begin{bmatrix} -3 \\ -1 \end{bmatrix} = 3v_2 - 1v_1$, the matrix of T in v_1, v_2 is $A = \begin{bmatrix} 5 & -1 \\ 1 & 3 \end{bmatrix}$. According to our theorem, A should equal

$$C^{-1}TC = \begin{bmatrix} 0 & 1 \\ -1 & 0 \end{bmatrix}\begin{bmatrix} 3 & -1 \\ 1 & 5 \end{bmatrix}\begin{bmatrix} 0 & -1 \\ 1 & 0 \end{bmatrix} = \begin{bmatrix} 5 & -1 \\ 1 & 3 \end{bmatrix}, \text{ as it is.}$$

We return to the example preceding the theorem. Since

$$v_1 = \begin{bmatrix} 0 \\ 1 \\ 1 \end{bmatrix}, \quad v_2 = \begin{bmatrix} 1 \\ 0 \\ 1 \end{bmatrix}, \quad v_3 = \begin{bmatrix} 1 \\ 1 \\ 0 \end{bmatrix},$$

the C of the theorem is $\begin{bmatrix} 0 & 1 & 1 \\ 1 & 0 & 1 \\ 1 & 1 & 0 \end{bmatrix}$. Our matrix T was defined as $\begin{bmatrix} 1 & 2 & 3 \\ -1 & 4 & 7 \\ 5 & 0 & 1 \end{bmatrix}$. As we saw, the matrix of T in v_1, v_2, v_3 is $A = \begin{bmatrix} \frac{7}{2} & 4 & \frac{5}{2} \\ -\frac{5}{2} & 2 & \frac{5}{2} \\ \frac{15}{2} & 2 & \frac{1}{2} \end{bmatrix}$. We want to verify that $A = C^{-1}TC$. Since computing the inverse of C is a little messy, we will check the slightly weaker relation $CA = TC$. So, is the equation

$$\begin{bmatrix} 0 & 1 & 1 \\ 1 & 0 & 1 \\ 1 & 1 & 0 \end{bmatrix}\begin{bmatrix} \frac{7}{2} & 4 & \frac{5}{2} \\ -\frac{5}{2} & 2 & \frac{5}{2} \\ \frac{15}{2} & 2 & \frac{1}{2} \end{bmatrix} = \begin{bmatrix} 1 & 2 & 3 \\ -1 & 4 & 7 \\ 5 & 0 & 1 \end{bmatrix}\begin{bmatrix} 0 & 1 & 1 \\ 1 & 0 & 1 \\ 1 & 1 & 0 \end{bmatrix}$$

correct? Doing the matrix multiplication, we see that both sides equal $\begin{bmatrix} 5 & 4 & 3 \\ 11 & 6 & 3 \\ 1 & 6 & 5 \end{bmatrix}$.

So the theorem checks out—no surprise—in this example.

We can use our theorem to relate the matrix of a linear transformation T of $F^{(n)}$ in one basis to its matrix in another, as we observe in the

Corollary 3.9.2. Let T be a linear transformation of $F^{(n)}$ and let $v_1, \ldots, v_n$ and $w_1, \ldots, w_n$ be two bases of $F^{(n)}$. Let A be the matrix of T in $v_1, \ldots, v_n$ and B the matrix of T in $w_1, \ldots, w_n$. Then $CAC^{-1} = DBD^{-1}$, where C is the matrix whose columns are $v_1, \ldots, v_n$ and D the matrix whose columns are $w_1, \ldots, w_n$.

Proof: By Theorem 3.9.1, both CAC^{-1} and DBD^{-1} equal T when we regard the linear transformation T as a matrix. ∎

Corollary 3.9.2 shows us that the matrix B of T in $w_1, \ldots, w_n$ is related to the matrix A of T in $v_1, \ldots, v_n$ by $CAC^{-1} = DBD^{-1}$. Solving for B, we get $B = D^{-1}CAC^{-1}D = U^{-1}AU$, where $U = C^{-1}D$. Since $CU = D$, this implies that $w_s = \sum_{r=1}^{n} u_{rs}v_r$ for all s. (Prove!)

Sec. 3.9] Matrices and Bases

Definition. We call U the *matrix of the change of basis* from the basis v_r to the basis w_s.

By these observations, we have proved the

Theorem 3.9.3. Let T be a linear transformation of $F^{(n)}$ and let $v_1, \ldots, v_n$ and $w_1, \ldots, w_n$ be two bases for $F^{(n)}$. Then the matrix B of T in $w_1, \ldots, w_n$ is $B = U^{-1}AU$, where A is the matrix of T in $v_1, \ldots, v_n$ and $U = (u_{rs})$ is the matrix of the change of basis, that is, $w_s = \sum_{r=1}^{n} u_{rs} v_r$ for all s.

EXAMPLE

Let $T = \begin{bmatrix} 1 & 0 & 1 \\ 1 & 0 & 0 \\ 2 & 0 & 0 \end{bmatrix}$ and $v_1 = \begin{bmatrix} 1 \\ 0 \\ 0 \end{bmatrix}$, $v_2 = \begin{bmatrix} 1 \\ 1 \\ 0 \end{bmatrix}$, $v_3 = \begin{bmatrix} 0 \\ 0 \\ 2 \end{bmatrix}$. According to our theorem, the matrix of T in v_1, v_2, v_3 is $A = C^{-1}TC$, where $C = \begin{bmatrix} 1 & 1 & 0 \\ 0 & 1 & 0 \\ 0 & 0 & 2 \end{bmatrix}$. Following the method of Section 3.8 for finding inverses, we find $C^{-1} = \begin{bmatrix} 1 & -1 & 0 \\ 0 & 1 & 0 \\ 0 & 0 & \frac{1}{2} \end{bmatrix}$, so we get

$$A = \begin{bmatrix} 1 & -1 & 0 \\ 0 & 1 & 0 \\ 0 & 0 & \frac{1}{2} \end{bmatrix} \begin{bmatrix} 1 & 0 & 1 \\ 1 & 0 & 0 \\ 2 & 0 & 0 \end{bmatrix} \begin{bmatrix} 1 & 1 & 0 \\ 0 & 1 & 0 \\ 0 & 0 & 2 \end{bmatrix} = \begin{bmatrix} 0 & 0 & 2 \\ 1 & 1 & 0 \\ 1 & 1 & 0 \end{bmatrix}.$$

Similarly, for

$$w_1 = \begin{bmatrix} 0 \\ 1 \\ 0 \end{bmatrix}, \quad w_2 = \begin{bmatrix} 0 \\ 0 \\ 1 \end{bmatrix}, \quad w_3 = \begin{bmatrix} 1 \\ 0 \\ 0 \end{bmatrix},$$

we let $D = \begin{bmatrix} 0 & 0 & 1 \\ 0 & 0 & 1 \\ 1 & 0 & 0 \end{bmatrix}$, so $D^{-1} = \begin{bmatrix} 0 & 1 & 0 \\ 0 & 0 & 1 \\ 1 & 0 & 0 \end{bmatrix}$ and the matrix of T in the basis w_1, w_2, w_3 is

$$B = D^{-1}TD = \begin{bmatrix} 0 & 1 & 0 \\ 0 & 0 & 1 \\ 1 & 0 & 0 \end{bmatrix} \begin{bmatrix} 1 & 0 & 1 \\ 1 & 0 & 0 \\ 2 & 0 & 0 \end{bmatrix} \begin{bmatrix} 0 & 0 & 1 \\ 1 & 0 & 0 \\ 0 & 1 & 0 \end{bmatrix}$$

$$= \begin{bmatrix} 0 & 0 & 1 \\ 0 & 0 & 2 \\ 0 & 1 & 1 \end{bmatrix}.$$

The change of basis matrix from v_1, v_2, v_3 to w_1, w_2, w_3 is $U = C^{-1}D =$
$\begin{bmatrix} 1 & -1 & 0 \\ 0 & 1 & 0 \\ 0 & 0 & \frac{1}{2} \end{bmatrix} \begin{bmatrix} 0 & 0 & 1 \\ 1 & 0 & 0 \\ 0 & 1 & 0 \end{bmatrix} = \begin{bmatrix} -1 & 0 & 1 \\ 1 & 0 & 0 \\ 0 & \frac{1}{2} & 0 \end{bmatrix}$, which checks out since

$$\begin{bmatrix} 0 \\ 1 \\ 0 \end{bmatrix} = -1 \begin{bmatrix} 1 \\ 0 \\ 0 \end{bmatrix} + 1 \begin{bmatrix} 1 \\ 1 \\ 0 \end{bmatrix} + 0 \begin{bmatrix} 0 \\ 0 \\ 2 \end{bmatrix}$$

$$\begin{bmatrix} 0 \\ 0 \\ 1 \end{bmatrix} = 0 \begin{bmatrix} 1 \\ 0 \\ 0 \end{bmatrix} + 0 \begin{bmatrix} 1 \\ 1 \\ 0 \end{bmatrix} + (\tfrac{1}{2}) \begin{bmatrix} 0 \\ 0 \\ 2 \end{bmatrix}$$

$$\begin{bmatrix} 1 \\ 0 \\ 0 \end{bmatrix} = 1 \begin{bmatrix} 1 \\ 0 \\ 0 \end{bmatrix} + 0 \begin{bmatrix} 1 \\ 1 \\ 0 \end{bmatrix} + 0 \begin{bmatrix} 0 \\ 0 \\ 2 \end{bmatrix}.$$

According to our theorem, the matrices $A = \begin{bmatrix} 0 & 0 & 2 \\ 1 & 1 & 0 \\ 1 & 1 & 0 \end{bmatrix}$ and $B = \begin{bmatrix} 0 & 0 & 1 \\ 0 & 0 & 2 \\ 0 & 1 & 1 \end{bmatrix}$ of T in the two bases should be related by $B = U^{-1}AU$, or by $UB = AU$. Checking, we find that this is in fact so:

$$\begin{bmatrix} -1 & 0 & 1 \\ 1 & 0 & 0 \\ 0 & \frac{1}{2} & 0 \end{bmatrix} \begin{bmatrix} 0 & 0 & 1 \\ 0 & 0 & 2 \\ 0 & 1 & 1 \end{bmatrix} = \begin{bmatrix} 0 & 1 & 0 \\ 0 & 0 & 1 \\ 0 & 0 & 1 \end{bmatrix} = \begin{bmatrix} 0 & 0 & 2 \\ 1 & 1 & 0 \\ 1 & 1 & 0 \end{bmatrix} \begin{bmatrix} -1 & 0 & 1 \\ 1 & 0 & 0 \\ 0 & \frac{1}{2} & 0 \end{bmatrix}.$$

The relation expressed in Theorem 3.9.3, namely, $B = U^{-1}AU$ is an important one in matrix theory. It is called *similarity*.

Definition. If $A, B \in M_n(F)$, then B is said to be *similar* to A if $B = C^{-1}AC$ for some invertible matrix C in $M_n(F)$. We denote this by $A \sim B$.

Similarity behaves very much like equality in that it obeys certain rules.

Theorem 3.9.4. If A, B, C are in $M_n(F)$, then

1. $A \sim A$.
2. $A \sim B$ implies that $B \sim A$.
3. $A \sim B, B \sim C$ implies that $A \sim C$.

Proof: To see that $A \sim A$ we need an invertible matrix M such that $A = M^{-1}AM$; well, $M = I$ certainly does the trick. So $A \sim A$.

If $A \sim B$, then $B = M^{-1}AM$, hence $A = (M^{-1})^{-1}B(M^{-1})$. Because M^{-1} is invertible we have that $B \sim A$.

Finally, if $A \sim B$ and $B \sim C$, then $A = M^{-1}BM$ and $B = N^{-1}CN$. Thus $A = M^{-1}(N^{-1}CN)M = (NM)^{-1}C(NM)$, and since NM is invertible, we get $A \sim C$. ∎

If $A \in M_n(F)$, then the set of all matrices B such that $A \sim B$ is a very important subset of $M_n(F)$.

Definition. The *similarity class* of A, written as $\text{cl}(A)$, is

$$\text{cl}(A) = \{B \in M_n(F) \mid A \sim B\}.$$

Note that *A is contained in its similarity class* $\text{cl}(A)$. Moreover, *A is contained in no other similarity class* by the important

Theorem 3.9.5. *If A and B are in $M_n(F)$, then either their similarity classes are equal or they have no element in common.*

Proof: Suppose that the class of A and that of B have some matrix G in common. So $A \sim G$ and $B \sim G$. By property (2) in Theorem 3.9.4, $G \sim B$, hence, by property (3) in Theorem 3.9.4, $A \sim B$. So if $X \in \text{cl}(A)$, then $X \sim A$ and since $A \sim B$, we get that $X \sim B$, that is, $X \in \text{cl}(B)$. Therefore, $\text{cl}(A)$ is contained in $\text{cl}(B)$. But the argument works when we interchange the roles of A and B, so we get that $\text{cl}(B)$ is contained in $\text{cl}(A)$. Therefore, $\text{cl}(A) = \text{cl}(B)$. ∎

If B and A are similar, if we interpret A as a linear transformation in the canonical basis, then we can interpret B as the matrix of the same linear transformation in the basis $w_1, \ldots, w_n$, where $w_j = C(e_j)$ and where $B = C^{-1}AC$. This is exactly what Theorem 3.9.1 tells us. So in a certain sense, A and B can be viewed as coming from the same linear transformation.

Given a linear transformation T, there is nothing sacred impelling us to view T as a matrix in the canonical basis. We can view it as a matrix in any basis of our choice. *Why not pick a basis in which the matrix of T is the simplest looking?* Let's take an example. If T is the linear transformation whose matrix in the canonical basis is $\begin{bmatrix} 1 & 1 \\ 0 & 2 \end{bmatrix}$, what is its matrix in the basis $w_1 = e_1$, $w_2 = e_1 + e_2$? Since $T(e_1) = e_1 = w_1 = T(w_1)$ and $T(w_2) = T(e_1) + T(e_2) = e_1 + e_1 + 2e_2 = 2(e_1 + e_2) = 2w_2$, we see that the matrix of T in the basis w_1, w_2 is $\begin{bmatrix} 1 & 0 \\ 0 & 2 \end{bmatrix}$, a diagonal matrix. In some sense this is a nicer-looking matrix than $\begin{bmatrix} 1 & 1 \\ 0 & 2 \end{bmatrix}$.

Since we are free to pick any basis we want to get the matrix of a linear transformation, and any two bases give us matrices which are similar for this transformation, it is desirable to find nice matrices as road signs for our similarity classes. There are several different types of such road signs used in matrix theory. These are called *canonical forms*. To check that two matrices are similar then becomes checking if these two matrices have the same canonical form. We shall discuss canonical forms in Chapter 10.

PROBLEMS

NUMERICAL PROBLEMS

1. Find the matrix B of T in the given basis, where T is the linear transformation whose matrix is given in the canonical basis.

   (a) $T = \begin{bmatrix} 1 & 2 \\ 0 & 1 \end{bmatrix}$ in the basis $w_1 = e_2, w_2 = e_1$.

   (b) $T = \begin{bmatrix} 1 & 1 & i \\ 0 & 0 & 1 \\ 2 & i & 0 \end{bmatrix}$ in the basis $w_1 = e_2, w_2 = e_3, w_3 = e_1$.

   (c) $T = \begin{bmatrix} -1 & 0 & 1 \\ 0 & 1 & -1 \\ -4 & 1 & 0 \end{bmatrix}$ in the basis $w_1 = e_1 + e_2 + e_3, w_2 = e_1 + e_2, w_3 = e_3$.

   (d) $T = \begin{bmatrix} 1 & 2 & 0 & 1 \\ 0 & 1 & 2 & 0 \\ 0 & 0 & 1 & 2 \\ 0 & 0 & 0 & 1 \end{bmatrix}$ in the basis $w_1 = e_1, w_2 = e_1 + e_2, w_3 = e_1 - e_2,$ $w_4 = e_4$.

2. In each part of Problem 1 find an invertible matrix C such that $B = C^{-1}AC$, where A is the matrix T as given.

3. Find a basis w_1, w_2 of $\mathbb{C}^{(2)}$ such that the matrix of $\begin{bmatrix} 1 & i \\ 0 & 1 \end{bmatrix}$ in this basis is $\begin{bmatrix} 1 & 1 \\ 0 & 1 \end{bmatrix}$.

4. Find a basis w_1, w_2, w_3 of $\mathbb{C}^{(3)}$ in which the matrix of $\begin{bmatrix} 1 & 2 & 3 \\ 0 & i & 4 \\ 0 & 0 & -i \end{bmatrix}$ is diagonal.

5. Show that $\begin{bmatrix} 1 & 2 & 3 \\ 4 & 5 & 6 \\ 7 & 8 & 9 \end{bmatrix}$ and $\begin{bmatrix} 1 & 2 & 3 \\ 6 & 5 & 4 \\ 9 & 8 & 7 \end{bmatrix}$ are not similar.   (*Hint*: Use traces.)

6. Show that all matrices $\begin{bmatrix} 1 & a & b \\ 0 & 2 & c \\ 0 & 0 & 3 \end{bmatrix}$ are similar to $\begin{bmatrix} 1 & 0 & 0 \\ 0 & 2 & 0 \\ 0 & 0 & 3 \end{bmatrix}$.

7. Show that $\begin{bmatrix} 1 & 7 \\ -6 & -1 \end{bmatrix}$ and $\begin{bmatrix} i & 10 \\ -4 & -i \end{bmatrix}$ are similar in $M_2(\mathbb{C})$.

8. Show that $\begin{bmatrix} 1 & 0 & 1 \\ 0 & 1 & 0 \\ 0 & 0 & 1 \end{bmatrix}$ is *not* similar to I.

9. If A is a scalar matrix, find all elements in the similarity class of A.

MORE THEORETICAL PROBLEMS

Easier Problems

10. What is the matrix of $\begin{bmatrix} a_1 & & & 0 \\ & a_2 & & \\ & & \ddots & \\ 0 & & & a_n \end{bmatrix}$ in the basis $w_1 = e_{r_1}, \ldots, w_n = e_{r_n}$, where $e_{r_1}, \ldots, e_{r_n}$ are $e_1, \ldots, e_n$ in some order?

11. If you know the matrix of a linear transformation T in a basis $v_1, \ldots, v_n$ of $F^{(n)}$, what is the matrix of T in the basis $e_1, \ldots, e_n$ of $F^{(n)}$?

12. If A and B are the matrices of a linear transformation T in two different bases, show that $\mathrm{tr}(A) = \mathrm{tr}(B)$.

13. If $v_1, \ldots, v_m \in F^{(n)}$ are characteristic vectors associated with $a_1, \ldots, a_m$ which are *distinct* characteristic roots of $A \in M_n(F)$, show that $v_1, \ldots, v_m$ are linearly independent over F.

14. If the matrix $A \in M_n(F)$ has n distinct characteristic roots in F, show that $C^{-1}AC$ is a diagonal matrix for some invertible C in $M_n(F)$.

15. Show that there is no matrix C, invertible in $M_4(F)$, such that $C^{-1} \begin{bmatrix} 1 & 2 & 0 & 1 \\ 0 & 1 & 2 & 0 \\ 0 & 0 & 1 & 2 \\ 0 & 0 & 0 & 1 \end{bmatrix} C$ is a diagonal matrix.

16. Prove that if $A \in M_3(F)$ satisfies $A^3 = 0$, then for some invertible C in $M_3(F)$, $C^{-1}AC$ is an upper triangular matrix. What is the diagonal of $C^{-1}AC$?

17. If you know the matrix of a linear transformation T in the basis $v_1, \ldots, v_n$ of $F^{(n)}$ what is the matrix of T in terms of this in the basis $v_n, \ldots, v_1$ of $F^{(n)}$?

Middle-Level Problems

18. If $C \in M_n(F)$ is invertible, then $\mathrm{cl}(C^{-1}AC) = \mathrm{cl}(A)$ for any $A \in M_n(F)$.

Harder Problems

19. Show that any $n \times n$ upper triangular matrix A is similar to a lower triangular matrix.

3.10. BASES AND INNER PRODUCTS

We saw in Section 3.9 that the nature of a given basis of $F^{(n)}$ over F has a strong influence on the form of a given linear transformation on $F^{(n)}$. If $F = \mathbb{C}$ (or $\mathbb{R}$) we saw earlier that if $v = \begin{bmatrix} x_1 \\ \vdots \\ x_n \end{bmatrix}$ and $w = \begin{bmatrix} y_1 \\ \vdots \\ y_n \end{bmatrix}$ are in $\mathbb{C}^{(n)}$, then their inner product $(v, w) = \sum_{r=1}^{n} x_r \bar{y}_r$ enjoys some very nice properties. How can we use the inner product on $\mathbb{C}^{(n)}$ or $\mathbb{R}^{(n)}$? This will be the major theme of this section.

Recall that two elements v and w of $\mathbb{C}^{(n)}$ are said to be *orthogonal* if $(v, w) = 0$. Suppose that $\mathbb{C}^{(n)}$ has a basis $v_1, \ldots, v_n$, where whenever $r \neq s$, $(v_r, v_s) = 0$. We give such a basis a name.

Definition. The basis $v_1, \ldots, v_n$ of $\mathbb{C}^{(n)}$ (or $\mathbb{R}^{(n)}$) is called an *orthogonal basis* if $(v_r, v_s) = 0$ for $r \neq s$.

What advantage does such a basis enjoy over just any old base? Well, for one thing, if we have a vector w in $\mathbb{C}^{(n)}$, say, we know that $w = a_1 v_1 + \cdots + a_n v_n$. Can we find a nice expression for these coefficients a_r? If we consider (w, v_r), we get

$$(w, v_r) = (a_1 v_1 + \cdots + a_n v_n, v_r) = a_1(v_1, v_r) + \cdots + a_n(v_n, v_r),$$

and since $(v_s, v_r) = 0$ if $s \neq r$, we end up with $a_r(v_r, v_r) = (w, v_r)$. Because $v_r \neq 0$, we know that $a_r = (w, v_r)/(v_r, v_r)$. So if we knew all the (v_r, v_r), we would have a nice, intrinsic expression for the a_r's in terms of w and the v_r's.

This expression would be even nicer if all the $(v_r, v_r) = 1$, that is, *if every v_j had length 1*. A vector of length 1 is called a *unit vector*.

Definition. An orthogonal basis $v_1, \ldots, v_n$ of $\mathbb{C}^{(n)}$ (or $\mathbb{R}^{(n)}$) is called an *orthonormal* basis of $\mathbb{C}^{(n)}$ (or $\mathbb{R}^{(n)}$) if each v_r is a unit vector.

So an orthonormal basis $v_1, \ldots, v_n$ is a basis of $C^{(n)}$ (or $\mathbb{R}^{(n)}$) such that $(v_r, v_s) = 0$ if $r \neq s$, and $(v_r, v_r) = 1$ for all $r = 1, 2, \ldots, n$. Note that the canonical basis $e_1, \ldots, e_n$ is an orthonormal basis of $\mathbb{R}^{(n)}$ over $\mathbb{R}$ and of $\mathbb{C}^{(n)}$ over $\mathbb{C}$.

If $v_1, \ldots, v_n$ is an orthonormal basis of $\mathbb{C}^{(n)}$, say over $\mathbb{C}$, the computation above showed that if $w = a_1 v_1 + \cdots + a_n v_n$, then $a_r = (w, v_r)/(v_r, v_r) = (w, v_r)$ since $(v_r, v_r) = 1$. Plugging this in the expression for w, we obtain

Lemma 3.10.1. *If $v_1, \ldots, v_n$ is an orthonormal basis of $\mathbb{C}^{(n)}$ (or $\mathbb{R}^{(n)}$), then given $w \in \mathbb{C}^{(n)}$ (or $\mathbb{R}^{(n)}$), $w = (w, v_1) v_1 + \cdots + (w, v_n) v_n$.*

Suppose that $v_1, \ldots, v_n$ is an orthonormal basis of $F^{(n)}$. Given $u = \sum_{r=1}^{n} a_r v_r$ and $w = \sum_{s=1}^{n} b_s v_s$, what does their inner product (u, w) look like in terms of the a's and b's?

Lemma 3.10.2. *If $v_1, \ldots, v_n$ is an orthonormal basis of $F^{(n)}$ and $u = \sum_{r=1}^{n} a_r v_r$, $w = \sum_{s=1}^{n} b_s v_r$ are in $F^{(n)}$, then $(u, w) = \sum_{s=1}^{n} a_s \bar{b}_s$.*

Proof: Before going to the general result, let's look at a simple case. Suppose that $u = a_1 v_1 + a_2 v_2$ and $w = b_1 v_1 + b_2 v_2$. Then

$$(u, w) = (a_1 v_1 + a_2 v_2, b_1 v_1 + b_2 v_2)$$
$$= (a_1 v_1, b_1 v_1 + b_2 v_2) + (a_2 v_2, b_1 v_1 + b_2 v_2)$$

$$= a_1(v_1, b_1v_1 + b_2v_2) + a_2(v_2, b_1v_1 + b_2v_2)$$
$$= a_1(v_1, b_1v_1) + a_1(v_1, b_2v_2) + a_2(v_2, b_1v_1) + a_2(v_2, b_2v_2)$$
$$= a_1\bar{b}_1(v_1, v_1) + a_1\bar{b}_2(v_1, v_2) + a_2\bar{b}_1(v_2, v_1) + a_2\bar{b}_2(v_2, v_2)$$
$$= a_1\bar{b}_1 + a_2\bar{b}_2$$

since $(v_1, v_1) = (v_2, v_2) = 1$ and $(v_1, v_2) = (v_2, v_1) = 0$. Notice that in carrying out this computation we have used the additive properties of the inner product and the rules for pulling a scalar out of an inner product.

The argument just given for the simple case is the basic clue of how the lemma should be proved. As we did above,

$$(u, w) = \left(\sum_{r=1}^{n} a_r v_r, \sum_{r=1}^{n} b_s v_s\right)$$
$$= \sum_{r=1}^{n} \sum_{s=1}^{n} a_r \bar{b}_s (v_r, v_s),$$

by the defining properties of the inner product. In this double sum, the only nonzero contribution comes if $r = s$, for otherwise $(v_r, v_s) = 0$. So the double sum reduces to a single sum, namely, $(u, w) = \sum_{s=1}^{n} a_s \bar{b}_s (v_s, v_s)$, and since $(v_s, v_s) = 1$ we get that $(u, w) = \sum_{s=1}^{n} a_s \bar{b}_s$. ∎

Notice that if $v_1, \ldots, v_n$ is the canonical basis $e_1, \ldots, e_n$, then the result shows that if $u = \sum_{r=1}^{n} a_r e_r$ and $w = \sum_{s=1}^{n} b_s e_s$, then $(u, w) = \sum_{s=1}^{n} a_s \bar{b}_s$. *This should come as no surprise because this is precisely the way in which we defined the inner product* (u, w).

Suppose that $v_1, \ldots, v_n$ is an orthonormal basis of $F^{(n)}$. We saw in Theorem 3.8.1 that the linear transformation T defined by $T(e_r) = v_r$ for $r = 1, 2, \ldots, n$ is invertible. However, both $e_1, \ldots, e_n$ and $v_1, \ldots, v_n$ are not any old bases of $F^{(n)}$ but are, in fact, both orthonormal. Surely this should force some further condition on T other than it be invertible. Indeed, it does!

Theorem 3.10.3. *If $v_1, \ldots, v_n$ is an orthonormal basis of $F^{(n)}$, then the linear transformation T, defined by $v_r = Te_r$ for $r = 1, 2, \ldots, n$ has the property that for all u, w in $F^{(n)}$, $(Tu, Tw) = (u, w)$.*

Proof: If $u = \sum_{r=1}^{n} a_r e_r$ and $w = \sum_{r=1}^{n} b_r e_r$ are in $F^{(n)}$, then $(u, w) = \sum_{s=1}^{n} a_s \bar{b}_s$. Now, by the definition of T and the fact that it is a linear transformation on $F^{(n)}$,

$$Tu = T\left(\sum_{r=1}^{n} a_r e_r\right) = \sum_{r=1}^{n} T(a_r e_r) = \sum_{r=1}^{n} a_r T e_r = \sum_{r=1}^{n} a_r v_r.$$

Similarly, $Tw = \sum_{s=1}^{n} b_s v_s$. By the result of Lemma 3.10.2, $(Tu, Tw) = \sum_{s=1}^{n} a_s \bar{b}_s = (u, w)$. ∎

In the theorem the condition $(Tu, Tw) = (u, w)$ for all $u, w \in F^{(n)}$ is equivalent to the condition $TT^* = I$, which we name in

Definition. A linear transformation T of $F^{(n)}$ is *unitary* if $TT^* = I$.

We illustrate Theorem 3.10.3. with an example. If $w_1 = e_1 + e_2$ and $w_2 = e_1 - e_2$, what is (w_1, w_2)? Calculating, we obtain

$$(w_1, w_2) = (e_1 + e_2, e_1 - e_2) = (e_1, e_1) - (e_1, e_2) + (e_2, e_1) - (e_2, e_2)$$
$$= 1 - 0 + 0 - 1 = 0.$$

So w_1 and w_2 are orthogonal. Also, $(w_1, w_1) = 2$ and $(w_2, w_2) = 2$ (Verify!), so the length of w_1, that is, $\sqrt{(w_1, w_1)}$, is $\sqrt{2}$, as is that of w_2. So if we let $v_1 = (1/\sqrt{2})w_1$ and $v_2 = (1/\sqrt{2})w_2$, then

$$(v_1, v_1) = ([1/\sqrt{2}]w_1, [1/\sqrt{2}]w_1) = \tfrac{1}{2}(w_1, w_1) = \tfrac{1}{2} \cdot 2 = 1.$$

Similarly, $(v_2, v_2) = 1$. So v_1 and v_2 form an orthonormal basis of $F^{(2)}$.

The change of basis linear transformation T is defined by $T(e_j) = v_j$ for $j = 1, 2$. So $T(e_1) = v_1 = (1/\sqrt{2})w_1 = (1/\sqrt{2})e_1 + (1/\sqrt{2})e_2$, and $T(e_2) = (1/\sqrt{2})e_1 - (1/\sqrt{2})e_2$. So the matrix of T in the basis e_1, e_2 is $\begin{bmatrix} 1/\sqrt{2} & 1/\sqrt{2} \\ 1/\sqrt{2} & -1/\sqrt{2} \end{bmatrix}$. Notice that

$$TT^* = \begin{bmatrix} 1/\sqrt{2} & 1/\sqrt{2} \\ 1/\sqrt{2} & -1/\sqrt{2} \end{bmatrix} \begin{bmatrix} 1/\sqrt{2} & 1/\sqrt{2} \\ 1/\sqrt{2} & -1/\sqrt{2} \end{bmatrix} = \begin{bmatrix} 1 & 0 \\ 0 & 1 \end{bmatrix};$$

that is, T is unitary.

Theorem 3.10.3 can be generalized. We leave the proof of the next result to the reader.

Theorem 3.10.4. *Let $v_1, \ldots, v_n$ and $w_1, \ldots, w_n$ be two orthonormal bases of $F^{(n)}$. Then the linear transformation T defined by $T(v_r) = w_r$ for $r = 1, 2, \ldots, n$ is unitary.*

What Theorem 3.10.4 says is that changing bases from one orthonormal basis to another is carried out by a unitary transformation.

What do these results translate into matricially? By Theorem 3.9.3, if A is a matrix, then the matrix of A—as a linear transformation—in a basis $v_1, \ldots, v_n$ is $B = C^{-1}AC$, where C is the matrix defined by $v_r = C(e_r)$ for some $r = 1, 2, \ldots, n$. If, furthermore, $v_1, \ldots, v_n$ is an orthonormal basis, what is implied for C? We have an exercise that $CC^* = I$; that is, C is unitary. That is,

Theorem 3.10.5. *If A is a matrix in $M_n(F)$, then A is transformed as a matrix, in*

Sec. 3.10] Bases and Inner Products 161

the orthonormal basis $v_1,\ldots,v_n$, into the matrix $B = C^{-1}AC$, where $v_r = C(e_r)$ and C is unitary.

In closing this section, we mention a useful theorem on independence of orthogonal vectors.

Theorem 3.10.6. If $v_1,\ldots,v_r$ are nonzero mutually orthogonal vectors in $F^{(n)}$, then they are linearly independent.

Proof: If $a_1v_1 + \cdots + a_rv_r = 0$ and $1 \leq s \leq r$, then $0 = (a_1v_1 + \cdots + a_rv_r, v_s) = a_s(v_s, v_s)$. Since v_s is nonzero, and $(v_s, v_s) \neq 0$, a_s must be 0. Since this is true for all s, the $v_1,\ldots,v_r$ are linearly independent. ∎

PROBLEMS

NUMERICAL PROBLEMS

1. Check which of the following are orthonormal bases of the given $F^{(n)}$.

   (a) $\begin{bmatrix} \frac{1}{2} \\ 0 \\ 1 \end{bmatrix}, \begin{bmatrix} 0 \\ 1 \\ 0 \end{bmatrix}, \begin{bmatrix} -1 \\ 1 \\ 1 \end{bmatrix}$ in $F^{(3)}$.

   (b) $\begin{bmatrix} i/\sqrt{2} \\ -i/\sqrt{2} \end{bmatrix}, \begin{bmatrix} -1/\sqrt{2} \\ 1/\sqrt{2} \end{bmatrix}$ in $\mathbb{C}^{(2)}$.

   (c) $\begin{bmatrix} 1 \\ 0 \\ 0 \end{bmatrix}, \begin{bmatrix} 0 \\ 1 \\ 0 \end{bmatrix}, \begin{bmatrix} \frac{1}{2} \\ 0 \\ \frac{\sqrt{3}}{2} \end{bmatrix}$ in $F^{(3)}$.

2. Find an element in $F^{(3)}$ orthogonal to both $\begin{bmatrix} 1 \\ 0 \\ 0 \end{bmatrix}$ and $\begin{bmatrix} 0 \\ 1 \\ 1 \end{bmatrix}$.

3. Find the matrix of the change of basis from

   $$\begin{bmatrix} 1 \\ 0 \\ 0 \end{bmatrix}, \begin{bmatrix} 0 \\ 1 \\ 0 \end{bmatrix}, \begin{bmatrix} 0 \\ 0 \\ 1 \end{bmatrix}$$

   to

   $$\begin{bmatrix} i/\sqrt{2} \\ 0 \\ -1/\sqrt{2} \end{bmatrix}, \begin{bmatrix} 0 \\ 1 \\ 0 \end{bmatrix}, \begin{bmatrix} 0 \\ 0 \\ i \end{bmatrix}$$

   in $\mathbb{C}^{(3)}$.

4. Verify that the matrix you obtain in Problem 3 is unitary.

5. For what value of a is the basis $\begin{bmatrix} a \\ -a \end{bmatrix}$ and $\begin{bmatrix} i/\sqrt{2} \\ i/\sqrt{2} \end{bmatrix}$ an orthonormal basis of $\mathbb{C}^{(2)}$?

MORE THEORETICAL PROBLEMS

Easier Problems

6. Give a complete proof of Theorem 3.10.4.
7. Give a complete proof of Theorem 3.10.5.
8. If $v \neq 0$ is in $F^{(n)}$, for what values of a in F is $\dfrac{v}{a}$ a unit vector?
9. If $v_1, \ldots, v_n$ is an orthogonal basis of $F^{(n)}$, construct from them an orthonormal basis of $F^{(n)}$.
10. If v_1, v_2, v_3 is a basis of $F^{(3)}$, use them to construct an orthogonal basis of $F^{(3)}$. How would you make this new basis into an orthonormal one?
11. If v_1, v_2, v_3 are in $F^{(4)}$, show that you can find a vector w in $F^{(4)}$ that is orthogonal to all of v_1, v_2, and v_3.
12. If T is a unitary linear transformation on $F^{(n)}$ such that $T(v_1), \ldots, T(v_n)$ is an orthonormal basis of $F^{(n)}$, show that $v_1, \ldots, v_n$ is an orthonormal basis of $F^{(n)}$.

Middle-Level Problems

13. If $T^* = T$ is a linear transformation on $F^{(n)}$ and if for $v \neq 0$, $w \neq 0$, $T(v) = av$ and $T(w) = bw$, where $a \neq b$ are in F, show that v and w are orthogonal.
14. If A is a symmetric matrix in $M_n(F)$ and if $v_1, \ldots, v_n$ in $F^{(n)}$ are nonzero and such that $Av_1 = a_1 v_1$, $Av_2 = a_2 v_2, \ldots, Av_n = a_n v_n$, where the a_j are *distinct* elements of F, show
    (a) $v_1, \ldots, v_n$ is an orthogonal basis of $F^{(n)}$.
    (b) We can construct from $v_1, \ldots, v_n$ an orthogonal basis of $F^{(n)}$, $w_1, \ldots, w_n$, such that $Aw_j = a_j w_j$ for $j = 1, 2, \ldots, n$.
    (c) There exists a matrix C in $M_n(F)$ such that $C^{-1}AC = \begin{bmatrix} a_1 & & & 0 \\ & a_2 & & \\ & & \ddots & \\ 0 & & & a_n \end{bmatrix}$.

15. In Part (c) of Problem 14, show that we can find a unitary C such that
$$C^{-1}AC = \begin{bmatrix} a_1 & & & 0 \\ & a_2 & & \\ & & \ddots & \\ 0 & & & a_n \end{bmatrix}.$$

Harder Problems

16. If $v_1, \ldots, v_n$ is a basis of $F^{(n)}$ and T a linear transformation on $F^{(n)}$ such that
$$Tv_1 = a_{11} v_1,$$
$$Tv_2 = a_{21} v_1 + a_{22} v_2,$$

$$Tv_j = a_{j1}v_1 + \cdots + a_{jj}v_j,$$
$$\vdots$$
$$Tv_n = a_{n1}v_1 + \cdots + a_{nn}v_n,$$

show that T satisfies $(T - a_{11}I)(T - a_{22}I)\cdots(T - a_{nn}I) = 0$.

17. If $A \in M_3(F)$ and A satisfies $A^3 = 0$, show that we can find a basis v_1, v_2, v_3 of $F^{(3)}$ such that $Av_1 = 0$, $Av_2 = av_1$, and $Av_3 = bv_1 + cv_2$ where a, b, c are in F.

18. For A in $M_3(F)$ such that $A^3 = 0$, show that we can find C invertible in $M_3(F)$ such that $C^{-1}AC = \begin{bmatrix} u & * & * \\ 0 & v & * \\ 0 & 0 & w \end{bmatrix}$.

CHAPTER 4

More on $n \times n$ Matrices

4.1. SUBSPACES

When we first considered $F^{(n)}$, we observed that $F^{(n)}$ enjoys two fundamental properties, namely, if $v, w \in F^{(n)}$, then $v + w$ is also in $F^{(n)}$, and if $a \in F$, then av is again in $F^{(n)}$. It is quite possible that a nonempty subset of $F^{(n)}$ could imitate $F^{(n)}$ in this regard and also satisfy these two properties. Such subsets are thus distinguished among all other subsets of $F^{(n)}$. We single them out in

Definition. A nonempty subset W of $F^{(n)}$ is called a *subspace* of $F^{(n)}$ if:

1. $v, w \in W$ implies that $v + w \in W$.
2. $a \in F, v \in W$ implies that $av \in W$.

A few simple examples of subspaces come to mind immediately. Clearly, $F^{(n)}$ itself is a subspace of $F^{(n)}$. Also, the set consisting of the element 0 alone is a subspace of $F^{(n)}$. These are called *trivial subspaces*.

A subspace $W \neq F^{(n)}$ of $F^{(n)}$ is called a *proper subspace* of $F^{(n)}$. What are some proper subspaces of $F^{(n)}$? Let $W = \left\{ \begin{bmatrix} x \\ y \\ z \end{bmatrix} \in F^{(3)} \mid x + y + z = 0 \right\}$. Then W is a proper subspace of $F^{(3)}$. Why? If $\begin{bmatrix} a \\ b \\ c \end{bmatrix}$ and $\begin{bmatrix} u \\ v \\ w \end{bmatrix}$ are in W, then $a + b + c = 0$ and $u + v + w = 0$, hence $(a + u) + (b + v) + (c + w) = 0$; thus the element

Sec. 4.1] Subspaces 165

$$\begin{bmatrix} a+u \\ b+v \\ c+w \end{bmatrix} = \begin{bmatrix} a \\ b \\ c \end{bmatrix} + \begin{bmatrix} u \\ v \\ w \end{bmatrix}$$ satisfies the requirements for membership in W, hence is in W. Similarly, $t\begin{bmatrix} a \\ b \\ c \end{bmatrix} = \begin{bmatrix} ta \\ tb \\ tc \end{bmatrix}$ is such that $ta + tb + tc = 0$ if $a + b + c = 0$.

Thus $t\begin{bmatrix} a \\ b \\ c \end{bmatrix}$ is in W for $t \in F$ and $\begin{bmatrix} a \\ b \\ c \end{bmatrix}$ in W. Therefore, W is a subspace of $F^{(3)}$.

The element $\begin{bmatrix} 1 \\ 0 \\ 0 \end{bmatrix}$ is not in W since $1 + 0 + 0 \neq 0$. So, W is also proper.

In fact, the vectors $\begin{bmatrix} u_1 \\ \vdots \\ u_n \end{bmatrix}$, where the elements $u_1, \ldots, u_n$ are the solutions of a system of homogeneous linear equations, form a subspace of $F^{(n)}$. (Prove!) The case above where $x + y + z = 0$ in $F^{(3)}$ is merely a special case of this.

If T is a linear transformation on $F^{(n)}$ and if $W = \{T(v) | v \in F^{(n)}\}$, then W is a subspace of $F^{(n)}$. Why? If $w_1 = Tv_1$ and $w_2 = Tv_2$, then $w_1 + w_2 = Tv_1 + Tv_2 = T(v_1 + v_2)$, so is in W. Similarly, if $a \in F$ and $w_1 = Tv_1$, then $aw_1 = aTv_1 = T(av_1)$, so aw_1 is in W. This subspace W, which we may denote as $T(F^{(n)})$, is called the *image space* of T. As we shall see, $T(F^{(n)})$ is a *proper* subspace of $F^{(n)}$ if and only if T is *not* invertible. For example, the image space of $T = \begin{bmatrix} 1 & 1 & 3 \\ 0 & 1 & 1 \\ 1 & 0 & 2 \end{bmatrix}$ is the plane

$$W = \left\{ u\begin{bmatrix} 1 \\ 0 \\ 1 \end{bmatrix} + v\begin{bmatrix} 1 \\ 1 \\ 0 \end{bmatrix} \middle| u, v \in \mathbb{R} \right\}$$

since the last column is 2 times the first plus 1 times the second, so that

$$\begin{bmatrix} 1 & 1 & 3 \\ 0 & 1 & 1 \\ 1 & 0 & 2 \end{bmatrix}\begin{bmatrix} a \\ b \\ c \end{bmatrix} = a\begin{bmatrix} 1 \\ 0 \\ 1 \end{bmatrix} + b\begin{bmatrix} 1 \\ 1 \\ 0 \end{bmatrix} + c\begin{bmatrix} 3 \\ 1 \\ 2 \end{bmatrix}$$

$$= (a + 2c)\begin{bmatrix} 1 \\ 0 \\ 1 \end{bmatrix} + (b + c)\begin{bmatrix} 1 \\ 1 \\ 0 \end{bmatrix} = u\begin{bmatrix} 1 \\ 0 \\ 1 \end{bmatrix} + v\begin{bmatrix} 1 \\ 1 \\ 0 \end{bmatrix} = \begin{bmatrix} u+v \\ v \\ u \end{bmatrix}$$

for $u = a + 2c$ and $v = b + c$.

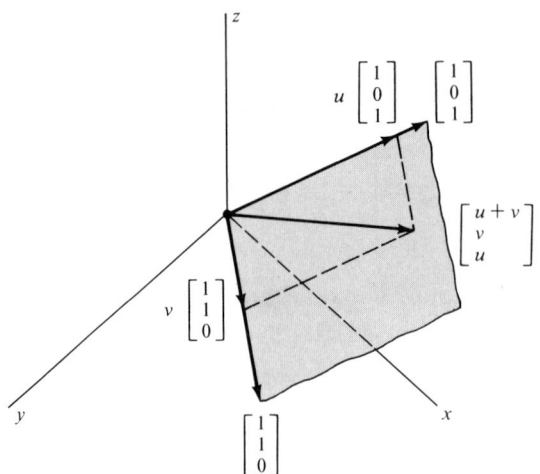

Finally, the span $\langle v_1, \ldots, v_r \rangle$ over F of vectors $v_1, \ldots, v_r$ in $F^{(n)}$ is a subspace of $F^{(n)}$. Since we have tremendous flexibility in forming such a span (we simply choose a finite number of vectors any way we want to), you may wonder whether every subspace of $F^{(n)}$ can be gotten as such a span. The answer is "yes." In fact, every subspace has a basis in the sense of the following generalization of our definition for basis of $F^{(n)}$.

Definition. A *basis* for a subspace V of $F^{(n)}$ is a set $v_1, \ldots, v_r$ of vectors of $F^{(n)}$ such that

1. The vectors $v_1, \ldots, v_r$ are linearly independent.
2. $v_1, \ldots, v_r$ span V over F, that is, $\langle v_1, \ldots, v_r \rangle = V$.

How can we get a basis for a subspace V of $F^{(n)}$? Easily! We know that if $v_1, \ldots, v_r$ are linearly independent elements of $F^{(n)}$, then $r \leq n$. Thus we can take a collection $v_1, \ldots, v_r$ of distinct linearly independent vectors of V over F which is *maximal* in the sense that any larger set $v_1, \ldots, v_r, v$ of vectors in V is linearly dependent over V. But then the vectors $v_1, \ldots, v_r$ form a basis. Why? We need only show that any vector v in V is a linear combination of $v_1, \ldots, v_r$. To show this, note that the set of vectors $v_1, \ldots, v_r, v$ must be linearly dependent by the maximality condition for $v_1, \ldots, v_r$. Let $a_1, \ldots, a_r, a$ be elements of F (not all zero) such that

$$a_1 v_1 + \cdots + a_r v_r + av = 0.$$

We can't have $a = 0$ since the vectors $v_1, \ldots, v_r$ are linearly independent. After all, if $a = 0$, then $a_1 v_1 + \cdots + a_r v_r = 0$ and so $a_1 = \cdots = a_r = 0$, since $v_1, \ldots, v_r$ are linearly independent. Hence we can solve for v as a linear combination

$$v = -\frac{1}{a}(a_1 v_1 + \cdots + a_r v_r) = -\frac{1}{a} a_1 v_1 - \cdots - \frac{1}{a} a_r v_r$$

of $v_1, \ldots, v_r$. Thus the linearly independent vectors $v_1, \ldots, v_r$ also span V, so form a basis for V over F. This proves the

Theorem 4.1.1. Every maximal linearly independent set $v_1, \ldots, v_r$ of vectors in a subspace V of $F^{(n)}$ is a basis for V over F.

By starting with any linearly independent set $v_1, \ldots, v_r$ of vectors in a subspace V of $F^{(n)}$ and enlarging it to a maximal linearly independent set $v_1, \ldots, v_s$ of vectors in V, we put the vectors $v_1, \ldots, v_r$ in a basis $v_1, \ldots, v_s$ for V over F, by Theorem 4.1.1. This procedure of *extending* the set $v_1, \ldots, v_r$ to a basis $v_1, \ldots, v_s$ is very useful, so we state our observations here in the

Corollary 4.1.2. Every linearly independent set $v_1, \ldots, v_r$ of vectors in a subspace V of $F^{(n)}$ can be extended to a basis of V over F.

We saw in Corollary 3.6.3 that if $v_1, \ldots, v_r$ and $w_1, \ldots, w_s$ are two linear independent sets of vectors in $F^{(n)}$ such that $\langle v_1, \ldots, v_r \rangle$ and $\langle w_1, \ldots, w_s \rangle$ are equal, then $r = s$. This proves the important

Theorem 4.1.3. Any two bases for a subspace V of $F^{(n)}$ over F have the same number of elements.

Thanks to Theorems 4.1.1 and 4.1.3, we now can associate a well-defined dimension to any subspace V of $F^{(n)}$.

Definition. The *dimension* of a subspace V of $F^{(n)}$ over F is the number of elements of a basis for V over F. We denote it by $\dim(V)$.

We now proceed to investigate the properties of this dimension function, starting with

Theorem 4.1.4. Let V and W be subspaces of $F^{(n)}$ and suppose that V is a proper subset of W. Then $\dim(V) < \dim(W)$.

Proof: Take a basis $v_1, \ldots, v_r$ for V over F. Then extend the basis $v_1, \ldots, v_r$ of V to a basis $v_1, \ldots, v_s$ for W. Since V and W are not equal, r does not equal s. But then $r < s$. ∎

The case of Theorem 4.1.4 with $W = F^{(n)}$ is the

Corollary 4.1.5. If V is a subspace of $F^{(n)}$ of dimension n, then $V = F^{(n)}$.

We leave it as an exercise for you to refine the proof of Theorem 4.1.4 to show that if V and W are subspaces of $F^{(n)}$ and V is a proper subset of W, then there are subspaces $V_d = V$, V_{d+1}, ..., $V_{d+c} = W$, where the V_k satisfy the following conditions:

1. V_k is contained in V_{k+1} for $d \leq k < d + c$.
2. $\dim(V_k) = k$ for $d \leq k \leq d + c$.

PROBLEMS

NUMERICAL PROBLEMS

1. In $F^{(3)}$ describe every element in $\langle w_1, w_2, w_3 \rangle$, where

$$w_1 = \begin{bmatrix} 1 \\ 0 \\ 0 \end{bmatrix}, \quad w_2 = \begin{bmatrix} 1 \\ 2 \\ 0 \end{bmatrix}, \quad w_3 = \begin{bmatrix} 1 \\ 2 \\ 3 \end{bmatrix}.$$

2. What is the dimension of $\langle w_1, w_2, w_3 \rangle$ in Problem 1?

3. In $\mathbb{C}^{(4)}$ find the form of every elements in $\langle w_1, w_2, w_3 \rangle$, where

$$w_1 = \begin{bmatrix} 1 \\ 1 \\ 0 \\ 0 \end{bmatrix}, \quad w_2 = \begin{bmatrix} i \\ 0 \\ 1 \\ 1 \end{bmatrix}, \quad w_3 = \begin{bmatrix} i \\ i \\ 0 \\ 0 \end{bmatrix}.$$

4. What is the dimension of the $\langle w_1, w_2, w_3 \rangle$ in Problem 3 over $\mathbb{C}$?

5. If $A = \begin{bmatrix} 1 & 2 & 3 \\ 0 & 4 & 2 \\ 2 & 4 & 6 \end{bmatrix}$, find the dimension of $A(F^{(3)})$ over F. Describe the general element in $A(F^{(3)})$.

MORE THEORETICAL PROBLEMS

Easier Problems

6. Prove Corollary 4.1.2.
7. Show that if $w_1, \ldots, w_r$ are in $F^{(n)}$, then $\langle w_1, \ldots, w_r \rangle$ is a subspace of $F^{(n)}$.
8. Show that $\dim(\langle w_1, \ldots, w_r \rangle) \leq r$ and equals r if and only if $w_1, \ldots, w_r$ are linearly independent over F.
9. If $z_1, \ldots, z_m$ are in $\langle w_1, \ldots, w_r \rangle$, prove that $\langle z_1, \ldots, z_m \rangle \subset \langle w_1, \ldots, w_r \rangle$.
10. If V and W are subspaces of $F^{(n)}$, show that $V \cap W$ is a subspace of $F^{(n)}$.
11. If $W_1, \ldots, W_k$ are subspaces of $F^{(n)}$, show that $W_1 \cap W_2 \cap \cdots \cap W_k$ is a subspace of $F^{(n)}$.
12. If $A \in M_n(F)$, let $W = \{v \in F^{(n)} \mid Av = 0\}$. Prove that W is a subspace of $F^{(n)}$.
13. If $A \in M_n(F)$, let

$$V = \{v \in F^{(n)} \mid A^k v = 0 \text{ for some positive integer } k \text{ depending on } v\}.$$

Show that V is a subspace of $F^{(n)}$.

14. If $A \in M_n(F)$, let $W_a = \{v \in F^{(n)} \mid Av = av\}$, where $a \in F$. Show that W_a is a subspace of $F^{(n)}$.

Sec. 4.1]   Subspaces   169

15. For $A \in M_n(F)$, let

$$V_a = \{v \in F^{(n)} \mid (A - aI)^k v = 0 \text{ for some positive integer } k \text{ depending on } v\},$$

where $a \in F$. Prove that V_a is a subspace of $F^{(n)}$.

16. If $a \neq b$ are in F, show that $V_a \cap V_b = \{0\}$, where V_a, V_b are as defined in Problem 15.

17. If V and W are subspaces of $F^{(n)}$, define the *sum*, $V + W$, of V and W by $V + W = \{v + w \mid v \in V, w \in W\}$. Show that $V + W$ is a subspace of $F^{(n)}$.

18. If $V_1, \ldots, V_k$ are subspaces of $F^{(n)}$, how would you define the sum, $V_1 + V_2 + \cdots + V_k$, of $V_1, \ldots, V_k$? Show that what you get is a subspace of $F^{(n)}$.

19. If V and W are subspaces of $F^{(n)}$ such that $V \cap W = \{0\}$, show that every element in $V + W$ has a *unique* representation in the form $v + w$, where v is in V and w is in W. (In this case we call $V + W$ the *direct sum* of V and W and write it as $V \oplus W$.)

20. If V_1, V_2, V_3 are subspaces of $F^{(n)}$ such that every element of $V_1 + V_2 + V_3$ has a *unique* representation in the form $v_1 + v_2 + v_3$, where $v_1 \in V_1$, $v_2 \in V_2$, $v_3 \in V_3$, show that $V_1 \cap (V_2 + V_3) = V_2 \cap (V_1 + V_3) = V_3 \cap (V_1 + V_2) = \{0\}$. (In this case, $V_1 + V_2 + V_3$ is called the *direct sum* of V_1, V_2, V_3 and is written as $V_1 \oplus V_2 \oplus V_3$.)

Middle-Level Problems

21. Generalize the result of Problem 20 to $V_1, \ldots, V_k$ subspaces of $F^{(n)}$.

22. If $A \in M_n(F)$ and V is a subspace of $F^{(n)}$, we say that V is *invariant* under A if $Av \in V$ for all $v \in V$, that is, if $AV \subset V$. Prove that the subspace V_a of Problem 15 is invariant under A.

23. If T is a linear transformation on $F^{(n)}$ and V is a subspace of $F^{(n)}$ invariant under T (i.e., $TV \subset V$), show that $\tilde{T}$ defined on V by $\tilde{T}v = Tv$ is a mapping from V to V such that $T(v + w) = Tv + Tw$ and $T(cv) = cT(v)$ for all $c \in F$.

24. If V and W are subspaces of $F^{(n)}$ such that $F^{(n)} = V \oplus W$ (see Problem 19) and $v_1, \ldots, v_k$ is a basis of V over F and $w_1, \ldots, w_m$ is a basis of W over F, show that $v_1, \ldots, v_k, w_1, \ldots, w_m$ is a basis of $F^{(n)}$ over F.

25. If V, W, $v_1, \ldots, v_k$, $w_1, \ldots, w_m$ are as in Problem 24 and if T is a linear transformation on $F^{(n)}$ such that $T(V) \subset V$ and $T(W) \subset W$, show that the matrix of T in the bases $v_1, \ldots, v_k, w_1, \ldots, w_m$ looks like $\begin{bmatrix} A & 0 \\ \hline 0 & B \end{bmatrix}$, where A is a $k \times k$ matrix and B is an $m \times m$ matrix over F.

26. How could you describe A and B in Problem 25 in terms of the linear transformations $\tilde{T}v = Tv (v \in V)$ of V and $\tilde{\tilde{T}}w = Tw (w \in W)$ of W?

27. If $v \in F^{(n)}$, let $v^\perp = \{w \in F^{(n)} \mid (v, w) = 0\}$. Prove that $v^\perp$ is a subspace of $F^{(n)}$.

28. If W is a subspace of $F^{(n)}$ and $w_1, \ldots, w_m$ is a basis of W over F, suppose that $v \in F^{(n)}$ is such that $(w_j, v) = 0$ for $j = 1, 2, \ldots, m$. Show that $(w, v) = 0$ for all $w \in W$.

29. If W is a subspace of $F^{(n)}$, let $W^\perp = \{v \in F^{(n)} \mid (v, w) = 0 \text{ for all } w \in W\}$. Show that $W^\perp$ is a subspace of $F^{(n)}$.

30. If W and $W^\perp$ are as in Problem 29, show that the sum $W + W^\perp$ is a direct sum.

Harder Problems

31. If V and W are subspaces of $F^{(n)}$ such that $V \cap W = 0$, then prove that $\dim(V + W) = \dim(V) + \dim(W)$.

32. Generalize Problem 31 as follows: If V and W are subspaces of $F^{(n)}$, show that $\dim(V + W) = \dim(V) + \dim(W) - \dim(V \cap W)$.

33. Let $v_1, \ldots, v_m$ be a set of orthogonal elements in $F^{(n)}$. Construct an element $w \in F^{(n)}$ such that $(w, v_j) = 0$ for all $j = 1, 2, \ldots, m$, assuming $m < n$.

34. If W is a subspace of $F^{(n)}$ of dimension m, show that we can find a $1 - 1$ mapping f of W onto $F^{(m)}$ such that f satisfies both
    (a) $f(w_1 + w_2) = f(w_1) + f(w_2)$
    (b) $\quad f(aw_1) = af(w_1)$
    for all w_1, w_2 in W and all a in F.

35. Suppose that you know that you have a mapping $f: F^{(m)} \to F^{(n)}$ which is *onto* and which satisfies
    (a) $f(u + v) = f(u) + f(v)$
    (b) $\quad f(au) = af(u)$
    for all $u, v \in F^{(n)}$ and all $a \in F$. Prove that $m \geq n$.

36. In Problem 35, when can you conclude that $m = n$?

37. Refine the proof of Theorem 4.1.4 to show that if V and W are subspaces of $F^{(n)}$ and V is a proper subset of W, then there are subspaces $V_d = V$, $V_{d+1}, \ldots, V_{d+c} = W$ where the V_k satisfy the following conditions:
    (a) V_k is contained in V_{k+1} for $d \leq k < d + c$;
    (b) $\dim V_k = k$ for $d \leq k \leq d + c$.

38. If W is a subspace of $F^{(n)}$ with $\dim(W) = s$, then given vectors $w_1, \ldots, w_s$ in W which are linearly independent over F, show that they form a basis for W over F.

NOTE TO THE READER. The problems from Section 4.1 on represent important ideas that we develop more formally in the next few sections. The results and techniques needed to solve them are not only important for matrices but will play a key role in discussing vector spaces. Don't give in to the temptation to peek ahead to solve them. Try to solve them on your own.

4.2. MORE ON SUBSPACES

We shall examine here some of the items that appeared as exercises in the problem set of the preceding section. If you haven't tried those problems as yet, don't read on. Go back and take a stab at the problems. Solving them on your own is n times as valuable for your understanding of the material than letting us solve them for you, with n a very large integer.

We begin with

Definition. If V and W are subspaces of $F^{(n)}$, then their *sum* $V + W$ is defined by $V + W = \{v + w \mid v \in V, w \in W\}$.

Sec. 4.2] More on Subspaces 171

Of course, $V + W$ is not merely a subset of $F^{(n)}$, it inherits the basic properties of $F^{(n)}$. In short, $V + W$ is a subspace of $F^{(n)}$. We proceed to prove it now. The proof is easy but is the prototype of many proofs that we might need to verify that a given subset of $F^{(n)}$ is a subspace of $F^{(n)}$.

Lemma 4.2.1. If V and W are subspaces of $F^{(n)}$, then $V + W$ is a subspace of $F^{(n)}$.

Proof: What must we show in order to establish the lemma? We need only demonstrate two things: first, if z_1 and z_2 are in $V + W$, then $z_1 + z_2$ is also in $V + W$, and second, if a is any element of F and z is in $V + W$, then az is also in $V + W$.

Suppose then that z_1 and z_2 are in $V + W$. By the definition of $V + W$ we have that $z_1 = v_1 + w_1$ and $z_2 = v_2 + w_2$, where v_1, v_2 are in V and w_1, w_2 are in W. Therefore,

$$z_1 + z_2 = v_1 + w_1 + v_2 + w_2 = (v_1 + v_2) + (w_1 + w_2).$$

Because V and W are subspaces of $F^{(n)}$, we know that $v_1 + v_2$ is in V and $w_1 + w_2$ is in W. Hence $z_1 + z_2$ is in $V + W$.

If $a \in F$ and $z = v + w$ is in $V + W$, then $az = a(v + w) = av + aw$. However, V and W are subspaces of $F^{(n)}$, so $av \in V$ and $aw \in W$. Thus $az \in V + W$.

Having shown that $V + W$ satisfies the criteria for a subspace of $F^{(n)}$, we know that $V + W$ is a subspace of $F^{(n)}$. ∎

Of course, having in hand the fact that $V + W$ is a subspace of $F^{(n)}$ allows us to prove that *the sum of any number of subspaces of $F^{(n)}$ is again a subspace of $F^{(n)}$*. We leave this as an exercise for the reader.

Some sums of subspaces are better than others. For instance, if

$$V_1 = \left\{ \begin{bmatrix} x \\ t \\ t \end{bmatrix} \,\middle|\, x, t \in \mathbb{R} \right\}$$

and

$$V_2 = \left\{ \begin{bmatrix} x \\ x \\ t \end{bmatrix} \,\middle|\, x, t \in \mathbb{R} \right\},$$

the sum $V_1 + V_2$ is $\mathbb{R}^{(3)}$, but elements $v_1 + v_2$ from this sum can be expressed in more than one way $\left(\text{e.g., } \begin{bmatrix} 1 \\ 2 \\ 2 \end{bmatrix} + \begin{bmatrix} 2 \\ 2 \\ 4 \end{bmatrix} = \begin{bmatrix} 3 \\ 4 \\ 6 \end{bmatrix} = \begin{bmatrix} 2 \\ 3 \\ 3 \end{bmatrix} + \begin{bmatrix} 1 \\ 1 \\ 3 \end{bmatrix} \right)$. A better sum is $V_1 + V_3$, where $V_3 = \left\{ \begin{bmatrix} x \\ x \\ x \end{bmatrix} \,\middle|\, x \in \mathbb{R} \right\}$. This sum is also $\mathbb{R}^{(3)}$, but it also has the property that elements $v_1 + v_3 (v_1 \in V_1, v_3 \in V_3)$ can be expressed in only one way. This sum is an example of the special sums which we introduce in

Definition. If $V_1, \ldots, V_m$ is a set of subspaces of $F^{(n)}$, then their sum $V_1 + V_2 + \cdots + V_m$ is called a *direct sum* if every element w in $V_1 + V_2 + \cdots + V_m$ has a *unique* representation as

$$w = v_1 + v_2 + \cdots + v_m,$$

where each v_j is in V_j for $j = 1, 2, \ldots, m$.

We shall denote the fact that the sum of $V_1, \ldots, V_m$ is a direct sum by writing it as $V_1 \oplus V_2 \oplus \cdots \oplus V_m$.

We consider an easier case first, that in which $m = 2$. When can we be sure that $V + W$ is a direct sum? We claim that all we need, in this special instance, is that $V \cap W = \{0\}$. Assuming that $V \cap W = \{0\}$, suppose that $z \in V + W$ has two representations $z = v_1 + w_1$ and $z = v_2 + w_2$, where v_1, v_2 are in V and w_1, w_2 are in W. Therefore, $v_1 + w_1 = v_2 + w_2$, so that $v_1 - v_2 = w_2 - w_1$. But $v_1 - v_2$ is in V and $w_2 - w_1$ is in W, and since they are equal, we must have that $v_1 - v_2 \in V \cap W = \{0\}$ and $w_2 - w_1 \in V \cap W = \{0\}$. These yield that $v_1 = v_2$ and $w_1 = w_2$. So z only has one representation in the form $v + w$. Thus the sum of V and W is direct.

In the opposite direction, we claim that if $V + W = V \oplus W$, that is, the sum is direct, then $V \cap W = \{0\}$. Why? Suppose that u is in both V and W. Then certainly $u = u + 0$, where the u and 0 on the right-hand side are viewed as in V and W, respectively. However, we also know that $u = 0 + u$, where $0, u$ are in V and W, respectively. So $u + 0 = 0 + u$ in $V \oplus W$; by the directness of the sum we conclude that $u = 0$. Hence $V \cap W = \{0\}$.

We have proved

Theorem 4.2.2. *If V and W are nonzero subspaces of $F^{(n)}$, their sum $V + W$ is a direct sum if and only if $V \cap W = \{0\}$.*

Whereas the criterion that the sum of two subspaces be direct is quite simple, that for determining the directness of *more* than two subspaces is much more complicated. *It is no longer true that the sum $V_1 + V_2 + \cdots + V_m$ is a direct sum if and only if $V_j \cap V_k = \{0\}$ for all $j \neq k$.* It is true that if the sum $V_1 + \cdots + V_m$ is direct, then $V_j \cap V_k = \{0\}$ for $j \neq k$. The proof of this is a slight adaptation of the argument given above showing $V \cap W = 0$. *But the condition is not sufficient.* The best way of demonstrating this is by an example. In $F^{(3)}$ let

$$V_1 = \left\{ \begin{bmatrix} a \\ b \\ 0 \end{bmatrix} \,\bigg|\, a, b \in F \right\}, \quad V_2 = \left\{ \begin{bmatrix} 0 \\ 0 \\ c \end{bmatrix} \,\bigg|\, c \in F \right\}, \quad V_3 = \left\{ \begin{bmatrix} d \\ d \\ d \end{bmatrix} \,\bigg|\, d \in F \right\}.$$

We easily see that

$$V_1 + V_2 = F^{(3)}$$

$$V_1 \cap V_2 = V_1 \cap V_3 = V_2 \cap V_3 = \{0\},$$

yet the sum $V_1 + V_2 + V_3$ is *not* direct. To see why, simply express some nonzero element $\begin{bmatrix} a \\ b \\ 0 \end{bmatrix}$ of V_1 in terms of elements $\begin{bmatrix} 0 \\ 0 \\ c \end{bmatrix}$ and $\begin{bmatrix} d \\ d \\ d \end{bmatrix}$. We can do this by taking $a = b = d = 1$ and $c = -1$, getting the two representations

$$1\begin{bmatrix} 1 \\ 1 \\ 0 \end{bmatrix} + 0\begin{bmatrix} 0 \\ 0 \\ -1 \end{bmatrix} + 0\begin{bmatrix} 1 \\ 1 \\ 1 \end{bmatrix} = \begin{bmatrix} 1 \\ 1 \\ 0 \end{bmatrix} = 1\begin{bmatrix} 0 \\ 0 \\ 0 \end{bmatrix} + 1\begin{bmatrix} 0 \\ 0 \\ -1 \end{bmatrix} + 1\begin{bmatrix} 1 \\ 1 \\ 1 \end{bmatrix}.$$

The necessary and sufficient condition for the directness of a sum is given in

Theorem 4.2.3. If $V_1, \ldots, V_m$ are nonzero subspaces of $F^{(n)}$, then their sum is a direct sum if and only if for each $j = 1, 2, \ldots, m$,

$$V_j \cap (V_1 + V_2 + \cdots + V_{j-1} + V_{j+1} + \cdots + V_m) = \{0\}.$$

We leave the proof of Theorem 4.2.3 to the reader with the reminder that all one must show is that every element in the sum $V_1 + \cdots + V_m$ has a unique representation of the form $v_1 + v_2 + \cdots + v_m$, where $v_j \in V_j$, if and only if the condition stated in the theorem holds.

We give an instance of when a sum is direct. If $W \neq \{0\}$ is a subspace of $F^{(n)}$, let $W^\perp = \{v \in F^{(n)} | (v, w) = 0$ for all $w \in W\}$. That $W^\perp$ is a subspace of $F^{(n)}$ we leave to the reader (see Problem 29 of Section 4.1). We claim that $F^{(n)} = W \oplus W^\perp$ *if we happen to know that W has an orthonormal basis*. (As we shall soon see, this is no restriction on W; every subspace W will be shown to have an orthonormal basis.)

We need the following remark to carry out the proof: If $w_1, \ldots, w_k$ is a basis of W and $v \in F^{(n)}$ is such that $(v, w_j) = 0$ for $j = 1, 2, \ldots, k$, then $v \in W^\perp$. We leave this as an exercise.

Theorem 4.2.4. Let $V_1, \ldots, V_r$ be mutually orthogonal subspaces of $F^{(n)}$. Then the sum $V_1 + \cdots + V_r$ is direct.

Proof: Suppose that $v \in V_1 \cap (V_2 + \cdots + V_r)$. Then v is orthogonal to v (Prove!), which implies that $v = 0$. So, $V_1 \cap (V_2 + \cdots + V_r) = \{0\}$. Similarly, for $s > 1$

$$V_2 \cap (V_{j_2} + \cdots + V_{j_r}) = \{0\},$$

where $j_2, \ldots, j_r$ are the integers from 1 to r leaving out s. So the sum $V_1 + \cdots + V_r$ is direct. ∎

Theorem 4.2.5. If $W \neq \{0\}$ is a subspace of $F^{(n)}$, and if W has an orthonormal basis $w_1, \ldots, w_k$, then $F^{(n)} = W \oplus W^\perp$.

Proof: The sum $W + W^\perp$ is direct by Theorem 4.2.4. So it remains only to show that $W + W^\perp = F^{(n)}$. For this, let v be any element of $F^{(n)}$ and consider the element

$$z = v - (v, w_1)w_1 - (v, w_2)w_2 - \cdots - (v, w_k)w_k.$$

Calculating (z, w_j) for $j = 1, 2, \ldots, k$, we get

$$(z, w_j) = (v - (v, w_1)w_1 - \cdots - (v, w_k)w_k, w_j)$$
$$= (v, w_j) - (v, w_1)(w_1, w_j) - \cdots - (v, w_k)(w_k, w_j).$$

Because $w_1, \ldots, w_k$ is an orthonormal basis of W, $(w_r, w_s) = 0$ if $r \neq s$ and $(w_r, w_r) = 1$ for all r, s. So the sum above only has nonzero contributions from the term (v, w_j) and $-(v, w_j)(w_j, w_j)$. In a word, the value we get is

$$(z, w_j) = (v, w_j) - (v, w_j) = 0.$$

Therefore, z is orthogonal to a basis of W; by the remark preceding the theorem, $z \in W^\perp$. But

$$z = v - (v, w_1)w_1 - \cdots - (v, w_k)w_k,$$

hence

$$v = z + ((v, w_1)w_1 - \cdots - (v, w_k)w_k),$$

where $z \in W^\perp$. So we have shown that every v in $F^{(n)}$ is in $W + W^\perp$. Thus $F^{(n)} = W \oplus W^\perp$. ∎

We repeat what we said earlier. The condition "W has an orthonormal basis" is no restriction whatsoever on W. Every nonzero subspace has an orthonormal basis. To show this is our next objective.

PROBLEMS

NUMERICAL PROBLEMS

1. If $A = \begin{bmatrix} 1 & -1 & 0 \\ 0 & 2 & 3 \\ 0 & 0 & 0 \end{bmatrix}$ and $B = \begin{bmatrix} 0 & 0 & 0 \\ 0 & 1 & 0 \\ 1 & 0 & 1 \end{bmatrix}$ are in $M_3(F)$, then

    (a) Find the form of the general element of $A(F^{(3)})$ and of $B(F^{(3)})$.
    (b) Show that from Part (a), $A(F^{(3)})$ and $B(F^{(3)})$ are subspaces of $F^{(3)}$.
    (c) Find the general form of the elements of $A(F^{(3)}) + B(F^{(3)})$.
    (d) From the result of Part (c), show that $A(F^{(3)}) + B(F^{(3)})$ is a subspace of $F^{(3)}$.
    (e) Is $A(F^{(3)}) + B(F^{(3)}) = (A + B)(F^{(3)})$?
    (f) What is $\dim(A(F^{(3)}) + B(F^{(3)}))$?

2. If $A = \begin{bmatrix} 1 & 2 \\ 0 & 0 \end{bmatrix}$ and $B = \begin{bmatrix} 0 & 0 \\ 1 & a \end{bmatrix}$, find conditions on a so that $A(F^{(2)}) + B(F^{(2)}) \neq (A + B)(F^{(2)})$.

3. In Problem 1 find the explicit form of each element of $A(F^{(3)}) \cap B(F^{(3)})$. What is its dimension?

4. In Problem 3, verify that $\dim(A(F^{(3)}) + B(F^{(3)})) = \dim(A(F^{(3)})) + \dim(B(F^{(3n)})) - \dim(A(F^{(3)}) \cap B(F^{(3)}))$.

5. If $v_1 = \begin{bmatrix} 0 \\ 1 \\ 1 \\ 1 \end{bmatrix}$, $v_2 = \begin{bmatrix} 0 \\ 1 \\ 1 \\ 0 \end{bmatrix}$, $v_3 = \begin{bmatrix} 0 \\ 1 \\ 0 \\ 0 \end{bmatrix}$, and $w_1 = \begin{bmatrix} 1 \\ 0 \\ 1 \\ 0 \end{bmatrix}$, $w_2 = \begin{bmatrix} 0 \\ 0 \\ 1 \\ 1 \end{bmatrix}$, find the form of the general elements in $\langle v_1, v_2, v_3 \rangle$ and $\langle w_1, w_2 \rangle$. From these find the general element in $\langle v_1, v_2, v_3 \rangle + \langle w_1, w_2 \rangle$.

6. What are $\dim(\langle v_1, v_2, v_3 \rangle)$, $\dim(\langle w_1, w_2 \rangle)$, and $\dim(\langle v_1, v_2, v_3 \rangle + \langle w_1, w_2 \rangle)$?

7. Is the sum $\langle v_1, v_2, v_3 \rangle + \langle w_1, w_2 \rangle$ in Problem 6 direct?

8. If $v_1 = \begin{bmatrix} 0 \\ 1 \\ 1 \\ 1 \end{bmatrix}$, $v_2 = \begin{bmatrix} 0 \\ 0 \\ 1 \\ 1 \end{bmatrix}$, $v_3 = \begin{bmatrix} 0 \\ 0 \\ 0 \\ 1 \end{bmatrix}$, and $w_1 = \begin{bmatrix} 1 \\ 0 \\ 0 \\ 0 \end{bmatrix}$, $w_2 = \begin{bmatrix} i \\ 0 \\ 0 \\ 0 \end{bmatrix}$ are in $\mathbb{C}^{(4)}$, find $V = \langle v_1, v_2, v_3 \rangle$, $W = \langle w_1, w_2 \rangle$, and $V + W$.

9. Is the sum $\langle v_1, v_2, v_3 \rangle + \langle w_1, w_2 \rangle$ in Problem 8 direct?

10. If v_1, v_2, v_3, w_1 are as in Problem 8, for what values of $w = \begin{bmatrix} a \\ b \\ c \\ d \end{bmatrix}$ is the sum $V + W$, where $V = \langle v_1, v_2, v_3 \rangle$ and $W = \langle w_1, w \rangle$, direct?

11. Find $\langle v_1, v_2 \rangle^\perp$ if $v_1 = \begin{bmatrix} 1 \\ 2 \\ -1 \end{bmatrix}$, $v_2 = \begin{bmatrix} 0 \\ 1 \\ 1 \end{bmatrix}$ are in $F^{(3)}$.

12. If $W = \langle e_1, e_2, \ldots, e_{n-1} \rangle$ in $F^{(n)}$, find $W^\perp$.

13. In Problem 11 show that $\langle v_1, v_2 \rangle + \langle v_1, v_2 \rangle^\perp = F^{(3)}$.

14. If $W = \langle e_1, \ldots, e_m \rangle$ in $F^{(n)}$, show that $(W^\perp)^\perp = W$.

MORE THEORETICAL PROBLEMS

Easier Problems

15. If $V_1, \ldots, V_m$ are subspaces of $F^{(n)}$, define $V_1 + V_2 + \cdots + V_m$ and show that it is a subspace of $F^{(n)}$.

16. If $V_1 + \cdots + V_m$ is a *direct* sum of the subspaces $V_1, \ldots, V_m$ of $F^{(n)}$, show that $V_j \cap V_k = \{0\}$ if $j \neq k$.

17. If $v \in F^{(n)}$ is such that $(v, w_j) = 0$ for all basis elements $w_1, \ldots, w_k$ of W, show that $v \in W^\perp$.

18. If $w_1, \ldots, w_k$ is a basis of orthogonal vectors of the subspace W [i.e., $(w_r, w_s) = 0$ if $r \neq s$], show how we can use $w_1, \ldots, w_k$ to get an orthonormal basis of W.

19. If v_1, v_2, v_3 in $F^{(n)}$ are linearly independent over F, construct an orthonormal basis for $\langle v_1, v_2, v_3 \rangle$.

Middle-Level Problems

20. If v_1, v_2, v_3, v_4 is a basis of $W \subset F^{(n)}$, construct an orthonormal basis for $\langle v_1, v_2, v_3, v_4 \rangle = W$.
21. If $W \neq F^{(5)}$ is a subspace of $F^{(5)}$, show that $W + W^\perp = F^{(5)}$.
22. If W is a subspace of $F^{(5)}$, show that $(W^\perp)^\perp = W$.
23. If W is any subspace of $F^{(n)}$, show that $(W^\perp)^\perp \supset W$.
24. If $A \in M_5(F)$ is Hermitian and if W is a subspace of $F^{(5)}$ such that $A(W) \subset W$, prove that $A(W^\perp) \subset W^\perp$.

4.3. GRAM–SCHMIDT ORTHOGONALIZATION PROCESS

In Theorem 4.2.5 we proved that if W is a subspace of $F^{(n)}$ which has an orthonormal basis, then $W \oplus W^\perp = F^{(n)}$. It would be aesthetically satisfying and highly useful if we could drop the assumption "W has an orthonormal basis" in this theorem and prove for *all* subspaces W of $F^{(n)}$, $F^{(n)} = W \oplus W^\perp$. One way of achieving this is to show that *any* subspace of $F^{(n)}$ has an orthonormal basis. Fortunately this is true, and we shall prove it in a short while. But first we need the procedure for constructing such an orthonormal basis. This is provided us by what is known as the *Gram–Schmidt orthogonalization process*.

Note first that if we could produce an orthogonal basis $w_1, \ldots, w_k$ for a subspace W of $F^{(n)}$, then it is extremely easy to adapt this basis to get an orthonormal one. If $a_j = \|w_j\|$, then for $v_j = w_j/a_j$ we have $\|v_j\| = \|w_j/a_j\| = (1/a_j)\|w_j\| = 1$. So $v_1, \ldots, v_k$ are unit vectors in W. They are also orthogonal since $(v_r, v_s) = \left(\dfrac{w_r}{a_r}, \dfrac{w_s}{a_s}\right) = (1/a_r a_s)(w_r, w_s) = 0$ for $r \neq s$. Thus $v_1, \ldots, v_k$ is an orthonormal basis of W.

So to get to where we want to get to, *we must merely produce an orthogonal basis* of W. Before jumping to the general case we try some simple (and sample) examples.

Suppose that v_1 and v_2 in $F^{(n)}$ are linearly independent over F and let, as usual, $\langle v_1, v_2 \rangle$ be the subspace of $F^{(n)}$ spanned by v_1 and v_2 over F. We want to find w_1 and w_2 in $\langle v_1, v_2 \rangle$ which span $\langle v_1, v_2 \rangle$ over F, and furthermore, satisfy $(w_1, w_2) = 0$. Our choice for w_1 is very easy; namely, we let

$$w_1 = v_1.$$

How to get w_2? Let $w_2 = v_2 + av_1 = v_2 + aw_1$, where a in F is to be chosen judiciously. We want $(w_2, w_1) = 0$, that is, $(v_2 + aw_1, w_1) = 0$. But $(v_2 + aw_1, w_1) = (v_2, w_1) + a(v_1, w_1)$; hence this is 0 if $a = -\dfrac{(v_2, w_1)}{(w_1, w_1)}$. Therefore, our choice for w_2 is

$$w_2 = v_2 - \frac{(v_2, w_1)}{(w_1, w_1)} w_1,$$

and since $v_1 = w_1$ and v_2 are linearly independent over F, we know that $w_2 \neq 0$. Note that w_1 and w_2 are in $\langle v_1, v_2 \rangle$ and since $v_1 = w_1$, $v_2 = w_2 + \dfrac{(v_2, w_1)}{(w_1, w_1)} w_1$, the span of

Sec. 4.3] Gram–Schmidt Orthogonalization Process

w_1 and w_2 contains the span of v_1 and v_2. Thus $\langle w_1, w_2 \rangle \supset \langle v_1, v_2 \rangle$, which together with $\langle w_1, w_2 \rangle \subset \langle v_1, v_2 \rangle$, yields $\langle v_1, v_2 \rangle = \langle w_1, w_2 \rangle$. Finally, $(w_1, w_2) = 0$ since this was forced by our choice of a.

Let us go on to the case of v_1, v_2, v_3 linearly independent over F. *We will use the result proved above for two vectors to settle this case.*
Let

$$w_1 = v_1,$$

$$w_2 = v_2 - \frac{(v_2, w_1)}{(w_1, w_1)} w_1,$$

as above. So $w_1 \neq 0$, $w_2 \neq 0$, and $(w_1, w_2) = 0$. We seek an element $w_3 \neq 0$ in $\langle v_1, v_2, v_3 \rangle$ such that $(w_3, w_1) = (w_3, w_2) = 0$ and $\langle v_1, v_2, v_3 \rangle = \langle w_1, w_2, w_3 \rangle$. Let $w_3 = v_3 + a w_2 + b w_1$. Since $(w_2, w_1) = 0$, to get $(w_3, w_1) = 0$, that is,

$$(v_3 + a w_2 + b w_1, w_1) = 0,$$

we have to satisfy $(v_3, w_1) + b(w_1, w_1) = 0$. So $b = -\dfrac{(v_3, w_1)}{(w_1, w_1)}$. To have $(w_3, w_2) = 0$, we want $(v_3 + a w_2 + b w_1, w_2) = 0$, so $(v_3, w_2) + a(w_2, w_2) = 0$. Thus we use $a = -\dfrac{(v_3, w_2)}{(w_2, w_2)}$. So our choice for w_3 is

$$w_3 = v_3 - \frac{(v_3, w_2)}{(w_2, w_2)} w_2 - \frac{(v_3, w_1)}{(w_1, w_1)} w_1.$$

Is $w_3 \neq 0$? Since w_1, w_2 only involve v_1 and v_2, and since v_1, v_2, v_3 are linearly independent over F, $w_3 \neq 0$. Finally, is $\langle w_1, w_2, w_3 \rangle$ all of $\langle v_1, v_2, v_3 \rangle$? From the construction in the paragraph above, v_1 and v_2 are linear combinations of w_1 and w_2. By construction $v_3 = w_3 + \dfrac{(v_3, w_2)}{(w_2, w_2)} w_2 + \dfrac{(v_3, w_1)}{(w_1, w_1)} w_1$, so v_3 is a linear combination of w_1, w_2, w_3. Therefore, $\langle v_1, v_2, v_3 \rangle \subset \langle w_1, w_2, w_3 \rangle \subset \langle v_1, v_2, v_3 \rangle$, yielding that $\langle v_1, v_2, v_3 \rangle = \langle w_1, w_2, w_3 \rangle$.

The discussion of the cases of two and three vectors has deliberately been long-winded and probably overly meticulous. However, its payoff will be a simple discussion of the general case. What we do is a bootstrap operation, going from one case to the next one involving one more vector.

Suppose we know that for some positive integer k for any set $v_1, \ldots, v_k$ of linearly independent elements of $F^{(n)}$ we can find *mutually orthogonal* nonzero elements $w_1, \ldots, w_k$ in $\langle v_1, \ldots, v_k \rangle$ such that

(1) $\langle v_1, \ldots, v_k \rangle = \langle w_1, \ldots, w_k \rangle$

(2) for each $j = 1, 2, \ldots, k$, $w_j \in \langle v_1, v_2, \ldots, v_j \rangle$. (*)

(We certainly did this explicitly for $k = 2$ and 3.) Given $v_1, v_2, \ldots, v_k, v_{k+1}$ in $F^{(n)}$, linearly independent over F, can we find $w_1, \ldots, w_{k+1}$ *mutually orthogonal* nonzero elements in $\langle v_1, \ldots, v_{k+1} \rangle$ such that

(1) $\langle v_1, \ldots, v_{k+1} \rangle = \langle w_1, \ldots, w_{k+1} \rangle$

(2) for each $j = 1, 2, \ldots, k, k+1$, $w_j \in \langle v_1, v_2, \ldots, v_j \rangle$? (**)

By (*) we have $w_1, \ldots, w_k$ already satisfying our requirements. How do we pick w_{k+1}? Guided by the cases done for $k = 2$ and $k = 3$, we simply let

$$w_{k+1} = v_{k+1} - \frac{(v_{k+1}, w_k)}{(w_k, w_k)} w_k - \frac{(v_{k+1}, w_{k-1})}{(w_{k-1}, w_{k-1})} w_{k-1} - \cdots - \frac{(v_{k+1}, w_1)}{(w_1, w_1)} w_1.$$

We leave the verification that w_{k+1} is nonzero and that $w_1, \ldots, w_{k+1}$ satisfy the requirements in (**) to the reader.

Knowing the result for $k = 3$, by the argument above, we know it to be true for $k = 4$. Knowing it for $k = 4$, we get it for $k = 5$, and so on, to get that the result is true for every positive integer n. We have proved

Theorem 4.3.1. If $v_1, \ldots, v_m$ in $F^{(n)}$ are linearly independent over F, then the subspace $\langle v_1, \ldots, v_m \rangle$ of $F^{(n)}$ has a basis $w_1, \ldots, w_m$ such that $(w_j, w_k) = 0$ for all $j \neq k$. That is, $\langle v_1, \ldots, v_m \rangle$ has an orthogonal basis, hence an orthonormal basis.

Theorem 4.3.1 has an immediate consequence:

Theorem 4.3.2. If $W = 0$ is a subspace of $F^{(n)}$, then W has an orthonormal basis.

Proof: W has a basis $v_1, \ldots, v_m$, so $W = \langle v_1, \ldots, v_m \rangle$. Thus, by Theorem 4.3.1, W has an orthonormal basis. ∎

In Theorem 4.2.5 we proved that if $W \neq 0$ is a subspace of $F^{(n)}$ and if W has an orthonormal basis, then $F^{(n)} = W \oplus W^\perp$. By Theorem 4.3.2, *every* nonzero subspace W of $F^{(n)}$ has an orthonormal basis. Thus we get the definitive form of Theorem 4.2.5,

Theorem 4.3.3. Given any subspace W of $F^{(n)}$, then $F^{(n)} = W \oplus W^\perp$.

Corollary 4.3.4. If W is a subspace of $F^{(n)}$, then $\dim(W^\perp) = n - \dim(W)$.

Proof: Let $w_1, \ldots, w_k$ be a basis of W over F, and $v_1, \ldots, v_r$ a basis of $W^\perp$. Then $w_1, \ldots, w_k, v_1, \ldots, v_r$ is a basis for $F^{(n)}$, so $n = k + r$ or $r = n - k$. This translates into $\dim(W^\perp) = n - \dim(W)$. ∎

From Theorem 4.3.3 we get a nice result; namely, given any set of nonzero mutually orthogonal elements of $F^{(n)}$, we can fill it out to an orthogonal basis of $F^{(n)}$. More explicitly,

Corollary 4.3.5. Given any subspace $W \neq \{0\}$ of $F^{(n)}$ and any subspace V of $F^{(n)}$ containing W, $V = W \oplus (V \cap (W^\perp))$.

Sec. 4.3]   Gram–Schmidt Orthogonalization Process   179

Proof: If $v \in V$, then $v = w + u$ with $w \in W$ and $u \in W^\perp$. But then $u = v - w \in V$, since $W \subset V$. So $V = W + V \cap (W^\perp)$. Obviously, $W \cap (V \cap (W^\perp)) = \{0\}$. ∎

Theorem 4.3.6. If $v_1, \ldots, v_k$ in $F^{(n)}$ are mutually orthogonal over F, then we can find mutually orthogonal elements $u_1, \ldots, u_{n-k}$ in $F^{(n)}$ such that $v_1, \ldots, v_k, u_1, \ldots, u_{n-k}$ is an orthogonal basis of $F^{(n)}$ over F.

Proof: Let $W = \langle v_1, \ldots, v_k \rangle$; so that $v_1, \ldots, v_k$ is an orthogonal basis of W, by Theorem 3.10.6. By Theorem 4.3.3, $F^{(n)} = W \oplus W^\perp$. Let $u_1, \ldots, u_r$ be an orthogonal basis of $W^\perp$ over F. Then $v_1, \ldots, v_k, u_1, \ldots, u_r$ is an orthogonal basis of $F^{(n)}$ over F. (Prove!) Since $\dim(F^{(n)}) = n$, and since $v_1, \ldots, v_k, u_1, \ldots, u_r$ is a basis of $F^{(n)}$ over F, $n = k + r$, so $r = n - k$. The theorem is proved. ∎

Note that we also get from Theorem 4.3.4 the result that given any set of linearly independent elements of $F^{(n)}$, we can fill it out to a basis for $F^{(n)}$. This is proved along the same lines as is the proof of Theorem 4.3.6.

We close this section with

Theorem 4.3.7. If $W \neq 0$ is a subspace of $F^{(n)}$, then $(W^\perp)^\perp = W$.

Proof: Let $V = W^\perp$. If $w \in W$ and $v \in V$, then $(v, w) = 0$. Hence $w \in V^\perp = (W^\perp)^\perp$. This says that $W \subset V^\perp = (W^\perp)^\perp$.

By the Corollary 4.3.4, $\dim(W^\perp) = n - \dim(W)$. Also, by Corollary 4.3.4, $\dim(W^\perp)^\perp = n - \dim(W^\perp) = n - (n - \dim(W)) = \dim(W)$.

Therefore, W is a subspace of $(W^\perp)^\perp$ and is of the same dimension as $(W^\perp)^\perp$. Thus $W = (W^\perp)^\perp$. ∎

PROBLEMS

NUMERICAL PROBLEMS

1. For the given $v_1, \ldots, v_k$, use the Gram–Schmidt process to get an orthonormal basis of $\langle v_1, \ldots, v_k \rangle$.

   (a) $v_1 = \begin{bmatrix} 1 \\ 2 \\ 3 \end{bmatrix}$, $v_2 = \begin{bmatrix} 0 \\ 1 \\ 2 \end{bmatrix}$.

   (b) $v_1 = \begin{bmatrix} 1 \\ i \\ 0 \\ 0 \end{bmatrix}$, $v_2 = \begin{bmatrix} 0 \\ i \\ 1 \\ 0 \end{bmatrix}$, $v_3 = \begin{bmatrix} 1 \\ i \\ 0 \\ 0 \end{bmatrix}$ in $\mathbb{C}^{(4)}$.

   (c) $v_1 = \begin{bmatrix} i \\ 2 \\ 3 \\ 4 \\ 5 \end{bmatrix}$, $v_2 = \begin{bmatrix} 1 \\ 1 \\ 1 \\ 1 \\ 1 \end{bmatrix}$, $v_3 = \begin{bmatrix} 0 \\ 1 \\ 1 \\ i \\ 0 \end{bmatrix}$ in $\mathbb{C}^{(5)}$.

(d) $v_1 = \begin{bmatrix} i \\ -1 \\ 1 \\ -1 \\ 1 \\ -1 \end{bmatrix}$, $v_2 = \begin{bmatrix} 0 \\ 1 \\ 2 \\ 0 \\ 1 \\ 1 \end{bmatrix}$, $v_3 = \begin{bmatrix} 1 \\ 0 \\ 1 \\ 0 \\ 1 \\ 0 \end{bmatrix}$, $v_4 = \begin{bmatrix} i \\ 0 \\ 0 \\ 0 \\ 0 \\ 1 \end{bmatrix}$ in $\mathbb{C}^{(6)}$.

2. For each part of Problem 1 find $\langle v_1, \ldots, v_k \rangle^\perp$ explicitly.
3. For each part of Problem 1 verify that $\langle v_1, \ldots, v_k \rangle + \langle v_1, \ldots, v_k \rangle^\perp = F^{(n)}$.
4. For each part of Problem 1, find $(\langle v_1, \ldots, v_k \rangle^\perp)^\perp$ and show that $(\langle v_1, \ldots, v_k \rangle^\perp)^\perp = \langle v_1, \ldots, v_k \rangle$.
5. For each part of Problem 1, fill out the given $v_1, \ldots, v_k$ to a basis for $F^{(n)}$.

MORE THEORETICAL PROBLEMS

Easier Problems

6. Fill in all the details in the proof of Theorem 4.3.1 to show that $w_1, \ldots, w_{k+1}$ satisfy (**).
7. What are $W^\perp$ if $W = 0$, and if $W = F^{(n)}$?
8. If $A \in M_n(\mathbb{C})$ is Hermitian and a is a real number, let $W = \{v \in \mathbb{C}^{(n)} \mid Av = av\}$. Prove that $(A - aI)(\mathbb{C}^{(n)}) \subset W^\perp$.
9. If $A \in M_n(\mathbb{C})$ is Hermitian, and W is a subspace of $\mathbb{C}^{(n)}$ such that $A(W) \subset W$, prove that $AW^\perp \subset W^\perp$.
10. In Problem 8 show that $A(W) \subset W$.

Harder Problems

11. If the W in Problem 8 is of dimension $k \geq 1$ (so that a is a characteristic root of A), show that we can find an orthonormal basis of $F^{(n)}$ in which the matrix of A is of the form $\begin{bmatrix} aI_k & 0 \\ 0 & B \end{bmatrix}$, where I_k is the $k \times k$ unit matrix and B is a Hermitian $(n-k) \times (n-k)$ matrix.
12. In Problem 11 show that there is a unitary matrix U such that
$$U^{-1}AU = \begin{bmatrix} aI_k & 0 \\ 0 & B \end{bmatrix}.$$
13. If W is a subspace of $F^{(n)}$ of dimension k, show that there is a *one-to-one* mapping, f, of W onto $F^{(k)}$ which satisfies $f(w_1 + w_2) = f(w_1) + f(w_2)$ and $f(aw_1) = af(w_1)$ for all w_1, w_2 in W and all $a \in F$.

4.4. RANK AND NULLITY

Given a matrix A in $M_n(F)$ (or, if you want, a linear transformation on $F^{(n)}$), we shall attach two integers to A which help to describe the nature of A. These integers are merely the dimensions of two subspaces of $F^{(n)}$ intimately related to A.

The set $N = \{v \in F^{(n)} \mid Av = 0\}$ is a subspace of $F^{(n)}$, because $Av = Aw = 0$ implies

Sec. 4.4] Rank and Nullity 181

that $A(v + w) = Av + Aw = 0$ and $A(av) = aA(v) = 0$ for all $a \in F$. It is called the *nullspace* of A.

Definition. The *nullity* of A, written $n(A)$, is the dimension of N over F.

Thus if the nullity of A is $n = \dim(F^{(n)})$, then $N = F^{(n)}$, so $Av = 0$ for all $v \in F^{(n)}$. In other words, $A = 0$. At the other end of the spectrum, if $n(A) = 0$, then $N = 0$, so $Av = 0$ forces $v = 0$. This tells us that A is invertible.

The set $A(F^{(n)}) = \{Av \mid v \in F^{(n)}\}$ is called the *range* of A. It is a subspace of $F^{(n)}$. The second integer we attach to A is the dimension of the range of A.

Definition. The *rank* of A, written $r(A)$, is the dimension of $A(F^{(n)})$ over F.

So if the rank of A is 0, then $A(F^{(n)}) = 0$, hence $A = 0$. On the other hand, if $r(A) = n$, then $A(F^{(n)}) = F^{(n)}$. This tells us that A is onto and so is invertible [so $n(A) = 0$]. Notice that in both these cases $n(A) + r(A) = n$. In fact, as the next theorem shows, this is always the case for any matrix A in $F^{(n)}$.

Theorem 4.4.1. Given $A \in M_n(F)$, then $n(A) + r(A) = n$. That is,

$$\text{rank} + \text{nullity} = \dim(F^{(n)}).$$

Proof: Letting $N = \{v \in F^{(n)} \mid Av = 0\}$, we find that N is a subspace of $F^{(n)}$ of dimension $n(A)$. By Theorem 4.3.3, $F^{(n)} = N \oplus N^\perp$. Therefore, since $A(N) = 0$, $A(F^{(n)}) = A(N \oplus N^\perp) = A(N^\perp)$. Hence $r(A) = \dim(A(F^{(n)})) = \dim(A(N^\perp))$.

Let $v_1, \ldots, v_k$ be a basis of $N^\perp$. We assert that $Av_1, \ldots, Av_k$ are linearly independent over F. Why? If $a_1 A(v_1) + \cdots + a_k A(v_k) = 0$, where $a_1, \ldots, a_k$ are in F, then $A(a_1 v_1 + \cdots + a_k v_k) = 0$. This puts $a_1 v_1 + \cdots + a_k v_k$ in N. However, since $v_1, \ldots, v_k$ are in $N^\perp$, $a_1 v_1 + \cdots + a_k v_k$ is in $N^\perp$. Because $N \cap N^\perp = 0$ and $a_1 v_1 + \cdots + a_k v_k \in N \cap N^\perp = 0$, we get that $a_1 v_1 + \cdots + a_k v_k = 0$. By the linear independence of $v_1, \ldots, v_k$ we conclude that $a_1 = a_2 = \cdots = a_k = 0$. Hence $Av_1, \ldots, Av_k$ are linearly independent over F.

Because $Av_1, \ldots, Av_k$ span $A(N^\perp)$ and are linearly independent over F, they form a basis of $A(N^\perp)$ over F. Therefore, $k = \dim(A(N^\perp)) = \dim(A(F^{(n)})) = r(A)$.

By the Corollary 4.3.4, $n = \dim(N) + \dim(N^\perp) = n(A) + k = n(A) + r(A)$. This proves the theorem. ∎

In Chapter 7 we give an effective and computable method for calculating the rank of a given matrix. For the present, we must find the form of all the elements in $A(F^{(n)})$ and calculate its dimension to obtain the value of $r(A)$.

PROBLEMS

NUMERICAL PROBLEMS

1. Find the rank of the following matrices, A, by finding $A(F^{(n)})$.

   (a) $\begin{bmatrix} 1 & 2 & 3 & 4 \\ 0 & -1 & 6 & 5 \\ 0 & 0 & 0 & 1 \\ 0 & 0 & 0 & 0 \end{bmatrix}$.

(b) $\begin{bmatrix} 1 & -1 & 0 \\ 6 & 6 & 1 \\ -.5 & .5 & 0 \end{bmatrix}$.

(c) $\begin{bmatrix} \frac{1}{4} & \frac{1}{4} & \frac{1}{4} & \frac{1}{4} \\ \frac{1}{4} & \frac{1}{4} & \frac{1}{4} & \frac{1}{4} \\ 0 & 0 & 0 & 0 \\ 0 & 0 & 0 & 0 \end{bmatrix}$.

(d) $\begin{bmatrix} 0 & 0 & 0 & 0 & 0 \\ 1 & 0 & 0 & 0 & 0 \\ 0 & 1 & 0 & 0 & 0 \\ 0 & 0 & 1 & 0 & 0 \\ 0 & 0 & 0 & 1 & 0 \end{bmatrix}$.

(e) $\begin{bmatrix} 1 & 0 & 1 \\ 0 & 1 & 0 \\ 1 & 0 & 0 \end{bmatrix}$.

(f) $\begin{bmatrix} 1 & 0 & 0 \\ 0 & 0 & 0 \\ 0 & 0 & 0 \end{bmatrix}$.

2. Find the nullity of the matrices in Problem 1 by finding N and its dimension.

3. Verify for the matrices in Problem 1 that $r(A) + n(A) = n$.

4. Find the rank of $\begin{bmatrix} 1 & 2 & 3 \\ 4 & 5 & 6 \\ 7 & 8 & 9 \end{bmatrix} \begin{bmatrix} 1 & 0 & 0 \\ 0 & 0 & 0 \\ 0 & 0 & 0 \end{bmatrix}$. How does it compare with the ranks of $\begin{bmatrix} 1 & 2 & 3 \\ 4 & 5 & 6 \\ 7 & 8 & 9 \end{bmatrix}$ and $\begin{bmatrix} 1 & 0 & 0 \\ 0 & 0 & 0 \\ 0 & 0 & 0 \end{bmatrix}$?

5. If $A = \begin{bmatrix} 1 & 2 & 0 \\ 0 & 1 & 0 \\ 0 & 0 & 0 \end{bmatrix} \begin{bmatrix} 1 & 0 & 0 \\ 0 & 0 & 0 \\ 0 & 0 & 1 \end{bmatrix}$, find the rank of A and those of $\begin{bmatrix} 1 & 2 & 0 \\ 0 & 1 & 0 \\ 0 & 0 & 0 \end{bmatrix}$ and $\begin{bmatrix} 1 & 0 & 0 \\ 0 & 0 & 0 \\ 0 & 0 & 1 \end{bmatrix}$.

MORE THEORETICAL PROBLEMS

Easier Problems

6. If $A = BC$, what can you say about $r(A)$ in comparison to $r(B)$ and $r(C)$?
7. If $A = B + C$, what can you say about $r(A)$ in terms of $r(B)$ and $r(C)$?
8. If $r(A) = 1$, prove that $A^2 = aA$ for some $a \in F$.

9. Show that any matrix can be written as the sum of matrices of rank 1.
10. If E_{jk} are the matrix units, what are the ranks $r(E_{jk})$?

Middle-Level Problems

11. If $E^2 = E \in M_n(F)$, give an explicit form for the typical element in the nullspace $N = \{v \in F^{(n)} \mid Ev = 0\}$.
12. For Problem 11, show that $N^\perp = E(F^{(n)})$ if $E^* = E = E^2$.
13. Use Problem 11 to find a basis of $F^{(n)}$ in which the matrix of E is of the form $\begin{bmatrix} I_k & 0 \\ 0 & 0 \end{bmatrix}$, where I_k is the $k \times k$ unit matrix.
14. How would you describe the k in Problem 13?
15. If $E^2 = E$, show that $\operatorname{tr}(E)$ is an integer.

Harder Problems

16. If $E^2 = E$ is of rank 1, prove that $EM_n(F)E = FE$ [i.e., $EAE = aE$ for $A \in M_n(F)$, $a \in F$].

4.5. CHARACTERISTIC ROOTS

Our intention now is to interrelate algebraic properties of a matrix A in $M_n(F)$ with the fact that $F^{(n)}$ is of dimension n over F.

If v is any vector in $F^{(n)}$, consider the vectors $v, Av, A^2v, \ldots, A^nv$. Here we have written down $n + 1$ vectors in $F^{(n)}$, which is of dimension n over F. By Theorem 3.6.2 these vectors must be linearly dependent over F. Thus there are elements $a_0, a_1, \ldots, a_n$ in F, not all 0, such that

$$a_0 A^n v + a_1 A^{n-1} v + \cdots + a_{n-1} Av + a_n v = 0.$$

So if $p(x)$ is the polynomial

$$p(x) = a_0 x^n + a_1 x^{n-1} + \cdots + a_{n-1} x + a_n,$$

replacing x by the matrix A, we obtain

$$p(A) = a_0 A^n + a_1 A^{n-1} + \cdots + a_{n-1} A + a_n I$$

and the statement on linear dependence translates into $p(A)(v) = 0$. Note that this polynomial $p(x)$ is not 0 and depends on v. Also note that $p(x)$ is of degree at most n.

Let $e_1, e_2, \ldots, e_n$ be the canonical basis of $F^{(n)}$. By the above, there exist nonzero polynomials $p_1(x), p_2(x), \ldots, p_n(x)$ of degree at most n such that $p_j(A)e_j = 0$ for $j = 1, 2, \ldots, n$. Consider the polynomial $p(x) = p_1(x)p_2(x)\ldots p_n(x)$; it is nonzero and of degree at most n^2. What can we say about $p(A)$? To begin with, $p(A)e_n = p_1(A)\ldots p_n(A)e_n = 0$, since $p_n(A)e_n = 0$. In fact, since $p(x)$ can also be written as a product with $p_j(x)$ at the end, for any j, and since $p_j(A)e_j = 0$, we get that $p(A)e_j = 0$ for $j = 1, 2, \ldots, n$. Since $p(A)$ is a matrix which annihilates every element of the basis

$e_1, e_2, \ldots, e_n$ of $F^{(n)}$ over F, $p(A)$ must annihilate *every* element v in $F^{(n)}$. But this says that $p(A) = 0$. We have proved

Theorem 4.5.1. Given $A \in M_n(F)$, then there exists a nonzero polynomial $p(x)$, of degree at most n^2 and with coefficients in F, such that $p(A) = 0$.

We shall see later—in the famous Cayley–Hamilton Theorem—that there is a specific and easy-to-construct polynomial $f(A)$ *of degree n* with coefficients in F such that $f(A) = 0$.

Since there is some nonzero polynomial $p(x)$ which has the property that $p(A) = 0$, there is a nonzero polynomial of lowest possible degree—call it $q_A(x)$—with the property that $q_A(A) = 0$. If we insist that the highest coefficient of q_A be 1—which we do—then this polynomial is unique. (Prove!) We call it the *minimal polynomial for A*.

Consider the matrix $A = \begin{bmatrix} 1 & 0 & 0 \\ 0 & 1 & 0 \\ 0 & 0 & 0 \end{bmatrix}$. Then as a calculation reveals, $A^2 = A$.

We quickly then see that $q_A(x) = x^2 - x$. If we go through the rigmarole preceding Theorem 4.5.1, what polynomial $p(x)$ do we get? Well, $Ae_1 = \begin{bmatrix} 1 & 0 & 0 \\ 0 & 1 & 0 \\ 0 & 0 & 0 \end{bmatrix} \begin{bmatrix} 1 \\ 0 \\ 0 \end{bmatrix} = \begin{bmatrix} 1 \\ 0 \\ 0 \end{bmatrix} = e_1$, $Ae_2 = \begin{bmatrix} 1 & 0 & 0 \\ 0 & 1 & 0 \\ 0 & 0 & 0 \end{bmatrix} \begin{bmatrix} 0 \\ 1 \\ 0 \end{bmatrix} = \begin{bmatrix} 0 \\ 1 \\ 0 \end{bmatrix} = e_2$, and $Ae_3 = \begin{bmatrix} 1 & 0 & 0 \\ 0 & 1 & 0 \\ 0 & 0 & 0 \end{bmatrix} \begin{bmatrix} 0 \\ 0 \\ 1 \end{bmatrix} = \begin{bmatrix} 0 \\ 0 \\ 0 \end{bmatrix}$. So the polynomials $p_1(x), p_2(x), p_3(x)$ used in the proof of Theorem 4.5.1 are, respectively, $p_1(x) = x - 1$, $p_2(x) = x - 1$, $p_3(x) = x$, and the polynomial

$$p(x) = p_1(x)p_2(x)p_3(x) = x(x-1)^2 = x^3 - 2x + x.$$

Note that $A^3 - 2A + A = 0$.

Let $q_A(x) = x^m + a_1 x^{m-1} + \cdots + a_{m-1} x + a_m$ be the minimum polynomial for A. We claim that A is invertible if and only if $a_m \neq 0$. To see this, note that since

$$0 = q_A(A) = A^m + a_1 A^{m-1} + \cdots + a_{m-1} A + a_m I,$$

if $a_m \neq 0$, then

$$\left(\frac{A^{m-1} + a_1 A^{m-2} + \cdots + a_{m-1} I}{-a_m} \right) A = I$$

and A is invertible.

On the other hand, if $a_m = 0$, then

$$(A^{m-1} + a_1 A^{m-2} + \cdots + a_{m-1} I)A = 0.$$

Because $h(x) = x^{m-1} + a_1 x^{m-2} + \cdots + a_{m-1}$ is of lower degree than $q_A(x)$, $h(A)$ *cannot* be 0. Now $h(A)A = 0$ and $h(A) \neq 0$. We leave it as an exercise to show that A cannot be invertible. Note, too, that since $h(A) \neq 0$, there is some element v in $F^{(n)}$ such that $w = h(A)(v) \neq 0$. However, $Aw = Ah(A)v = q_A(A)v = 0$.

Summarizing this long discussion, we have

Theorem 4.5.2. Given A in $M_n(F)$, then A is invertible if and only if the constant term of its minimum polynomial is nonzero. Moreover:

1. If A is invertible, the inverse of A is a polynomial expression over F in A.
2. If A is not invertible, then
   (a) $Aw = 0$ for some $w \neq 0$ in $F^{(n)}$;
   (b) $AF^{(n)} \neq F^{(n)}$.

Proof: (1) is clear from our discussion above, as is Part (a) of (2). For Part (b) of (2), we've observed in our discussion above that $h(A)A = 0$ and $h(A)$ is not 0. Thus $AF^{(n)}$ cannot be all of $F^{(n)}$, since otherwise $h(A)F^{(n)} = h(A)AF^{(n)} = 0$. ∎

We have talked several times earlier about the characteristic roots of a matrix. One time we said that $a \in F$ is a characteristic root of $A \in M_n(F)$ if $A - aI$ is *not* invertible. Another time we said that a is a characteristic root of A if for some $v \neq 0$ in $F^{(n)}$, $Av = av$; and we called such a vector v a characteristic vector associated with a.

Note that both these descriptions—in light of Theorem 4.5.2—are the same. For if $Av = av$, then $(A - aI)v = 0$; hence $A - aI$ certainly cannot be invertible. In the other direction, if $B = A - aI$ is not invertible, by Part (2) of Theorem 4.5.2 there exists $w \neq 0$ in $F^{(n)}$ such that $Bw = 0$, that is, $(A - aI)w = 0$. This yields $Aw = aw$. So, recalling that $a \in F$ is a characteristic root of $A \in M_n(F)$ if there exists $v \neq 0$ in $F^{(n)}$ such that $Av = av$, we have then

Theorem 4.5.3. The element a in F is a characteristic root of $A \in M_n(F)$ if and only if $(A - aI)$ is not invertible.

If we consider $A = \begin{bmatrix} 0 & 1 \\ -1 & 0 \end{bmatrix}$ in $M_2(\mathbb{R})$, then $A^2 + I = 0$ and A has *no real characteristic roots*. However, viewing A as sitting in $M_2(\mathbb{C})$, we have that A has $+i$ and $-i$ as characteristic roots in $\mathbb{C}$.

Note that $A - iI = \begin{bmatrix} -i & 1 \\ -1 & -i \end{bmatrix}$ and $A - iI$ is not invertible. (Prove.) Also,

$A \begin{bmatrix} 1 \\ i \end{bmatrix} = \begin{bmatrix} 0 & 1 \\ -1 & 0 \end{bmatrix} \begin{bmatrix} 1 \\ i \end{bmatrix} = \begin{bmatrix} i \\ -1 \end{bmatrix} = i \begin{bmatrix} 1 \\ i \end{bmatrix}$, so $(A - iI) \begin{bmatrix} 1 \\ i \end{bmatrix} = 0$.

What the example above shows is that whether or not a given matrix has characteristic roots in F depends strongly on whether $F = \mathbb{R}$ or $F = \mathbb{C}$.

Until now—except for the rare occasion—our theorems were equally true for $M_n(\mathbb{R})$ and $M_n(\mathbb{C})$. However, in speaking about characteristic roots, a real difference between the two, $M_2(\mathbb{R})$ and $M_2(\mathbb{C})$, crops up.

We elaborate on this. What we need, and what we shall use—although we do not prove it—is a very important theorem due to Gauss which is known as the

Fundamental Theorem of Algebra. This theorem can be stated in several ways. One way says that given a polynomial $p(x)$ of degree at least 1, having complex coefficients, then $p(x)$ has a root in $\mathbb{C}$; that is, there is an $a \in \mathbb{C}$ such that $p(a) = 0$. Another—and equivalent—version states that given a polynomial $p(x)$ of degree $n \geq 1$, having complex coefficients (and leading coefficient 1), then $p(x)$ factors in the form

$$p(x) = (x - a_1)(x - a_2) \cdots (x - a_n),$$

where $a_1, \ldots, a_n \in \mathbb{C}$ are the roots of $p(x)$. (It may happen that a given root appears several times.)

We shall use both versions of this important theorem of Gauss. Unfortunately, to prove it would require the development of a rather elaborate mathematical machinery.

Let $a \in \mathbb{C}$ be a characteristic root of $A \in M_n(\mathbb{C})$. So there is an element $v \neq 0$ in $\mathbb{C}^{(n)}$ such that $Av = av$. Therefore,

$$A^2 v = A(Av) = A(av) = aAv = a(av) = a^2 v.$$

Continuing in this way, we get

$$A^k v = a^k v \qquad \text{for } k \geq 1,$$

so that if $f(x)$ is any polynomial having complex coefficients, then

$$f(A)v = f(a)(v).$$

(Prove!) Thus if $f(A) = 0$, then $f(a) = 0$, that is, if a is a characteristic root of A and if $f(x)$ is a polynomial such that $f(A) = 0$, then $f(a) = 0$, that is, a is a root of $f(x)$. So any characteristic root of A must be a root of $q_A(x)$, the minimum polynomial for A over $\mathbb{C}$. However, $q_A(x)$ has only a finite number of roots in $\mathbb{C}$—this finite number being at most the degree of $q_A(x)$. Therefore, A has only a finite number of characteristic roots and these must be the roots of $q_A(x)$.

We want to sharpen this last statement even further. This is

Theorem 4.5.4. *If $A \in M_n(\mathbb{C})$, then every characteristic root of A is a root of $q_A(x)$, the minimal polynomial for A. Furthermore, any root of $q_A(x)$ is a characteristic root of A.*

Proof: In the discussion preceding the statement of Theorem 4.5.4, we showed the first half, namely, that a characteristic root of A must be a root of $q_A(x)$.

So all we have to do is to prove the second half, that part beginning with "Furthermore." Let $a_1, \ldots, a_k$ be the roots of $q_A(x)$. Then, by Gauss's theorem,

$$q_A(x) = (x - a_1)(x - a_2) \cdots (x - a_k).$$

Therefore,

$$0 = q_A(A) = (A - a_1 I)(A - a_2 I) \cdots (A - a_k I).$$

Given j, let $h_j(x)$ be the product of all the $(x - a_t)$ for $t \neq j$. So $q_A(x) = h_j(x)(x - a_j)$. Since the degree of $h_j(x)$ = degree of $q_A(x) - 1$, $h_j(A) \neq 0$ by the minimality of $q_A(x)$. So for some $w \neq 0$ in $\mathbb{C}^{(n)}$, $v = h_j(A)w \neq 0$. However,

$$(A - a_j I)(v) = (A - a_j I)h_j(A)(w) = q_A(A)w = 0.$$

This says that $Av = a_j v$, with $v \neq 0$. Therefore, a_j is a characteristic root of A. ∎

Let's take a closer look at some of this in the special case where our $n \times n$ matrix A is upper triangular. Then any polynomial $f(A)$ in A is also an upper triangular matrix. (Prove!) So we get a sharp corollary to Theorem 4.5.2, namely

Corollary 4.5.5. *Given an invertible upper triangular A in $M_n(\mathbb{C})$, its diagonal entries are all nonzero and its inverse A^{-1} is upper triangular.*

Proof: Since the inverse of A is a polynomial in A, by Theorem 4.5.2, it is also upper triangular. (Prove!) Letting the diagonal elements of A and A^{-1} be $a_1, \ldots, a_n$ and $a_1^*, \ldots, a_n^*$, respectively, the diagonal elements of the product AA^{-1} of these two upper triangular matrices are $a_1 a_1^*, \ldots, a_n a_n^*$ (Prove!). Since $I = AA^{-1}$, all these diagonal entries are 1, so that the diagonal entries a_r of A must be nonzero. ∎

If A is a diagonal matrix with all entries on the diagonal nonzero, then A is invertible and its inverse is a diagonal matrix with diagonal entries inverses of those of A. (Prove!) If A is an upper triangular *unipotent* $n \times n$ matrix, that is, one all of whose diagonal entries are 1, then A can be expressed as $A = I - N$, where N is an upper triangular *nilpotent* $n \times n$ matrix, that is, one all of whose diagonal entries are 0. We call N nilpotent because $N^k = 0$ for some k; and we call A unipotent because $(A - I)^k = 0$ for some k. Thus the matrices

$$\begin{bmatrix} 0 & 3 & 6 \\ 0 & 0 & 4 \\ 0 & 0 & 0 \end{bmatrix}, \begin{bmatrix} 0 & 5 \\ 0 & 0 \end{bmatrix}$$

are nilpotent and the matrices

$$\begin{bmatrix} 1 & 3 & 6 \\ 0 & 1 & 4 \\ 0 & 0 & 1 \end{bmatrix}, \begin{bmatrix} 1 & 5 \\ 0 & 1 \end{bmatrix}$$

are unipotent. *An upper triangular $n \times n$ matrix A with $a_{11} = 0, \ldots, a_{nn} = 0$ is nilpotent* since, for $k \geq 1$, the entries $b_{1,1+t}, b_{2,2+t}, \ldots, b_{n-t,n}$ of $B = A^k$ are 0 for $0 \leq t \leq k - 1$. For example, if $n = 5$ and $k = 2$, t can be 0 or 1 and the entries $b_{11}, \ldots, b_{55}$ and $b_{12}, \ldots, b_{45}$ are 0. Why is it always true? It is true if $k = 1$, by hypothesis. In general, we go by induction on k. Assuming that it is true for k, it is also true for $k + 1$, since

A^{k+1} equals $AA^k = AB$ and the entries $c_{r,r+t} = \sum_{j=1}^{n} a_{rj} b_{j,r+t}$ of A^{k+1} are 0 for $1 \leq r \leq$

$n - t$ and $0 \leq t \leq k$. Thus it is true for all k. If A is an upper triangular unipotent matrix, then A is again invertible. In fact, letting $N = I - A$, we get

$$A^{-1} = (I - N)^{-1} = I + N + N^2 + \cdots + N^n,$$

as you see by showing that

$$(I - N)(I + N + N^2 + \cdots + N^n) = I$$

(Do so!). The upshot of this is the

Theorem 4.5.6. Let $A \in M_n(F)$ be upper triangular. Then
1. A is invertible if and only if all of its diagonal entries are nonzero;
2. The diagonal entries of A are its characteristic roots (with possible duplications).

Proof: If A is invertible, we've seen above that its diagonal entries are all nonzero. Suppose, conversely, that all diagonal entries of A are nonzero and let D be the diagonal matrix with the same diagonal entries as A. Then $U = D^{-1}A$ is unipotent, so U is invertible by what we showed in our earlier discussion. But then $A = DU$ is the product of invertible matrices, so is itself invertible. This proves (1).

Now let's prove (2). Since the diagonal entries of A are just those scalars a such that $A - aI$ has 0 as a diagonal entry, the diagonal entries of A are just those scalars a such that $A - aI$ is not invertible, by (1). But then, by Theorem 4.5.3, the diagonal entries of A are just its characteristic roots. ∎

This section—so far—has been replete with important results, many of them not too easy. So we break the discussion here to give the reader a chance to digest the material. We pick up this theme again in the next section.

PROBLEMS

NUMERICAL PROBLEMS

1. For the given v and A find a polynomial $p(x)$ such that $p(A)(v) = 0$.

   (a) $v = \begin{bmatrix} 1 \\ 1 \end{bmatrix}$, $A = \begin{bmatrix} 1 & 2 \\ 2 & 1 \end{bmatrix}$.

   (b) $v = \begin{bmatrix} 1 \\ 1 \\ 0 \end{bmatrix}$, $A = \begin{bmatrix} 1 & 2 & 0 \\ 2 & 1 & 0 \\ 0 & 0 & 0 \end{bmatrix}$.

   (c) $v = \begin{bmatrix} 0 \\ 1 \\ -1 \end{bmatrix}$, $A = \begin{bmatrix} 1 & 0 & 0 \\ 0 & 1 & 0 \\ 1 & 2 & 3 \end{bmatrix}$.

(d) $v = \begin{bmatrix} 0 \\ -1 \\ 1 \\ -1 \end{bmatrix}$, $A = \begin{bmatrix} 0 & 1 & 0 & 0 \\ 0 & 0 & 1 & 0 \\ 0 & 0 & 0 & 1 \\ 0 & 0 & 1 & 0 \end{bmatrix}$.

(e) $v = \begin{bmatrix} 1 \\ 2 \\ 1 \end{bmatrix}$, $A = \begin{bmatrix} 0 & 0 & 0 \\ 1 & 0 & 0 \\ 0 & 1 & 0 \end{bmatrix}$.

2. (a) For $A = \begin{bmatrix} 0 & 1 & 0 \\ 0 & 0 & 1 \\ 1 & 0 & 0 \end{bmatrix}$, find polynomials $p_1(x)$, $p_2(x)$, $p_3(x)$ such that
$p_j(A)(e_j) = 0$ for $j = 1, 2, 3$.

   (b) By directly calculating $p(A)$, where $p(x) = p_1(x)p_2(x)p_3(x)$, show that $p(A) = 0$.

3. From the result of Problem 2, express A^{-1} as a polynomial expression in A.

4. For the A in Problem 2, find its minimal polynomial over $\mathbb{C}$.

5. For the A in Problem 2 and the minimal polynomial $q_A(x)$ found in Problem 4, show, by explicitly exhibiting a $v \neq 0$, that if a is a root of $q_A(x)$, then a is a characteristic root of A.

6. In Problem 5, show directly that if a is a root of $q_A(x)$, then $(A - aI)$ is not invertible.

7. (a) If $A = \begin{bmatrix} 1 & 2 & 3 \\ 0 & 4 & 5 \\ 0 & 0 & 6 \end{bmatrix}$, prove that $(A - I)(A - 4I)(A - 6I) = 0$.

   (b) What are the characteristic roots of A?
   (c) What is $q_A(x)$?

8. (a) If $A = \begin{bmatrix} 1 & 0 & 0 \\ 0 & 1 & 2 \\ 0 & 2 & 4 \end{bmatrix}$, find $q_A(x)$.

   (b) Show that A has three distinct, real, characteristic roots.
   (c) If a_1, a_2, a_3 are the three characteristic roots of A, find $v_1 \neq 0, v_2 \neq 0, v_3 \neq 0$ such that $Av_j = a_j v_j$ for $j = 1, 2, 3$.
   (d) Verify that $(v_j, v_k) = 0$, for $j \neq k$, for the v_1, v_2, v_3 found in Part (c).

9. (a) If $A = \begin{bmatrix} 0 & 0 & 0 \\ 0 & 0 & 1 \\ 0 & -1 & 0 \end{bmatrix}$, find $q_A(x)$.

   (b) Find the roots of $q_A(x)$ in $\mathbb{C}$.
   (c) Verify that each root of $q_A(x)$ found in Part (b) is a characteristic root of A.
   (d) Find v_1, v_2, v_3 in $\mathbb{C}^{(3)}$ such that $Av_j = a_j v_j$, $j = 1, 2, 3$, where a_1, a_2, a_3 are the characteristic roots of A.

(e) Verify that $(v_j, v_k) = 0$, for $j \neq k$, for v_1, v_2, v_3 of Part (d).
(f) By a direct calculation show that $q_A(A) = 0$.

10. For the A in Problem 8, find an invertible matrix U such that $U^{-1}AU = \begin{bmatrix} a_1 & 0 & 0 \\ 0 & a_2 & 0 \\ 0 & 0 & a_3 \end{bmatrix}$.

11. For the A in Problem 9, find an invertible matrix U such that $U^{-1}AU = \begin{bmatrix} a_1 & 0 & 0 \\ 0 & a_2 & 0 \\ 0 & 0 & a_3 \end{bmatrix}$.

12. For the matrices A in Problems 8 and 9, can you exhibit *unitary* matrices such that $U^{-1}AU = \begin{bmatrix} a_1 & 0 & 0 \\ 0 & a_2 & 0 \\ 0 & 0 & a_3 \end{bmatrix}$?

MORE THEORETICAL PROBLEMS

Easier Problems

13. If $p(x)$ is a polynomial of least positive degree such that $p(A) = 0$, for $A \in M_n(F)$, and if the highest coefficient of $p(x)$ is 1, show that $p(x)$ is unique. [That is, if $q(x)$ is possibly another polynomial of least positive degree with highest coefficient equal to 1 such that $q(A) = 0$, then $p(x) = q(x)$].

14. If $Av = av$, show that for any polynomial $f(x)$ having coefficients in F, $f(A)(v) = f(a)v$.

15. Let $A = \begin{bmatrix} 0 & 1 & 0 & 0 \\ 0 & 0 & 1 & 0 \\ 0 & 0 & 0 & 1 \\ 1 & 0 & 0 & 0 \end{bmatrix}$. Show that A has four distinct characteristic roots in $\mathbb{C}$, and find these roots.

16. What is $q_A(x)$ for the A in Problem 15?

17. If a_1, a_2, a_3, a_4 are the characteristic roots of the A in Problem 15, find $v_1 \neq 0$, $v_2 \neq 0$, $v_3 \neq 0$, $v_4 \neq 0$ in $\mathbb{C}^{(4)}$ such that $Av_j = a_j v_j$ for $j = 1, 2, 3, 4$.

18. Show that v_1, v_2, v_3, v_4 in Problem 17 form a orthogonal basis of $\mathbb{C}^{(4)}$.

19. Show that we can find a unitary matrix U such that $U^{-1}AU = \begin{bmatrix} a_1 & 0 & 0 & 0 \\ 0 & a_2 & 0 & 0 \\ 0 & 0 & a_3 & 0 \\ 0 & 0 & 0 & a_4 \end{bmatrix}$ for the A in Problem 15.

20. If $A \in M_n(\mathbb{C})$ and $B = A^*$, the Hermitian adjoint of A, express $q_B(x)$ in terms of $q_A(x)$.

21. Express the relationship between the characteristic roots of A and of A^*.

22. If $A \in M_n(\mathbb{C})$ and $f(A)v = 0$ for some $v \neq 0$ in $\mathbb{C}^{(n)}$, is every root of $f(x)$ a characteristic root of A?

Middle-Level Problems

23. If $q_A(x)$ is the minimal polynomial for A and if $p(x)$ is such that $p(A) = 0$, show that $p(x) = q_A(x)t(x)$ for some polynomial $t(x)$.

24. Let $A \in M_n(\mathbb{C})$ and $v \neq 0 \in \mathbb{C}^{(n)}$. Suppose that $u(x)$ is the polynomial of least positive degree, with highest coefficient 1, such that $u(A)v = 0$. Prove:
  (a) Degree $u(x) \leq$ degree $q_A(x)$.
  (b) $q_A(x) = u(x)s(x)$ for some polynomial $s(x)$ with complex coefficients.
  (c) Every root of $u(x)$ is a characteristic root of A.

25. If A is *not* invertible, show that there is a nonzero matrix B such that $AB = BA = 0$.

26. If $A = \begin{bmatrix} a_1 & & & 0 \\ & a_2 & & \\ & & \ddots & \\ * & & & a_n \end{bmatrix}$ is a lower triangular matrix, show that A satisfies $p(x) = 0$, where $p(x) = (x - a_1)(x - a_2) \cdots (x - a_n)$.

27. If A is as in Problem 26, is the rank of A always equal to the number of elements $a_1, a_2, \ldots, a_n$ which are nonzero?

Harder Problems

28. If $A \neq 0$ is not invertible, show that there is a matrix B such that $AB = 0$ but $BA \neq 0$.

29. If A is a matrix such that $AC = I$ for some matrix C, prove that $CA = I$ (thus A is invertible).

30. If $A \in M_n(\mathbb{C})$, prove that A can have *at most* n distinct characteristic roots. (*Hint*: Show that if $v_1, \ldots, v_r$ are characteristic vectors corresponding to distinct characteristic roots, then they are linearly independent.)

31. If A is a Hermitian matrix and $A^2 v = 0$ for some $v \in \mathbb{C}^{(n)}$, prove that $Av = 0$.

32. Generalize Problem 31 as follows: If A is Hermitian and $A^k v = 0$ for $k \geq 1$, then $Av = 0$.

33. If A is Hermitian and b is a characteristic root of A, show that if $(A - bI)^k v = 0$ for some $k \geq 1$, then $Av = bv$.

Very Hard Problems

34. If A is Hermitian and $q_A(x)$ is its minimal polynomial, prove that $q_A(x)$ is of the form $q_A(x) = (x - a_1) \cdots (x - a_k)$, where $a_1, \ldots, a_k$ are *distinct* elements of $\mathbb{R}$.

35. If A is a *real* skew-symmetric matrix in $M_n(\mathbb{R})$, where n is *odd*, show that A cannot be invertible.

36. If A is Hermitian, show that $p(A) = 0$ for some nonzero polynomial of degree at most n. (*Hint*: Try to combine Problems 30 and 34.)

37. If $A \in M_n(\mathbb{C})$ is nilpotent (i.e., $A^k = 0$ for some $k \geq 1$) prove that $A^n = 0$.

4.6. HERMITIAN MATRICES

Although one could not tell from the section title, we continue here in the spirit of Section 4.5. Our objective is to study the relationship between characteristic roots and the property of being Hermitian. We had earlier obtained one piece of this relationship; namely, we showed in Theorem 3.5.7 that the characteristic roots of a Hermitian matrix must be *real*. We now fill out the picture. Some of the results which we shall obtain came up in the problems of the preceding section. But we want to do all this officially, here.

We begin with the following property of Hermitian matrices, which, as you may recall, are the matrices A in $M_n(\mathbb{C})$ such that $A = A^*$.

Lemma 4.6.1. If A in $M_n(\mathbb{C})$ is Hermitian and v in $\mathbb{C}^{(n)}$ is such that $A^k v = 0$ for some $k \geq 1$, then $Av = 0$.

Proof: Consider the easiest case first — that for $k = 2$, that is, where $A^2 v = 0$. But then $(A^2 v, v) = 0$ and, since $A = A^*$, we get $0 = (A^2 v, v) = (Av, A^* v) = (Av, Av)$. Since $(Av, Av) = 0$, the defining relations for inner product allow us to conclude that $Av = 0$.

For the case $k > 2$, note that $A^2(A^{k-2} v) = 0$ implies that $A(A^{k-2} v) = 0$, that is, $A^{k-1} v = 0$. So for $k \geq 2$ we can go from $A^k v = 0$ to $A^{k-1} v = 0$ and we can safely say "continue this procedure" to end up with $Av = 0$. ∎

If A is Hermitian and a is a characteristic root of A, then because a is real, $A - aI$ is Hermitian [since $(A - aI)^* = A^* - \bar{a}I = A - aI$ from the reality of a]. So as an immediate consequence of the lemma, we get

Corollary 4.6.2. If a is a characteristic root of the Hermitian matrix A and v is such that $(A - aI)^k v = 0$ (for $k \geq 1$), then $Av = av$.

Proof: Because $A - aI$ is Hermitian, our lemma assures us that $(A - aI)v = 0$, that is, $Av = av$. ∎

A fundamental implication of this corollary is to specify quite sharply the nature of the minimal polynomial, $q_A(x)$, of the Hermitian matrix A. This is

Theorem 4.6.3. If A is Hermitian with minimal polynomial $q_A(x)$, then $q_A(x)$ is of the form $q_A(x) = (x_1 - a_1) \cdots (x - a_t)$, where the $a_1, \ldots, a_t$ are *real* and *distinct*.

Proof: By the Fundamental Theorem of Algebra, we can factor $q_A(x)$ as

$$q_A(x) = (x_1 - a_1)^{m_1} \cdots (x - a_t)^{m_t},$$

where the m_r are positive integers and the $a_1, \ldots, a_t$ are distinct. By Theorem 4.5.4, $a_1, \ldots, a_t$ are *all* of the distinct characteristic roots of A, and by Theorem 3.5.7 they are real. So our job is to show that m_r is precisely 1 for each r.

Since $q_A(x)$ is the minimal polynomial for A, we have

$$0 = q_A(A) = (A - a_1 I)^{m_1} \cdots (A - a_t I)^{m_t}.$$

Applying these mappings to any v in $\mathbb{C}^{(n)}$, we have

$$0 = q_A(A)v = (A - a_1 I)^{m_1} \cdots (A - a_t I)^{m_t} v$$
$$= (A - a_1 I)^{m_1}((A - a_2 I)^{m_2} \cdots (A - a_t I)^{m_t} v).$$

Letting w denote $(A - a_2 I)^{m_2} \cdots (A - a_t I)^{m_t} v$, this says that

$$(A - a_1 I)^{m_1} w = 0.$$

By Corollary 4.6.2. we get that

$$Aw = a_1 w.$$

So

$$(A - a_1 I)[(A - a_2 I)^{m_2} \cdots (A - a_t I)^{m_t} v] = 0$$

for all v in $\mathbb{C}^n$, whence

$$(A - a_1 I)(A - a_2 I)^{m_2} \cdots (A - a_t I)^{m_t} = 0.$$

Now, rearranging the order of the terms to put the second one first, we can proceed to give the same treatment to m_2 as we gave to m_1. Continuing with the other terms in turn, we end up with

$$(A - a_1 I)(A - a_2 I) \cdots (A - a_t I) = 0,$$

as claimed. ∎

Since the minimum polynomial of A has real roots if A is Hermitian, we can prove

Theorem 4.6.4. *Let F be $\mathbb{R}$ or $\mathbb{C}$ and let $A \in M_n(F)$ be Hermitian. If W is a nonzero subspace of $F^{(n)}$ and $AW \subset W$, then W contains a characteristic vector of A.*

Proof: Let the minimum polynomial $q_A(x)$ of A be $(x - a_1) \cdots (x - a_r)$ and choose the first integer $r \geq 1$ such that $(A - a_1 I) \cdots (A - a_r I)W = 0$. If $r = 1$, then any nonzero element of W is a characteristic vector of A. If $r > 1$, then any nonzero vector of $(A - a_1 I) \cdots (A - a_{r-1} I)W$ is a characteristic vector of A associated with the characteristic root a_r. ∎

The proof just given actually shows a little more, namely, if $F = \mathbb{C}$ and $AW \subset W$, then W contains a characteristic vector of A.

The results we have already obtained allow us to pass to an extremely important theorem. It is difficult to overstate the importance of this result, not only in all of mathematics, but even more so, in physics. In physics, an infinite-dimensional version of what we have proved and are about to prove plays a crucial role. There the characteristic roots of a certain Hermitian linear transformation called the *Hamiltonian* are the *energies* of the physical state. The fact that such a linear transformation has, in a change of basis, such a simple form is vital in the quantum mechanics. In addition to all of this, it really is a very pretty result.

First, some preliminary steps.

Lemma 4.6.5. If A is Hermitian and a_1, a_2 are distinct characteristic roots of A, then, if $Av = a_1 v$ and $Aw = a_2 w$, we have $(v, w) = 0$.

Proof: Consider (Av, w). On one hand, since $Av = a_1 v$, we have

$$(Av, w) = (a_1 v, w) = a_1(v, w).$$

On the other hand,

$$(Av, w) = (v, A^*w) = (v, Aw) = (v, a_2 w) = a_2(v, w),$$

since a_2 is real. Therefore,

$$a_1(v, w) = (Av, w) = a_2(v, w) \text{ and } (a_1 - a_2)(v, w) = 0.$$

Because a_1 and a_2 are distinct, $a_1 - a_2$ is nonzero. Canceling, the outcome is that $(v, w) = 0$, as claimed in the lemma. ∎

The next result is really a corollary to Lemma 4.6.5, and we leave its proof as an exercise for the reader.

Lemma 4.6.6. If A is a Hermitian matrix in $M_n(\mathbb{C})$ and if $a_1, \ldots, a_t$ are distinct characteristic roots of A with corresponding characteristic vectors $v_1, \ldots, v_t \neq 0$ (i.e., $Av_r = a_r v_r$ for $1 \leq r \leq t$), then the vectors $v_1, \ldots, v_t$ are pairwise orthogonal (and therefore linearly independent) over $\mathbb{C}$.

Since $\mathbb{C}^{(n)}$ has at most n linearly independent vectors, by the result of Lemma 4.6.6, we know that $t \leq n$. So A has *at most n* distinct characteristic roots. Together with Theorem 4.6.3, we then have that $q_A(x)$ is of *degree at most n*. So A satisfies a polynomial of degree n or less over $\mathbb{C}$. This is a weak version of the Cayley–Hamilton Theorem, which will be proved in due course. We summarize these remarks into

Lemma 4.6.7. If A in $M_n(\mathbb{C})$ is Hermitian, then A has at most n distinct characteristic roots, and $q_A(x)$ is of degree at most n.

Let A be a Hermitian matrix in $M_n(\mathbb{C})$ acting on the vector space $V = \mathbb{C}^{(n)}$. If $a_1, \ldots, a_k$ are distinct characteristic roots of A, then we know that the a_r are real and the minimal polynomial of A is $q_A(x) = (x - a_1) \cdots (x - a_k)$.

Let $V_{a_r} = \{v \in V \mid Av = a_r v\}$ for $r = 1, \ldots, k$. Clearly, $A(V_{a_r}) \subseteq V_{a_r}$ for all r; in fact, A acts as a_r times the identity map on $V_{a_r} \neq \{0\}$.

Theorem 4.6.8. $V = V_{a_1} \oplus \cdots \oplus V_{a_k}$.

Proof: Let $Z = V_{a_1} + \cdots + V_{a_k}$. The subspaces V_{a_r} are mutually orthogonal by

Lemma 4.6.6. By Theorem 4.2.4, their sum is direct. So $Z = V_{a_1} \oplus \cdots \oplus V_{a_k}$. Also, because $A(V_{a_r}) \subseteq V_{a_r}$, we have $A(Z) \subset Z$.

Now it suffices to show that $V = Z$. For this, let $Z^\perp$ be the orthogonal complement of Z in V. We claim that $A(Z^\perp) \subset Z^\perp$. Why? Since $A^* = A$, $A(Z) \subset Z$ and $(Z, Z^\perp) = 0$, we have

$$0 = (A(Z), Z^\perp) = (Z, A^*(Z^\perp)) = (Z, A(Z^\perp)).$$

But this tells us that $A(Z^\perp) \subset Z^\perp$. So if $Z^\perp \neq 0$, by Theorem 4.6.4 there is a characteristic vector of A in $Z^\perp$. But this is impossible since all the characteristic roots are used up in Z and $Z^\perp \cap V_{a_s} = \{0\}$ for all s. It follows that $Z^\perp = \{0\}$, hence $Z = Z^{\perp \perp} = \{0\}^\perp = V$, by Theorem 4.3.7. This, in turn, says that

$$V = Z = V_{a_1} \oplus \cdots \oplus V_{a_s}. \quad \blacksquare$$

An immediate and important consequence of Theorem 4.6.8 is

Theorem 4.6.9. Let A be a Hermitian matrix in $M_n(\mathbb{C})$. Then there exists an orthonormal basis of $\mathbb{C}^{(n)}$ in which the matrix of A is diagonal. Equivalently, there is a unitary matrix U in $M_n(\mathbb{C})$ such that the matrix $U^{-1}AU$ is diagonal.

Proof: For each V_{a_r}, $r = 1, \ldots, k$, we can find an orthonormal basis. These, put together, form an orthonormal basis of V since

$$V = V_{a_1} \oplus \cdots \oplus V_{a_k}.$$

By the definition of V_{a_r}, the matrix A acts like multiplication by the scalar a_r on V_{a_r}. So, in this put-together basis of V, the matrix of A is the diagonal matrix

$$\begin{bmatrix} a_1 I_{m_1} & & & 0 \\ & a_2 I_{m_2} & & \\ & & \ddots & \\ 0 & & & a_k I_{m_k} \end{bmatrix},$$

where I_{m_r} is the $m_r \times m_r$ identity matrix and $m_r = \dim(V_{a_r})$ for all r. Since a change of basis from one orthonormal basis to another is achieved by a unitary matrix U, we have that

$$U^{-1}AU = \begin{bmatrix} a_1 I_{m_1} & & & 0 \\ & a_2 I_{m_2} & & \\ & & \ddots & \\ 0 & & & a_k I_{m_k} \end{bmatrix}$$

for such a U and the theorem is proved. $\blacksquare$

This theorem is usually described as: *A Hermitian matrix can be brought to diagonal form, or can be diagonalized, by a unitary matrix.*

It is not true that every matrix can be brought to diagonal form. We leave as an exercise that there is no invertible matrix $\begin{bmatrix} a & b \\ c & d \end{bmatrix}$ such that

$$\begin{bmatrix} a & b \\ c & d \end{bmatrix}^{-1} \begin{bmatrix} 0 & 1 \\ 0 & 0 \end{bmatrix} \begin{bmatrix} a & b \\ c & d \end{bmatrix}$$

is diagonal. So, in this regard, Hermitian matrices are remarkable.

There is an easy passage from Hermitian to skew-Hermitian matrices, and vice versa. If A is Hermitian, then $B = iA$ is skew-Hermitian, since $B^* = (iA)^* = -iA = -B$. Similarly, if A is skew-Hermitian, then iA is Hermitian. Thus we can easily get analogues, for the theorems we proved for Hermitian matrices, for skew-Hermitian matrices. As a sample:

Corollary 4.6.10. If A in $M_n(\mathbb{C})$ is skew-Hermitian, then for some unitary matrix U, $U^{-1}AU$ is a diagonal matrix, with pure imaginary entries on the diagonal.

Proof: Let $B = iA$. Then B is Hermitian. Therefore, $U^{-1}BU$ is diagonal for some unitary U and has real entries on its diagonal. That is, the matrix $U^{-1}(iA)U$ is diagonal with real diagonal entries. Hence $U^{-1}AU$ is diagonal with pure imaginary entries on its diagonal. ■

A final word in the spirit of what was done in this section. Suppose that A is a *symmetric* matrix with *real entries*; thus A is Hermitian. As such, we know that $U^{-1}AU$ is diagonal for some unitary matrix in $M_n(\mathbb{C})$, that is, one having complex entries. A natural question then presents itself: Can we find a *unitary matrix B* with *real* entries such that $B^{-1}AB$ is diagonal? The answer is yes. While working in the larger set of matrices with complex entries, we showed in Theorem 4.6.3 that the minimum polynomial of A is $q_A(x) = (x - a_1)\cdots(x - a_k)$, where the $a_1, \ldots, a_k$ are distinct and real. So we can get this result here by referring back to what was done for the complex case and outlining how to proceed.

Theorem 4.6.11. Let A be a symmetric matrix *with real entries*. Then there is a unitary matrix U *with real entries* such that $U^{-1}AU$ is diagonal.

Proof: If we look back at what we already have done, and check the proofs, we see that very often we did not use the fact that we were working over the complex numbers. The arguments go through, as given, even when we are working over the real numbers. Specifically,

1. The Gram–Schmidt orthogonalization process goes over.
2. So any nonzero subspace has an orthonormal basis.
3. Theorems 4.6.4, 4.2.4, and 4.3.6 and Lemma 4.6.6 go over.

Now we can repeat the proofs of Theorems 4.6.8 and 4.6.9 word for word, with $\mathbb{R}$ in place of $\mathbb{C}$. ■

PROBLEMS

NUMERICAL PROBLEMS

1. By a direct calculation show that if $A = \begin{bmatrix} a & b \\ b & c \end{bmatrix}$, $a, b, c \in \mathbb{R}$, is such that $A^2 = 0$ then $A = 0$.

2. (a) Find the characteristic roots of $A = \begin{bmatrix} 0 & 2 \\ 2 & 0 \end{bmatrix}$.
   (b) From Part (a), find $q_A(x)$.
   (c) For each characteristic root s of A, find a nonzero element v in $\mathbb{R}^{(2)}$ such that $Av = sv$.
   (d) Verify that if v, w are as in Part (c), for distinct characteristic roots of A, then $(v, w) = 0$.

3. Do Problem 2 for the matrix $A = \begin{bmatrix} 1 & 0 & 0 \\ 0 & 0 & 2 \\ 0 & 2 & 0 \end{bmatrix}$.

4. Let $A = \begin{bmatrix} 0 & 1 \\ -1 & 0 \end{bmatrix}$. Show that A is invertible and that $\begin{bmatrix} 0 & -1 \\ 1 & 0 \end{bmatrix} A^{-1}$ is unitary.

5. (a) Show that $A = \begin{bmatrix} i & 0 & 0 \\ 0 & 0 & 1 \\ 0 & -1 & 0 \end{bmatrix}$ in $M_3(\mathbb{C})$ is skew-Hermitian and find its characteristic roots.
   (b) Calculate A^2 and show that A^2 is Hermitian.
   (c) Find A^{-1} and show that it is skew-Hermitian.

MORE THEORETICAL PROBLEMS

Easier Problems

6. Prove Lemma 4.6.6.
7. If A is skew-Hermitian, show that A^k is Hermitian if k is even and is skew-Hermitian if k is odd.
8. If A is invertible, show that
   (a) If A is Hermitian, then A^{-1} is Hermitian.
   (b) If A is skew-Hermitian, then A^{-1} is skew-Hermitian.
   (c) A^* is invertible (find its inverse).
9. If A is such that $AA^* = A^*A$ and A is invertible, then $U = A^*A^{-1}$ is unitary.
10. If $A \in M_n(\mathbb{C})$ and $V = \{v \in \mathbb{C}^{(n)} | (A - aI)^k v = 0 \text{ for some } k\}$, show that if $BA = AB$, then $B(V) \subseteq V$.
11. Prove Theorem 4.6.11 in detail.

Middle-Level Problems

12. If A is a skew-Hermitian matrix with *real* entries, show that if a is a characteristic root of A, then $\bar{a}$, the complex conjugate of a, is a characteristic root of A.

13. If A is skew-Hermitian with real entries, in $M_n(\mathbb{R})$, where n is odd, show tht A cannot be invertible.

14. If $A \in M_n(\mathbb{R})$ is skew-Hermitian and n is even, show that

$$q_A(x) = p_1(x) \cdots p_k(x),$$

where $p_1(x), \ldots, p_k(x)$ are polynomials of degree 2 with real coefficients.

15. What is the analogue of the result of Problem 14 if n is odd?

16. If $A \in M_n(\mathbb{R})$ is skew-Hermitian, show that it must have even rank.

17. If A is Hermitian and has only one distinct characteristic root, show that A must be a scalar matrix.

18. If $E^2 = E$ is Hermitian, for what values of $a \in \mathbb{C}$ is $I - aE$ unitary?

19. Let A be a skew-Hermitian matrix in $M_n(\mathbb{C})$.
  (a) Show that $aI - A$ is invertible for all *real* $a \neq 0$.
  (b) Show that $B = (I - A)(I + A)^{-1}$ is unitary.
  (c) Show that for some unitary matrix, U, the matrix $U^{-1}BU$ is diagonal.

Harder Problems

20. If $A, B \in M_n(\mathbb{C})$ are such that $AB = BA$, show that if

$$W = \{w \in \mathbb{C}^{(n)} \mid Aw = aw\},$$

then there is an element $w \neq 0$ in W such that $Bw = bw$ for some $b \in \mathbb{C}$.

21. Use the result of Problem 20 to prove that if A and B are two Hermitian matrices in $M_n(\mathbb{C})$ such that $AB = BA$, then there exists a unitary matrix U such that *both* $U^{-1}AU$ and $U^{-1}BU$ are *diagonal*.

22. If $A \in M_n(\mathbb{C})$ is such that $AA^* = A^*A$ and $v \in \mathbb{C}^{(n)}$ is such that $Av = 0$, prove that $A^*v = 0$.

23. If $A \in M_n(\mathbb{C})$ is such that $AA^* = A^*A$ and $v \in \mathbb{C}^{(n)}$ is such that $Av = av$, prove that $A^*v = \bar{a}v$, where $\bar{a}$ is the complex conjugate of a.

Very Hard Problems

24. If $A \in M_n(\mathbb{C})$ is such that $AA^* = A^*A$, prove:
  (a) If $A^2v = 0$ for $v \in \mathbb{C}^{(n)}$, then $Av = 0$.
  (b) If $A^kv = 0$ for $v \in \mathbb{C}^{(n)}$ for some $k \geq 1$, then $Av = 0$.
  (c) $q_A(x) = (x - a_1) \cdots (x - a_t)$, where $a_1, \ldots, a_t$ are all the distinct characteristic roots of A.

25. If $A \in M_n(\mathbb{C})$ is such that $AA^* = A^*A$, prove that there exists a unitary matrix, U, in $M_n(\mathbb{C})$, such that $U^{-1}AU$ is diagonal.

26. If $A \in M_n(\mathbb{C})$ is such that $AA^* = A^*A$, prove:
  (a) A is Hermitian if and only if all its characteristic roots are real.
  (b) A is skew-Hermitian if and only if all its characteristic roots are pure imaginary.

(c) A is unitary if and only if all its characteristic roots, a, satisfy $|a| = 1$ (i.e., $a\bar{a} = 1$).

27. If $A \in M_n(\mathbb{C})$ is unitary, prove that $U^{-1}AU$ is diagonal for some unitary U in $M_n(\mathbb{C})$.

4.7. TRIANGULARIZING MATRICES WITH COMPLEX ENTRIES

The subject matter we are about to take up is somewhat more subtle—consequently somewhat more difficult—than the material treated hitherto. This difficulty is not so much in the technical nature of the proofs, but rather, is in the appreciation, the first time one sees these things, of the importance and role that the results play in matrix theory.

Until now, we have used mathematical induction only informally, contenting ourselves with phrases such as "and so on" and "repeat the procedure." Many, if not most readers will have had some exposure to mathematical induction. However, for them and for those who have had no experience with induction, we give a quick synopsis of that topic.

Mathematical induction is a technique for proving theorems about integers, or in situations where integers measure some phenomenon. We state the *principle of mathematical induction* now:

Suppose that $P(n)$ is a proposition such that

1. *$P(1)$ is true.*
2. *Whenever we know that $P(k)$ is true for some integer $k \geq 1$ then we know that $P(k + 1)$ is also true.*

Then we know that $P(n)$ is true for all positive integers n.

Another form—equivalent to the one above—runs as follows:

Suppose that $P(1)$ is true and that for any $m > 1$, the truth of $P(k)$ for all positive integers $k < m$ implies the truth of $P(m)$. Then $P(n)$ is true for all positive integers n.

What is the rationale for this? Knowing $P(1)$ to be valid by (1) tells us, by (2), that $P(2)$ is valid. The validity of $P(2)$, again by (2), assures that of $P(3)$. That of $P(3)$ gives us $P(4)$, and so on.

We illustrate this technique with an example, one that is overworked in almost all discussions of induction. We want to prove that

$$1 + 2 + \cdots + n = \frac{n(n + 1)}{2}.$$

So let the proposition $P(n)$ be that $1 + 2 + \cdots + n = \frac{n(n + 1)}{2}$. Clearly, if $n = 1$, then, since $1 = \dfrac{1(1+1)}{2}$, $P(1)$ is true. Suppose that we happen to know that $P(k)$ is true for some integer $k \geq 1$, that is,

$$1 + 2 + \cdots + k = \frac{k(k+1)}{2}.$$

We want to show from this that $P(k+1)$ is true, that is, that

$$1 + 2 + \cdots + (k+1) = \frac{(k+1)((k+1)+1)}{2}.$$

Now, by our assumption,

$$1 + 2 + \cdots + k + (k+1) = \frac{k(k+1)}{2} + (k+1) = \frac{(k+1)(k+2)}{2}$$

$$= \frac{(k+1)((k+1)+1)}{2}.$$

So the validity of $P(k)$ implies that of $P(k+1)$. Therefore, by the principle of mathematical induction, $P(n)$ is true for all positive integers n; that is, $1 + 2 + \cdots + n = \dfrac{n(n+1)}{2}$ for all positive integers n.

In the proof we are about to give we will go by induction on the dimension of V. Precisely what does this mean? It means that we will show two things:

1. The theorem we want to prove is true if $\dim V = 1$.
2. The truth of the theorem for all subspaces W of $\mathbb{C}^{(n)}$ of the same kind for which $\dim W$ is less than $\dim V$ implies the truth of the theorem for V.

We shall describe this procedure by saying: *We go by induction on the dimension of V.*

Theorem 4.7.1. Let V be a subspace of $\mathbb{C}^{(n)}$, T and element of $M_n(\mathbb{C})$ such that $T(V) \subset V$. Then we can find a basis $v_1, \ldots, v_t$ of V such that $Tv_s = \sum_{r=1}^{s} a_{rs}v_r$, that is, Tv_s is a linear combination of v_s and its predecessors, for $1 \leq s \leq t$. In other words, we can find a basis of V in which the $t \times t$ matrix (a_{rs}) of T on V is upper triangular.

Proof: We go by induction on $\dim V$. If $\dim V = 1$, then $V = Fv$ and $Tv = av$ for some scalar a. Since v is a basis for V, it does the trick and we have proved the result for $\dim V = 1$.

Next, let V be a subspace of $\mathbb{C}^{(n)}$ and suppose that the theorem is correct for all proper subspaces W of V such that $T(W) \subset W$. So, for such a subspace W, we can find a basis $w_1, \ldots, w_u$ such that Tw_s is a linear combination of w_s and its predecessors for $s = 1, 2, \ldots, u$.

We can factor $q_T(x)$ as a product $q_T(x) = (x - b_1)\cdots(x - b_k)$, where $b_1,\ldots,b_k$ are elements (not necessarily all different) of $\mathbb{C}$.

Since $0 = q_T(T) = (T - b_1 I)\cdots(T - b_k I)$ we can't have $(T - b_p I)(V) = V$ for all p, since otherwise $V = (T - b_k I)(V) = (T - b_{k-1} I)(T - b_k I)(V) = \cdots = (T - b_1 I)\cdots(T - b_k I)(V) = q_T(T)(V) = 0$. Choose p such that $(T - b_p I)(V) \neq V$ and let $W = (T - b_p I)(V)$; hence W is a proper subspace of V. Because

$$T(W) = T((T - b_p I)(V)) = (T - b_p I)(T(V)) \subset (T - b_p I)(V) = W$$

[using the fact that $T(V) \subset V$], we have $T(W) \subset W$. Furthermore, since W is a proper subspace of V, $\dim W < \dim V$. By induction, we therefore can find a basis $w_1,\ldots,w_u$ such that for each $s = 1, 2, \ldots, u$, $Tw_s = \sum_{r=1}^{s} a_{rs} w_r$.

By Corollary 4.1.2, we can find elements $w_{u+1},\ldots,w_t$ such that $w_1,\ldots,w_u$, $w_{u+1},\ldots,w_t$ form a basis of V. We know from the above that for $s \leq u$, $Tw_s = \sum_{r=1}^{s} a_{rs} w_r$. What is the action of T on $w_{u+1},\ldots,w_t$? Because $W = (T - b_p I)(V)$, each of the vectors $(T - b_p I)w_s$, for $s > u$, is in W, so is a linear combination of $w_1,\ldots,w_u$. That is, we can express them as

$$(T - b_p I)w_s = \sum_{r=1}^{u} a_{rs} w_r$$

for $s > u$. But this implies that even for $s > u$, we can express them as

$$(T - b_p I)w_s = \sum_{r=1}^{s} a_{rs} w_r$$

(take the coefficients to be 0 for $u < r \leq s$). So here, too, Tw_s is a linear combination of w_s and its predecessors. Thus this is true for every element in the basis we created, namely, $w_1,\ldots,w_u, w_{u+1},\ldots,w_t$. In other words, the desired result holds for V. Since $a_{rs} = 0$ for $r > s$, the matrix A of T in this basis is upper triangular. ∎

We single out a special case of the theorem, which is of special interest.

Theorem 4.7.2. Let $T \in M_n(\mathbb{C})$. Then there exists a basis $v_1,\ldots,v_n$ of $\mathbb{C}^{(n)}$ over $\mathbb{C}$ such that for each $s = 1, 2, \ldots, n$, $Tv_s = \sum_{r=1}^{s} a_{rs} v_r$. In other words, there exists a basis for $\mathbb{C}^{(n)}$ in which the matrix A of T is upper triangular.

Proof: After all, $\mathbb{C}^{(n)}$ is a subspace of itself, so by a direct application of Theorem 4.7.1, we obtain Theorem 4.7.2. ∎

Theorem 4.7.2 has important implications for matrices. We begin to catalog such implications now.

Theorem 4.7.3. If $T \in M_n(\mathbb{C})$, then there exists an invertible matrix C in $M_n(\mathbb{C})$ such that $C^{-1}TC$ is an upper triangular matrix $\begin{bmatrix} a_{11} & & * \\ & \ddots & \\ 0 & & a_{nn} \end{bmatrix}$.

Proof: Let $v_1, \ldots, v_n$ be a basis of $\mathbb{C}^{(n)}$ over $\mathbb{C}$ in which the matrix A of T is upper triangular. If C is the matrix whose columns are the vectors $v_1, \ldots, v_n$, then C is invertible and $A = C^{-1}TC$ by Theorem 3.9.1. ∎

We should point out that Theorem 4.7.3 does not hold if we are merely working over $\mathbb{R}$. The following example of $T = \begin{bmatrix} 0 & 1 \\ -1 & 0 \end{bmatrix}$ reveals this. We claim that there is no invertible matrix V with real entries such that $V^{-1}TV = \begin{bmatrix} a & b \\ 0 & c \end{bmatrix}$, that is, such that $\begin{bmatrix} 0 & 1 \\ -1 & 0 \end{bmatrix} V = V \begin{bmatrix} a & b \\ 0 & c \end{bmatrix}$. For suppose that $V = \begin{bmatrix} u & v \\ w & x \end{bmatrix}$. Then

$$\begin{bmatrix} 0 & 1 \\ -1 & 0 \end{bmatrix} \begin{bmatrix} u & v \\ w & x \end{bmatrix} = \begin{bmatrix} u & v \\ w & x \end{bmatrix} \begin{bmatrix} a & b \\ 0 & c \end{bmatrix}$$

yields $w = au$, $-u = aw$. But then $w \neq 0$ (otherwise, $u = w = 0$ and the first basis vector is 0) and $w = au = a(-aw) = -a^2 w$ and so $(1 + a^2)w = 0$, which is impossible since a is real.

Theorem 4.7.3 is one of the key results in matrix theory. It is usually described as:

Any matrix in $M_n(\mathbb{C})$ can be brought to triangular form in $M_n(\mathbb{C})$.

Theorem 4.7.3 has many important consequences for us. We begin with an example with $n = 3$. Suppose that T is in $M_3(\mathbb{C})$ and that we take a basis C to triangularize the matrix A of T in C, that is, $A = C^{-1}TC = \begin{bmatrix} a & b & c \\ 0 & d & e \\ 0 & 0 & f \end{bmatrix}$. Then the columns v_1, v_2, v_3 of C form a basis of $\mathbb{C}^{(3)}$ such that

$$Av_1 = av_1, \quad Av_2 = bv_1 + dv_2, \quad Av_3 = cv_1 + ev_2 + fv_3.$$

Therefore,

$$(A - aI)v_1 = 0, \quad (A - dI)v_2 = bv_1, \quad (A - fI)v_3 = cv_1 + ev_2.$$

Thus we can knock out the basis vectors one at a time:

$$(A - aI)v_1 = 0,$$
$$(A - aI)(A - dI)v_2 = (A - aI)(bv_1) = 0,$$
$$(A - aI)(A - dI)(A - fI)v_3 = (A - aI)(A - dI)(cv_1 + ev_2) = 0. \quad \text{(Why?)}$$

Sec. 4.7]   **Triangularizing Matrices with Complex Entries**

If we let $D = (A - aI)(A - dI)(A - fI)$, it follows that D knocks out all three basis vectors at the same time, so that $Dv = 0$ for all v in $\mathbb{C}^{(3)}$. (Prove!) In other words, $D = 0$ and $(A - aI)(A - dI)(A - fI) = 0$. Expressing A as $C^{-1}TC$, we have

$$(C^{-1}TC - aI)(C^{-1}TC - dI)(C^{-1}TC - fI) = 0.$$

But

$$(C^{-1}TC - aI)(C^{-1}TC - dI)(C^{-1}TC - fI) = C^{-1}(T - aI)(T - dI)(T - fI)C$$

(Prove!), so that

$$C^{-1}(T - aI)(T - dI)(T - fI)C = 0.$$

Multiplying the equation on the left by C and on the right by C^{-1}, we see that

$$(T - aI)(T - dI)(T - fI) = 0.$$

What does the equation $(T - aI)(T - dI)(T - fI) = 0$ tell us? Several things:

(1) Any matrix in $M_3(\mathbb{C})$ satisfies a polynomial of degree 3 over $\mathbb{C}$.

(2) This polynomial is the product of terms of the form $T - uI$, where the u are all of the characteristic roots of T, in this case, a, b, and c. (Prove!) It may happen that these characteristic roots are not distinct, that is, that some are included more than once in this polynomial.

What we did for T in $M_3(\mathbb{C})$ we shall do—in exactly the same way—for matrices in $M_n(\mathbb{C})$, to obtain results like (1) and (2) above. This is why we did the 3×3 case in so painstaking a way.

Theorem 4.7.4. Given T in $M_n(\mathbb{C})$, there is a polynomial, $p(x)$, *of degree n* with complex coefficients such that $p(T) = 0$. If T is an upper triangular matrix $\begin{bmatrix} a_1 & & * \\ & \ddots & \\ 0 & & a_n \end{bmatrix}$, one such polynomial $p(x)$ is $(x - a_1)\cdots(x - a_n)$, and the $a_1, \ldots, a_n$ are the characteristic roots of T (all of them with possible duplication).

Proof: The scheme for proving the result is just what we did for the 3×3 case, with the obvious modification needed for the general situation.

By Theorem 4.7.2 we have a basis $v_1, \ldots, v_n$ of $\mathbb{C}^{(n)}$ such that for each s, $Tv_s = \sum_{r=1}^{s} a_{rs}v_r$. Take any such basis, let V be the invertible matrix whose columns are the v_r and write the entries a_{kk} as a_k. Thus $C^{-1}TC$ is the upper triangular matrix $A = \begin{bmatrix} a_1 & & * \\ & \ddots & \\ 0 & & a_n \end{bmatrix}$. Note that $(T - a_sI)v_s$ is a linear combination of $v_1, \ldots, v_{s-1}$ for all s. By an induction—or going step by step as for the 3×3 case—we have that

$$(T - a_1 I)(T - a_2 I) \cdots (T - a_{s-1} I) v_k = 0$$

for $k = 1, 2, \ldots, s - 1$. But then

$$(T - a_1 I)(T - a_2 I) \cdots (T - a_{s-1} I)(T - a_s I) v_k = 0$$

not only for $k = 1, 2, \ldots, s - 1$, but also for $k = s$. (Why?) The vector $(T - a_s I) v_k$ is a linear combination of $v_1, \ldots, v_{s-1}$. This is true for all s, therefore, and so, in particular,

$$(T - a_1 I) \cdots (T - a_n I) v_k = 0$$

for $k = 1, 2, \ldots, n$. Because $(T - a_1 I) \cdots (T - a_n I)$ annihilates a basis of $\mathbb{C}^{(n)}$, it must annihilate all of $\mathbb{C}^{(n)}$, that is, $(T - a_1 I) \cdots (T - a_n I) v = 0$ for all v in $\mathbb{C}^{(n)}$. In short, the linear transformation $(T - a_1 I) \cdots (T - a_n I)$ is 0.

It follows that the minimum polynomial $q_T(x)$ divides $(x - a_1) \cdots (x - a_n)$. Since the characteristic roots of T are just the roots of its minimum polynomial, it follows that every characteristic root occurs as some a_r. Conversely, for any $1 \le r \le n$, the matrix $A - a_r I$ is not invertible, by Theorem 4.5.3. So $T - a_r I = C(A - a_r I) C^{-1}$ is also not invertible (prove!), so that a_r is a characteristic root of T. ∎

This sharpens considerably our previous result, namely, that T satisfies a polynomial of degree n^2, for we have cut down from n^2 to n. We could call this theorem the *weak Cayley–Hamilton Theorem*. Why weak? Later, we prove this again and, at the same time, give Cayley–Hamilton's explicit description of this polynomial of degree n.

Suppose that T is a *nilpotent* matrix, that is, that $T^k = 0$ for some k. If $a \in \mathbb{C}$ is a characteristic root of T, then $Tv = av$ for some nonzero vector v in $\mathbb{C}^{(n)}$. Thus

$$T^2 v = T(Tv) = T(av) = aT(v) = aav = a^2 v,$$
$$T^3 v = T^2(Tv) = T^2(av) = aT^2(v) = aa^2 v = a^3 v,$$

and so on. By induction we can show that $T^k v = a^k v$ for all positive integers k (Do so!). Since $T^k = 0$ for some k and v is nonzero, and since $0 = T^k v = a^k v$, it follows that a is 0. Therefore, the only characteristic root of T is 0. By Theorem 4.7.3 it follows that there is an invertible matrix C such that $A = C^{-1} T C$ is an upper triangular matrix with 0's on the diagonal, proving

Theorem 4.7.5. If T in $M_n(\mathbb{C})$ is nilpotent, then there is an invertible C in $M_n(\mathbb{C})$ such that the matrix $C^{-1} T C$ is an upper triangular matrix $\begin{bmatrix} 0 & & * \\ & \ddots & \\ 0 & & 0 \end{bmatrix}$ with 0's on the diagonal.

In particular, the trace of the matrix $C^{-1} T C = \begin{bmatrix} 0 & & * \\ & \ddots & \\ 0 & & 0 \end{bmatrix}$ of the nilpotent

T in Theorem 4.7.5 is 0. But then also $0 = \text{tr}(C^{-1}TC) = \text{tr}(T)$. This proves

Corollary 4.7.6. If T in $M_n(\mathbb{C})$ is nilpotent, $\text{tr}(T) = 0$.

There is a sort of converse to Theorem 4.7.5. It asserts that if T in $M_n(\mathbb{C})$ is such that $\text{tr}(T^k) = 0$ for $k = 1, 2, \ldots, n$, then $T^k = 0$ for some k. This will appear as a very (very) hard problem for general n. We do the case $n = 2$ here—it is not too hard, and it gives the spirit of the proof for all n.

Suppose that T in $M_2(\mathbb{C})$ is such that $\text{tr}(T) = 0$ and $\text{tr}(T^2) = 0$. By Theorem 4.7.3, $A = C^{-1}TC = \begin{bmatrix} a_1 & b \\ 0 & a_2 \end{bmatrix}$ for some invertible matrix C. Since the diagonal entries of the upper triangular matrix A are a_1 and a_2, we have

$$a_1 + a_2 = \text{tr}\, A = \text{tr}(C^{-1}TC) = \text{tr}(T) = 0.$$

As for A^2, we have

$$A^2 = (C^{-1}TC)^2 = (C^{-1}TC)(C^{-1}TC) = C^{-1}T^2C.$$

Since the diagonal entries of the upper triangular matrix A^2 are a_1^2 and a_2^2, we have

$$a_1^2 + a_2^2 = \text{tr}(A^2) = \text{tr}(C^{-1}T^2C) = \text{tr}(T^2) = 0.$$

The equations $a_1 + a_2 = 0$ and $a_1^2 + a_2^2 = 0$ imply that a_1 and a_2 are both 0. (Why?) Thus $A = \begin{bmatrix} 0 & b \\ 0 & 0 \end{bmatrix}$. But then $A^2 = 0$. From this it follows that $C^{-1}T^2C = 0$, hence $T^2 = 0$.

We can combine Theorems 4.7.3 and 4.7.4 to obtain an important property of the trace function.

Theorem 4.7.7. If $T \in M_n(\mathbb{C})$, then $\text{tr}(T) = a_1 + \cdots + a_n$, where $a_1, a_2, \ldots, a_n$ are the characteristic roots of T (with possible duplications).

Proof: By Theorem 4.7.3 we know that for some invertible C, $A = C^{-1}TC = \begin{bmatrix} a_1 & & * \\ & \ddots & \\ 0 & & a_n \end{bmatrix}$. By Theorem 4.7.4 the diagonal entries are the characteristic roots of T. Now $\text{tr}\, T = \text{tr}\, C^{-1}TC = \text{tr}\, A = a_1 + \cdots + a_n$. ∎

Of course, the a_r need not be distinct. In the simplest possible case, namely, $T = I = \begin{bmatrix} 1 & & 0 \\ & \ddots & \\ 0 & & 1 \end{bmatrix}$ in $M_n(\mathbb{C})$, T has trace n, $A = I$ and this n is the sum of n duplicates of the characteristic root 1 appearing on the diagonal of A.

How many duplicates of the various characteristic roots of T are needed in the trace? At this point we can only say that every characteristic root is needed at least once and the *total number* needed is the number of characteristic roots on the diagonal of A, including duplicates, which of course is n. Later, however, we answer this question definitively by saying that

The characteristic roots (including duplicates) are the roots of the characteristic polynomial.

The *characteristic polynomial* is a very important player in the world of linear algebra, whose identity we reveal in Chapter 5.

PROBLEMS

NUMERICAL PROBLEMS

1. For the given T, find a C in $M_n(\mathbb{C})$, invertible, such that $C^{-1}TC$ is triangular.

   (a) $T = \begin{bmatrix} 1 & 2 \\ 3 & 4 \end{bmatrix}$.

   (b) $T = \begin{bmatrix} 0 & 1 \\ 1 & 1 \end{bmatrix}$.

   (c) $T = \begin{bmatrix} 0 & i \\ -i & 0 \end{bmatrix}$.

   (d) $T = \begin{bmatrix} 1 & 0 & 0 \\ 0 & 0 & i \\ 0 & -i & 0 \end{bmatrix}$.

   (e) $T = \begin{bmatrix} 1 & 1 & 1 \\ 1 & 1 & 1 \\ 0 & 0 & 0 \end{bmatrix}$.

   (f) $T = \begin{bmatrix} 0 & 1 & 0 \\ 0 & 0 & 1 \\ 1 & 0 & 0 \end{bmatrix}$.

2. For the matrices in Problem 1, find bases $v_1, \ldots, v_n$ such that Tv_s is a linear combination of its predecessors, for $s = 1, 2, 3, \ldots, n$.

3. Let $t = 2$, $V = \left\{ \begin{bmatrix} a \\ 0 \\ b \end{bmatrix} \middle| a, b \in \mathbb{C} \right\}$, and $T = \begin{bmatrix} 0 & 1 & 0 \\ 0 & 2 & 0 \\ 1 & 2 & 3 \end{bmatrix}$. Find a polynomial $P_{T,V}(x)$ of degree t such that $P_{T,V}(T)(V) = 0$.

4. Let $t = 2$, $V = \left\{ \begin{bmatrix} a \\ b \\ c \end{bmatrix} \middle| a, b, c \in \mathbb{C}, a + b + c = 0 \right\}$, and $T = \begin{bmatrix} 1 & 0 & 0 \\ 0 & 0 & 1 \\ 0 & 1 & 0 \end{bmatrix}$. Find a polynomial $P_{T,V}(x)$ of degree t such that $P_{T,V}(T)(V) = 0$.

5. If $T = \begin{bmatrix} a & b \\ c & d \end{bmatrix}$ satisfies $T^2 = 0$, show that $a + d = 0$.

6. Find an invertible matrix C such that $C \begin{bmatrix} 0 & 1 \\ 0 & 0 \end{bmatrix} = \begin{bmatrix} 0 & 0 \\ 1 & 0 \end{bmatrix} C$.

MORE THEORETICAL PROBLEMS
Easier Problems

7. Suppose that $TT' = T$. Then show that $T^2 = T$.
8. Suppose that
$$(A - aI)v_1 = 0,$$
$$(A - dI)v_2 = bv_1,$$
$$(A - fI)v_3 = cv_1 + ev_2,$$
and let $D = (A - aI)(A - dI)(A - fI)$. Prove that $(A - aI)(A - dI)(A - fI)v = 0$ for every linear combination v of v_1, v_2, v_3.
9. Prove that the expression $(C^{-1}TC - aI)(C^{-1}TC - dI)(C^{-1}TC - fI)$ equals $C^{-1}(T - aI)(T - dI)(T - fI)C$.
10. If T in $M_n(\mathbb{C})$ is such that $T^2 = T$, find the characteristic roots of T.
11. If T in $M_n(\mathbb{C})$ is such that $T^2 = T$, prove that $\text{tr}(T) = \text{rank } T$.
12. If T in $M_n(\mathbb{C})$ is such that $T^2 = T$, show that for some invertible C in $M_n(\mathbb{C})$, $C^{-1}TC$ is a matrix of the form $\begin{bmatrix} I_k & 0 \\ 0 & 0 \end{bmatrix}$ (I_k is the $k \times k$ identity matrix and the 0's represent blocks of the right sizes all of whose entries are 0), where k is the rank of T.
13. Show that there exists an invertible matrix C such that $C^{-1} \begin{bmatrix} 0 & 0 & 0 \\ 1 & 0 & 0 \\ 0 & 1 & 0 \end{bmatrix} C = \begin{bmatrix} 0 & 1 & 0 \\ 0 & 0 & 1 \\ 0 & 0 & 0 \end{bmatrix}$.

Middle-Level Problems

14. If T in $M_3(\mathbb{C})$ is such that $\text{tr}(T) = \text{tr}(T^2) = \text{tr}(T^3) = 0$, prove that T cannot be invertible.
15. If T in $M_3(\mathbb{C})$ is such that $\text{tr}(T) = \text{tr}(T^2) = \text{tr}(T^3) = 0$, prove that $T^3 = 0$.
16. If T in $M_n(\mathbb{C})$ is the matrix $T = \begin{bmatrix} 0 & 0 & \cdots & 0 \\ 1 & 0 & \ddots & \vdots \\ \vdots & \ddots & \ddots & 0 \\ 0 & \cdots & 1 & 0 \end{bmatrix}$, show that you can find an invertible C in $M_n(\mathbb{C})$ such that $C^{-1}TC = \begin{bmatrix} 0 & 1 & \cdots & 0 \\ 0 & 0 & \ddots & \vdots \\ \vdots & \ddots & \ddots & 1 \\ 0 & \cdots & 0 & 0 \end{bmatrix}$.

17. Show that there is no invertible matrix C such that

$$C^{-1}\begin{bmatrix} 0 & 1 & 0 & 0 \\ 0 & 0 & 1 & 0 \\ 0 & 0 & 0 & 1 \\ 0 & 0 & 0 & 0 \end{bmatrix} C = \begin{bmatrix} 0 & 1 & 0 & 0 \\ 0 & 0 & 1 & 0 \\ 0 & 0 & 0 & 0 \\ 0 & 0 & 0 & 0 \end{bmatrix}.$$

18. If T is in $M_n(\mathbb{C})$ and $p(x)$ is a polynomial such that $p(T) = 0$, show that $p(C^{-1}TC) = 0$ for all invertible C in $M_n(\mathbb{C})$.

19. If A, B are elements of $M_3(\mathbb{C})$ such that $A(AB - BA) = (AB - BA)A$ and $B(AB - BA) = (AB - BA)B$, show that $(AB - BA)^3 = 0$.

Harder Problems

20. If A, B are elements of $M_3(\mathbb{C})$ such that $A(AB - BA) = (AB - BA)A$, show that $(AB - BA)^3 = 0$.

21. If T is an element of $M_n(\mathbb{R})$ all of whose characteristic roots are real, show that there is an invertible C in $M_n(\mathbb{R})$ such that $C^{-1}TC$ is upper triangular.

22. If A is an element of $M_n(\mathbb{C})$ such that $\text{tr}(A^k) = 0$ for $k = 1, 2, \ldots, n$, prove that A cannot be invertible.

23. If T is an element of $M_3(\mathbb{R})$, prove that for some invertible C in $M_3(\mathbb{R})$,

$$C^{-1}TC = \begin{bmatrix} a & d & e \\ 0 & b & f \\ 0 & 0 & c \end{bmatrix} \text{ or } C^{-1}TC = \begin{bmatrix} a & d & e \\ 0 & 0 & b \\ 0 & 1 & c \end{bmatrix}.$$

Very Hard Problems

24. If A is an element of $M_n(C)$ such that $\text{tr}(A^k) = 0$ for $k = 1, 2, \ldots, n$, prove that $A^n = 0$.

25. If A and B are elements of $M_n(\mathbb{C})$ such that $A(AB - BA) = (AB - BA)A$, prove that $(AB - BA)^n = 0$. (*Hint:* Try to bring the result of Problem 24 into play.)

26. Show that we could have proved a variation of Theorem 4.7.2 in which the matrix of T in the constructed basis is upper triangular with its diagonal entries arranged in an order $a_1, \ldots, a_1, \ldots, a_k, \ldots, a_k$ so that equal entries are grouped together.

4.8. TRIANGULARIZING MATRICES WITH REAL ENTRIES (OPTIONAL)

Theorem 4.7.3 assures us that if T is an $n \times n$ matrix with complex entries, then for some invertible matrix C in $M_n(\mathbb{C})$, the matrix $C^{-1}TC$ is triangular. Moreover, the entries on the main diagonal are the characteristic roots of $C^{-1}TC$, and so, of T.

If T is a matrix with real entries, it may happen that its characteristic roots are not necessarily real—as $T = \begin{bmatrix} 0 & 1 \\ -1 & 0 \end{bmatrix}$ shows us. Of course, T does have its characteristic roots in $\mathbb{C}$. For such a matrix, it is clear that there is no invertible matrix C *with real entries* such that $C^{-1}TC$ is triangular. If there were such, since C^{-1}, T, C all have real entries, $C^{-1}TC$ would have real entries. So $C^{-1}TC$ would not have the characteristic roots of T—which are not real—on its main diagonal. So $C^{-1}TC$ could not possibly be triangular.

Fine, what cannot be cannot be. But is there some substitute result, something akin to triangularization, that does hold when we restrict ourselves to working with matrices with real entries? We shall obtain such a result as one of the theorems in this section. But first we need the real number version of the Fundamental Theorem of Algebra.

Theorem 4.8.1 (Fundamental Theorem of Algebra of R). Let $f(x)$ be a nonconstant polynomial with real coefficients. Then $f(x) = f_1(x) \cdots f_t(x)$, where each factor is a real polynomial of degree 1 or 2.

Proof: Let $z_1, \ldots, z_u$ be the complex roots of $f(x)$. Since $f(z) = 0$ implies that $f(\bar{z}) = 0$, where $\bar{z}$ is the complex conjugate of z (Prove!), *the complex roots of the real polynomial $f(x)$ occur in conjugate pairs.* Thus we can list these roots as $z_1, \ldots, z_s$ (the real roots), $z_{s+1}, \bar{z}_{s+1}, \ldots, z_t, \bar{z}_t$ (the roots that are not real). We can assume that $f(x)$ is monic (i.e., that the coefficient of the highest power of x is 1). Then, by the Fundamental Theorem of Algebra over $\mathbb{C}$,

$$f(x) = (x - z_1) \cdots (x - z_s)(x - z_{s+1})(x - \bar{z}_{s+1}) \cdots (x - z_t)(x - \bar{z}_t).$$

Thus

$$f(x) = f_1(x) f_s(x) f_{s+1}(x) \cdots f_t(x),$$

where the factors $f_r(x)$ are degree 1 polynomials $x - z_r$ or degree 2 polynomials $(x - z_r)(x - \bar{z}_r) = x^2 - (z + \bar{z})x + z\bar{z}$ with real coefficients. ∎

We now go to the real number version of Theorem 4.7.1.

Theorem 4.8.2. Let V be a nonzero subspace of $\mathbb{R}^{(n)}$, $T \in M_n(\mathbb{R})$ such that $T(V) \subseteq V$. Then we can find a basis $v_1, \ldots, v_t$ of V such that $v_s = \sum_{r=1}^{t} a_{rs} v_r$, where the $t \times t$ matrix

$A = (a_{rs})$ is $\begin{bmatrix} A_1 & & * \\ & \ddots & \\ 0 & & A_k \end{bmatrix}$, where A_p is either a 1×1 matrix or a 2×2 matrix of

the form $\begin{bmatrix} 0 & -a_p \\ 1 & -b_p \end{bmatrix}$ for $1 \leq p \leq t$.

Proof: We go by induction on dim V. If dim $V = 1$, then $V = Fv$ and $Tv = av$ for some $a \in F$, and v does the trick. Next, let V be a subspace of $\mathbb{R}^{(n)}$ of dimension greater than 1 such that $T(V) \subseteq V$ and suppose that the theorem is correct for all proper subspaces W of V such that $T(W) \subseteq W$. So for such a subspace W, either W is $\{0\}$ or we can find a basis $w_1, \ldots, w_u$ such that $Tw_s = \sum_{r=1}^{u} a_{rs} w_s$ and the $u \times u$ matrix (a_{rs}) has the desired form.

We can factor the minimum polynomial $q_T(x)$ as $q_T(x) = q_1(x) \cdots q_t(x)$, where the $q_r(x)$ are real polynomials of degrees 1 or 2, by the real number version of the Fundamental Theorem of Algebra. Since $0 = q_T(T) = q_1(T) \cdots q_t(T)$, we cannot have $q_p(T)(V) = V$ for all p, since otherwise $V = q_T(T)(V) = 0$. (Prove!) Choose p such that $q_p(T)(V) \neq V$. Then there are proper subspaces W of V such that $q_p(T)(V) \subseteq W$ and $T(W) \subseteq W$. For example, $q_p(T)(V)$ itself is a proper subspace of V such that $T(q_p(T)(V)) = q_p(T)(V)(T(V)) \subseteq q_p(T)(V)$. Among all proper subspaces W of V such that $q_p(T)(V) \subseteq W$ and $T(W) \subseteq W$, take one whose dimension is as big as possible. By our induction hypothesis, either W is $\{0\}$, and we set $u = 0$, or we can find a basis $w_1, \ldots, w_u$ of W such that $Tw_s = \sum_{r=1}^{u} a_{rs} w_r$ and the $u \times u$ matrix (a_{rs}) has the desired form. Take any element w' that is in V and *not* in W. This is possible, of course, since W is a proper subspace of V. We know that the degree of $q_p(x)$ is 1 or 2 and we consider these cases separately.

Case 1. If $q_p(x)$ has degree 1, then $q_p(x) = x - a$ for some a in $\mathbb{R}$. The condition $q_p(T)(V) \subseteq W$ implies that $(T - aI)w' \in W$ and $Tw' - aw'$ is a linear combination of $w_1, \ldots, w_n$. Letting $w_{u+1} = w'$ and $a_{u+1\,u+1} = a$, we have

$$Tw_{u+1} = a_{u+1\,u+1} w_{u+1} + \sum_{r=1}^{u} a_{r\,u+1} w_r$$

for suitable $a_{r\,u+1} \in F$. Moreover, the subspace $W + Fw_{u+1}$ contains $q_p(T)(V)$, satisfies the condition $T(W + Fw_{u+1}) \subseteq W + Fw_{u+1}$, and has dimension greater than dim W. Thus $W + Fw_{u+1} = V$ has the desired kind of basis $w_1, \ldots, w_{u+1}$.

Case 2. If, on the other hand, $q_p(x)$ has degree 2, then $q_p(x) = x^2 + bx + a$ for suitable a, b in $\mathbb{R}$. The condition $q_p(T)(V) \subseteq W$ implies that $(T^2 + bT + aI)(w') \in W$, so $T^2 w' + bTw' + aw'$ is a linear combination of $w_1, \ldots, w_n$. Thus

$$T^2 w' = -aw' - bTw' + \sum_{r=1}^{u} a_r w_r$$

for suitable scalars a_r in $\mathbb{R}$. Letting $w_{u+1} = w'$, $w_{u+2} = Tw'$, and $a_{r\,u+2} = a_r$, we have

$$Tw_{u+1} = Tw' = 0 w_{u+1} + 1 w_{u+2}$$

$$Tw_{u+2} = T^2 w' = -aw' - bTw' + \sum_{r=1}^{u} a_r w_r = -a w_{u+1} - b w_{u+2} + \sum_{r=1}^{u} a_{r\,u+2} w_r.$$

Moreover, the subspace $W + Fw_{u+1} + Fw_{u+2}$ contains $q_p(T)(V)$, satisfies the condition $T(W + Fw_{u+1} + Fw_{u+2}) \subseteq W + Fw_{u+1} + Fw_{u+2}$, and has dimension greater than that of W. Thus $W + Fw_{u+1} + Fw_{u+2} = V$ and V has the desired kind of basis $w_1, \ldots, w_{u+1}$. ∎

Of course, the following counterparts of Theorems 4.7.2 and 4.7.3 follow from Theorem 4.8.2. We omit the proofs, which are similar to their complex counterparts.

Theorem 4.8.3. Let $T \in M_n(\mathbb{R})$. Then there exists a basis $v_1, \ldots, v_n$ of $\mathbb{R}^{(n)}$ over $\mathbb{R}$ in which the matrix of T is $\begin{bmatrix} A_1 & & * \\ & \ddots & \\ 0 & & A_k \end{bmatrix}$, where A_p is either a 1×1 matrix or a 2×2 matrix of the form $\begin{bmatrix} 0 & -a_p \\ 1 & -b_p \end{bmatrix}$ for $1 \leq p \leq t$.

Theorem 4.8.4. If $T \in M_n(\mathbb{R})$, then there exists an invertible matrix C in $M_n(\mathbb{R})$ such that $C^{-1}TC$ is $\begin{bmatrix} A_1 & & * \\ & \ddots & \\ 0 & & A_k \end{bmatrix}$, where A_p is either a 1×1 matrix or a 2×2 matrix of the form $\begin{bmatrix} 0 & -a_p \\ 1 & -b_p \end{bmatrix}$ for $1 \leq p \leq t$.

Let $T \in M_n(\mathbb{R})$. Then T sits in $M_n(\mathbb{C})$, and so, as such, by Theorem 4.7.4, $p(T) = 0$ for some polynomial $p(x)$ of degree n having *complex* coefficients. It doesn't seem too unreasonable to ask: Is there a polynomial $h(x)$ of degree n having *real* coefficients such that $h(T) = 0$? The answer to this is "yes," as the next theorem shows.

Theorem 4.8.5. If $T \in M_n(\mathbb{R})$, then there is a polynomial $h(x)$ of degree n, with *real* coefficients, such that $h(T) = 0$.

Proof: By Theorem 4.7.4 there is a polynomial $p(x) = x^n + a_1 x^{n-1} + \cdots + a_n$, with $a_1, \ldots, a_n \in \mathbb{C}$, such that $p(T) = 0$. Each a_k is $u_k + v_k i$ where u_k, v_k are real. Thus

$$p(x) = (x^n + u_1 x^{n-1} + \cdots + u_n) + (v_1 x^{n-1} + \cdots + v_n)i,$$

whence

$$0 = p(T) = (T^n + u_1 T^{n-1} + \cdots + u_n I) + (v_1 T^{n-1} + \cdots + v_n I)i.$$

Since T has real entries, all powers of T have real entries, and so, since $u_1, \ldots, u_n$ are real, all the entries of $T^n + u_1 T^{n-1} + \cdots + u_n I$ are real. Similarly, all the entries of $v_1 T^{n-1} + \cdots + v_n I$ are real, hence all the entries of $(v_1 T^{n-1} + \cdots + v_n I)i$ are *pure imaginary*. The equality of

$$T^n + u_1 T^{n-1} + \cdots + u_n I$$

with

$$-(v_1 T^{n-1} + \cdots + v_n I)i$$

then forces each of these to be 0. So $T^n + u_1 T^{n-1} + \cdots + u_n I = 0$. Therefore, if we let $h(x) = x^n + u_1 x^{n-1} + \cdots + u_n$, then $h(x)$ is of degree n, has real coefficients, and $h(T) = 0$. This is exactly what the theorem asserts. ∎

Maybe it would be more transparent if we repeat the argument just given for a specific matrix. Let $T = \begin{bmatrix} 1 & 0 \\ 2 & 3 \end{bmatrix}$. Then $T^2 = \begin{bmatrix} 1 & 0 \\ 8 & 9 \end{bmatrix}$. If $a_1 = b_1 + c_1 i$, $a_2 = b_2 + c_2 i$, are such that b_1, c_1, b_2, c_2 are real and $T^2 + a_1 T + a_2 I = 0$, then

$$0 = \begin{bmatrix} 1 & 0 \\ 8 & 9 \end{bmatrix} + \begin{bmatrix} b_1 + c_1 i & 0 \\ 2(b_1 + c_1 i) & 3(b_1 + c_1 i) \end{bmatrix} + \begin{bmatrix} b_2 + c_2 i & 0 \\ 0 & b_2 + c_2 i \end{bmatrix}.$$

So

$$0 = \begin{bmatrix} 1 & 0 \\ 8 & 9 \end{bmatrix} + \begin{bmatrix} b_1 & 0 \\ 2b_1 & 3b_1 \end{bmatrix} + \begin{bmatrix} b_2 & 0 \\ 0 & b_2 \end{bmatrix} + i\left(\begin{bmatrix} c_1 & 0 \\ 2c_1 & 3c_1 \end{bmatrix} + \begin{bmatrix} c_2 & 0 \\ 0 & c_2 \end{bmatrix}\right).$$

Because the b's and c's are real, looking at each entry, we get

$$\begin{bmatrix} 1 & 0 \\ 8 & 9 \end{bmatrix} + \begin{bmatrix} b_1 & 0 \\ 2b_1 & 3b_1 \end{bmatrix} + \begin{bmatrix} b_2 & 0 \\ 0 & b_2 \end{bmatrix} = 0,$$

that is, $T^2 + b_1 T + b_2 I = 0$.

PROBLEMS

NUMERICAL PROBLEMS

1. Find a basis in which the matrix $\begin{bmatrix} 1 & 0 & 0 \\ 7 & 3 & 0 \\ 8 & 2 & 5 \end{bmatrix}$ is upper triangular.

2. Find a basis in which the matrix $\begin{bmatrix} 0 & 1 & 0 & 0 \\ 0 & 0 & 1 & 0 \\ 0 & 0 & 0 & 1 \\ 1 & 0 & 0 & 0 \end{bmatrix}$ is upper block triangular with real entries of the form described in Theorem 4.8.2.

3. Find a basis in which the matrix $\begin{bmatrix} 0 & 1 & 0 & 0 \\ 1 & 0 & 1 & 0 \\ 0 & 1 & 0 & 1 \\ 0 & 0 & 1 & 0 \end{bmatrix}$ is upper triangular with real entries.

4. Find a basis in which the matrix $\begin{bmatrix} 0 & 1 & 0 & 0 \\ 0 & 0 & 1 & 0 \\ 0 & 0 & 0 & 1 \\ 0 & -1 & 1 & 1 \end{bmatrix}$ is upper triangular with real entries.

MORE THEORETICAL PROBLEMS

Easier Problems

5. Show that an upper triangular matrix A has 0 as one of its diagonal entries if and only if there is a nonzero solution v in $F^{(n)}$ to the equation $Av = 0$.

Middle-Level Problems

6. Let a, b, c be distinct real numbers. Let T be the matrix $\begin{bmatrix} 0 & 0 & -abc \\ 1 & 0 & -bc-ac-ab \\ 0 & 1 & -a-b-c \end{bmatrix}$.

   Show for a that if $v_a = \begin{bmatrix} bc \\ b+c \\ 1 \end{bmatrix}$, then $T(v_a) = -a(v_a)$. Do the same for b and c and find a basis in which the matrix for T is diagonal. Finally, show that

   $(T + aI)(T + bI)(T + cI) = 0.$

CHAPTER 5

Determinants

5.1. INTRODUCTION

In our discussion of 2×2 matrices in Chapter 1, we introduced the notion of the determinant of the matrix $\begin{bmatrix} a & b \\ c & d \end{bmatrix}$ as $\begin{vmatrix} a & b \\ c & d \end{vmatrix} = ad - bc$. This notion turned out to have many uses and consequences, even in so easy a case as the 2×2 matrices. For instance, the determinant cropped up in solving two linear equations in two unknowns, in determining the invertibility of a matrix, in finding the characteristic roots of a matrix, and in other contexts.

We shall soon define the determinant of an $n \times n$ matrix for any $n \geq 2$. The procedure will be a step-by-step one: Knowing what a 2×2 determinant is will allow us to define what a 3×3 determinant is. Then, knowing what a 3×3 determinant is, we will be able to define a 4×4 one, and so on. In other words, we are defining the determinant *inductively*.

Whereas the proofs for the 2×2 case of all the properties of the determinant were very easy, merely involving rather simple computations, the proofs of these properties for the general case of n are much more subtle and difficult. Knowing the results for the 2×2 case will always be the starting point for the argument—an argument by induction—and then the $n \times n$ case will be handled by knowing it for the $(n-1) \times (n-1)$ case.

To try to clarify the arguments used, we shall use these arguments for the special case of 3×3 matrices. Here everything is quite visible and so should give the reader an inkling of what is going on in general.

Definition. If $A = [a_{rs}] = \begin{bmatrix} a_{11} & a_{12} & & a_{1n} \\ a_{21} & a_{22} & \cdots & a_{2n} \\ \vdots & \vdots & & \vdots \\ a_{n1} & a_{n2} & & a_{nn} \end{bmatrix} \in M_n(F)$, then the (r,s) *minor sub-*

Sec. 5.1] Introduction

matrix A_{rs} of A is that matrix which remains when we cross out row r and column s of A.

For example, if $A = \begin{bmatrix} 1 & 2 & 3 & 4 \\ 4 & 3 & 2 & 1 \\ 0 & 1 & 2 & 3 \\ -1 & 4 & 1 & 1 \end{bmatrix}$, then the $(3,2)$ minor submatrix A_{32} is given by $A_{32} = \begin{bmatrix} 1 & 3 & 4 \\ 4 & 2 & 1 \\ -1 & 1 & 1 \end{bmatrix}$, that matrix which remains on eliminating row 3 and column 2 of A.

The matrices A_{rs} are, of course, $(n-1) \times (n-1)$ matrices, *so we proceed by assuming that what is meant by their determinant is known* (as we outlined a few paragraphs above).

Definition. The (r,s) *minor* of A is given by the determinant M_{rs} of the minor submatrix A_{rs}.

Since this notion is so important, to try to understand it completely, let's look at some minors for 3×3 matrices.

If $A = \begin{bmatrix} 1 & 2 & 3 \\ 4 & 5 & 6 \\ 7 & 8 & 9 \end{bmatrix}$, then

$$A_{11} = \begin{bmatrix} 5 & 6 \\ 8 & 9 \end{bmatrix}$$

and

$$M_{11} = \begin{vmatrix} 5 & 6 \\ 8 & 9 \end{vmatrix} = 5 \cdot 9 - 8 \cdot 6 = 45 - 48 = -3,$$

while

$$A_{32} = \begin{bmatrix} 1 & 3 \\ 4 & 6 \end{bmatrix}$$

and

$$M_{23} = \begin{vmatrix} 1 & 3 \\ 4 & 6 \end{vmatrix} = 1 \cdot 6 - 4 \cdot 3 = 6 - 12 = -6.$$

We are now ready to define the determinant of any $n \times n$ matrix.

Definition. If $A = [a_{rs}] \in M_n(F)$, then the *determinant* of A, written as $\det(A)$, or

$\det(a_{rs})$, or $\begin{vmatrix} a_{11} & a_{12} & & a_{1n} \\ a_{21} & a_{22} & \cdots & a_{2n} \\ \vdots & \vdots & & \vdots \\ a_{n1} & a_{n2} & & a_{nn} \end{vmatrix}$, is defined as

$$a_{11}M_{11} - a_{21}M_{21} + a_{31}M_{31} - \cdots + (-1)^{r+1}a_{r1}M_{r1} + \cdots + (-1)^{n+1}a_{n1}M_{n1}.$$

So in summation notation,

$$\det(A) = \sum_{r=1}^{n} (-1)^{r+1} a_{r1} M_{r1}.$$

Before seeing what this definition says for a 3 × 3 matrix, let's note some aspects built into the definition.

1. The emphasis is given to the *first column* of A—an unsymmetrical fact that will cause us some annoyance later—since we are using the minors of the first column of A. For this reason we call this expansion the *expansion by the minors of the first column of A*.
2. The signs attached to the minors *alternate*, going as +, −, +, −, and so on.
3. We are not restricting the rows of A in the definition. This will be utterly essential for us in the next section.

EXAMPLES

1. $\begin{vmatrix} 1 & 2 & 3 \\ 3 & -1 & 4 \\ 0 & 1 & 1 \end{vmatrix} = 1 \begin{vmatrix} -1 & 4 \\ 1 & 1 \end{vmatrix} - 3 \begin{vmatrix} 2 & 3 \\ 1 & 1 \end{vmatrix} + 0 \begin{vmatrix} 2 & 3 \\ -1 & 4 \end{vmatrix} = -5 - 3(-1) + 0 = -2.$

2. $\begin{vmatrix} 1 & 2 & 3 & 4 \\ 0 & 1 & 2 & 3 \\ 0 & 0 & 1 & 2 \\ 0 & 0 & 0 & 1 \end{vmatrix} = 1 \begin{vmatrix} 1 & 2 & 3 \\ 0 & 1 & 2 \\ 0 & 0 & 1 \end{vmatrix} - 0 \begin{vmatrix} 2 & 3 & 4 \\ 0 & 1 & 2 \\ 0 & 0 & 1 \end{vmatrix} + 0 \begin{vmatrix} 2 & 3 & 4 \\ 1 & 2 & 3 \\ 0 & 0 & 1 \end{vmatrix} - 0 \begin{vmatrix} 2 & 3 & 4 \\ 1 & 2 & 3 \\ 0 & 1 & 2 \end{vmatrix} = 1.$

3. $\begin{vmatrix} 1 & 0 & 0 & 0 \\ 2 & 1 & 0 & 0 \\ 3 & 2 & 1 & 0 \\ 4 & 3 & 2 & 1 \end{vmatrix} = 1 \begin{vmatrix} 1 & 0 & 0 \\ 2 & 1 & 0 \\ 3 & 2 & 1 \end{vmatrix} - 2 \begin{vmatrix} 0 & 0 & 0 \\ 2 & 1 & 0 \\ 3 & 2 & 1 \end{vmatrix} + 3 \begin{vmatrix} 0 & 0 & 0 \\ 1 & 0 & 0 \\ 3 & 2 & 1 \end{vmatrix} - 4 \begin{vmatrix} 0 & 0 & 0 \\ 1 & 0 & 0 \\ 2 & 1 & 0 \end{vmatrix} = ?$

(Fill in the "?".)

Notice in Examples 2 and 3, which involve 3 × 3 determinants, that the calculations can be made because we know (in terms of 2 × 2 determinants) what a 3 × 3 determinant is. We leave it to the reader to show that both determinants in Examples 2 and 3 are equal to 1.

Sec. 5.1] Introduction

Examples 2 and 3 are examples of what we called triangular matrices. Recall that an *upper triangular matrix* is one all of whose entries *below* the main diagonal are 0, and that a *lower triangular matrix* is one all of whose entries *above* the main diagonal are 0.

For triangular matrices (upper or lower) the determinant is extremely easy to evaluate. Notice that in Examples 2 and 3 the determinant is just the product of the diagonal entries. This is the story in general, as our next theorem shows. The proof will show how we exploit knowing the result for the $(n-1) \times (n-1)$ case.

Theorem 5.1.1. The determinant of a triangular (upper or lower) matrix A is the product of its diagonal entries.

Proof: We first do the easier case of the upper triangular matrix. Let

$$A = \begin{bmatrix} a_{11} & a_{12} & & a_{1n} \\ 0 & a_{22} & \cdots & a_{2n} \\ \vdots & \vdots & & \vdots \\ 0 & 0 & & a_{nn} \end{bmatrix}.$$

Since the entries $a_{21}, a_{31}, \ldots, a_{n1}$ of the first column of A are all 0, our definition of the determinant of A gives us

$$\begin{vmatrix} a_{11} & a_{12} & & a_{1n} \\ 0 & a_{22} & \cdots & a_{2n} \\ \vdots & \vdots & & \vdots \\ 0 & 0 & & a_{nn} \end{vmatrix} = a_{11} M_{11} = a_{11} \begin{vmatrix} a_{22} & a_{23} & & a_{2n} \\ 0 & a_{33} & \cdots & a_{3n} \\ \vdots & \vdots & & \vdots \\ 0 & 0 & & a_{nn} \end{vmatrix}.$$

The $(1, 1)$ minor submatrix of A is itself an upper triangular matrix, but an $(n-1) \times (n-1)$ one. Hence by our induction procedure,

$$\begin{vmatrix} a_{22} & a_{23} & & a_{2n} \\ 0 & a_{33} & \cdots & a_{3n} \\ \vdots & \vdots & & \vdots \\ 0 & 0 & & a_{nn} \end{vmatrix} = a_{22} a_{33} \cdots a_{nn},$$

the product of its diagonal entries. Therefore,

$$\det(A) = a_{11} M_{11} = a_{11}(a_{22} a_{33} \cdots a_{nn}) = a_{11} a_{22} a_{33} \cdots a_{nn},$$

the product of the diagonal entries of A.

Now to the case of a lower triangular matrix. Suppose that

$$A = \begin{bmatrix} a_{11} & 0 & & 0 \\ a_{21} & a_{22} & \cdots & 0 \\ \vdots & \vdots & & \vdots \\ a_{n1} & a_{n2} & & a_{nn} \end{bmatrix}.$$

Then

$$\begin{vmatrix} a_{11} & 0 & & 0 \\ a_{21} & a_{22} & \cdots & 0 \\ \vdots & \vdots & & \vdots \\ a_{n1} & a_{n2} & & a_{nn} \end{vmatrix} = a_{11}M_{11} - a_{21}M_{21} + a_{31}M_{31} - \cdots + (-1)^{n+1}M_{n1}.$$

Notice that in computing M_{r1} for $r > 1$ the $0 \ 0 \ \cdots \ 0$ part of the first row stays in M_{r1}. So M_{r1} is the determinant of a lower triangular matrix with a 0 as its first diagonal entry. Since it is an $(n - 1) \times (n - 1)$ determinant, by induction it is the product of the diagonal entries. Since the first diagonal entry is 0, this product is 0. So $M_{r1} = 0$ for $r > 1$. So all that is left to compute in $\det(A)$ is

$$a_{11}M_{11} = a_{11}\begin{vmatrix} a_{22} & 0 & & 0 \\ a_{32} & a_{33} & \cdots & 0 \\ \vdots & \vdots & & \vdots \\ a_{n2} & a_{n2} & & a_{nn} \end{vmatrix}.$$

By induction, the determinant M_{11} of this $(n - 1) \times (n - 1)$ matrix is $a_{22}a_{33} \cdots a_{nn}$, the product of its diagonal entries. So

$$\det(A) = a_{11}M_{11} = a_{11}a_{22} \cdots a_{nn},$$

the product of the diagonal entries of A. This completes the proof of Theorem 5.1.1.

PROBLEMS

NUMERICAL PROBLEMS

1. Compute the following determinants.

(a) $\begin{vmatrix} 1 & 3 & 4 \\ 2 & 0 & 2 \\ 3 & 0 & 3 \end{vmatrix}.$

(b) $\begin{vmatrix} 1 & 0 & 5 & 6 \\ 2 & 0 & 6 & 5 \\ 3 & 0 & 5 & 6 \\ 4 & 0 & 6 & 5 \end{vmatrix}.$

(c) $\begin{vmatrix} 9 & 0 & 0 \\ 2 & 5 & 6 \\ -1 & -2 & 0 \end{vmatrix}.$

(d) $\begin{vmatrix} 9 & 2 & -1 \\ 0 & 5 & -2 \\ 0 & 6 & 0 \end{vmatrix}.$

(e) $\begin{vmatrix} -1 & 2 & -1 & 3 \\ 2 & 3 & 2 & 1 \\ 3 & 4 & 3 & 0 \\ 0 & 1 & 0 & 1 \end{vmatrix}.$

(f) $\begin{vmatrix} 3 & 0 & 0 & 0 \\ 4 & 2 & 1 & 0 \\ 0 & 1 & 2 & 3 \\ -1 & -1 & 1 & 1 \end{vmatrix}.$

2. Compute the determinants of the transposes of the matrices in Problem 1.

3. Show that $\begin{vmatrix} 1 & 3 & 4 \\ 8 & x & 2 \\ 3 & r & 3 \end{vmatrix}$ and $\begin{vmatrix} 1+3 & 3 & 4 \\ 8+x & x & 2 \\ 3+r & r & 3 \end{vmatrix}$ are equal.

4. Evaluate $\begin{vmatrix} 1 & 4 & 6 & 8 \\ 0 & 1 & 2 & 3 \\ 0 & 2 & 3 & 1 \\ 0 & 5 & 6 & -1 \end{vmatrix}$ and compare it with $\begin{vmatrix} 1 & 2 & 3 & 4 \\ 0 & 1 & 2 & 0 \\ 0 & 2 & 3 & 1 \\ 0 & 5 & 6 & -1 \end{vmatrix}$.

5. Show that $\begin{vmatrix} 4 & 5 & 1 \\ 2 & f & s \\ 3 & d & 6 \end{vmatrix} + \begin{vmatrix} 1 & 5 & 4 \\ s & f & 2 \\ 6 & d & 3 \end{vmatrix} = 0$.

6. Compute $\begin{vmatrix} 2 & 3 & 4 \\ r & r & 2+r \\ t & t & 3+t \end{vmatrix}$.

7. Compute $\begin{vmatrix} 1 & 1 & 1 \\ a & b & c \\ a^2 & b^2 & c^2 \end{vmatrix}$.

8. Compute the determinant $\begin{vmatrix} e & e & g \\ u & u & w+r \\ t & t & b+t \end{vmatrix}$ and simplify.

9. Compute $\begin{vmatrix} 1 & 1 & 1 & 1 \\ 3 & 0 & 1 & 2 \\ 0 & 1 & 2 & 3 \\ 1 & 5 & 6 & 7 \end{vmatrix}$ and $\begin{vmatrix} 0 & 1 & 2 & 3 \\ 3 & 0 & 1 & 2 \\ 1 & 1 & 1 & 1 \\ 1 & 5 & 6 & 7 \end{vmatrix}$ and compare the answers.

MORE THEORETICAL PROBLEMS

Easier Problems

10. Show that $\begin{vmatrix} 1 & 1 & 1 & 1 \\ a & b & c & d \\ a^2 & b^2 & c^2 & d^2 \\ a^3 & b^3 & c^3 & d^3 \end{vmatrix}$ is 0 if and only if some two of a, b, c, d are equal.

11. Show that $\begin{vmatrix} a+r & b+s & c+t \\ d & e & f \\ g & h & k \end{vmatrix} = \begin{vmatrix} a & b & c \\ d & e & f \\ g & h & k \end{vmatrix} + \begin{vmatrix} r & s & t \\ d & e & f \\ g & h & k \end{vmatrix}$.

12. Prove that $\begin{vmatrix} qa_{11} & a_{12} & & a_{1n} \\ qa_{21} & a_{22} & \cdots & a_{2n} \\ \vdots & \vdots & & \vdots \\ qa_{n1} & a_{n2} & & a_{nn} \end{vmatrix} = q \begin{vmatrix} a_{11} & a_{12} & & a_{1n} \\ a_{21} & a_{22} & \cdots & a_{2n} \\ \vdots & \vdots & & \vdots \\ a_{n1} & a_{n2} & & a_{nn} \end{vmatrix}$.

13. Prove that
$$\begin{vmatrix} qa_{11} & qa_{12} & \cdots & qa_{1n} \\ a_{21} & a_{22} & & a_{2n} \\ \vdots & \vdots & & \vdots \\ a_{n1} & a_{n2} & & a_{nn} \end{vmatrix} = q \begin{vmatrix} a_{11} & a_{12} & \cdots & a_{1n} \\ a_{21} & a_{22} & & a_{2n} \\ \vdots & \vdots & & \vdots \\ a_{n1} & a_{n2} & & a_{nn} \end{vmatrix}.$$

Middle-Level Problems

14. If A and B are upper triangular $n \times n$ matrices are such that
$$\det(aA + bB) = a\det(A) + b\det(B)$$
for all $a, b \in F$, show that $\det(A) = \det(B) = 0$.

15. If A and B are lower triangular matrices $n \times n$ such that
$$\det(aA + bB) = a\det(A) + b\det(B)$$
for all $a, b \in F$, show that $\det(A) = \det(B) = 0$.

16. If A and B are upper triangular $n \times n$ matrices, show that
$$\det(A + xB) = \det(A) + b_1 x + b_2 x^2 + \cdots + b_{n-1} x^{n-1} + \det(B) x^n$$
for suitable b_r.

Harder Problems

17. Evaluate
$$\begin{vmatrix} 1 & 1 & & 1 \\ a_1 & a_1^2 & & a_1^n \\ a_2 & a_2^2 & \cdots & a_2^n \\ \vdots & \vdots & & \vdots \\ a_{n-1} & a_{n-1}^2 & & a_{n-1}^n \end{vmatrix} \text{ and show } \begin{vmatrix} 1 & 1 & & 1 \\ a_1 & a_1^2 & & a_1^n \\ a_2 & a_2^2 & \cdots & a_2^n \\ \vdots & \vdots & & \vdots \\ a_{n-1} & a_{n-1}^2 & & a_{n-1}^n \end{vmatrix} = 0$$
if some two of $a_1, \ldots, a_{n-1}$ are equal.

18. If A, B are 3×3 matrices, show that $\det(AB) = \det(A)\det(B)$.

19. If A and B are 3×3 matrices and B is invertible, show that $\det(A + xB) = 0$ for some x in F.

20. Is the result of Problem 19 correct if $B \neq 0$ is not invertible? Either prove or give a counterexample.

21. If A is an invertible 3×3 matrix, show that $\det(A) \neq 0$ and $\det(A^{-1}) = \dfrac{1}{\det(A)}$.

(*Hint:* Use Problem 18.)

22. For a 3×3 matrix A, show that $\det(A) = \det(A')$.

23. If A and C are 3×3 matrices and C is invertible, show that $\det(C^{-1}AC) = \det(A)$.

24. If A is an upper triangular 3×3 matrix with real entries *whose diagonal entries are all equal*, then $A^m = I$ for some m if and only if A is I or $-I$.

5.2. PROPERTIES OF DETERMINANTS: ROW OPERATIONS

In defining the determinant of a matrix we did it by the expansion by the minors of the first *column*. In this way we did not interface very much with the *rows*. Fortunately, this will allow us to show some things that can be done by playing around with the *rows* of a determinant. We begin with

Theorem 5.2.1. If a row of an $n \times n$ matrix A consists only of zeros, then $\det(A) = 0$.

Proof: We go by induction on n. The result is trivially true for $n = 2$. Next let $n > 2$ and suppose that row r of A consists of zeros. Then every minor M_{k1}, for $k \neq r$, retains a row of zeros. So by induction, each $M_{k1} = 0$ if $k \neq r$. On the other hand, $a_{r1} = 0$, whence $a_{r1} M_{r1} = 0$. Thus

$$\det(A) = \sum_{k=1}^{n} (-1)^{k+1} a_{k1} M_{k1} = 0. \quad \blacksquare$$

Another property that is fairly easy to establish, using our induction type of argument, is

Theorem 5.2.2. Suppose that a matrix B is obtained from A by multiplying each entry of some row of A by a constant u. Then $\det(B) = u \det(A)$.

Proof: Before doing the general case, let us consider the story for the general 3×3 matrix

$$A = \begin{bmatrix} a_{11} & a_{12} & a_{13} \\ a_{21} & a_{22} & a_{23} \\ a_{31} & a_{32} & a_{33} \end{bmatrix}$$

and

$$B = \begin{bmatrix} a_{11} & a_{12} & a_{13} \\ ua_{21} & ua_{22} & ua_{23} \\ a_{31} & a_{32} & a_{33} \end{bmatrix}.$$

Let M_{r1} be the $(r, 1)$ minor of A and N_{r1} that of B. So

$$\det(B) = a_{11} N_{r1} - u a_{21} N_{21} + a_{31} N_{31}.$$

Since the minor N_{21} does not involve the second row, we see that $N_{21} = M_{21}$. The other two minors, N_{11} and N_{31}, which are 2×2, have a row multiplied by u. So we have $N_{11} = u M_{11}$, $N_{31} = u M_{31}$. Thus

$$\det(B) = a_{11}(u M_{11}) - u a_{21} M_{21} + a_{31}(u M_{31})$$
$$= u(a_{11} M_{11} - a_{21} M_{21} + a_{31} M_{31})$$
$$= u \det(A).$$

The general case goes in a similar fashion. Let B be obtained from A by multiplying every entry in row r of A by u. So if $B = (b_{st})$, then $b_{st} = a_{st}$ if $s \neq r$ and $b_{rt} = ua_{rt}$. If N_{k1} is the $(k, 1)$ minor of B, then

$$\det(B) = \sum_{k=1}^{n} (-1)^{k+1} b_{k1} N_{k1}.$$

But N_{k1}, for $k \neq r$, is the same as M_{k1}, the $(k, 1)$ minor of A, except that one of its rows is that of M_{k1} multiplied by u. Since N_{k1} is $(n-1) \times (n-1)$, by induction $N_{k1} = uM_{k1}$ for $k \neq r$. On the other hand, since in forming N_{r1} we strike out row r of A, we see that $N_{r1} = M_{r1}$. Thus for $k \neq r$, $b_{k1}N_{k1} = a_{k1}N_{k1}$ while $b_{r1}N_{r1} = (ua_{r1})M_{r1}$, hence

$$\det(B) = \sum_{k=1}^{n} (-1)^{k+1} b_{k1} N_{k1} = \sum_{k=1}^{n} (-1)^{k+1} ua_{k1} M_{k1}$$

$$= u \sum_{k=1}^{n} (-1)^{k+1} a_{k1} M_{k1} = u \det(A). \quad \blacksquare$$

We come to some theorems of key importance in the theory of determinants.

Theorem 5.2.3. If the matrix B is obtained from A by interchanging two *adjacent* rows of A, then $\det(B) = -\det(A)$.

Proof: Again we illustrate the proof for 3×3 matrices. Suppose that in

$$A = \begin{bmatrix} a_{11} & a_{12} & a_{13} \\ a_{21} & a_{22} & a_{23} \\ a_{31} & a_{32} & a_{33} \end{bmatrix}$$

we interchange rows 2 and 3 to obtain the matrix

$$B = \begin{bmatrix} a_{11} & a_{12} & a_{13} \\ a_{31} & a_{32} & a_{33} \\ a_{21} & a_{22} & a_{23} \end{bmatrix} = \begin{bmatrix} b_{11} & b_{12} & b_{13} \\ b_{21} & b_{22} & b_{23} \\ b_{31} & b_{32} & b_{33} \end{bmatrix}.$$

Then

$$\det(B) = b_{11}\begin{vmatrix} b_{22} & b_{23} \\ b_{32} & b_{33} \end{vmatrix} - b_{21}\begin{vmatrix} b_{12} & b_{13} \\ b_{32} & b_{33} \end{vmatrix} + b_{31}\begin{vmatrix} b_{12} & b_{13} \\ b_{22} & b_{23} \end{vmatrix}$$

$$= a_{11}\begin{vmatrix} a_{32} & a_{33} \\ a_{22} & a_{23} \end{vmatrix} - a_{31}\begin{vmatrix} a_{12} & a_{13} \\ a_{22} & a_{23} \end{vmatrix} + a_{21}\begin{vmatrix} a_{12} & a_{13} \\ a_{32} & a_{33} \end{vmatrix}.$$

Notice in the first minor $N_{11} = \begin{vmatrix} a_{32} & a_{33} \\ a_{22} & a_{23} \end{vmatrix}$ of B, the rows of the minor $M_{11} = \begin{vmatrix} a_{22} & a_{23} \\ a_{32} & a_{33} \end{vmatrix}$ corresponding to the second and third rows of A are interchanged.

Sec. 5.2] Properties of Determinants: Row Operations 223

So by the 2×2 case, $N_{11} = -M_{11}$. Notice also that $N_{21} = \begin{vmatrix} a_{12} & a_{13} \\ a_{22} & a_{23} \end{vmatrix} = M_{31}$

and $N_{31} = \begin{vmatrix} a_{12} & a_{13} \\ a_{32} & a_{33} \end{vmatrix} = M_{21}$. So

$$\begin{aligned} \det(B) &= b_{11}N_{11} - b_{21}N_{21} + b_{31}N_{31} \\ &= -a_{11}M_{11} - a_{31}M_{31} + a_{21}M_{21} \\ &= -(a_{11}M_{11} - a_{21}M_{21} + a_{31}M_{31}) \\ &= -\det(A), \end{aligned}$$

since $b_{31} = a_{21}$ and $b_{21} = a_{31}$.

Now to the general case. Suppose that we interchange rows r and $r+1$ of A to get B. Thus if $B = (b_{uv})$, then for $u \neq r, r+1$ we have $b_{uv} = a_{uv}$, while $b_{rv} = a_{r+1,v}$ and $b_{r+1,v} = a_{rv}$. So if N_{u1} is the $(u, 1)$ minor of B, then

$$\det(B) = \sum_{u=1}^{n} (-1)^{u+1} b_{u1} N_{u1}.$$

If $u \neq r$ or $r+1$, then in obtaining the minor N_{u1} from M_{u1}, the $(u, 1)$ minor of A, we have interchanged two adjacent rows of M_{u1}. So, for these, by the $(n-1) \times (n-1)$ case, $N_{u1} = -M_{u1}$. Also, $N_{r1} = M_{r+1,1}$ and $N_{r+1,1} = M_{r1}$ (Why? Look at the 3×3 case!). So $b_{r1}N_{r1} = a_{r+1,1}M_{r+1,1}$ and $b_{r+1,1}N_{r+1,1} = a_{r1}M_{r1}$. Hence

$$\begin{aligned} \det(B) &= b_{11}N_{11} - b_{21}N_{21} + \cdots + (-1)^{r+1}b_{r1}N_{r1} \\ &\quad + (-1)^{r+2}b_{r+1,1}N_{r+1,1} + \cdots + (-1)^{n+1}b_{n1}N_{n1} \\ &= -a_{11}M_{11} + a_{21}M_{21} - \cdots + (-1)^{r+1}a_{r+1,1}M_{r+1,1} \\ &\quad + (-1)^{r+2}a_{r1}M_{r1} - \cdots - (-1)^{n+1}a_{n1}M_{n1} \\ &= -(a_{11}M_{11} - a_{21}M_{21} + \cdots + (-1)^{r+1}a_{r1}M_{r1} \\ &\quad + (-1)^{r+2}a_{r+1,1}M_{r+1,1} + \cdots + (-1)^{n+1}a_{n1}M_{n1}) \\ &= -\det(A). \end{aligned}$$

So $\det(B) = -\det(A)$, which proves the theorem. ∎

We realize that the proof just given may be somewhat intricate. We advise the reader to follow its steps for the 4×4 matrices. This should help clarify the argument.

The next result has nothing to do with matrices or determinants, but we shall make use of it in those contexts.

Suppose that n balls, numbered from the left as $1, 2, \ldots, n$, are lined up in a row. If $s > r$, we claim that we can interchange balls r and s by a succession of an *odd* number of interchanges of adjacent ones. How? First interchange ball s with ball $s - 1$, then with ball $s - 2$, and so on. To bring ball s to position r thus requires $s - r$ interchanges of adjacent balls. Also, ball r is now in position $r + 1$, so to move it to position s requires $s - (r + 1)$ interchanges of adjacent ones. So, in all, to interchange balls r and s, we have made $(s - r) + (s - (r + 1)) = 2(s - r) - 1$, an *odd* number, of interchanges of adjacent ones.

Let's see this in the case of four balls **(1) (2) (3) (4)**, where we want to interchange the first and fourth balls. The sequence pictorially is

(1)	(2)	(3)	(4)	(The initial configuration)
(1)	(2)	(4)	(3)	(After the first interchange)
(1)	(4)	(2)	(3)	(After the second interchange)
(4)	(1)	(2)	(3)	(After the third interchange)
(4)	(2)	(1)	(3)	(After the fourth interchange)
(4)	(2)	(3)	(1)	(After the fifth interchange)

So altogether it took $5 = 2(4 - 1) - 1$ adjacent interchanges to effect the interchange of balls 1 and 4.

We note the result proved above, not for balls, but for rows of a matrix, as

Lemma 5.2.4. We can interchange any two rows of a matrix by an *odd* number of interchanges of adjacent rows.

This lemma, together with Theorem 5.2.3, has the very important consequence,

Theorem 5.2.5. If we interchange two rows of a determinant, then the sign changes.

Proof: According to Lemma 5.2.4, we can effect the interchange of these two rows by an odd number of interchanges of adjacent rows. By Theorem 5.2.3, each such change of adjacent rows changes the sign of the determinant. Since we do this an odd number of times, the sign of the determinant is changed in the interchange of two of its rows. ■

Suppose that two rows of the matrix A are the same. So if we interchange these two rows, we do not change it, hence we don't change $\det(A)$. But by Theorem 5.2.5, on interchanging these two rows of A, the determinant changes sign. This can happen only if $\det(A) = 0$. Hence

Theorem 5.2.6. If two rows of A are equal, then $\det(A) = 0$.

Before going formally to the next result, we should like to motivate the theorem we shall prove with an example.

Consider the 3×3 matrix

$$C = \begin{bmatrix} a_{11} & a_{12} & a_{13} \\ a_{21} + b_{21} & a_{22} + b_{22} & a_{23} + b_{23} \\ a_{31} & a_{32} & a_{33} \end{bmatrix}.$$

We should like to interrelate $\det(C)$ with two other determinants, $\det(A)$ and $\det(B)$, where

$$A = \begin{bmatrix} a_{11} & a_{12} & a_{13} \\ a_{21} & a_{22} & a_{23} \\ a_{31} & a_{32} & a_{33} \end{bmatrix}$$

Sec. 5.2] Properties of Determinants: Row Operations

and

$$B = \begin{bmatrix} a_{11} & a_{12} & a_{13} \\ b_{21} & b_{22} & b_{23} \\ a_{31} & a_{32} & a_{33} \end{bmatrix}.$$

Of course, since 3 is not a very large integer, we could expand $\det(A)$, $\det(B)$, $\det(C)$ by brute force and compare the answers. But this would not be very instructive. Accordingly, we'll carry out the discussion along the lines that we shall follow in the general case.

What do we hope to show? Our objective is to prove that $\det(C) = \det(A) + \det(B)$. Now

$$\det(C) = a_{11} \begin{vmatrix} a_{22}+b_{22} & a_{23}+b_{23} \\ a_{32} & a_{33} \end{vmatrix}$$

$$- (a_{21}+b_{21}) \begin{vmatrix} a_{12} & a_{13} \\ a_{32} & a_{33} \end{vmatrix}$$

$$+ a_{31} \begin{vmatrix} a_{12} & a_{13} \\ a_{22}+b_{22} & a_{23}+b_{23} \end{vmatrix}.$$

For the 2×2 case we readily have that

$$\begin{vmatrix} a_{22}+b_{22} & a_{23}+b_{23} \\ a_{32} & a_{33} \end{vmatrix} = \begin{vmatrix} a_{22} & a_{23} \\ a_{32} & a_{33} \end{vmatrix} + \begin{vmatrix} b_{22} & b_{23} \\ a_{32} & a_{33} \end{vmatrix}$$

and

$$\begin{vmatrix} a_{12} & a_{13} \\ a_{22}+b_{22} & a_{23}+b_{23} \end{vmatrix} = \begin{vmatrix} a_{12} & a_{13} \\ a_{22} & a_{23} \end{vmatrix} + \begin{vmatrix} a_{12} & a_{13} \\ b_{22} & b_{23} \end{vmatrix}.$$

Substituting these values in the expression for $\det(C)$, we obtain

$$\det(C) = a_{11} \left(\begin{vmatrix} a_{22} & a_{23} \\ a_{32} & a_{33} \end{vmatrix} + \begin{vmatrix} b_{22} & b_{23} \\ a_{32} & a_{33} \end{vmatrix} \right)$$

$$- (a_{21}+b_{21}) \begin{vmatrix} a_{12} & a_{13} \\ a_{32} & a_{33} \end{vmatrix}$$

$$+ a_{31} \left(\begin{vmatrix} a_{12} & a_{13} \\ a_{22} & a_{23} \end{vmatrix} + \begin{vmatrix} a_{12} & a_{13} \\ b_{22} & b_{23} \end{vmatrix} \right).$$

So regrouping terms, we get

$$\det(C) = \left(a_{11} \begin{vmatrix} a_{22} & a_{23} \\ a_{32} & a_{33} \end{vmatrix} - a_{21} \begin{vmatrix} a_{12} & a_{13} \\ a_{32} & a_{33} \end{vmatrix} + a_{31} \begin{vmatrix} a_{12} & a_{13} \\ a_{22} & a_{23} \end{vmatrix} \right)$$

$$+ \left(a_{11} \begin{vmatrix} b_{22} & b_{23} \\ a_{32} & a_{33} \end{vmatrix} - b_{21} \begin{vmatrix} a_{12} & a_{13} \\ a_{32} & a_{33} \end{vmatrix} + a_{31} \begin{vmatrix} a_{12} & a_{13} \\ b_{22} & b_{23} \end{vmatrix} \right).$$

We recognize the first expression

$$a_{11}\begin{vmatrix} a_{22} & a_{23} \\ a_{32} & a_{33} \end{vmatrix} - a_{21}\begin{vmatrix} a_{12} & a_{13} \\ a_{32} & a_{33} \end{vmatrix} + a_{31}\begin{vmatrix} a_{12} & a_{13} \\ a_{22} & a_{23} \end{vmatrix}$$

as $\det(A)$ and the second expression

$$a_{11}\begin{vmatrix} b_{22} & b_{23} \\ a_{32} & a_{33} \end{vmatrix} - b_{21}\begin{vmatrix} a_{12} & a_{13} \\ a_{32} & a_{33} \end{vmatrix} + a_{31}\begin{vmatrix} a_{12} & a_{13} \\ b_{22} & b_{23} \end{vmatrix}$$

as $\det(B)$. Consequently, $\det(C) = \det(A) + \det(B)$.

The pattern just exhibited will be that used in the proof of

Theorem 5.2.7. If the matrix $C = (c_{uv})$ is such that for $u \neq r$, $c_{uv} = a_{uv}$ but for $u = r$, $c_{rv} = a_{rv} + b_{rv}$, then $\det(C) = \det(A) + \det(B)$, where

$$A = \begin{bmatrix} a_{11} & a_{12} & & a_{1n} \\ \vdots & \vdots & & \vdots \\ a_{r-1,1} & a_{r-1,2} & & a_{r-1,n} \\ a_{r1} & a_{r2} & \cdots & a_{rn} \\ a_{r+1,1} & a_{r+1,2} & & a_{r+1,n} \\ \vdots & \vdots & & \vdots \\ a_{n1} & a_{n2} & & a_{nn} \end{bmatrix} \text{ and } B = \begin{bmatrix} a_{11} & a_{12} & & a_{1n} \\ \vdots & \vdots & & \vdots \\ a_{r-1,1} & a_{r-1,2} & & a_{r-1,n} \\ b_{r1} & b_{r2} & \cdots & b_{rn} \\ a_{r+1,1} & a_{r+1,2} & & a_{r+1,n} \\ \vdots & \vdots & & \vdots \\ a_{n1} & a_{n2} & & a_{nn} \end{bmatrix}.$$

Proof: The matrix C is such that in all the rows, except for row r, the entries are a_{uv} but in row r the entries are $a_{rv} + b_{rv}$. Let M_{k1}, N_{k1}, Q_{k1} denote the $(k, 1)$ minors of A, B, and C, respectively. Thus

$$\det(C) = \sum_{u=1}^{n} (-1)^{k+1} c_{k1} Q_{k1}.$$

Note that the minor Q_{r1} is exactly the same as M_{r1} and N_{r1}, since in finding these minors row r has been crossed out, so the a_{rs} and b_{rs} do not enter the picture. So $Q_{r1} = M_{r1} = N_{r1}$. What about the other minors Q_{k1}, where $k \neq r$? Since row r is retained in this minor, one of its rows has the entries $a_{rv} + b_{rv}$. Because Q_{k1} is $(n - 1) \times (n - 1)$, by induction we have that $Q_{k1} = M_{k1} + N_{k1}$. So the formula for $\det(C)$ becomes

$$a_{11}(M_{11} + N_{11}) - a_{21}(M_{21} + N_{21}) + \cdots + (-1)^r a_{r-1,1}(M_{r-1,1} + N_{r-1,1})$$
$$+ (-1)^{r+1}(a_{r1} + b_{r1})Q_{r1}$$
$$+ (-1)^{r+2} a_{r+1,1}(M_{r+1,1} + N_{r+1,1}) + \cdots + (-1)^{n+1} a_{n1}(M_{n1} + N_{n1}).$$

Using that $Q_{r1} = M_{r1} = N_{r1}$, we can replace the term $(-1)^{r+1}(a_{r1} + b_{r1})Q_{r1}$ by $(-1)^{r+1} a_{r1} M_{r1} + (-1)^{r+1} b_{r1} N_{r1}$. Regrouping then gives us

$$\det(C) = (a_{11}M_{11} - a_{21}M_{21} + \cdots + (-1)^{r+1} a_{r1} M_{r1} + \cdots + (-1)^{n+1} a_{n1} M_{n1})$$
$$+ (a_{11}N_{11} - a_{21}N_{21} + \cdots + (-1)^{r+1} b_{r1} N_{r1} + \cdots + (-1)^{n+1} a_{n1} N_{n1}).$$

Sec. 5.2] Properties of Determinants: Row Operations

We recognize the first expression

$$(a_{11}M_{11} - a_{21}M_{21} + \cdots + (-1)^{r+1}a_{r1}M_{r1} + \cdots + (-1)^{n+1}a_{n1}M_{n1})$$

as $\det(A)$ and the second expression

$$(a_{11}M_{11} - a_{21}M_{21} + \cdots + (-1)^{r+1}a_{r1}M_{r1} + \cdots + (-1)^{n+1}a_{n1}M_{n1})$$

as $\det(B)$. In consequence, $\det(C) = \det(A) + \det(B)$, as claimed in the theorem. ∎

For instance, from the theorem, we know that

$$\begin{vmatrix} 1 & 2 & 3 & 4 \\ 1 & 1 & 1 & 1 \\ 2 & 5 & 6 & 2 \\ 0 & 1 & 2 & 3 \end{vmatrix} = \begin{vmatrix} 1 & 2 & 3 & 4 \\ 1 & 1 & 1 & 1 \\ 1 & 4 & 5 & 1 \\ 0 & 1 & 2 & 3 \end{vmatrix} + \begin{vmatrix} 1 & 2 & 3 & 4 \\ 1 & 1 & 1 & 1 \\ 1 & 1 & 1 & 1 \\ 0 & 1 & 2 & 3 \end{vmatrix},$$

and the second determinant is 0 by Theorem 5.2.6, since two of its rows are equal. So

$$\begin{vmatrix} 1 & 2 & 3 & 4 \\ 1 & 1 & 1 & 1 \\ 2 & 5 & 6 & 2 \\ 0 & 1 & 2 & 3 \end{vmatrix} = \begin{vmatrix} 1 & 2 & 3 & 4 \\ 1 & 1 & 1 & 1 \\ 1 & 4 & 5 & 1 \\ 0 & 1 & 2 & 3 \end{vmatrix}.$$

We could proceed with this to get that

$$\begin{vmatrix} 1 & 2 & 3 & 4 \\ 1 & 1 & 1 & 1 \\ 2 & 5 & 6 & 2 \\ 0 & 1 & 2 & 3 \end{vmatrix} = \begin{vmatrix} 1 & 2 & 3 & 4 \\ 1 & 1 & 1 & 1 \\ 0 & 3 & 4 & 0 \\ 0 & 1 & 2 & 3 \end{vmatrix}$$

and even to get

$$\begin{vmatrix} 1 & 2 & 3 & 4 \\ 1 & 1 & 1 & 1 \\ 2 & 5 & 6 & 2 \\ 0 & 1 & 2 & 3 \end{vmatrix} = \begin{vmatrix} 0 & 1 & 2 & 3 \\ 1 & 1 & 1 & 1 \\ 0 & 3 & 4 & 0 \\ 0 & 1 & 2 & 3 \end{vmatrix}.$$

Can you see how this comes about? Notice that the last determinant has all but one entry equal to 0 in its first column, so it is relatively easy to evaluate, for we need but calculate the (2, 1) minor. By the way, what are the exact values of $\begin{vmatrix} 0 & 1 & 2 & 3 \\ 1 & 1 & 1 & 1 \\ 0 & 3 & 4 & 0 \\ 0 & 1 & 2 & 3 \end{vmatrix}$

and of $\begin{vmatrix} 1 & 2 & 3 & 4 \\ 1 & 1 & 1 & 1 \\ 2 & 5 & 6 & 2 \\ 0 & 1 & 2 & 3 \end{vmatrix}$? Evaluate them by expanding as is given in the definition and compare the results. If you make no computational mistake, you should find that they are equal.

The next result is an important consequence of Theorem 5.2.7. Before doing it we again look at a 3×3 example. Let

$$A = \begin{bmatrix} a_{11} & a_{12} & a_{13} \\ a_{21} & a_{22} & a_{23} \\ a_{31} & a_{32} & a_{33} \end{bmatrix}$$

and

$$B = \begin{bmatrix} a_{11} & a_{12} & a_{13} \\ a_{21}+qa_{31} & a_{22}+qa_{32} & a_{23}+qa_{33} \\ a_{31} & a_{32} & a_{33} \end{bmatrix}.$$

By Theorem 5.2.7,

$$\det(B) = \det(A) + \begin{vmatrix} a_{11} & a_{12} & a_{13} \\ qa_{31} & qa_{32} & qa_{33} \\ a_{31} & a_{32} & a_{33} \end{vmatrix}.$$

Moreover, by Theorem 5.2.2 we have

$$\begin{vmatrix} a_{11} & a_{12} & a_{13} \\ qa_{31} & qa_{32} & qa_{33} \\ a_{31} & a_{32} & a_{33} \end{vmatrix} = q \begin{vmatrix} a_{11} & a_{12} & a_{13} \\ a_{31} & a_{32} & a_{33} \\ a_{31} & a_{32} & a_{33} \end{vmatrix}$$

and then by Theorem 5.2.6 we have

$$\begin{vmatrix} a_{11} & a_{12} & a_{13} \\ a_{31} & a_{32} & a_{33} \\ a_{31} & a_{32} & a_{33} \end{vmatrix} = 0$$

since the last two rows of the latter determinant are equal. So we find that

$$\det(B) = \det(A) + \begin{vmatrix} a_{11} & a_{12} & a_{13} \\ qa_{31} & qa_{32} & qa_{33} \\ a_{31} & a_{32} & a_{33} \end{vmatrix} = \det(A) + 0$$

and, wonder of wonders, $\det(B) = \det(A)$.

The argument given is really the general argument but it never hurts to repeat, or almost repeat.

Theorem 5.2.8. If the matrix B is obtained from A by adding a constant times one row of A to another row of A, then $\det(B) = \det(A)$.

Proof: Suppose that we add q times row s of A to row r of A to get B. Then all the entries of B, except those in row r, are the same as the corresponding entries of A. But the (r, v) entry of B is $a_{rv} + qa_{sv}$. Thus by Theorem 5.2.7,

$$\det(B) = \det(A) + \begin{vmatrix} a_{11} & a_{12} & & a_{1n} \\ \vdots & \vdots & & \vdots \\ qa_{s1} & qa_{s2} & \cdots & qa_{sn} \\ \vdots & \vdots & & \vdots \\ a_{s1} & a_{s2} & & a_{sn} \\ \vdots & \vdots & & \vdots \\ a_{n1} & a_{n2} & & a_{nn} \end{vmatrix} \begin{matrix} \\ \\ \text{(row } r\text{)} \\ \\ \text{(row } s\text{)} \\ \\ \end{matrix}$$

By Theorem 5.2.2 we have

$$\begin{vmatrix} a_{11} & a_{12} & & a_{1n} \\ \vdots & \vdots & & \vdots \\ qa_{s1} & qa_{s2} & \cdots & qa_{sn} \\ \vdots & \vdots & & \vdots \\ a_{s1} & a_{s2} & & a_{sn} \\ \vdots & \vdots & & \vdots \\ a_{n1} & a_{n2} & & a_{nn} \end{vmatrix} = q \begin{vmatrix} a_{11} & a_{12} & & a_{1n} \\ \vdots & \vdots & & \vdots \\ a_{s1} & a_{s2} & \cdots & a_{sn} \\ \vdots & \vdots & & \vdots \\ a_{s1} & a_{s2} & & a_{sn} \\ \vdots & \vdots & & \vdots \\ a_{n1} & a_{n2} & & a_{nn} \end{vmatrix},$$

and by Theorem 5.2.7 we have

$$\begin{vmatrix} a_{11} & a_{12} & & a_{1n} \\ \vdots & \vdots & & \vdots \\ a_{s1} & a_{s2} & \cdots & a_{sn} \\ \vdots & \vdots & & \vdots \\ a_{s1} & a_{s2} & & a_{sn} \\ \vdots & \vdots & & \vdots \\ a_{n1} & a_{n2} & & a_{nn} \end{vmatrix} = 0,$$

since two rows of the latter determinant are equal. So we have

$$\det(B) = \det(A) + \begin{vmatrix} a_{11} & a_{12} & & a_{1n} \\ \vdots & \vdots & & \vdots \\ qa_{s1} & qa_{s2} & \cdots & qa_{sn} \\ \vdots & \vdots & & \vdots \\ a_{s1} & a_{s2} & & a_{sn} \\ \vdots & \vdots & & \vdots \\ a_{n1} & a_{n2} & & a_{nn} \end{vmatrix} = \det(A) + 0$$

and $\det(B) = \det(A)$, which proves the theorem. ∎

This last theorem (5.2.8) is very useful in computing determinants. We illustrate its use with a 4×4 example. Suppose that we want to compute the determinant
$\begin{vmatrix} 5 & -1 & 6 & 7 \\ 1 & 3 & -1 & 2 \\ 4 & 5 & 0 & 1 \\ -1 & 6 & 2 & 2 \end{vmatrix}$. If we add -5 times the second row to the first row, we get

$$\begin{vmatrix} 0 & -16 & 11 & -3 \\ 1 & 3 & -1 & 2 \\ 4 & 5 & 0 & 1 \\ -1 & 6 & 2 & 2 \end{vmatrix}.$$

Now add the second row to the fourth, which yields

$$\begin{vmatrix} 0 & -16 & 11 & -3 \\ 1 & 3 & -1 & 2 \\ 4 & 5 & 0 & 1 \\ 0 & 9 & 1 & 4 \end{vmatrix}.$$

Finally, adding -4 times the second row to the third results in

$$\begin{vmatrix} 0 & -16 & 11 & -3 \\ 1 & 3 & -1 & 2 \\ 0 & -7 & 4 & -7 \\ 0 & 9 & 1 & 4 \end{vmatrix}.$$

All these determinants are equal by Theorem 5.2.8. But what have we obtained by all this jockeying around? Because all the entries, except for the (2, 1) one, in the first column are 0, the determinant merely becomes

$$(-1)^{2+1} \begin{vmatrix} -16 & 11 & -3 \\ -7 & 4 & -7 \\ 9 & 1 & 4 \end{vmatrix} = - \begin{vmatrix} -16 & 11 & -3 \\ -7 & 4 & -7 \\ 9 & 1 & 4 \end{vmatrix}.$$

In this way we have reduced the evaluation of a 4×4 determinant to that of a 3×3 one. We could carry out a similar game for the 3×3 matrix $\begin{vmatrix} -16 & 11 & -3 \\ -7 & 4 & -7 \\ 9 & 1 & 4 \end{vmatrix}$ to obtain zeros in its first column. For instance, adding $\frac{16}{9}$ times row 3 to row 1 gives us a 0 in the corner:

$$\begin{vmatrix} 0 & 11+\frac{16}{9} & -3+\frac{64}{9} \\ -7 & 4 & -7 \\ 9 & 1 & 4 \end{vmatrix} = \begin{vmatrix} 0 & \frac{115}{9} & \frac{37}{9} \\ -7 & 4 & -7 \\ 9 & 1 & 4 \end{vmatrix}.$$

Then adding $\frac{7}{9}$ times row 3 to row 2 gives us another 0 in the first column:

$$\begin{vmatrix} 0 & \frac{115}{9} & \frac{37}{9} \\ 0 & 4+\frac{7}{9} & -7+\frac{28}{9} \\ 9 & 1 & 4 \end{vmatrix} = \begin{vmatrix} 0 & \frac{115}{9} & \frac{37}{9} \\ 0 & \frac{43}{9} & -\frac{35}{9} \\ 9 & 1 & 4 \end{vmatrix}.$$

Again, all these determinants are equal by Theorem 5.2.8. And again, since all the entries, except for the (3, 1) one, in the first column are 0, the determinant merely becomes

$$(-1)^{3+1} 9 \begin{vmatrix} \frac{115}{9} & \frac{37}{9} \\ \frac{43}{9} & -\frac{35}{9} \end{vmatrix} = 9((\tfrac{115}{9})(-\tfrac{35}{9}) - (\tfrac{37}{9})(\tfrac{43}{9})) = -\tfrac{5616}{9}.$$

Of course, this playing around with the rows can almost be done visually, so it is a rather fast way of evaluating determinants. For a computer this is relatively child's play, and easy to program.

The material in this section has not been easy. If you don't understand it all the first time around, go back to it and try out what is being said for specific examples.

PROBLEMS

NUMERICAL PROBLEMS

1. Evaluate the given determinants directly from the definition, and then evaluate them using the row operations.

   (a) $\begin{vmatrix} 1 & 4 & -1 \\ 7 & 6 & 2 \\ \pi & -\pi & 1 \end{vmatrix}.$

   (b) $\begin{vmatrix} 1+i & i & -i & 0 \\ 1 & 1 & 1 & 1 \\ i & i-1 & -i-1 & -1 \\ 5 & \pi & \pi^2 & 17 \end{vmatrix}.$

   (c) $\begin{vmatrix} 5 & -1 & 6 & 7 \\ 1 & 3 & -1 & 2 \\ 4 & 5 & 0 & 1 \\ -1 & 6 & 2 & 2 \end{vmatrix}.$

   (d) $\begin{vmatrix} 5 & -1 & 6 & 7 \\ 1 & 3 & -1 & 2 \\ 3 & 4 & 1 & -1 \\ 0 & 9 & 1 & 4 \end{vmatrix}.$

2. Verify by a direct calculation or by row operations that

   (a) $\begin{vmatrix} 1 & -1 & 6 \\ 4 & 3 & -11 \\ 6 & 2 & -1 \end{vmatrix} = \begin{vmatrix} 1 & -1 & 6 \\ 3 & 2 & -12 \\ 6 & 2 & -1 \end{vmatrix} + \begin{vmatrix} 1 & -1 & 6 \\ 1 & 1 & 1 \\ 6 & 2 & -1 \end{vmatrix}.$

(b) $\begin{vmatrix} 1 & 2 & 4 \\ 6 & 7 & 9 \\ -11 & 5 & 10 \end{vmatrix} + \begin{vmatrix} 2 & 1 & 4 \\ 7 & 6 & 9 \\ 5 & -11 & 10 \end{vmatrix} = 0.$

(c) $\begin{vmatrix} 1 & 2 & 4 \\ 6 & 7 & 9 \\ -11 & 5 & 10 \end{vmatrix} = \begin{vmatrix} 2 & 4 & 1 \\ 7 & 9 & 6 \\ 5 & 10 & -11 \end{vmatrix}.$

(d) $\begin{vmatrix} 0 & 5 & 0 & 0 \\ 1 & 2 & -1 & 0 \\ 3 & 4 & 1 & 6 \\ 2 & 1 & 5 & 1 \end{vmatrix} = -5 \begin{vmatrix} 1 & -1 & 1 \\ 3 & 1 & 6 \\ 2 & 5 & 1 \end{vmatrix}.$

3. Compute all the minors of

(a) $\begin{vmatrix} 1 & 3 & 0 \\ 3 & 4 & 5 \\ 0 & 0 & 1 \end{vmatrix}.$

(b) $\begin{vmatrix} 1 & 2 & 0 \\ 0 & 1 & 1 \\ 6 & 1 & 2 \end{vmatrix}.$

(c) $\begin{vmatrix} 1 & 2 & 3 & 4 \\ 0 & 3 & 0 & 4 \\ 1 & 0 & 1 & 0 \\ 0 & 1 & 0 & 1 \end{vmatrix}.$

MORE THEORETICAL PROBLEMS

Easier Problems

4. If A is a triangular matrix, what are the necessary and sufficient conditions on A that $\det(A) \neq 0$?

5. If A and B are both upper triangular matrices, show that

$$\det(AB) = \det(A)\det(B).$$

6. If A is an upper triangular matrix, prove that $\det(A) = \det(A')$.
7. If A is a 3×3 matrix, show that A is invertible if $\det(A) \neq 0$.
8. If A is a 3×3 matrix and A is invertible, show that $\det(A) \neq 0$.
9. In Problem 8 show that $\det(A^{-1}) = \dfrac{1}{\det(A)}.$

10. Verify for the matrix $A = \begin{bmatrix} 2 & 2 & 2 \\ 1 & 0 & 2 \\ 4 & 1 & 2 \end{bmatrix}$ that $|AA'| = |A|^2$.

Harder Problems

11. Prove the statement

There is one and only one scalar-valued function f on 3×3 matrices which satisfies the properties:

D1. $f\left(\begin{bmatrix} a+a' & b & c \\ d+d' & e & f \\ g+g' & h & k \end{bmatrix}\right) = f\left(\begin{bmatrix} a & b & c \\ d & e & f \\ g & h & k \end{bmatrix}\right) + f\left(\begin{bmatrix} a' & b & c \\ d' & e & f \\ g' & h & k \end{bmatrix}\right);$

D2. $f\left(\begin{bmatrix} ua & b & c \\ ud & e & f \\ ug & h & k \end{bmatrix}\right) = uf\left(\begin{bmatrix} a & b & c \\ d & e & f \\ g & h & k \end{bmatrix}\right);$

D3. $f\left(\begin{bmatrix} a & b & c \\ d & e & f \\ g & h & k \end{bmatrix}\right) = -f\left(\begin{bmatrix} b & a & c \\ e & d & f \\ h & g & k \end{bmatrix}\right) = -f\left(\begin{bmatrix} a & c & b \\ d & f & e \\ g & k & h \end{bmatrix}\right);$

D4. $f\left(\begin{bmatrix} 1 & 0 & 0 \\ 0 & 1 & 0 \\ 0 & 0 & 1 \end{bmatrix}\right) = 1;$

namely the determinant function.

12. For 3×3 matrices A, show that the function $f(A) = \det(A')$ satisfies the conditions of Problem 11, so that $f(A)$ must be the determinant of A for all 3×3 matrices A.

13. For $A = \begin{bmatrix} a & b & c \\ d & e & f \\ g & h & k \end{bmatrix}$, define $f(A) = -b \begin{vmatrix} d & f \\ g & k \end{vmatrix} + e \begin{vmatrix} a & c \\ g & k \end{vmatrix} - h \begin{vmatrix} a & c \\ d & f \end{vmatrix}$ and show

directly that the function $f(A)$ satisfies the conditions of Problem 11 and, therefore, must be the determinant function.

Very Hard Problems

14. If A is an upper triangular matrix, show that for all B, $\det(AB) = \det(A)\det(B)$, using row operations.

15. If A is a lower triangular matrix, show that for all B, $\det(AB) = \det(A)\det(B)$.

5.3. PROPERTIES OF DETERMINANTS: COLUMN OPERATIONS

In Section 5.2 we saw what operations can be carried out on the rows of a given determinant. We now want to establish the analogous results for the columns of a given determinant. The style of many of the proofs will be similar to those given in Section 5.2. But there is a difference. Because the determinant was defined in terms of the expansion by minors of the first column, we shall often divide the argument into two parts: one, in which the first column enters in the argument, and two, in which the first column is not involved. For the results about the rows we didn't need such a case division, since no row was singled out among all the other rows.

With the experience in the manner of proof used in Section 5.2 under our belts, we can afford, at times, to be a little sketchier in our arguments than we were in Section 5.2. As before, the proof will be by induction, assuming the results to be correct for $(n-1) \times (n-1)$ matrices, and passing from this to the validity of the results for the $n \times n$ ones. We shall also try to illustrate what is going on by resorting to 3×3 examples.

We begin the discussion with

Theorem 5.3.1. *If a column of A consists of zeros, then $\det(A) = 0$.*

Proof: If the column of zeros is the first column, then the proof is very easy, for $\det(A) = \sum_{r=1}^{n} (-1)^{r+1} a_{r1} M_{r1}$ and since each $a_{r1} = 0$ we get that $\det(A) = 0$.

What happens if this column of zeros is not the first one? In that case, since in expressing M_{r1} for $r > 1$, we see that each M_{r1} has a column of zeros. Because M_{r1} is an $(n-1) \times (n-1)$ determinant, we know that $M_{r1} = 0$. Hence $\det(A) = 0$ by the formula

$$\det(A) = \sum_{r=1}^{n} (-1)^{r+1} a_{r1} M_{r1}. \quad \blacksquare$$

In parallel with what we did in Section 5.2 we now prove

Theorem 5.3.2. *If the matrix B is obtained from A by multiplying each entry of some column of A by a constant, u, then $\det(B) = u \det(A)$.*

Proof: Let's see the situation first for the 3×3 case. If B is obtained from A by multiplying the first column of A by u, then

$$B = \begin{bmatrix} ua_{11} & a_{12} & a_{13} \\ ua_{21} & a_{22} & a_{23} \\ ua_{31} & a_{32} & a_{33} \end{bmatrix},$$

where

$$A = \begin{bmatrix} a_{11} & a_{12} & a_{13} \\ a_{21} & a_{22} & a_{23} \\ a_{31} & a_{32} & a_{33} \end{bmatrix}.$$

Thus

$$\det(B) = ua_{11} \begin{vmatrix} a_{22} & a_{23} \\ a_{32} & a_{33} \end{vmatrix} - ua_{21} \begin{vmatrix} a_{12} & a_{13} \\ a_{32} & a_{33} \end{vmatrix} + ua_{31} \begin{vmatrix} a_{12} & a_{13} \\ a_{22} & a_{23} \end{vmatrix}$$

$$= u \left(a_{11} \begin{vmatrix} a_{22} & a_{23} \\ a_{32} & a_{33} \end{vmatrix} - a_{21} \begin{vmatrix} a_{12} & a_{13} \\ a_{32} & a_{33} \end{vmatrix} + a_{31} \begin{vmatrix} a_{12} & a_{13} \\ a_{22} & a_{23} \end{vmatrix} \right)$$

$$= u \det(A),$$

the desired result.

If a column other than the first is multiplied by u, then in each minor N_{k1} of B a column in N_{k1} is multiplied by u. Hence $N_{k1} = uM_{k1}$, by the 2×2 case, where M_{k1} is the $(k, 1)$ minor of $\det(A)$. So

$$\begin{aligned}\det(B) &= a_{11}N_{11} - a_{21}N_{21} + a_{31}N_{31} \\ &= ua_{11}M_{11} - ua_{21}M_{21} + ua_{31}M_{31} \\ &= u(a_{11}M_{11} - a_{21}M_{21} + a_{31}M_{31}) \\ &= u\det(A).\end{aligned}$$

We leave the details of the proof for the case of $n \times n$ matrices to the readers. In that way they can check if they have this technique of proof under control. The proof runs along pretty well as it did for 3×3 matrices. ∎

At this point the parallelism with Section 5.2 breaks. (It will be picked up later.) The result we now prove depends heavily on the results for row operations, and is a little tricky. So before embarking on the proof of the general result let's try the argument to be given on a 3×3 matrix.

Consider the 3×3 matrix

$$A = \begin{bmatrix} a & d & a \\ b & e & b \\ c & f & c \end{bmatrix}.$$

Notice that the first and third columns of A are equal. Suppose, for the sake of argument, that $a \neq 0$ (this really is not essential). By Theorem 5.2.8 we can add any multiple of the first row to any other row without changing the determinant. Add $-\frac{b}{a}$ times the first row to the second one and $-\frac{c}{a}$ times the first row to the third one. The resulting matrix B is

$$\begin{bmatrix} a & d & a \\ 0 & u & 0 \\ 0 & v & 0 \end{bmatrix},$$

where $u = e - \frac{b}{a}d$ and $v = f - \frac{c}{a}d$ (and are of no importance). Thus

$$\det(A) = \det(B) = a \begin{vmatrix} u & 0 \\ v & 0 \end{vmatrix} = 0$$

since $\begin{vmatrix} u & 0 \\ v & 0 \end{vmatrix}$ has a column of zeros.

This simple example suggests

Theorem 5.3.3. If two columns of A are equal, then $\det(A) = 0$.

Proof: If neither column in question is the first, then for each minor M_{k1} we have two columns equal. So by the $(n-1) \times (n-1)$ case, $M_{k1} = 0$. Therefore, $\det(A) = 0$.

Suppose then that the two columns involved are the first and the rth. If the first column of A consists only of zeros, then $\det(A) = 0$ by Theorem 5.3.1, and we are done. So we may assume that $a_{s1} \neq 0$ for some s. If we interchange row s with the first row, we get a matrix B whose $(1, 1)$ entry is nonzero. Moreover, by Theorem 5.2.5, doing this interchange merely changes the sign of the determinant. So if the determinant of B is 0, since $\det(B) = -\det(A)$ we get the desired result, $\det(A) = 0$.

So it is enough to carry out the argument for B. In other words, we may assume that the $(1, 1)$ entry a_{11} of A is nonzero.

By our assumption

$$A = \begin{bmatrix} a_{11} & & \overset{(r)}{a_{11}} & \\ a_{21} & & a_{21} & \\ \vdots & * & \vdots & * \\ a_{n1} & & a_{n1} & \end{bmatrix},$$

where the *'s indicate entries in which we have no interest, and where the second $\begin{bmatrix} a_{11} \\ a_{21} \\ \vdots \\ a_{n1} \end{bmatrix}$ occurs in column r.

By Theorem 5.2.8 we can add any multiple of the first row to any other row without changing the determinant. Since $a_{11} \neq 0$, if we add $-\dfrac{a_{k1}}{a_{11}}$ times the first row to row k, for $k = 2, \ldots, n$, the resulting matrix C looks like

$$\begin{bmatrix} a_{11} & & a_{11} & \\ 0 & & 0 & \\ \vdots & * & \vdots & * \\ 0 & & 0 & \end{bmatrix},$$

and $\det(C) = \det(A)$. But what is $\det(C)$? Since all the entries in the first column are 0 except for a_{11}, $\det(C) = a_{11}$ times the $(1, 1)$ minor of C. But look at this minor! Its column $r - 1$, which comes from column r of A, consists only of zeros. Thus this minor is 0! Therefore, $\det(C) = 0$, and since $\det(C) = \det(A)$, we get the desired result that $\det(A) = 0$. ∎

We should like to prove that interchanging two columns of a determinant merely changes its sign. This is the counterpart of Theorem 5.2.5. We are almost there, but we first need

Theorem 5.3.4. If $C = (c_{uv})$ is a matrix such that for $v \neq r$, $c_{uv} = a_{uv}$ and for $v = r$,

$c_{ur} = a_{ur} + b_{ur}$, then $\det(C) = \det(A) + \det(B)$, where

$$A = \begin{bmatrix} a_{11} & a_{12} & \cdots & a_{1r} & \cdots & a_{1n} \\ a_{21} & a_{22} & \cdots & a_{2r} & \cdots & a_{2n} \\ \vdots & \vdots & & \vdots & & \vdots \\ a_{n1} & a_{n2} & \cdots & a_{nr} & \cdots & a_{nn} \end{bmatrix}$$

and

$$B = \begin{bmatrix} a_{11} & a_{12} & \cdots & b_{1r} & \cdots & a_{1n} \\ a_{21} & a_{22} & \cdots & b_{2r} & \cdots & a_{2n} \\ \vdots & \vdots & & \vdots & & \vdots \\ a_{n1} & a_{n2} & \cdots & b_{nr} & \cdots & a_{nn} \end{bmatrix}.$$

$\left(\text{That is, } B \text{ is the same as } A \text{ except that its column } r \text{ is replaced by } \begin{bmatrix} b_{1r} \\ b_{2r} \\ \vdots \\ b_{nr} \end{bmatrix}. \right)$

Proof: If $r = 1$, then the result is easy. Why? Because

$$\det(C) = \sum_{k=1}^{n} (-1)^{k+1} c_{k1} Q_{k1}$$
$$= \sum_{k=1}^{n} (-1)^{k+1} (a_{k1} + b_{k1}) Q_{k1},$$

where Q_{k1} is the $(k, 1)$ minor of $\det(C)$. But what is Q_{k1}? Since the columns, other than the first, do not involve the b's, we have that $Q_{k1} = M_{k1}$, the $(k, 1)$ minor of $\det(A)$. Thus

$$\det(C) = \sum_{k=1}^{n} (-1)^{k+1} a_{k1} M_{k1} + \sum_{k=1}^{n} (-1)^{k+1} b_{k1} M_{k1}.$$

We recognize the first sum as $\det(A)$ and the second one as $\det(B)$. Therefore, $\det(C) = \det(A) + \det(B)$.

Suppose, then, that $r > 1$. Then we have that in column $r - 1$ of Q_{k1} that each entry is of the form $c_{ur} = a_{ur} + b_{ur}$, while all the other entries come from A. Since Q_{k1} is an $(n - 1) \times (n - 1)$ matrix we have, by induction, that $Q_{k1} = M_{k1} + N_{k1}$, where N_{k1} is the $(k, 1)$ minor of $\det(B)$. Thus

$$\det(C) = \sum_{k=1}^{n} (-1)^{k+1} c_{k1} Q_{k1} = \sum_{k=1}^{n} (-1)^{k+1} a_{k1} (M_{k1} + N_{k1})$$
$$= \sum_{k=1}^{n} (-1)^{k+1} a_{k1} M_{k1} + \sum_{k=1}^{n} (-1)^{k+1} a_{k1} N_{k1}$$
$$= \det(A) + \det(B). \blacksquare$$

If all the $\sum$'s in the proof confuse you, try the argument out for the 3 × 3 situation. Let's see that the result is correct for a specific 3 × 3 matrix. Let

$$A = \begin{bmatrix} 1 & 2 & 3 \\ 0 & 4 & 5 \\ 0 & 1 & 3 \end{bmatrix} = \begin{bmatrix} 1 & 2 & 1+2 \\ 0 & 4 & 1+4 \\ 0 & 1 & 2+1 \end{bmatrix}.$$

According to the theorem,

$$\det \begin{bmatrix} 1 & 2 & 3 \\ 0 & 4 & 5 \\ 0 & 1 & 3 \end{bmatrix} = \det \begin{bmatrix} 1 & 2 & 1 \\ 0 & 4 & 1 \\ 0 & 1 & 2 \end{bmatrix} + \det \begin{bmatrix} 1 & 2 & 2 \\ 0 & 4 & 4 \\ 0 & 1 & 1 \end{bmatrix}$$

$$= \det \begin{bmatrix} 1 & 2 & 1 \\ 0 & 4 & 1 \\ 0 & 1 & 2 \end{bmatrix} + 0$$

$$= \det \begin{bmatrix} 1 & 2 & 1 \\ 0 & 4 & 1 \\ 0 & 1 & 2 \end{bmatrix}.$$

Is that the case? Well,

$$\det \begin{bmatrix} 1 & 2 & 3 \\ 0 & 4 & 5 \\ 0 & 1 & 3 \end{bmatrix} = 1 \begin{vmatrix} 4 & 5 \\ 1 & 3 \end{vmatrix} = 12 - 5 = 7$$

and

$$\det \begin{bmatrix} 1 & 2 & 1 \\ 0 & 4 & 1 \\ 0 & 1 & 2 \end{bmatrix} = 1 \begin{vmatrix} 4 & 1 \\ 1 & 2 \end{vmatrix} = 8 - 1 = 7.$$

So they are indeed equal. No surprise!

We now can prove that interchanging two columns of $\det(A)$ changes the sign of $\det(A)$.

We do a specific case first. Let

$$A = \begin{bmatrix} a_{11} & a_{12} & a_{13} \\ a_{21} & a_{22} & a_{23} \\ a_{31} & a_{32} & a_{33} \end{bmatrix},$$

our old friend, and

$$B = \begin{bmatrix} a_{12} & a_{11} & a_{13} \\ a_{22} & a_{21} & a_{23} \\ a_{32} & a_{31} & a_{33} \end{bmatrix},$$

the matrix obtained from A by interchanging the first two columns of A. Let

$$C = \begin{bmatrix} a_{11} + a_{12} & a_{11} + a_{12} & a_{13} \\ a_{21} + a_{22} & a_{21} + a_{22} & a_{23} \\ a_{31} + a_{32} & a_{31} + a_{32} & a_{33} \end{bmatrix}.$$

Since the first two columns of C are equal, $\det(C) = 0$ by Theorem 5.3.3. By Theorem 5.3.4 used several times, this determinant can also be written as

$$0 = \begin{vmatrix} a_{11} + a_{12} & a_{11} + a_{12} & a_{13} \\ a_{21} + a_{22} & a_{21} + a_{22} & a_{23} \\ a_{31} + a_{32} & a_{31} + a_{32} & a_{33} \end{vmatrix}$$

$$= \begin{vmatrix} a_{11} & a_{11} + a_{12} & a_{13} \\ a_{21} & a_{21} + a_{22} & a_{23} \\ a_{31} & a_{31} + a_{32} & a_{33} \end{vmatrix} + \begin{vmatrix} a_{12} & a_{11} + a_{12} & a_{13} \\ a_{22} & a_{21} + a_{22} & a_{23} \\ a_{32} & a_{31} + a_{32} & a_{33} \end{vmatrix}$$

$$= \begin{vmatrix} a_{11} & a_{11} & a_{13} \\ a_{21} & a_{21} & a_{23} \\ a_{31} & a_{31} & a_{33} \end{vmatrix} + \begin{vmatrix} a_{11} & a_{12} & a_{13} \\ a_{21} & a_{22} & a_{23} \\ a_{31} & a_{32} & a_{33} \end{vmatrix} + \begin{vmatrix} a_{12} & a_{11} & a_{13} \\ a_{22} & a_{21} & a_{23} \\ a_{32} & a_{31} & a_{33} \end{vmatrix} + \begin{vmatrix} a_{12} & a_{12} & a_{13} \\ a_{22} & a_{22} & a_{23} \\ a_{32} & a_{32} & a_{33} \end{vmatrix}.$$

Notice that the first and last of these four determinants have two equal columns, so are 0 by Theorem 5.3.3. Thus we are left with

$$0 = \begin{vmatrix} a_{11} & a_{12} & a_{13} \\ a_{21} & a_{22} & a_{23} \\ a_{31} & a_{32} & a_{33} \end{vmatrix} + \begin{vmatrix} a_{12} & a_{11} & a_{13} \\ a_{22} & a_{21} & a_{23} \\ a_{32} & a_{31} & a_{33} \end{vmatrix}.$$

Hence $\det(B) = -\det(A)$.

The proof for the general case will be along the lines of the proof for the 3×3 matrix.

Theorem 5.3.5. If B is obtained from A by the interchange of two columns of A, then $\det(B) = -\det(A)$.

Proof: Suppose that $A = (a_{rs})$ and that B is obtained from A by interchanging columns r and s. Consider the matrix $C = (c_{uv})$, where $c_{uv} = a_{uv}$ if $v \neq r$ or $v \neq s$ and where $c_{ur} = a_{ur} + a_{us}$ and $c_{us} = a_{ur} + a_{us}$. Then columns r and s of C are equal, therefore, by Theorem 5.3.3, $\det(C) = 0$. Now, by several uses of Theorem 5.3.4, we get that

$$0 = \det(C) = \begin{vmatrix} & & (\text{col. } r) & (\text{col. } s) & \\ a_{11} & a_{1r} + a_{1s} & a_{1r} + a_{1s} & a_{11} \\ a_{21} & a_{2r} + a_{2s} & a_{2r} + a_{2s} & a_{21} \\ \vdots & \cdots & \vdots & \vdots & \cdots & \vdots & \cdots & \vdots \\ a_{n1} & a_{nr} + a_{ns} & a_{nr} + a_{ns} & a_{n1} \end{vmatrix}$$

$$= \begin{vmatrix} a_{11} & a_{1s} & a_{1r}+a_{1s} & a_{11} \\ a_{21} & a_{2s} & a_{2r}+a_{2s} & a_{21} \\ \vdots & \vdots & \vdots & \vdots \\ a_{n1} & a_{nr} & a_{nr}+a_{ns} & a_{n1} \end{vmatrix} \begin{matrix} (r) & (\text{col. } s) \\ \\ \\ \\ \end{matrix}$$

$$+ \begin{vmatrix} a_{11} & a_{1s} & a_{1r}+a_{1s} & a_{11} \\ a_{21} & a_{2s} & a_{2r}+a_{2s} & a_{21} \\ \vdots & \vdots & \vdots & \vdots \\ a_{n1} & a_{ns} & a_{nr}+a_{ns} & a_{n1} \end{vmatrix} \begin{matrix} (r) & (\text{col. } s) \\ \\ \\ \\ \end{matrix}$$

$$= \det(A) + \begin{vmatrix} a_{11} & a_{1r} & a_{1r} & a_{11} \\ a_{21} & a_{2r} & a_{2r} & a_{21} \\ \vdots & \vdots & \vdots & \vdots \\ a_{n1} & a_{nr} & a_{nr} & a_{n1} \end{vmatrix} \begin{matrix} (r) & (s) \\ \\ \\ \\ \end{matrix}$$

$$+ \det(B) + \begin{vmatrix} a_{11} & a_{1s} & a_{1s} & a_{11} \\ a_{21} & a_{2s} & a_{2s} & a_{21} \\ \vdots & \vdots & \vdots & \vdots \\ a_{n1} & a_{ns} & a_{ns} & a_{n1} \end{vmatrix} \begin{matrix} (r) & (s) \\ \\ \\ \\ \end{matrix}.$$

Since the second and last determinants on the right-hand side above have two equal columns, by Theorem 5.3.3 they are both 0. This leaves us with $0 = \det(A) + \det(B)$. Thus $\det(B) = -\det(A)$ and the theorem is proved. ∎

We close this section with the column analog of Theorem 5.2.8.

If D is an $n \times n$ matrix, let's use the shorthand $D = (d_1, d_2, \ldots, d_n)$ to represent D, where the vector d_k is column k of D.

Theorem 5.3.6. *If a multiple of one column is added to another in $\det(A)$, the determinant does not change.*

Proof: Let $A = (a_1, \ldots, a_n)$ and suppose that we add q times column s to column r. The resulting matrix is $C = (a_1, \ldots, a_r + qa_s, \ldots, a_n)$, where only column r has been changed, as indicated.

Thus, by Theorem 5.3.4, we have that

$$\det(C) = \det(a_1, \ldots, a_r, \ldots, a_s, \ldots, a_n) + \det(a_1, \ldots, qa_s, \ldots, a_s, \ldots, a_n).$$

However, by Theorem 5.3.2,

$$\det(a_1, \ldots, qa_s, \ldots, a_s, \ldots, a_n) = q \det(a_1, \ldots, a_s, \ldots, a_s, \ldots, a_n),$$

and since two of the columns of $(a_1,\ldots,a_s,\ldots,a_s,\ldots,a_n)$ are equal,

$$\det(a_1,\ldots,a_s,\ldots,a_s,\ldots,a_n) = 0.$$

Thus we get that

$$\det(C) = \det(a_1,\ldots,a_r,\ldots,a_s,\ldots,a_n) = \det(A). \qquad \blacksquare$$

We do the proof just done in detail for 3×3 matrices. Suppose that

$$A = \begin{bmatrix} a_{11} & a_{12} & a_{13} \\ a_{21} & a_{22} & a_{23} \\ a_{31} & a_{32} & a_{33} \end{bmatrix}.$$

Then $C = (a_1, a_2 + qa_3, a_3)$ is the matrix

$$C = \begin{bmatrix} a_{11} & a_{12} + qa_{13} & a_{13} \\ a_{21} & a_{22} + qa_{23} & a_{23} \\ a_{31} & a_{32} + qa_{33} & a_{33} \end{bmatrix}$$

and using Theorems 5.3.4 and 5.3.2 as in the proof above, we have

$$\begin{aligned}
\det(C) &= \begin{vmatrix} a_{11} & a_{12} + qa_{13} & a_{13} \\ a_{21} & a_{22} + qa_{23} & a_{23} \\ a_{31} & a_{32} + qa_{33} & a_{33} \end{vmatrix} \\
&= \begin{vmatrix} a_{11} & a_{12} & a_{13} \\ a_{21} & a_{22} & a_{23} \\ a_{31} & a_{32} & a_{33} \end{vmatrix} + \begin{vmatrix} a_{11} & qa_{13} & a_{13} \\ a_{21} & qa_{23} & a_{23} \\ a_{31} & qa_{33} & a_{33} \end{vmatrix} \\
&= \det(A) + q \begin{vmatrix} a_{11} & a_{13} & a_{13} \\ a_{21} & a_{23} & a_{23} \\ a_{31} & a_{33} & a_{33} \end{vmatrix} \\
&= \det(A) + 0 \\
&= \det(A).
\end{aligned}$$

PROBLEMS

NUMERICAL PROBLEMS

1. Verify by a direct computation that

   (a) $\begin{vmatrix} 1 & -1 & 6 \\ 4 & \frac{1}{2} & 0 \\ 3 & -1 & 2 \end{vmatrix} = -\begin{vmatrix} 6 & -1 & 1 \\ 0 & \frac{1}{2} & 4 \\ 2 & -1 & 3 \end{vmatrix}.$

(b) $\begin{vmatrix} 1 & -1 & 6 \\ 4 & \frac{1}{2} & 0 \\ 3 & -1 & 2 \end{vmatrix} = \begin{vmatrix} 1 & 0 & 0 \\ 4 & 4+\frac{1}{2} & -24 \\ 3 & 2 & -16 \end{vmatrix}$.

(c) $\begin{vmatrix} 1 & 2 & 3 & 4 \\ 4 & 3 & 2 & 1 \\ 3 & 2 & 1 & 4 \\ 5 & 10 & 15 & 20 \end{vmatrix} = \begin{vmatrix} 1 & 3 & 4 & 5 \\ 4 & 7 & 6 & 5 \\ 3 & 5 & 4 & 7 \\ 5 & 15 & 20 & 25 \end{vmatrix}$.

(d) $\begin{vmatrix} 1 & -2 & 6 \\ 4 & \frac{1}{2} & 0 \\ 3 & -1 & 2 \end{vmatrix} = \begin{vmatrix} 1 & 4 & 3 \\ -2 & \frac{1}{2} & -1 \\ 6 & 0 & 2 \end{vmatrix}$.

(e) $\begin{vmatrix} 1 & 2 & 3 \\ 4 & 5 & 6 \\ 7 & 8 & 9 \end{vmatrix} = \begin{vmatrix} 3 & 2 & 3 \\ 6 & 5 & 6 \\ 9 & 8 & 9 \end{vmatrix} + \begin{vmatrix} -2 & 2 & 3 \\ -2 & 5 & 6 \\ -2 & 8 & 9 \end{vmatrix}$.

2. Evaluate $\begin{vmatrix} 1 & a & b & c \\ 1 & a^2 & b^2 & c^2 \\ 1 & a^3 & b^3 & c^3 \\ 1 & a^4 & b^4 & c^4 \end{vmatrix}$.

3. When is the determinant in Problem 2 equal to 0?

MORE THEORETICAL PROBLEMS

Middle-Level Problems

4. Complete the proof of Theorem 5.3.2 for $n \times n$ matrices.

5. Prove that if you carry out the operation on A of putting its first row into the second, the second row into the third, and the third row into the first, then the determinant remains unchanged.

6. Show that if the columns of A are linearly dependent as vectors, then $\det(A) \leq 0$.

7. Show that if the columns of A are linearly independent, then A is invertible.

8. Let $V = F^{(n)}$ and $W = F^{(m)}$. Suppose that f is a function of two variables such that
   (a) $f(x, y) \in W$ for all $x, y \in V$;
   (b) $f(ax, y) = f(x, ay) = af(x, y)$ for all $a \in F$, $x, y \in V$;
   (c) $f(x + x', y) = f(x, y) + f(x', y)$ for all $x, x', y \in V$;
   (d) $f(x, y + y') = f(x, y) + f(x, y')$ for all $x, y, y' \in V$.
   (e) $f(x, x) = 0$ for all $x \in V$.
   Show that $f(x, y) = -f(y, x)$ for all $x, y \in V$.

9. If A is an upper triangular matrix, show that a scalar s is a characteristic root of A if $\det(A - sI) = 0$, where sI denotes s times the identity matrix.

10. Show that if the rows or columns of A are linearly dependent, then $|A| = 0$.

5.4. CRAMER'S RULE

The rules for operations with columns and rows of a determinant allow us to make use of determinants in solving systems of linear equations. Consider the system of equations

$$
\begin{aligned}
a_{11}x_1 + \cdots + a_{1n}x_n &= y_1 \\
a_{21}x_1 + \cdots + a_{2n}x_n &= y_2 \\
\vdots \quad\quad\quad\quad\quad \vdots &\quad \vdots \\
a_{n1}x_1 + \cdots + a_{nn}x_n &= y_n.
\end{aligned}
\tag{1}
$$

We can represent this system as a matrix-vector equation $Ax = y$, where A is the $n \times n$ matrix (a_{rs}) and $x, y \in F^{(n)}$, that is,

$$
\begin{bmatrix} a_{11} & & a_{1n} \\ \vdots & \cdots & \vdots \\ a_{n1} & & a_{nn} \end{bmatrix} \begin{bmatrix} x_1 \\ \vdots \\ x_n \end{bmatrix} = \begin{bmatrix} y_1 \\ \vdots \\ y_n \end{bmatrix}.
$$

Consider $x_r \det(A)$. By Theorem 5.3.2 we can absorb this x_r in column r, that is,

$$
x_r \det(A) = \begin{vmatrix} a_{11} & a_{12} & & x_r a_{1r} & & a_{1n} \\ \vdots & \vdots & \cdots & \vdots & \cdots & \vdots \\ a_{n1} & a_{n2} & & x_r a_{nr} & & a_{nn} \end{vmatrix}.
$$

If we add x_1 times the first column, x_2 times the second,..., x_k times column k to column r of this determinant, for $k \neq r$, by Theorem 5.3.6 we do not change the determinant. Thus

$$
x_r \det(A) = \begin{vmatrix} a_{11} & a_{12} & \overset{(\text{column } r)}{a_{11}x_1 + \cdots + a_{1n}x_n} & a_{1n} \\ \vdots & \vdots & \cdots \quad\quad \vdots \quad\quad \cdots & \vdots \\ a_{n1} & a_{n2} & a_{n1}x_1 + \cdots + a_{nn}x_n & a_{nn} \end{vmatrix}.
$$

Hence if $x_1, \ldots, x_n$ is a solution to the system (1), we have

$$
x_r \det(A) = \begin{vmatrix} a_{11} & a_{12} & \overset{(r)}{y_1} & a_{1n} \\ \vdots & \vdots & \cdots \vdots \cdots & \vdots \\ a_{n1} & a_{n2} & y_n & a_{nn} \end{vmatrix} = \det(A_r),
$$

where the matrix A_r is obtained from A by replacing column r of A by the vector $\begin{bmatrix} y_1 \\ \vdots \\ y_n \end{bmatrix}$, that is, the vector of the values on the right-hand side of (1).

In particular, if $\det(A) \neq 0$, we have

Theorem 5.4.1 (Cramer's Rule) If $\det(A) \neq 0$, the solution to the system (1) of linear equations is given by

$$x_r = \frac{\det(A_r)}{\det(A)} \quad \text{for } r = 1, 2, \ldots, n,$$

where A_r is as described above.

To illustrate the result, let's use it to solve the three linear equations in three unknowns:

$$\begin{aligned} x_1 + x_2 + x_3 &= 1 \\ 2x_1 - x_2 + 2x_3 &= 2 \\ 3x_2 - 4x_3 &= 3. \end{aligned}$$

Thus here the matrix A is

$$A = \begin{bmatrix} 1 & 1 & 1 \\ 2 & -1 & 2 \\ 0 & 3 & -4 \end{bmatrix}$$

and

$$A_1 = \begin{bmatrix} 1 & 1 & 1 \\ 2 & -1 & 2 \\ 3 & 3 & -4 \end{bmatrix}, \quad A_2 = \begin{bmatrix} 1 & 1 & 1 \\ 2 & 2 & 2 \\ 0 & 3 & -4 \end{bmatrix}, \quad A_3 = \begin{bmatrix} 1 & 1 & 1 \\ 2 & -1 & 2 \\ 0 & 3 & 3 \end{bmatrix}.$$

Evaluating all these determinants, we obtain

$$\det(A) = 12, \quad \det(A_1) = 21, \quad \det(A_2) = 0, \quad \det(A_3) = -9.$$

Consequently, by Cramer's rule,

$$x_1 = \tfrac{21}{12} = \tfrac{7}{4}, \quad x_2 = \tfrac{0}{12} = 0, \quad x_3 = -\tfrac{9}{12} = -\tfrac{3}{4}.$$

Checking, we see that we have indeed solved the system:

$$\begin{aligned} (\tfrac{7}{4}) + (0) + (-\tfrac{3}{4}) &= 1 \\ 2(\tfrac{7}{4}) - (0) + 2(-\tfrac{3}{4}) &= 2 \\ 3(0) - 4(-\tfrac{3}{4}) &= 3. \end{aligned}$$

Aside from its role in solving systems of n linear equations in n unknowns, Cramer's rule has a very important consequence interrelating the behavior of a matrix and its determinant. In the next result we get a very powerful and useful criterion for the invertibility of a matrix.

Theorem 5.4.2. If $\det(A) \neq 0$, then A is invertible.

Proof: If $\det(A) \neq 0$, then, by Cramer's rule, we can solve the matrix-vector equation $Ax = v$ for any vector $v = \begin{bmatrix} y_1 \\ \vdots \\ y_n \end{bmatrix}$. In particular, if $e_1, e_2, \ldots, e_n$ is the canonical basis of $F^{(n)}$, we can find vectors $X_1 = \begin{bmatrix} x_{11} \\ \vdots \\ x_{n1} \end{bmatrix}$, $X_2 = \begin{bmatrix} x_{12} \\ \vdots \\ x_{n2} \end{bmatrix}, \ldots, X_n = \begin{bmatrix} x_{1n} \\ \vdots \\ x_{nn} \end{bmatrix}$ such that for each $r = 1, 2, \ldots, n$, $AX_r = e_r$. Thus if $B = \begin{bmatrix} x_{11} & & x_{1n} \\ \vdots & \cdots & \vdots \\ x_{n1} & & x_{nn} \end{bmatrix}$, from $AX_r = e_r$ we obtain

$$AB = A(X_1, \ldots, X_n) = (e_1, \ldots, e_n) = \begin{bmatrix} 1 & 0 & \cdots & 0 \\ 0 & 1 & \ddots & \vdots \\ \vdots & & 1 & 0 \\ 0 & \cdots & 0 & 1 \end{bmatrix} = I.$$

In short, A is invertible. ∎

We shall later see that the converse of Theorem 5.4.2 is true, namely, that if A is invertible, then $\det(A) \neq 0$.

PROBLEMS

NUMERICAL PROBLEMS

Use Cramer's rule.

1. Solve for x_1, x_2, x_3, x_4 in the system

$$x_1 + 2x_2 + 3x_3 + 4x_4 = 1$$
$$x_1 + 2x_2 + 3x_3 + 3x_4 = 2$$
$$x_1 + 2x_2 + 2x_3 + 2x_4 = 3$$
$$x_1 + x_2 + x_3 + x_4 = 4.$$

2. Find the solution $\begin{bmatrix} a \\ b \\ c \\ d \end{bmatrix}$ to the equation $\begin{bmatrix} 5 & 5 & 5 & 5 \\ 4 & 4 & 5 & 5 \\ 4 & 5 & 4 & 5 \\ 4 & 5 & 0 & 0 \end{bmatrix} \begin{bmatrix} a \\ b \\ c \\ d \end{bmatrix} = \begin{bmatrix} 2 \\ \pi \\ 1 \\ 7 \end{bmatrix}$.

3. Solve for x_1, x_2, x_3 in the system

$$3x_1 + 2x_2 - 3x_3 = 4$$
$$3x_1 + 8x_2 + x_3 = 9$$
$$21x_1 + 22x_2 + x_3 = -1.$$

4. Find, for the vector $\begin{bmatrix} a \\ b \\ c \end{bmatrix}$, a vector $\begin{bmatrix} u \\ v \\ w \end{bmatrix}$ such that $\begin{bmatrix} 1 & 5 & -1 \\ 2 & 0 & 1 \\ 3 & 1 & 1 \end{bmatrix} \begin{bmatrix} u \\ v \\ w \end{bmatrix} = \begin{bmatrix} a \\ b \\ c \end{bmatrix}$.

MORE THEORETICAL PROBLEMS

Harder Problems

5. If $A = (a_{rs})$ is an $n \times n$ matrix such that the vectors

$$v_1 = \begin{bmatrix} a_{11} \\ \vdots \\ a_{n1} \end{bmatrix}, \quad v_2 = \begin{bmatrix} a_{12} \\ \vdots \\ x_{n2} \end{bmatrix}, \quad \ldots, \quad v_n = \begin{bmatrix} a_{1n} \\ \vdots \\ a_{nn} \end{bmatrix}$$

are linearly independent, using the column operations of Section 5.3, show that $\det(A) \neq 0$.

5.5. PROPERTIES OF DETERMINANTS: OTHER EXPANSIONS

After the short respite from row and column operations, which was afforded us in Section 5.4, we return to a further examination of such operations.

In our initial definition of the determinant of A, the determinant was defined in terms of the expansion by minors of the first column. In this way we favored the first column over all the others. Is this really necessary? Can we define the determinant in terms of the expansion by the minors of any column? The answer to this is "yes," which we shall soon demonstrate.

To get away from the annoying need to alternate the signs of the minors, we introduce a new notion, closely allied to that of a minor.

Definition. If A is an $n \times n$ matrix, then the (r, s) *cofactor* A_{rs} of $\det(A)$ is defined by $A_{rs} = (-1)^{r+s} M_{rs}$, where M_{rs} is the (r, s) minor of $\det(A)$. (We used A_{rs} in another context in Section 5.1. It is not the same as the A_{rs} we now use to denote a cofactor.)

So, for example,

$$A_{57} = (-1)^{5+7} M_{57} = M_{57},$$

while

$$A_{34} = (-1)^{3+4} M_{34} = -M_{34}.$$

Sec. 5.5]   Properties of Determinants: Other Expansions   247

Note that in terms of cofactors we have that

$$\det(A) = \sum_{r=1}^{n} a_{r1} A_{r1}.$$

As we so often have done, let us look at a particular matrix, say,

$$A = \begin{bmatrix} 1 & 2 & 3 \\ 4 & 5 & 6 \\ 7 & 8 & 9 \end{bmatrix}.$$

What are the cofactors of the entries of the second column of A? They are

$$A_{12} = (-1)^3 M_{12} = -\begin{vmatrix} 4 & 6 \\ 7 & 9 \end{vmatrix} = 6$$

$$A_{22} = (-1)^4 M_{22} = +\begin{vmatrix} 1 & 3 \\ 7 & 9 \end{vmatrix} = -12$$

$$A_{32} = (-1)^5 M_{32} = -\begin{vmatrix} 1 & 3 \\ 4 & 6 \end{vmatrix} = 6.$$

So

$$a_{12}A_{12} + a_{22}A_{22} + a_{32}A_{32} = 2(6) + 5(-12) + 8(6) = 0.$$

If we expand $\det(A)$ by the first column, we get that $\det(A) = 0$. So what we might call the "expansion by the second column of A,"

$$\sum_{r=1}^{3} a_{r2} A_{r2},$$

in this particular instance turns out to be the same as $\det(A)$. Do you think this is happenstance?

The next theorem shows that the expansion by *any* column of A—using cofactors in this expansion—gives us $\det(A)$. In other words, *the first column is no longer favored over its colleagues.*

Theorem 5.5.1. For any s, $\det(A) = \sum_{r=1}^{n} a_{rs} A_{rs}$.

Proof: Let $A = (a_{rs})$ and let B be the matrix

$$B = \begin{bmatrix} a_{1s} & a_{11} & a_{12} & & a_{1n} \\ a_{2s} & a_{21} & a_{22} & & a_{2n} \\ \vdots & \vdots & \vdots & \cdots & \vdots \\ a_{ns} & a_{n1} & a_{n2} & & a_{nn} \end{bmatrix}$$

obtained from A by moving column s to the first column. Notice that this is accomplished by $s-1$ interchanges of adjacent columns, that is, by moving column across column $s-1$, then across column $s-2$, and so on. Since each such interchange changes the sign of the determinant (Theorem 5.3.5), we have that

$$\det(B) = (-1)^{s-1}\det(A).$$

But what is $\det(B)$? Using the expansion by the first column of B, we have that

$$(-1)^{s-1}\det(A) = \det(B)$$
$$= a_{1s}N_{11} - a_{2s}N_{21} + a_{31}N_{31} + \cdots + (-1)^{n+1}a_{ns}N_{n1},$$

where N_{r1} is the $(r,1)$ minor of $\det(B)$. However, we claim that $N_{r1} = M_{rs}$, the (r,s) minor of $\det(A)$. Why? To construct N_{r1} we eliminate the first column of B and its row r. What is left is precisely what we get by eliminating column s and row r of A. So $N_{r1} = M_{rs}$. Thus

$$(-1)^{s-1}\det(A) = \det(B) = a_{1s}M_{1s} - a_{2s}M_{2s} + a_{31}M_{3s} + \cdots + (-1)^{n+1}a_{ns}M_{ns}.$$

Because $A_{1s} = (-1)^{1+s}M_{1s}$, $A_{2s} = (-1)^{2+s}M_{2s}$ and so on, we can translate our expression for $(-1)^{s-1}\det(A) = \det(B)$ into

$$(-1)^{1+s}a_{1s}A_{1s} - (-1)^{2+s}a_{2s}A_{2s} + (-1)^{3+s}a_{31}A_{3s} + \cdots + (-1)^{n+s}(-1)^{n+1}a_{ns}A_{ns}.$$

To get $\det(A)$, we multiply through by $(-1)^{s-1}$, ending up with

$$\det(A) = (-1)^{2s}a_{1s}A_{1s} - (-1)^{1+2s}a_{2s}A_{2s} + (-1)^{2+2s}a_{31}A_{3s} + \cdots + (-1)^{2s+2n}a_{ns}A_{ns}$$
$$= (-1)^{2s}a_{1s}A_{1s} + (-1)^{2+2s}a_{2s}A_{2s} + (-1)^{2+2s}a_{31}A_{3s} + \cdots + (-1)^{2s+2n}a_{ns}A_{ns}$$
$$= a_{1s}A_{1s} + a_{2s}A_{2s} + a_{31}A_{3s} + \cdots + a_{ns}A_{ns},$$

since the powers of -1 in the expression for $\det(A)$ above are *even*. This completes the proof. ∎

Theorem 5.5.1 allows us to say something about the expansion of a determinant of A by the minors or cofactors of the first *row* of A. Consider the particular matrix
$A = \begin{bmatrix} 0 & a & 0 \\ b & c & d \\ e & f & g \end{bmatrix}$; by Theorem 5.5.1 we can expand A by the cofactors of the second column. Thus

$$\det(A) = a_{12}A_{12} + a_{22}A_{22} + a_{32}A_{32}$$
$$= (-1)^3 a_{12}M_{12} + (-1)^4 a_{22}M_{22} + (-1)^5 a_{32}M_{32}$$
$$= -a\begin{vmatrix} b & d \\ e & g \end{vmatrix} + c\begin{vmatrix} 0 & 0 \\ e & g \end{vmatrix} - f\begin{vmatrix} 0 & 0 \\ b & d \end{vmatrix} = -a\begin{vmatrix} b & d \\ e & g \end{vmatrix}$$
$$= -aM_{12},$$

since in the last two minors we have a row of zeros.

Sec. 5.5]   **Properties of Determinants: Other Expansions**   249

The argument just given is the model for that needed in the proof of

Theorem 5.5.2.

$$\begin{vmatrix} 0 & 0 & & 0 & a_{1s} & 0 & & 0 \\ a_{21} & a_{22} & & a_{2s-1} & a_{2s} & a_{2s+1} & & a_{2n} \\ \vdots & \vdots & \cdots & \vdots & \vdots & \vdots & \cdots & \vdots \\ a_{n1} & a_{n2} & & a_{ns-1} & a_{ns} & a_{ns-1} & & a_{nn} \end{vmatrix} = a_{1s}A_{1s}.$$

Proof: By Theorem 5.5.1 we may expand the determinant by cofactors of column s. We thus get

$$\begin{vmatrix} 0 & 0 & & 0 & a_{1s} & 0 & & 0 \\ a_{21} & a_{22} & & a_{2s-1} & a_{2s} & a_{2s+1} & & a_{2n} \\ \vdots & \vdots & \cdots & \vdots & \vdots & \vdots & \cdots & \vdots \\ a_{n1} & a_{n2} & & a_{ns-1} & a_{ns} & a_{ns-1} & & a_{nn} \end{vmatrix} = a_{1s}A_{1s} + a_{2s}A_{2s} + \cdots + a_{ns}A_{ns}.$$

Notice, however, that in each of the cofactors $A_{2s}, \ldots, A_{ns}$ we have a row of zeros. Hence $A_{2s} = \cdots = A_{ns} = 0$. This leaves us with

$$\begin{vmatrix} 0 & 0 & & 0 & a_{1s} & 0 & & 0 \\ a_{21} & a_{22} & & a_{2s-1} & a_{2s} & a_{2s+1} & & a_{2n} \\ \vdots & \vdots & \cdots & \vdots & \vdots & \vdots & \cdots & \vdots \\ a_{n1} & a_{n2} & & a_{ns-1} & a_{ns} & a_{ns-1} & & a_{nn} \end{vmatrix} = a_{1s}A_{1s}. \quad \blacksquare$$

Given the matrix $A = (a_{rs})$, then

$$\det(A) = \begin{vmatrix} a_{11} & a_{12} & & a_{1n} \\ a_{21} & a_{22} & & a_{2n} \\ \vdots & \vdots & \cdots & \vdots \\ a_{n1} & a_{n2} & & a_{nn} \end{vmatrix}$$

can be expressed as

$$\begin{vmatrix} a_{11} & 0 & & 0 \\ a_{21} & a_{22} & & a_{2n} \\ \vdots & \vdots & \cdots & \vdots \\ a_{n1} & a_{n2} & & a_{nn} \end{vmatrix} + \begin{vmatrix} 0 & a_{12} & 0 & & 0 \\ a_{21} & a_{22} & & & a_{2n} \\ \vdots & \vdots & & \cdots & \vdots \\ a_{n1} & a_{n2} & & & a_{nn} \end{vmatrix} + \cdots + \begin{vmatrix} 0 & 0 & & 0 & a_{1n} \\ a_{21} & a_{22} & & & a_{2n} \\ \vdots & \vdots & & \cdots & \vdots \\ a_{n1} & a_{n2} & & & a_{nn} \end{vmatrix}$$

by Theorem 5.3.4. But by Theorem 5.5.2, we have

$$\begin{vmatrix} 0 & 0 & & 0 & a_{1s} & 0 & & 0 \\ a_{21} & a_{22} & & & a_{2s} & & & a_{2n} \\ \vdots & \vdots & \cdots & & \vdots & & \cdots & \vdots \\ a_{n1} & a_{n2} & & & a_{ns} & & & a_{nn} \end{vmatrix} = a_{1s}A_{1s}$$

for all $s = 1, \ldots, n$. Thus

$$\det(A) = a_{11}A_{11} + a_{12}A_{12} + \cdots + a_{1s}A_{1s} + \cdots + a_{1n}A_{1n}.$$

In other words, $\det(A)$ can be evaluated by expanding it by the cofactors of the first row.

In summation notation, this result is $\det(A) = \sum_{s=1}^{n} a_{1s}A_{1s}$. Using the definition $A_{1s} = (-1)^{1+s}M_{1s}$, it translates to *the expansion*

$$\det(A) = \sum_{s=1}^{n} (-1)^{1+s} a_{1s} M_{1s}$$

of $\det(A)$ by the minors of the first row.

We record this very important result as

Theorem 5.5.3. $\det(A) = \sum_{s=1}^{n} (-1)^{1+s} a_{1s} M_{1s} = \sum_{s=1}^{n} a_{1s}A_{1s}$; that is, we can evaluate $\det(A)$ by an expansion by the minors or cofactors of the *first row* of A.

An example is now in order. Let $A = \begin{bmatrix} 1 & 2 & 3 \\ -1 & 1 & -1 \\ 2 & 4 & 1 \end{bmatrix}$. The minors of the first row are

$$\begin{vmatrix} 1 & -1 \\ 4 & 1 \end{vmatrix} = 5, \quad \begin{vmatrix} -1 & -1 \\ 2 & 1 \end{vmatrix} = 1, \quad \begin{vmatrix} -1 & 1 \\ 2 & 4 \end{vmatrix} = -6,$$

so that

$$\begin{vmatrix} 1 & 2 & 3 \\ -1 & 1 & -1 \\ 2 & 4 & 1 \end{vmatrix} = 1 \cdot 5 - 2 \cdot 1 + 3(-6) = -15.$$

Evaluate $\det(A)$ by the minors of the first column and you will get the same answer, namely, -15.

Theorem 5.5.3 itself has a very important consequence. Let A be a matrix on A' its transpose. Then

Theorem 5.5.4. $\det(A) = \det(A')$.

Proof: Once again we look at the situation of a 3×3 matrix. Let

$$A = \begin{bmatrix} a_{11} & a_{12} & a_{13} \\ a_{21} & a_{22} & a_{23} \\ a_{31} & a_{32} & a_{33} \end{bmatrix}.$$

Sec. 5.5]   Properties of Determinants: Other Expansions   251

Then
$$A' = \begin{bmatrix} a_{11} & a_{21} & a_{31} \\ a_{12} & a_{22} & a_{32} \\ a_{13} & a_{23} & a_{33} \end{bmatrix}.$$

Therefore, expanding $\det(A')$ by the minors of its first row, we have

$$\det(A') = a_{11}\begin{vmatrix} a_{22} & a_{32} \\ a_{23} & a_{33} \end{vmatrix} - a_{21}\begin{vmatrix} a_{12} & a_{32} \\ a_{13} & a_{33} \end{vmatrix} + a_{31}\begin{vmatrix} a_{12} & a_{22} \\ a_{13} & a_{23} \end{vmatrix}.$$

Expanding $\det(A)$ by the minors of its first column, we get

$$\det(A) = a_{11}\begin{vmatrix} a_{22} & a_{23} \\ a_{32} & a_{33} \end{vmatrix} - a_{21}\begin{vmatrix} a_{12} & a_{13} \\ a_{32} & a_{33} \end{vmatrix} + a_{31}\begin{vmatrix} a_{12} & a_{13} \\ a_{22} & a_{23} \end{vmatrix}.$$

Notice that each (r, s) minor in the expansion of $\det(A')$ is the 2×2 determinant of the transpose of the (s, r) minor submatrix of A (see Section 5.1). By the result for 2×2 matrices we then know that these minors are equal. This gives us that $\det(A') = \det(A)$.

Now to the general case, assuming the result known for $(n - 1) \times (n - 1)$ matrices. The first row of A' is the first column of A, from the very definition of transpose. If U_{rs} is the (r, s) minor submatrix of A and V_{rs} the (r, s) minor submatrix of A', then $V_{rs} = U'_{sr}$. (Prove!) By induction, N_{rs}, the (r, s) minor of A', is given by

$$N_{rs} = \det(V_{rs}) = \det(U'_{sr}) = \det(U_{sr}) = M_{sr},$$

where M_{sr} is the (s, r) minor of A.

If b_{1s} is the $(1, s)$ entry of A', then $b_{1s} = a_{s1}$. Also, by Theorem 5.5.3,

$$\det(A') = \sum_{s=1}^{n}(-1)^{s+1}b_{1s}N_{1s} = \sum_{s=1}^{n}(-1)^{s+1}a_{s1}M_{s1} = \det(A).$$

This proves the theorem. ∎

For the columns we showed in Theorem 5.5.1 that we can calculate $\det(A)$ by the cofactors of any column of A. We would hope that a similar result holds for the rows of A. In fact, this is true; it is

Theorem 5.5.5. For any $r \geq 1$, $\det(A) = \sum_{s=1}^{n} a_{rs}A_{rs}$, that is, we can expand $\det(A)$ by the cofactors of any row of A.

Proof: We leave the proof to the reader, but we do provide a few hints of how to go about it. By Theorem 5.5.4, $|A| = |A'|$ and, by Theorem 5.5.1, $|A'|$ can be expanded by the cofactors of its rth column. Since the (s, r) cofactor of A' equals the (r, s) cofactor of A (Prove!), it follows that the expansion of $|A'|$ by the cofactors of its rth column equals the expansion of $|A|$ by the cofactors of its rth row. ∎

PROBLEMS

NUMERICAL PROBLEMS

1. Verify that the expansion of the determinant of

$$A = \begin{bmatrix} 1 & 2 & 3 \\ 7 & -1 & 0 \\ 6 & 1 & 2 \end{bmatrix}$$

by the cofactors of the second column equals $\det(A)$.

2. Verify by a direct computation that $\begin{vmatrix} 1 & 0 & -5 \\ 6 & 2 & -1 \\ 5 & -11 & 6 \end{vmatrix}$, we evaluated by expansion by the minors of the first row and as evaluated by expansion by the minors of the first column are equal.

3. Evaluate by the expansion of the minors of the first row:

(a) $\begin{vmatrix} 1 & 2 & 3 \\ -3 & 2 & 1 \\ 3 & -1 & 2 \end{vmatrix} + \begin{vmatrix} 2 & 1 & 3 \\ -3 & 2 & 1 \\ 2 & -1 & 2 \end{vmatrix}$.

(b) $\begin{vmatrix} 1 & 2 & 3 \\ -3 & 2 & 1 \\ 3 & 0 & 4 \end{vmatrix} + \begin{vmatrix} 2 & 1 & 3 \\ -3 & 2 & 1 \\ 6 & -5 & 7 \end{vmatrix}$.

(c) $\begin{vmatrix} 1 & 1 & 1 & 1 \\ 2 & 2 & 2 & 2 \\ 3 & 3 & 3 & 3 \\ 4 & 4 & 4 & 4 \end{vmatrix}$.

(d) $\begin{vmatrix} 0 & 1 & 2 & 3 \\ 4 & 5 & 6 & 7 \\ 0 & 8 & 9 & 10 \\ 0 & 1 & 1 & 1 \end{vmatrix}$.

(e) Compare the result of Part (d) with the result obtained using the expansion by minors of the first column.

MORE THEORETICAL PROBLEMS

Easier Problems

4. Show that if A has two proportional columns, then $\det(A) = 0$.

5. For what value of a is $\begin{vmatrix} 1 & 5 & 9 & 0 \\ 2 & 6 & 10 & 0 \\ 3 & 7 & 0 & a \\ 4 & 8 & 0 & a \end{vmatrix} = 0$?

Middle-Level Problems

6. Completely prove Theorem 5.5.5.

5.6. THE CLASSICAL ADJOINT (OPTIONAL)

We have seen that the determinant of an $n \times n$ matrix A can be expanded by the cofactors of any row or column. In this section we use this to derive an interesting formula for the inverse of an invertible matrix based on

Definition. For any $n \times n$ matrix A, the *classical adjoint* of A is the $n \times n$ matrix $A^@$ whose (r, s)th entry is the (s, r)th cofactor A_{sr} for $1 \leq r, s \leq n$.

Let's write out the expansions of $|A|$ by the cofactors of the rows and columns of A:

$$|A| = \sum_{s=1}^{n} a_{rs} A_{rs} \quad \text{for all } r$$

$$|A| = \sum_{r=1}^{n} a_{rs} A_{rs} \quad \text{for all } s.$$

The first set of equations says that the (r, r) entry of the product $AA^@$ of A and the classical adjoint $A^@$ is the value $|A|$ for all r; that is, *the diagonal entries of $AA^@$ are all equal to the determinant of A*. The second says that the (s, s) entry of the product $A^@A$ of the classical adjoint $A^@$ and A is the value $|A|$ for all s; that is, *the diagonal entries of $A^@A$ are all equal to the determinant of A*. Using this we can now prove the following theorem.

Theorem 5.6.1. The product of an $n \times n$ matrix A and its classical adjoint in either order is its determinant times the identity matrix; that is, $AA^@ = A^@A = (\det A)I$.

Proof: It suffices to show that the off-diagonal entries of $AA^@$ and $A^@A$ are all 0, since we already know from row and column expansions of the determinant of A that their diagonal entries are all equal to $\det A$. Let $r \neq r'$ and note that the (r, r') entry of $AA^@$ is $\sum_{s=1}^{n} a_{rs} A_{r's}$. This is the determinant of the matrix that we would get from A by replacing the row r' of A by a second copy of the row r of A, as in the following case for $n = 3, r = 2, r' = 3$:

$$A = \begin{bmatrix} a_{11} & a_{12} & a_{13} \\ a_{21} & a_{22} & a_{23} \\ a_{31} & a_{32} & a_{33} \end{bmatrix}, \quad \sum_{s=1}^{n} a_{rs} A_{r's} = \begin{vmatrix} a_{11} & a_{12} & a_{13} \\ a_{21} & a_{22} & a_{23} \\ a_{21} & a_{22} & a_{23} \end{vmatrix}.$$

Since rows r and r' of the resulting matrix are equal, this determinant is 0. It follows that $\sum_{s=1}^{n} a_{rs} A_{r's} = 0$. Since this is true for all r, r' with $r \neq r'$, the off-diagonal entries of $AA^@$ are all 0. The off-diagonal entries of $A^@A$ also are all 0, by a corresponding argument using column properties rather than row properties. ■

Of course, if det A is nonzero, then Theorem 5.6.1 implies that A is invertible, as we now state in

Corollary 5.6.2. If the determinant of an $n \times n$ matrix A is nonzero, then A is invertible and its inverse is $A^{-1} = \dfrac{A^@}{|A|}$.

Proof: Since $AA^@ = A^@A = |A|I$, it must also be true that $A\dfrac{A^@}{|A|} = \dfrac{A^@}{|A|}A = I$. It follows that A is invertible and $A^{-1} = \dfrac{A^@}{|A|}$. ∎

EXAMPLE

The classical adjoint $A^@$ of the matrix $A = \begin{bmatrix} 1 & 6 & 7 \\ 4 & 3 & 4 \\ 2 & 3 & 9 \end{bmatrix}$ is the transpose of the array of values

$$A_{11} = \begin{vmatrix} 3 & 5 \\ 3 & 9 \end{vmatrix} = 12; \quad A_{12} = -\begin{vmatrix} 4 & 5 \\ 2 & 9 \end{vmatrix} = -26; \quad A_{13} = \begin{vmatrix} 4 & 3 \\ 2 & 3 \end{vmatrix} = 6;$$

$$A_{21} = -\begin{vmatrix} 6 & 7 \\ 3 & 9 \end{vmatrix} = -33; \quad A_{22} = \begin{vmatrix} 1 & 7 \\ 2 & 9 \end{vmatrix} = -5; \quad A_{23} = -\begin{vmatrix} 1 & 6 \\ 2 & 3 \end{vmatrix} = 9;$$

$$A_{31} = \begin{vmatrix} 6 & 7 \\ 3 & 5 \end{vmatrix} = 9; \quad A_{32} = -\begin{vmatrix} 1 & 7 \\ 4 & 5 \end{vmatrix} = 23; \quad A_{33} = \begin{vmatrix} 1 & 6 \\ 4 & 3 \end{vmatrix} = -21;$$

So $A^@ = \begin{bmatrix} 12 & -33 & 9 \\ -26 & -5 & 23 \\ 6 & 9 & -21 \end{bmatrix}$. If we compute $AA^@$, we get

$$\begin{bmatrix} -102 & 0 & 0 \\ 0 & -102 & 0 \\ 0 & 0 & -102 \end{bmatrix} = -102 \begin{bmatrix} 1 & 0 & 0 \\ 0 & 1 & 0 \\ 0 & 0 & 1 \end{bmatrix}.$$

In view of Theorem 5.6.1, this implies that $|A| = -102$. It also implies that

$$A^{-1} = -\frac{1}{102}\begin{bmatrix} 12 & -33 & 9 \\ -26 & -5 & 23 \\ 6 & 9 & -21 \end{bmatrix} = \frac{1}{|A|}\begin{bmatrix} 12 & -33 & 9 \\ -26 & -5 & 23 \\ 6 & 9 & -21 \end{bmatrix} = \frac{A^@}{|A|},$$

which is what we proved in Corollary 5.6.2.

Corollary 5.6.2 provides a way to solve an equation $Ax = y$ for $x = \begin{bmatrix} x_1 \\ \vdots \\ x_n \end{bmatrix}$ when A is an $n \times n$ matrix with nonzero determinant and y is a given column

Sec. 5.6] The Classical Adjoint

vector $\begin{bmatrix} y_1 \\ \vdots \\ y_n \end{bmatrix}$. Since A is invertible, by Corollary 5.6.2, letting x be $x = A^{-1}y$, we get $Ax = AA^{-1}y = y$, as desired. Nothing could be simpler: There is a solution x to $Ax = y$, and the only solution is $\begin{bmatrix} x_1 \\ \vdots \\ x_n \end{bmatrix} = A^{-1}y$; and, since $A^{-1} = \dfrac{A^@}{|A|}$, this solution is

$$\begin{bmatrix} x_1 \\ \vdots \\ x_n \end{bmatrix} = \frac{A^@}{|A|} \begin{bmatrix} y_1 \\ \vdots \\ y_n \end{bmatrix} = \frac{1}{|A|} A^@ \begin{bmatrix} y_1 \\ \vdots \\ y_n \end{bmatrix}.$$

So to get the individual entries x_s of x from this expression, we simply let

$$x_s = \frac{1}{|A|}(A_{1s}y_1 + \cdots + A_{ns}y_n),$$

since the (s,r)th entry of the matrix $A^@$ is the cofactor A_{rs} of A. But magically,

$$A_{1s}y_1 + \cdots + A_{ns}y_n = y_1 A_{1s} + \cdots + y_n A_{ns}$$

is the expansion of the determinant of A by column s of the matrix A_s obtained from A by putting $\begin{bmatrix} y_1 \\ \vdots \\ y_n \end{bmatrix}$ in place of its column s. This enables us to write the expression

$$x_s = \frac{1}{|A|}(A_{1s}y_1 + \cdots + A_{ns}y_n)$$

as

$$x_s = \frac{1}{|A|}|A^y| = \frac{|A_s|}{|A|}.$$

You may recognize this formula for the entries x_s as Cramer's rule, which we derived in Section 5.4.

PROBLEMS

NUMERICAL PROBLEMS

1. Find the solution $\begin{bmatrix} x_1 \\ x_2 \\ x_3 \\ x_4 \end{bmatrix}$ to the equation $\begin{bmatrix} 1 & 2 & 3 & 4 \\ 1 & 2 & 3 & 3 \\ 1 & 2 & 2 & 2 \\ 1 & 1 & 1 & 1 \end{bmatrix} \begin{bmatrix} x_1 \\ x_2 \\ x_3 \\ x_4 \end{bmatrix} = \begin{bmatrix} 1 \\ 2 \\ 3 \\ 4 \end{bmatrix}$ by computing the determinant, adjoint, and inverse of the coefficient matrix.

2. Do Problem 1 by, instead, using Cramer's rule.

MORE THEORETICAL PROBLEMS

Easier Problems

3. Show that $A'^{@} = (A^{@})'$; that is, the classical adjoint of the transpose of A equals the transpose of the classical adjoint of A.
4. Show that $|AA'^{@}| = |A|^n$ for any $n \times n$ matrix A.

5.7. ELEMENTARY MATRICES

We have seen in the preceding sections that there are certain operations which we can carry out on the row, or on the columns, of a matrix A which result in no change, or in a change of a very easy and specific sort, in $\det(A)$. Let's recall which they were. We list them for the rows, but remember, *the same thing holds for the columns.*

1. If we add the multiple of one row of A to another row of A, we do not change the determinant.
2. If we interchange two rows of A, then the determinant changes sign.
3. If we multiply every entry in a given row of A by a constant q, then the determinant is merely multiplied by q.

Since we are dealing with matrices, it is natural for us to ask whether these three operations can be achieved by matrix multiplication of A by three specific types of matrices. The answer is indeed "yes"! Before going to the general situation note, for the 3×3 matrices that:

1. $\begin{bmatrix} 1 & 0 & 0 \\ 0 & 1 & q \\ 0 & 0 & 1 \end{bmatrix} \begin{bmatrix} a_{11} & a_{12} & a_{13} \\ a_{21} & a_{22} & a_{23} \\ a_{31} & a_{32} & a_{33} \end{bmatrix} = \begin{bmatrix} a_{11} & a_{12} & a_{13} \\ a_{21} + qa_{31} & a_{22} + qa_{32} & a_{23} + qa_{33} \\ a_{31} & a_{32} & a_{33} \end{bmatrix}$, so this

multiplication adds q times the third row to the second.

2. $\begin{bmatrix} 1 & 0 & 0 \\ 0 & 0 & 1 \\ 0 & 1 & 0 \end{bmatrix} \begin{bmatrix} a_{11} & a_{12} & a_{13} \\ a_{21} & a_{22} & a_{23} \\ a_{31} & a_{32} & a_{33} \end{bmatrix} = \begin{bmatrix} a_{11} & a_{12} & a_{13} \\ a_{31} & a_{32} & a_{33} \\ a_{21} & a_{22} & a_{23} \end{bmatrix}$, so this multiplication inter-

changes rows 2 and 3.

3. $\begin{bmatrix} 1 & 0 & 0 \\ 0 & q & 0 \\ 0 & 0 & 1 \end{bmatrix} \begin{bmatrix} a_{11} & a_{12} & a_{13} \\ a_{21} & a_{22} & a_{23} \\ a_{31} & a_{32} & a_{33} \end{bmatrix} = \begin{bmatrix} a_{11} & a_{12} & a_{13} \\ qa_{21} & qa_{22} & qa_{23} \\ a_{31} & a_{32} & a_{33} \end{bmatrix}$, so this multiplication re-

sults in multiplying the second row by q.

As you can readily verify, *multiplying A on the right by*

$$\begin{bmatrix} 1 & 0 & 0 \\ 0 & 1 & q \\ 0 & 0 & 1 \end{bmatrix}, \begin{bmatrix} 1 & 0 & 0 \\ 0 & 0 & 1 \\ 0 & 1 & 0 \end{bmatrix}, \begin{bmatrix} 1 & 0 & 0 \\ 0 & q & 0 \\ 0 & 0 & 1 \end{bmatrix}$$

gives us the same story for the columns, except that the role of the indices is reversed namely

adds q times the *second* column to the *third*;
interchanges columns 3 and 2; and
multiplies column 2 by q, respectively.

So in the particular instance for 3×3 matrices, multiplying A on the *left* by the three matrices above achieves the three operations (1), (2), (3) above, and doing it on the *right* achieves the corresponding thing for the columns.

Fortunately, the story is equally simple for the $n \times n$ matrices. But first recall that the matrices E_{rs} introduced earlier in the book were defined by: E_{rs} is the matrix whose (r, s) entry is 1 and all of whose other entries are 0. The basic facts about these matrices were:

1. If $A = (a_{rs})$, then $A = \sum_{r=1}^{n} \sum_{s=1}^{n} a_{rs} E_{rs}$.
2. If $s \neq t$, then $E_{rs} E_{tu} = 0$.
3. $E_{rs} E_{st} = E_{rt}$.

Note that in the example above,

$$\begin{bmatrix} 1 & 0 & 0 \\ 0 & 1 & q \\ 0 & 0 & 1 \end{bmatrix} = I + qE_{23}, \quad \begin{bmatrix} 1 & 0 & 0 \\ 0 & q & 0 \\ 0 & 0 & 1 \end{bmatrix} = I + (q-1)E_{22},$$

and $\begin{bmatrix} 1 & 0 & 0 \\ 0 & 0 & 1 \\ 0 & 1 & 0 \end{bmatrix}$, which is obtained by interchanging the second and third columns of the matrix I, can be written in the (awkward) form $I + E_{23} + E_{32} - E_{22} - E_{23}$.

We are now ready to pass to the general context.

Definition. We define three types of matrices:

1. $A(r, s; q) = I + qE_{rs}$ for $r \neq s$.
2. $M(r; q) = I + (q-1)E_{rs}$ for $q \neq 0$.
3. $I(r, s)$, the matrix obtained from the unit matrix I by interchanging columns r and s of I. [We then know that $I(r, s) = I + E_{rs} + E_{sr} - E_{rr} - E_{ss}$ (Prove!), but we will not use this clumsy form of representing $I(r, s)$.]

We call these three types of matrices the *elementary matrices* and shall often represent them by the letter E or by E with *one* subscript.

What are the basic properties of these elementary matrices? We leave the proof of the next theorem to the reader.

Theorem 5.7.1. *For any matrix B,*

1. $A(r, s; q)B$, *for $r \neq s$, is that matrix obtained for B by adding q times row s to row r.*

1'. $BA(r, s; q)$, for $r \neq s$, is that matrix obtained from B by adding q times column r to column s.
2. $M(r; q)B$ is that matrix obtained from B by multiplying each entry in row r by q.
2'. $BM(r; q)$ is that matrix obtained from B by multiplying each entry in column r by q.
3. $I(r, s)B$ is that matrix obtained from B by interchanging rows r and s of B.
3'. $BI(r, s)$ is that matrix obtained from B by interchanging columns s and r of B.

Using the fact that $A(r, s; q)$ for $r \neq s$, is triangular, and that $M(r; q)$ is diagonal, with 1's on the diagonal except for the (r, r) entry which is q, we have:

$$\det(A(r, s; q)) = 1 \text{ for } r \neq s \text{ and } \det(M(r; q)) = q.$$

Finally, by Theorem 5.2.5, since we obtain $I(r, s)$, for $r \neq s$, by interchanging two columns of I,

$$\det(I(r, s)) = -\det(I) = -1.$$

Consider the effect on the determinant of multiplying a given matrix B by an elementary matrix E.

1. If $E = A(r, s; q)$, then $A(r, s; q)B$ is obtained from B by adding q times row s to row r. By Theorem 5.2.8 we have that

$$\det(A(r, s; q)B) = \det(B) = \det(A(r, s; q))\det(B).$$

And similarly, $BA(r, s; q)$ is obtained from B by adding q times column r to column s. By Theorem 5.3.6 we get that

$$\det(BA(r, s; q)) = \det(B) = \det(B)\det(A(r, s; q)).$$

2. If $E = M(r; q)$, then by Theorems 5.3.2 and 5.2.2 we know that

$$\det(BM(r; q)) = q\det(B) = \det(B)\det(M(r; q))$$
$$\det(M(r; q)B) = q\det(B) = \det(M(r; q))\det(B).$$

3. Finally, if $E = I(r, s)$, by Theorems 5.2.5 and 5.3.5 we know that

$$\det(I(r, s)B) = -\det(B) = \det(I(r, s))\det(B),$$
$$\det(BI(r, s)) = -\det(B) = \det(B)\det(I(r, s)).$$

We summarize this longish paragraph in

Theorem 5.7.2. For any matrix B and any elementary matrix E,

$$\det(EB) = \det(E)\det(B)$$

and

$$\det(BE) = \det(B)\det(E).$$

By iteration we can extend Theorem 5.7.2 tremendously. Take, for example, two elementary matrices E_1 and E_2, and consider both $E_1 E_2 B$ and $E_1 B E_2$. What are the determinants of these matrices? Now $E_1 E_2 B = E_1(E_2 B)$, so by Theorem 5.7.2,

$$\det(E_1 E_2 B) = \det(E_1)\det(E_2 B) = \det(E_1)\det(E_2)\det(B).$$

Similarly,

$$\det(E_1 B E_2) = \det(E_1)\det(B)\det(E_2).$$

We can continue this game and prove

Theorem 5.7.3. If $E_1, E_2, \ldots, E_m, E_{m+1}, \ldots, E_k$ are elementary matrices, then

1. $\det(E_1 E_2 \cdots E_m E_{m+1} \cdots E_k B)$
   $= \det(E_1)\det(E_2) \cdots \det(E_m)\det(E_{m+1}) \cdots \det(E_k)\det(B)$; and
2. $\det(E_1 E_2 \cdots E_m B E_{m+1} \cdots E_k)$
   $= \det(E_1)\det(E_2) \cdots \det(E_m)\det(B)\det(E_{m+1}) \cdots \det(E_k)$.

Proof: Either go back to the paragraph before the statement of the theorem and say: "Continuing in this way," or better still (and more formally), prove the result by induction. ∎

A simple but important corollary to Theorem 5.7.3 is

Theorem 5.7.4. If $E_1, \ldots, E_k$ are elementary matrices, then

$$\det(E_1 E_2 \cdots E_k) = \det(E_1)\det(E_2) \cdots \det(E_k).$$

Proof: In Part (a) of Theorem 5.7.3, put $B = I$. Then

$\det(E_1 E_2 \cdots E_k I)$
$= \det(E_1)\det(E_2) \cdots \det(E_k)\det(I) = \det(E_1)\det(E_2) \cdots \det(E_k).$ ∎

Theorem 5.7.4 will be of paramount importance to us. It is the key to proving the most basic property of the determinant. That property is that

$$\det(AB) = \det(A)\det(B)$$

for all $n \times n$ matrices A and B. This will be shown in the next section.

Theorem 5.7.5. If E is an elementary matrix, then E is invertible and E^{-1} is an elementary matrix.

Proof: We run through the three types of elementary matrices:

1. If $E = A(r,s;q)$, with $r \neq s$, then $E = I + qE_{rs}$. So

$$E(I - qE_{rs}) = (I + qE_{rs})(I - qE_{rs}) = I + qE_{rs} - qE_{rs} + q^2 E_{rs}^2 = I$$

Since $E_{rs}^2 = 0$ (because $r \neq s$). So $E^{-1} = I - qE_{rs} = A(r,s;-q)$.

2. If $E = M(r;q) = \begin{bmatrix} 1 & & & & \\ & \ddots & & 0 & \\ & & q & & \\ & 0 & & \ddots & \\ & & & & 1 \end{bmatrix}$ with $q \neq 0$, then

$$E^{-1} = \begin{bmatrix} 1 & & & & \\ & \ddots & & 0 & \\ & & q^{-1} & & \\ & 0 & & \ddots & \\ & & & & 1 \end{bmatrix} = M(r;q^{-1}).$$

3. If $E = I(r,s)$ with $r \neq s$, then interchanging columns r and s_1 twice brings us back to the original. That is, $E^2 = I$. Hence $E^{-1} = E = I(r,s)$. ∎

PROBLEMS

NUMERICAL PROBLEMS

1. Compute the product of the 3×3 elementary matrices $A(2,3;1)$, $A(2,3;2)$, $A(2,3;3)$, $A(2,3;4)$, $A(2,3;5)$ and show that the order of these factors does not affect your answer.

2. Compute the product of the 3×3 elementary matrices $A(2,3;1)$, $A(2,3;2)$, $A(1,3;3)$ in every possible order.

3. Compute the product $B = I(3,4)A(1,2;5)\begin{bmatrix} 5 & 0 & 0 \\ 0 & 3 & 0 \\ 0 & 0 & 2 \end{bmatrix} A(1,2;-5)I(4,3)$. Then show that the determinant of B is 30, that is, the determinants of $\begin{bmatrix} 5 & 0 & 0 \\ 0 & 3 & 0 \\ 0 & 0 & 2 \end{bmatrix}$ and B are equal.

4. Compute the inverses of the matrices

$$B, I(3,4)A(1,2;5), \begin{bmatrix} 5 & 0 & 0 \\ 0 & 3 & 0 \\ 0 & 0 & 2 \end{bmatrix}, A(1,2;-5), I(2,3)$$

in Problem 3 and show that the inverse of B equals

$$I(3,2)A(1,2;5) \begin{bmatrix} 5 & 0 & 0 \\ 0 & 3 & 0 \\ 0 & 0 & 2 \end{bmatrix}^{-1} A(1,2;-5)I(2,3).$$

5. Compute $A(1,3;5)A(1,3;8)$.
6. Compute $A(1,3;-5)A(1,3;8)A(1,3;5)$.
7. Compute $I(1,3)A(1,2;3)$ and $A(1,2;3)I(1,3)$.
8. Compute $I(1,3)A(1,2;3)I(1,3)$ and $A(1,2;-3)I(1,3)A(1,2;3)$.
9. Compute $I(1,3)M(3;3)I(1,3)$ and $M(3;1/3)I(1,3)M(3;3)$.

MORE THEORETICAL PROBLEMS

Easier Problems

10. Verify that multiplying B by $A(r,s;q)$ on the left adds q times row s to row r.
11. Verify that multiplying B by $A(r,s;q)$ on the right adds q times column r to column s.
12. Give a formula for the powers E^k of the elementary matrices E in the following cases:
    (a) $E = I(a,b)$.
    (b) $E = A(a,b;u)$.
    (c) $E = M(a;u)$.
13. Describe the entries of the matrix

    $$J = A(a,b;u)I(a,b)A(a,b;-u) \text{ (with } a \text{ not equal to } b\text{)}$$

    and verify that $J^2 = I$ (the identity matrix).
14. Show that transposes of elementary matrices are also elementary matrices by describing them explicitly in each case.

Middle-Level Problems

15. Compute $A(a,b;c)A(a,b;v)$ for a not equal to b for any v.
16. Compute $A(a,b;u)^k$ for a not equal to b for any k.
17. Using that $I(r,s) = I + E_{rs} + E_{sr} - E_{rr} - E_{ss}$, for $r \neq s$ prove, using the multiplication rules of the E_{uv}'s, that $I(r,s)^2 = I$.

5.8. THE DETERMINANT OF THE PRODUCT

Several results will appear in this section. One theorem stands out far and above all of the other results. We show that

$$\det(AB) = \det(A)\det(B)$$

for all matrices A and B in $M_n(F)$. With this result in hand we shall be able to do many things. For example, given a matrix A, using determinants we shall construct a polynomial, $P_A(x)$, whose roots are the characteristic roots of A. We'll be able to prove the important Cayley–Hamilton Theorem and a host of other nice theorems.

Our first objective is to show that an invertible matrix is the product of elementary matrices, that is, those introduced in Section 5.7.

Theorem 5.8.1. *If B is an invertible matrix, then B is the product of elementary matrices.*

Proof: The proof will be a little long, but it gives you an actual algorithm for factoring B as a product of elementary matrices.

Since B is invertible, it has no column of zeros, by Theorem 5.3.1. Hence the first column cannot consist only of zeros; in other words, $b_{r1} \neq 0$ for some s. If $r = 1$, that is, $b_{11} \neq 0$, fine. If $b_{11} = 0$, consider the matrix $B^{(1)} = I(r, 1)B$. In it the $(1, 1)$ entry is b_{r1} since we obtain $B^{(1)}$ from B by interchanging row r with the first row. We want to show that $B^{(1)}$ is the product of elementary matrices. If so, then $I(r, 1)B = E_1 \cdots E_k$, where $E_1, \ldots, E_k$ are elementary matrices and $B = I(r, 1)E_1 \cdots E_k$ would be a product of elementary matrices as well. We therefore proceed assuming that $b_{11} \neq 0$ (in other words, we are really carrying out the argument for $B^{(1)}$).

Consider $A\left(s, 1; -\dfrac{b_{s1}}{b_{11}}\right) B$ for $s > 1$. By the property of $A\left(s, 1; -\dfrac{b_{s1}}{b_{11}}\right)$, this matrix $A\left(s, 1; -\dfrac{b_{s1}}{b_{11}}\right) B$ is obtained from B by adding $-\dfrac{b_{s1}}{b_{11}}$ times the first row of B to row s of B. Consequently, the $(s, 1)$ entry of $A\left(s, 1; -\dfrac{b_{s1}}{b_{11}}\right) B$ is $b_{s1} + \left(-\dfrac{b_{s1}}{b_{11}}\right) b_{11} = 0$. So if we do this for $s = 2, 3, \ldots, n$—that is, we act on B by the product of

$$A\left(n, 1; -\dfrac{b_{n1}}{b_{11}}\right), \quad \ldots, \quad A\left(2, 1; -\dfrac{b_{21}}{b_{11}}\right)$$

—we arrive at a matrix all of whose entries in the first column are 0 except for the $(1, 1)$ entry. If we let

$$E_s = A\left(s, 1; -\dfrac{b_{s1}}{b_{11}}\right),$$

this last statement becomes

$$C = E_n E_{n-1} \cdots E_2 B = \begin{bmatrix} b_{11} & b_{12} & b_{13} & \cdots & b_{1n} \\ 0 & c_{22} & c_{23} & \cdots & c_{2n} \\ 0 & c_{32} & c_{33} & \cdots & c_{3n} \\ & \vdots & \vdots & & \vdots \\ 0 & c_{n2} & c_{n3} & \cdots & c_{nn} \end{bmatrix}.$$

Now consider what happens to C when we multiply it on the right by the elementary matrix $F_s = A\left(1, s; -\dfrac{b_{1s}}{b_{11}}\right)$. Since CF_s is obtained from C by adding $-\dfrac{b_{1s}}{b_{11}}$ times the first column of C to column s of C, the $(1, s)$ entry of CF_s, for $s > 1$, is $b_{1s} + \left(-\dfrac{b_{1s}}{b_{11}}\right) b_{11} = 0$. So

$$D = E_n \cdots E_2 B F_2 \cdots F_n = CF_2 \cdots F_n = \begin{bmatrix} b_{11} & 0 & 0 & \cdots & 0 \\ 0 & c_{22} & c_{23} & \cdots & c_{2n} \\ 0 & c_{32} & c_{33} & \cdots & c_{3n} \\ & \vdots & \vdots & & \vdots \\ 0 & c_{n2} & c_{n3} & \cdots & c_{nn} \end{bmatrix}.$$

Now D is invertible, since the E_s, F_r, and B are invertible; in fact,

$$D^{-1} = F_n^{-1} \cdots F_2^{-1} B^{-1} E_2^{-1} \cdots E_n^{-1}.$$

It follows from the invertibility of D that b_{11} is nonzero, the $(n-1) \times (n-1)$ matrix

$$G = \begin{bmatrix} c_{22} & c_{23} & \cdots & c_{2n} \\ c_{32} & c_{33} & \cdots & c_{3n} \\ \vdots & \vdots & & \vdots \\ c_{n2} & c_{n3} & \cdots & c_{nn} \end{bmatrix}$$

is invertible, and the inverse of

$$D = \left[\begin{array}{c|c} b_{11} & 0 \cdots 0 \\ \hline 0 & \\ \vdots & G \\ 0 & \end{array}\right]$$

is the matrix

$$D^{-1} = \left[\begin{array}{c|c} b_{11}^{-1} & 0 \cdots 0 \\ \hline 0 & \\ \vdots & G^{-1} \\ 0 & \end{array}\right].$$

(Prove!) Moreover, we can express D as

$$D = M(1; b_{11}) \begin{bmatrix} 1 & 0 \cdots 0 \\ \hline 0 & \\ \vdots & G \\ 0 & \end{bmatrix}.$$

By induction, G is a product $G = G_1 \cdots G_k$ of elementary $(n-1) \times (n-1)$ matrices $G_1, \ldots, G_k$. For each G_j of these, we define

$$\bar{G}_j = \begin{bmatrix} 1 & 0 \cdots 0 \\ \hline 0 & \\ \vdots & G_j \\ 0 & \end{bmatrix},$$

which is an $n \times n$ elementary matrix. We leave it as an exercise for the reader to show that $G = G_1 \cdots G_k$ implies that

$$\begin{bmatrix} 1 & 0 \cdots 0 \\ \hline 0 & \\ \vdots & G \\ 0 & \end{bmatrix} = \begin{bmatrix} 1 & 0 \cdots 0 \\ \hline 0 & \\ \vdots & G_1 \\ 0 & \end{bmatrix} \cdots \begin{bmatrix} 1 & 0 \cdots 0 \\ \hline 0 & \\ \vdots & G_k \\ 0 & \end{bmatrix}.$$

Multiplying by $M(1; b_{11})$, the left-hand side becomes D and the right-hand side becomes $M(1; b_{11}) \bar{G}_1 \cdots \bar{G}_k$. So this equation becomes

$$D = M(1; b_{11}) \bar{G}_1 \cdots \bar{G}_k.$$

Since $D = E_n \cdots E_2 B F_2 \cdots F_n$, we can now express B as

$$B = E_2^{-1} \cdots E_n^{-1} D F_n^{-1} \cdots F_2^{-1}$$
$$= E_2^{-1} \cdots E_n^{-1} M(1; b_{11}) \bar{G}_1 \cdots \bar{G}_k F_n^{-1} \cdots F_2^{-1},$$

which is a product of elementary matrices. This proves the theorem. ∎

If the proof just given strikes you as long and hard, try it out for 3×3 and 4×4 matrices, following the proof we just gave step by step. You will see that each step is fairly easy. What may confuse you in the general proof was the need for a certain amount of notation.

This last theorem gives us readily

Theorem 5.8.2. *If A and B are invertible $n \times n$ matrices, the $\det(AB) = \det(A)\det(B)$.*

Proof: By Theorem 5.8.1 we can write

$$A = E_1 \cdots E_k \quad \text{and} \quad B = F_1 \cdots F_m,$$

where the E_r and F_s are elementary matrices. So, by Theorem 5.7.4,

$$\det(AB) = \det(E_1 \cdots E_k F_1 \cdots F_m)$$
$$= \det(E_1) \cdots \det(E_k) \det(F_1) \cdots \det(F_m)$$

and again by Theorem 5.7.4,

$$\det(E_1) \cdots \det(E_k) = \det(E_1 \cdots E_k) = \det(A)$$

and

$$\det(F_1) \cdots \det(F_m) = \det(F_1 \cdots F_m) = \det(B).$$

Putting this all together gives us $\det(AB) = \det(A)\det(B)$. ∎

From Theorem 5.8.2 we get the immediate consequence

Theorem 5.8.3. *A is invertible if and only if $\det(A) \neq 0$. Moreover, if A is invertible, then $\det(A^{-1}) = \det(A)^{-1}$.*

Proof: If $\det(A) \neq 0$, we have that A is invertible; this is merely Theorem 5.4.2. On the other hand, if A is invertible, then from $AA^{-1} = I$, using Theorem 5.8.2, we get

$$1 = \det(I) = \det(AA^{-1}) = \det(A)\det(A^{-1}).$$

This give us at the same time that $\det(A)$ is nonzero and that $\det(A^{-1}) = \det(A)^{-1}$. ∎

As a consequence of Theorem 5.8.3 we have that if A is not invertible, then $\det(A) = 0$. But if A is not invertible, then AB is not invertible for any B by Corollary 3.8.2; so $\det(AB) = 0 = \det(A)\det(B)$. Therefore, Theorem 5.8.2 can be sharpened to

Theorem 5.8.4. *If A and B are in $M_n(F)$, then $\det(AB) = \det(A)\det(B)$.*

Proof: If both A and B are invertible, this result is just Theorem 5.8.2. If one of A or B is not invertible, then AB is not invertible, so $\det(AB) = 0 = \det(A)\det(B)$ since one of $\det(A)$ or $\det(B)$ is 0. ∎

An important corollary to these last few theorems is

Theorem 5.8.5. *If C is an invertible matrix, then $\det(C^{-1}AC) = \det(A)$ for all A.*

Proof: By Theorem 5.8.4,

$$\det(C^{-1}AC) = \det((C^{-1}A)C) = \det(C(C^{-1}A))$$
$$= \det(CC^{-1}A) = \det(A). \qquad \blacksquare$$

With this, we close this section. We cannot exaggerate the importance of Theorems 5.8.4 and 5.8.5. As you will see, we shall use them to good effect in what is to come.

PROBLEMS

NUMERICAL PROBLEMS

1. Evaluate $\det\left(\begin{bmatrix} 1 & 0 & 0 \\ 1 & 2 & 0 \\ 1 & 2 & 3 \end{bmatrix}\begin{bmatrix} 1 & 4 & 4 \\ 0 & 2 & 1 \\ 0 & 0 & 3 \end{bmatrix}\right)$ by multiplying the matrices out and computing the determinant.

2. Do Problem 1 by making use of Theorem 5.8.4.

3. Express the following as products of elementary matrices:

   (a) $\begin{bmatrix} 1 & 2 \\ 3 & 4 \end{bmatrix}$.

   (b) $\begin{bmatrix} 0 & 2 & 3 \\ 1 & -1 & 6 \\ 5 & 3 & 1 \end{bmatrix}$.

   (c) $\begin{bmatrix} 1 & 0 & 0 & 0 \\ 2 & 3 & 0 & 0 \\ 4 & 5 & 6 & 0 \\ 7 & 8 & 9 & 10 \end{bmatrix}$. (Follow the steps of the proof of Theorem 5.8.1.)

More Theoretical Problems

4. Prove that for any matrix A, $\det(A^k) = (\det(A))^k$ for all $k \geq 1$.

5. Prove that $\det(ABA^{-1}B^{-1}) = 1$ if A and B are invertible $n \times n$ matrices.

6. If A is a skew-symmetric matrix in $M_n(\mathbb{R})$, prove that if n is odd, then $\det(A) = 0$.

5.9. THE CHARACTERISTIC POLYNOMIAL

Given a matrix A we shall now construct a polynomial $p_A(x)$ whose roots are the characteristic roots of A. We do this immediately in the

Definition. The *characteristic polynomial*, $p_A(x)$, of A is defined by $p_A(x) = \det(xI - A)$.

So, for example, if $A = \begin{bmatrix} 1 & 2 & 3 \\ 4 & 5 & 6 \\ 7 & 8 & 9 \end{bmatrix}$, then

$$p_A(x) = \det(xI - A) = \begin{vmatrix} x-1 & -2 & -3 \\ -4 & x-5 & -6 \\ -7 & -8 & x-9 \end{vmatrix} = x^3 - 15x^2 - 18x. \text{ (Verify!)}$$

The key property that the characteristic polynomial enjoys is

Sec. 5.9] The Characteristic Polynomial 267

Theorem 5.9.1. The complex number a is a characteristic root of A if and only if a is a root of $p_A(x)$.

Proof: Recall that a number a is a characteristic root of A in $M_n(\mathbb{C})$ if and only if $aI - A$ is not invertible or, equivalently, if $Av = av$ for some $v \neq 0$ in $\mathbb{C}^{(n)}$. But by Theorem 5.8.3, $aI - A$ is not invertible if and only if $\det(aI - A) = 0$. Since

$$0 = \det(aI - A) = p_A(a),$$

we have proven the theorem. ■

Given $A, C \in M_n(F)$, and C invertible, then

$$C^{-1}(xI - A)C = C^{-1}(xI)C - C^{-1}AC = xI - C^{-1}AC.$$

Therefore, by Theorem 5.8.5,

$$\det(xI - A) = \det(C^{-1}(xI - A)C) = \det(xI - C^{-1}AC).$$

What this says is precisely

Theorem 5.9.2. If A, C are in $M_n(F)$ and C is invertible, then $p_A(x) = p_{C^{-1}AC}(x)$; that is, A and $C^{-1}AC$ have the same characteristic polynomial. Thus A and $C^{-1}AC$ have the same characteristic roots.

Proof: By definition,

$$p_A(x) = \det(xI - A)$$

and

$$p_{C^{-1}AC}(x) = \det(xI - C^{-1}AC).$$

Since

$$\det(xI - A) = \det(C^{-1}(xI - A)C) = \det(xI - C^{-1}AC),$$

as we saw above, it follows that

$$p_A(x) = p_{C^{-1}AC}(x)$$

and A and $C^{-1}AC$ have the same characteristic polynomials. But then, by Theorem 5.9.1, A and $C^{-1}AC$ also have the same characteristic roots. ■

Going back to the example of $A = \begin{bmatrix} 1 & 2 & 3 \\ 4 & 5 & 6 \\ 7 & 8 & 9 \end{bmatrix}$, for which we saw that $p_{C^{-1}AC}(x) = x^3 - 15x^2 - 18x$, we have that the characteristic roots of A, namely the roots of $p_{C^{-1}AC}(x) = x^3 - 15x^2 - 18x$, are 0 and the roots of $x^2 - 15x - 18$. The roots of this quadratic polynomial are $\dfrac{15 + 3\sqrt{33}}{2}$ and $\dfrac{15 - 3\sqrt{33}}{2}$. So the characteristic roots of A are 0, $\dfrac{15 + 3\sqrt{33}}{2}$, and $\dfrac{15 - 3\sqrt{33}}{2}$.

Note one further property of $p_A(x)$. If A' denotes the transpose of A, then $(xI - A)' = xI - A'$. But we know by Theorem 5.5.4 that the determinant of a matrix equals that of its transpose. This allows us to prove

Theorem 5.9.3. For any $A \in M_n(F)$, $p_A(x) = p_{A'}(x)$, that is, A and A' have the same characteristic polynomial.

Proof: By definition,

$$p_{A'}(x) = \det(xI - A') = \det((xI - A)') = \det(xI - A) = p_A(x). \quad \blacksquare$$

Given A in $M_n(F)$ and $p_A(x)$, we know that the roots of $p_A(x)$ all lie in $\mathbb{C}$ and

$$p_A(x) = (x - r_1) \cdots (x - r_n)$$

with $r_1, \ldots, r_n \in \mathbb{C}$, where the $r_1, \ldots, r_n$ are the roots of $p_A(x)$, and the characteristic roots of A (the roots need not be distinct).

We ask: What is $p_A(0)$? Setting $x = 0$ in the equation above, we get

$$p_A(0) = (0 - r_1) \cdots (0 - r_n) = (-1)^n r_1 \cdots r_n.$$

On the other hand, $p_A(x) = \det(xI - A)$, hence

$$p_A(0) = \det(-A) = \det((-I)A) = \det(-I)\det(A)$$
$$= (-1)^n \det(A).$$

Comparing these two evaluations of $p_A(0)$, we come up with

$$(-1)^n r_1 \cdots r_n = (-1)^n \det(A).$$

Thus we get the

Theorem 5.9.4. If $A \in M_n(F)$, then $\det(A)$ is the product of the characteristic roots of A.

PROBLEMS

NUMERICAL PROBLEMS

1. Compute $p_A(x)$ for $A = 0$ and for $A = I$.
2. Compute $p_A(x)$ for A a lower triangular matrix.

3. Compute $p_A(x)$ for

(a) $A = \begin{bmatrix} 1 & 0 & 1 \\ 2 & 0 & 2 \\ 5 & 6 & -1 \end{bmatrix}$.

(b) $A = \begin{bmatrix} 1 & 0 & 0 \\ 0 & a & b \\ 0 & c & d \end{bmatrix}$.

(c) $A = \begin{bmatrix} 0 & 1 & 0 \\ 1 & 0 & 0 \\ 0 & 0 & 1 \end{bmatrix}$.

(d) $A = \begin{bmatrix} 1 & 2 & 0 & 0 \\ 3 & 4 & 0 & 0 \\ 0 & 0 & 5 & 6 \\ 0 & 0 & 7 & 8 \end{bmatrix}$.

4. Find the characteristic roots for the matrices in Problem 3, and show directly that $\det(A)$ is the product of the roots of the characteristic polynomial of A.

MORE THEORETICAL PROBLEMS

Easier Problems

5. If at least one of A or B is invertible, prove that $p_{AB}(x) = p_{BA}(x)$.

6. Find $p_A(x)$ for $A = \begin{bmatrix} 0 & 0 & 0 & -c_0 \\ 1 & 0 & 0 & -c_1 \\ 0 & 1 & 0 & -c_2 \\ 0 & 0 & 1 & -c_3 \end{bmatrix}$.

7. If $A \in M_n(F)$ is nilpotent ($A^k = 0$ for some positive integer k), show that $p_A(x) = x^n$.

8. If $A = I + N$, where N is nilpotent, find $p_A(x)$.

9. If A^* is the Hermitian adjoint of A, express $p_{A^*}(x)$ in terms of $p_A(x)$.

10. If A is invertible, express $p_{A^{-1}}(x)$ in terms of $p_A(x)$ if $p_A(x) = x^n + a_1 x^{n-1} + \cdots + a_n$.

11. If A and B in $M_n(F)$ are upper triangular, prove that $p_{AB}(x) = p_{BA}(x)$.

Note: It is a fact that for all A, B in $M_n(F)$, $p_{AB}(x) = p_{BA}(x)$. However, if neither A nor B is invertible, it is difficult to prove. So although we alert you to this fact, we do not prove it for you.

12. Show that if an invertible $n \times n$ matrix D is of the form

$$D = \begin{bmatrix} b_{11} & 0 \cdots 0 \\ \hline 0 & \\ \vdots & G \\ 0 & \end{bmatrix},$$

then b_{11} is nonzero, G is invertible, and the inverse of D is

$$D^{-1} = \begin{bmatrix} b_{11}^{-1} & 0 \cdots 0 \\ \hline 0 & \\ \vdots & G^{-1} \\ 0 & \end{bmatrix}.$$

For any elementary $(n-1) \times (n-1)$ matrix G, we define $\bar{G}$ to be the matrix

$$\bar{G} = \begin{bmatrix} 1 & 0 \cdots 0 \\ \hline 0 & \\ \vdots & G \\ 0 & \end{bmatrix}$$

obtained by adorning G with one more row and column as displayed. Show that $\bar{G}$ is an elementary matrix. Then prove the following product rule:

$$\begin{bmatrix} 1 & 0 \cdots 0 \\ \hline 0 & \\ \vdots & G \\ 0 & \end{bmatrix} \begin{bmatrix} 1 & 0 \cdots 0 \\ \hline 0 & \\ \vdots & H \\ 0 & \end{bmatrix} = \begin{bmatrix} 1 & 0 \cdots 0 \\ \hline 0 & \\ \vdots & GH \\ 0 & \end{bmatrix}.$$

5.10. THE CAYLEY–HAMILTON THEOREM

We saw earlier that if A is in $M_n(F)$, then A *satisfies* a polynomial $p(x)$ of degree at most n^2 whose coefficients are in F, that is,

$$p(A) = 0.$$

What we shall show here is that A *actually satisfies a specific polynomial of degree n having coefficients in F*. In fact, this polynomial will turn out to be the characteristic polynomial of A. This is a famous theorem in the subject and is known as the *Cayley–Hamilton Theorem*.

Before proving the Cayley–Hamilton Theorem in its full generality, we prove it for the special case in which A is an upper triangular matrix. From there, using a result on triangularization that we proved earlier (Theorem 4.7.3), we shall be able to establish the full Cayley–Hamilton Theorem.

Theorem 5.10.1. *If A is an upper triangular matrix, then A satisfies $p_A(x)$.*

Proof: If

$$A = \begin{bmatrix} a_{11} & a_{12} & a_{13} & \cdots & a_{1n} \\ 0 & a_{22} & a_{23} & \cdots & a_{2n} \\ 0 & 0 & a_{33} & \cdots & a_{3n} \\ \vdots & \vdots & & \ddots & \vdots \\ 0 & 0 & \cdots & 0 & a_{nn} \end{bmatrix},$$

Sec. 5.10]   The Cayley–Hamilton Theorem   271

then

$$xI - A = \begin{bmatrix} x - a_{11} & -a_{12} & -a_{13} & \cdots & -a_{1n} \\ 0 & x - a_{22} & -a_{23} & \cdots & -a_{2n} \\ 0 & 0 & x - a_{33} & \cdots & -a_{3n} \\ \vdots & \vdots & & \ddots & \vdots \\ 0 & 0 & \cdots & 0 & x - a_{nn} \end{bmatrix},$$

so

$$p_A(x) = \det(xI - A) = (x - a_{11}) \cdots (x - a_{nn}).$$

If you look back at Theorem 4.7.4, we proved there that A satisfies

$$(x - a_{11}) \cdots (x - a_{nn}).$$

So we get that A satisfies $p_A(x) = (x - a_{11}) \cdots (x - a_{nn})$. ∎

In Theorem 4.7.3 it was shown that if $A \in M_n(\mathbb{C})$, then for some invertible matrix, C, $C^{-1}AC$ is upper triangular. Hence $C^{-1}AC$ satisfies $p_{C^{-1}AC}(x)$. By Theorem 5.9.2, $p_{C^{-1}AC}(x) = p_A(x)$. So $C^{-1}AC$ satisfies $p_A(x)$. If

$$p_A(x) = x^n + a_1 x^{n-1} + \cdots + a_n,$$

this means that

$$0 = p_A(C^{-1}AC) = (C^{-1}AC)^n + a_1(C^{-1}AC)^{n-1} + \cdots + a_n I.$$

However,

$$(C^{-1}AC)^k = C^{-1}A^k C$$

for all $k > 1$. Using this, it follows from the preceding equation that

$$0 = C^{-1}A^n C + a_1 C^{-1}A^{n-1} C + \cdots + a_n I$$
$$= C^{-1}(A^n + a_1 A^{n-1} + \cdots + a_n I)C.$$

Multiplying from the left by C and from the right by C^{-1}, we then obtain

$$0 = A^n + a_1 A^{n-1} + \cdots + a_n I = p_A(A),$$

which proves

Theorem 5.10.2 (Cayley–Hamilton). Every $A \in M_n(F)$ satisfies its characteristic polynomial.

PROBLEMS

NUMERICAL PROBLEMS

1. Prove that the following matrices A satisfy $p_A(x)$ by directly computing $p_A(A)$.

   (a) $A = \begin{bmatrix} 1 & 2 & 3 \\ 4 & 5 & 6 \\ 7 & 8 & 9 \end{bmatrix}$.

   (b) $A = \begin{bmatrix} 1 & 0 & 0 & 0 \\ 0 & 1 & 2 & 3 \\ 0 & 4 & 5 & 6 \\ 0 & 7 & 8 & 9 \end{bmatrix}$.

   (c) $A = \begin{bmatrix} 1 & 1 & 0 & 0 \\ 2 & 3 & 0 & 0 \\ 0 & 0 & 4 & 5 \\ 0 & 0 & 6 & 7 \end{bmatrix}$.

2. If $C = \begin{bmatrix} 1 & 0 & 0 \\ 0 & 2 & 0 \\ 0 & 1 & 3 \end{bmatrix}$, find C^{-1} and then by a direct computation show that $p_{C^{-1}AC}(x) = p_A(x)$ for $A = \begin{bmatrix} 1 & 2 & 3 \\ 4 & 5 & 6 \\ 7 & 8 & 9 \end{bmatrix}$.

MORE THEORETICAL PROBLEMS

Harder Problems

3. If $A = \begin{bmatrix} a & *\cdots* \\ \hline 0 & \\ \vdots & B \\ 0 & \end{bmatrix}$, where B is an $(n-1) \times (n-1)$ matrix, prove that
$p_A(x) = (x - a)p_B(x)$.

4. In Problem 3 prove by an induction argument that $p_A(A) = 0$, if A is upper triangular.

5. If $A = \begin{bmatrix} a & b & 0\cdots 0 \\ c & d & 0\cdots 0 \\ 0 & 0 & \\ \vdots & \vdots & B \\ 0 & 0 & \end{bmatrix}$, where B is an $(n-2) \times (n-2)$ matrix, prove that
$p_A(x) = p_C(x)p_B(x)$, where $C = \begin{bmatrix} a & b \\ c & d \end{bmatrix}$.

CHAPTER
6

Rectangular Matrices. More on Determinants

6.1. RECTANGULAR MATRICES

Even as we study square matrices, rectangular ones come into the picture from time to time. Now, we pause to introduce them as a tool.

An $m \times n$ *matrix* is an array $A = (a_{rs})$ of scalars having m rows and n columns. As for square matrices, a_{rs} denotes the (r, s) *entry* of A, that is, the scalar located in row r and column s. A *rectangular matrix* is just an $m \times n$ matrix, where m and n are positive integers. Thus every square matrix is a rectangular matrix. Row vectors are $1 \times n$ matrices, column vectors are $m \times 1$ matrices, and scalars are 1×1 matrices so that all these are rectangular matrices as well.

We now generalize our operations on square matrices to operations on rectangular matrices. Whenever operations were defined before (e.g., product of an $n \times n$ matrix and an $n \times 1$ column vector, product of two 1×1 scalars, sum of two $n \times 1$ column vectors), you should verify that they are special cases of the operations that we are about to define. We sometimes refer to $m \times n$ as the *shape* of an $m \times n$ matrix. So we will define addition for matrices of the same shape and multiplication UC for a matrix U of shape $b \times m$ and a matrix C of shape $m \times n$.

Definition. The *product* of a $b \times m$ matrix (u_{rs}) and an $m \times n$ matrix (v_{rs}) is the $b \times n$ matrix (w_{rs}) whose (r, s) entry w_{rs} is given by the expression $w_{rs} = \sum_{t=1}^{m} u_{rt}v_{ts}$.

Definition. Let A and B be the $m \times n$ matrices (a_{rs}), (b_{rs}) and let c be any scalar. Then $A + B$ is the $m \times n$ matrix $(a_{rs} + b_{rs})$, cA is the matrix $(ca_{rs}) - A$ is the matrix $(-a_{rs})$, $A - B$ is the matrix $A + (-B) = (a_{rs} - b_{rs})$, 0 denote the $m \times n$ matrix all of whose entries are 0, I or I_m denote the $m \times m$ identity matrix.

Whenever a concept for square matrices makes just as good sense in the case of rectangular matrices, we use it freely without formally introducing it. So, for instance, the transpose of the matrix $\begin{bmatrix} 1 & 2 \\ 3 & 5 \\ 6 & 7 \end{bmatrix}$ is the matrix $\begin{bmatrix} 1 & 3 & 6 \\ 2 & 5 & 7 \end{bmatrix}$.

Some properties of operations on square matrices carry over routinely to rectangular matrices, as long as their shapes are compatible; that is, *always multiply a $b \times m$ matrix by an $m \times n$ matrix for some n*, and *only add two matrices if they both have the same shape*. We list some of these properties in the following theorem, in which 0 denotes the matrix of 0's of the right shape, I_m or I_n denote square identity matrices, and a and b denote scalars.

Theorem 6.1.1. Rectangular $m \times m$ matrices satisfy the following properties when their shapes are compatible:

1. $(A + B) + C = A + (B + C)$;
2. $A + 0 = A$;
3. $A + (-A) = 0$;
4. $A + B = B + A$;
5. $(AB)C = A(BC)$;
6. $I_m A = A$ and $AI_n = A$;
7. $(A + B)C = AC + BC$;
8. $C(A + B) = CA + CB$;
9. $(a + b)C = aC + bC$;
10. $(ab)C = a(bC)$;
11. $1C = C$.

PROBLEMS

NUMERICAL PROBLEMS

1. Compute the product $\begin{bmatrix} 1 & 2 & 3 & 2 & 1 \\ 2 & 3 & 4 & 3 & 2 \\ 3 & 4 & 5 & 4 & 3 \end{bmatrix} \begin{bmatrix} 1 & 1 \\ 2 & 1 \\ 3 & 2 \\ 4 & 1 \\ 5 & 1 \end{bmatrix}$.

2. Find matrices A, B, C, D such that the C is 6×6, the product $ABCD$ makes sense, and $ABCD = \begin{bmatrix} 1 & 2 & 3 \\ 3 & 2 & 1 \end{bmatrix}$.

3. Compute the determinants of $\begin{bmatrix} 1 & 1 & 1 \\ 3 & 2 & 1 \end{bmatrix} \begin{bmatrix} 1 & 3 \\ 1 & 2 \\ 1 & 1 \end{bmatrix}$ and $\begin{bmatrix} 1 & 3 \\ 1 & 2 \\ 1 & 1 \end{bmatrix} \begin{bmatrix} 1 & 1 & 1 \\ 3 & 2 & 1 \end{bmatrix}$.

4. Are the determinants of

$$\begin{bmatrix} 1 & 1 & 1 \\ 3 & 2 & 1 \end{bmatrix} \begin{bmatrix} 1 & 3 \\ 1 & 2 \\ 1 & 1 \end{bmatrix} \begin{bmatrix} 2 & 1 & 1 \\ 3 & 2 & 1 \end{bmatrix} \begin{bmatrix} 1 & 3 \\ 1 & 2 \\ 1 & 1 \end{bmatrix}$$

and

$$\begin{bmatrix} 2 & 1 & 1 \\ 3 & 2 & 1 \end{bmatrix} \begin{bmatrix} 1 & 3 \\ 1 & 2 \\ 1 & 1 \end{bmatrix} \begin{bmatrix} 1 & 1 & 1 \\ 3 & 2 & 1 \end{bmatrix} \begin{bmatrix} 1 & 3 \\ 1 & 2 \\ 1 & 1 \end{bmatrix}$$

equal?

MORE THEORETICAL PROBLEMS

Easier Problems

5. Give an example of an $m \times n$ matrix A and $n \times m$ matrix B such that $AB = 0$ and the rows of A and the columns of B are linearly independent.

Middle-Level Problems

6. Show that if $m < n$ and A is an $m \times n$ matrix, the determinant of $A'A$ is 0.

Harder Problems

7. Show that the determinant of $A'A$ is nonzero for any real $m \times n$ matrix A whose columns are linearly independent.

6.2. BLOCK MULTIPLICATION (OPTIONAL)

Now that we have defined addition and multiplication of rectangular matrices, we sometimes can simplify multiplication of matrices C and D by breaking them up into blocks of compatible shapes and multiplying the blocks. This is useful if the blocks can be chosen so that some of them multiply easily.

EXAMPLE

To multiply $P = \begin{bmatrix} 1 & 2 & 3 & 4 & 5 \\ 4 & 3 & 2 & 1 & 0 \\ 1 & 2 & 3 & 4 & 3 \\ 0 & 0 & 0 & 1 & 2 \\ 0 & 0 & 1 & 0 & 1 \end{bmatrix}$ and $Q = \begin{bmatrix} 5 & 2 & 5 & 4 & 5 \\ 4 & 3 & 4 & 1 & 4 \\ 0 & 0 & 0 & 4 & 3 \\ 0 & 0 & 0 & 1 & 2 \\ 0 & 0 & 0 & 0 & 1 \end{bmatrix}$, we divide them

up as $P = \left[\begin{array}{cc|ccc} 1 & 2 & 3 & 4 & 5 \\ 4 & 3 & 2 & 1 & 0 \\ 1 & 2 & 3 & 4 & 3 \\ \hline 0 & 0 & 0 & 1 & 2 \\ 0 & 0 & 1 & 0 & 1 \end{array}\right]$ and $Q = \left[\begin{array}{ccc|cc} 5 & 2 & 5 & 4 & 5 \\ 4 & 3 & 4 & 1 & 4 \\ \hline 0 & 0 & 0 & 4 & 3 \\ 0 & 0 & 0 & 1 & 2 \\ 0 & 0 & 0 & 0 & 1 \end{array}\right]$ so that they can be represented by $P = \left[\begin{array}{c|c} A & B \\ \hline 0 & D \end{array}\right]$ and $Q = \left[\begin{array}{c|c} E & F \\ \hline 0 & H \end{array}\right]$ and their product by

$$PQ = \left[\begin{array}{c|c} AE + BG & AF + BH \\ \hline 0 & CF + DH \end{array}\right],$$

which is of the form $\begin{bmatrix} a & b & c & d & e \\ f & g & h & j & k \\ l & m & n & o & p \\ 0 & 0 & 0 & t & u \\ 0 & 0 & 0 & y & z \end{bmatrix}$.

PROBLEMS

NUMERICAL PROBLEMS

1. Compute the product

$$\begin{bmatrix} -1 & 2 & 3 & 4 & 5 \\ 0 & 3 & 2 & 1 & 0 \\ 0 & 0 & 3 & 4 & 3 \\ 0 & 0 & 1 & 0 & 0 \\ 0 & 0 & 0 & 1 & 0 \end{bmatrix} \begin{bmatrix} -1 & 0 & 0 & 1 & 0 \\ 0 & 1 & 0 & 0 & 1 \\ 0 & 0 & 0 & 0 & 0 \\ 0 & 0 & 0 & 1 & 0 \\ 0 & 0 & 0 & 0 & 1 \end{bmatrix}$$

by block multiplication.

2. Compute the product

$$\begin{bmatrix} 1 & 2 & 3 & 4 & 5 \\ 0 & 3 & 2 & 1 & 0 \\ 0 & 0 & 3 & 4 & 3 \\ 0 & 0 & 1 & 0 & 0 \\ 0 & 0 & 0 & 1 & 0 \end{bmatrix} \begin{bmatrix} 1 & 0 & 0 & 1 & 0 \\ 0 & 1 & 0 & 0 & 1 \\ 0 & 0 & 0 & 0 & 0 \\ 0 & 0 & 0 & 1 & 0 \\ 0 & 0 & 0 & 0 & 1 \end{bmatrix}$$

by block multiplication.

3. Compute the product

$$\begin{bmatrix} 1 & 2 & 3 & 4 & 5 \\ 3 & 3 & 2 & 1 & 0 \\ 0 & 0 & 3 & 4 & 3 \\ 0 & 0 & 1 & 3 & 0 \\ 0 & 0 & 0 & 0 & 2 \end{bmatrix} \begin{bmatrix} 1 & 0 & 0 & 1 & 0 \\ 0 & 1 & 0 & 0 & 1 \\ 0 & 0 & 1 & 0 & 0 \\ 0 & 0 & 0 & 1 & 0 \\ 0 & 0 & 0 & 0 & 7 \end{bmatrix}$$

by block multiplication.

Sec. 6.2] Block Multiplication 277

4. Decide on the blocks, but do not carry out the product for the product

$$\begin{bmatrix} h & 2 & 3 & 4 & 5 \\ 0 & 3 & 2 & 1 & 0 \\ 0 & 2 & 3 & 4 & 3 \\ 0 & 0 & 0 & r & 2 \\ 0 & 0 & 0 & 0 & k \end{bmatrix} \begin{bmatrix} 5 & 2 & 5 & 4 & 5 \\ 0 & 3 & 4 & 1 & 4 \\ 0 & 2 & 3 & 4 & 3 \\ 0 & 0 & 0 & 9 & 2 \\ 0 & 0 & 0 & 0 & 8 \end{bmatrix}.$$ Then compute the determinant of the product by any method.

MORE THEORETICAL PROBLEMS

Easier Problems

5. Show that if $P = \begin{bmatrix} A & B & C \\ 0 & D & E \\ 0 & 0 & F \end{bmatrix}$ and $Q = \begin{bmatrix} G & H & J \\ 0 & K & L \\ 0 & 0 & M \end{bmatrix}$, where the entries are rectangular matrices and the diagonal entries are invertible 15×15 matrices, then

$$PQ = \begin{bmatrix} AG & * & * \\ 0 & DK & * \\ 0 & 0 & FM \end{bmatrix},$$ where the *'s are *wild cards*; that is, matrices can be found to put in their places to make it into the right matrix.

6. Show that the determinant of the product AB of a 3×2 matrix A and a 2×3 matrix B is 0. Give an example to show that the determinant of BA may be nonzero.

Middle-Level Problems

7. Show that the determinant of $\begin{bmatrix} A & B & C \\ 0 & D & E \\ 0 & 0 & F \end{bmatrix} \begin{bmatrix} G & H & J \\ 0 & K & L \\ 0 & 0 & M \end{bmatrix}$ is $|A||D||F||G||K||M|$.

8. Show that the determinant of AB is 0 if A is $m \times n$, B is $n \times m$, and $m > n$.

9. Show that the determinant of $\begin{bmatrix} a & b & 3 & 4 & 5 \\ c & d & 2 & 1 & 0 \\ \hline e & f & 3 & 4 & 3 \\ 0 & 0 & 1 & 0 & r \\ 0 & 0 & 0 & 1 & s \end{bmatrix} \begin{bmatrix} g & h & j & 1 & 0 \\ k & m & n & 0 & 1 \\ 0 & 0 & 0 & p & q \\ \hline 0 & 0 & 0 & 1 & 0 \\ 0 & 0 & 0 & 0 & 1 \end{bmatrix}$ is 0 for all possible values of the variables a, b, c, and so on.

10. Show that the product $PQ = \begin{bmatrix} AE + BG & AF + BH \\ 0 & CF + DH \end{bmatrix} = \begin{bmatrix} a & b & c & d & e \\ f & g & h & j & k \\ l & m & n & o & p \\ 0 & 0 & 0 & t & u \\ 0 & 0 & 0 & y & z \end{bmatrix}$ of the matrices P and Q defined in our example cannot be invertible.

11. Show by block multiplication that if $B = (B_1, \ldots, B_n)$ describes an $m \times n$ matrix B in terms of its n columns in $F^{(m)}$ and if A is a $b \times m$ matrix, then the product AB is

$(AB_1, \ldots, AB_n)$, written in terms of its n columns AB_s (product of A and the column vector B_s).

12. State and prove the generalization of the theorem on $n \times n$ matrices A and B that the transpose of AB is $B'A'$. Use it to formulate a transposed version of Problem 11.

Harder Problems

13. Let A be *block upper triangular*; that is, A is an $n \times n$ matrix of the form

$$A = \begin{bmatrix} A_1 & & * \\ & \ddots & \\ \square & & A_k \end{bmatrix}, \text{ where the } A_r \text{ are } n_r \times n_r \text{ matrices and } n_1 + \cdots + n_k = n.$$

Show, using mathematical induction, that $|A| = |A_1| \cdots |A_k|$.

6.3. ELEMENTARY BLOCK MATRICES (OPTIONAL)

For $n = 2$, the elementary matrices are

$$\begin{bmatrix} 1 & 0 \\ 0 & 1 \end{bmatrix}, \begin{bmatrix} 0 & 1 \\ 1 & 0 \end{bmatrix}, \begin{bmatrix} 1 & u \\ 0 & 1 \end{bmatrix}, \begin{bmatrix} 1 & 0 \\ u & 1 \end{bmatrix}, \begin{bmatrix} u & 0 \\ 0 & 1 \end{bmatrix} (u \neq 0), \begin{bmatrix} 1 & 0 \\ 0 & u \end{bmatrix} (u \neq 0).$$

If we use $n \times n$ blocks in place of the scalars in these matrices, we get $(n + n) \times (n + n)$ matrices

$$\begin{bmatrix} I & \square \\ \square & I \end{bmatrix}, \begin{bmatrix} \square & I \\ I & \square \end{bmatrix}, \begin{bmatrix} I & U \\ \square & I \end{bmatrix}, \begin{bmatrix} I & \square \\ U & I \end{bmatrix},$$

$$\begin{bmatrix} U & \square \\ \square & I \end{bmatrix} (|U| \neq 0), \begin{bmatrix} I & \square \\ \square & U \end{bmatrix} (|U| \neq 0),$$

where I denotes the identity matrix and $\square$ denotes the zero matrix. These are just examples of a larger class consisting of all *elementary block matrices* where no restriction is made on the size of the blocks. In this section, we illustrate how elementary block matrices behave by using these elementary $(n + n) \times (n + n)$ matrices to give a conceptual proof of the multiplication property of the determinant.

Let A and B be $n \times n$ matrices. Since we want to prove that $|AB| = |A||B|$, one reasonable starting point is

Theorem 6.3.1. $\begin{vmatrix} A & \square \\ C & B \end{vmatrix} = |A||B|$ for any $m \times m$ matrix A, $n \times n$ matrix B, and $n \times m$ matrix C.

Proof: We expand $\begin{vmatrix} A & \square \\ C & B \end{vmatrix}$ by the minors of the first row. A typical term in the resulting expression is $(-1)^{1+s} a_{1s} M_{1s} |B|$, where M_{1s} is the $(1, s)$ minor of A.

Sec. 6.3]   Elementary Block Matrices   279

Why? When we look at the $(1, s)$ minor of $\begin{vmatrix} A & \square \\ C & B \end{vmatrix}$, it has the same *block triangular form* and therefore can be expressed, by mathematical induction, as the product of the determinants of its two diagonal blocks, namely the product of M_{1s} and $|B|$. How does this help? The terms $(-1)^{1+s}a_{1s}M_{1s}|B|$ all have the factor $|B|$. So when we add them to get $\begin{vmatrix} A & \square \\ C & B \end{vmatrix}$, we get

$$((-1)^{1+1}a_{11}M_{11} + \cdots + (-1)^{1+m}a_{1m}M_{1m})|B| = |A||B|. \quad \blacksquare$$

Does this help us prove $|AB| = |A||B|$ for $n \times n$ matrices A and B? Greatly! It shows us that the determinant of $\begin{bmatrix} AB & \square \\ \square & I \end{bmatrix}$ is $|AB|$ and the determinant of $\begin{bmatrix} A & \square \\ I & B \end{bmatrix}$ is $|A||B|$. So now we need only get from $\begin{bmatrix} AB & \square \\ \square & I \end{bmatrix}$ to $\begin{bmatrix} A & \square \\ I & B \end{bmatrix}$ by operations which, taken collectively, have no net effect on the determinant. What operations can we use? By the row and column properties of the determinant, multiplying (on either side) by an elementary matrix $A(a, b; u)$ has no effect on the determinant and multiplying by $I(a, b)$ changes its sign. We leave it for you as an interesting exercise to generalize this to

Theorem 6.3.2. The determinant of an $(n + n) \times (n + n)$ matrix is not affected if it is multiplied by $\begin{bmatrix} I & \square \\ U & I \end{bmatrix}$ or $\begin{bmatrix} I & U \\ \square & I \end{bmatrix}$ (on either side) and changes by a factor of $(-1)^n$ if it is multiplied by $\begin{bmatrix} \square & I \\ I & \square \end{bmatrix}$ (on either side).

Given this background, we now reprove the multiplication property.

Theorem 6.3.3. The determinant of the product of two $n \times n$ matrices A and B is the determinant of A times the determinant of B.

Proof:

$$|AB| = \det\begin{bmatrix} AB & \square \\ \square & I \end{bmatrix} = \det\left(\begin{bmatrix} AB & \square \\ \square & I \end{bmatrix}\begin{bmatrix} I & \square \\ B & I \end{bmatrix}\right)$$

$$= \det\begin{bmatrix} AB & \square \\ B & I \end{bmatrix} = \det\left(\begin{bmatrix} I & -A \\ \square & I \end{bmatrix}\begin{bmatrix} AB & \square \\ B & I \end{bmatrix}\right)$$

$$= \det\begin{bmatrix} AB - AB & -A \\ B & I \end{bmatrix} = \det\begin{bmatrix} \square & -A \\ B & I \end{bmatrix} = \det\left(\begin{bmatrix} -A & \square \\ I & B \end{bmatrix}\begin{bmatrix} \square & I \\ I & \square \end{bmatrix}\right)$$

$$= (-1)^n \det\begin{bmatrix} -A & \square \\ I & B \end{bmatrix} = (-1)^n(-1)^n \det\begin{bmatrix} A & \square \\ I & B \end{bmatrix} = \det\begin{bmatrix} A & \square \\ I & B \end{bmatrix} = |A||B|.$$

$\blacksquare$

PROBLEMS

NUMERICAL PROBLEMS

1. Illustrate the proof of Theorem 6.3.3 for $A = \begin{bmatrix} 1 & 2 \\ 3 & 4 \end{bmatrix}$, $B = \begin{bmatrix} 3 & 4 \\ 5 & 6 \end{bmatrix}$.

MORE THEORETICAL PROBLEM

Easier Problems

2. Prove Theorem 6.3.2 in detail.

6.4. A CHARACTERIZATION OF THE DETERMINANT FUNCTION (OPTIONAL)

The determinant function on $n \times n$ matrices plays such an important role in linear algebra that you may ask whether there are similar functions that can be defined and exploited. The answer to this question is "no," in the sense of

Theorem 6.4.1. Let f be a scalar-valued function on the set $M_n(F)$ of $n \times n$ matrices and suppose that f satisfies the following properties:

1. $f(I) = 1$;
2. $f(B) = -f(A)$ if B is obtained from A by interchanging two rows;
3. $f(B) = f(A)$ if B is obtained from A by adding a multiple of one row to another;
4. $f(B) = u(f(A))$ if B is obtained from A by multiplying some row by u.

Then $f(A) = |A|$ for all $n \times n$ matrices A.

Proof: The conditions (2)–(4) simply say that $f(EA) = |E|f(A)$ for every elementary matrix E. Thus if A is invertible and we factor A as a product $A = E_1 \cdots E_q$ of elementary matrices, we have that

$$|A| = |E_1 \cdots E_q| = |E_1| \cdots |E_q|.$$

But then

$$|E_1| \cdots |E_q| \cdot 1 = |E_1| \cdots |E_q| f(I) = \cdots = f(E_1 \cdots E_q I) = f(A),$$

by using $F(EA) = |E|f(A)$ over and over again. On the other hand, suppose that A is not invertible. If A has a zero row, then $|A| = 0$, and $f(A) = 0$ by property (4) (Prove!), so that $|A| = 0 = f(A)$. Otherwise, choose elementary matrices $E_1, \ldots, E_p$ such that $U = E_p \cdots E_1 A$ is an upper triangular matrix with leading entry of 1 on each nonzero row. Then U is not invertible since A is not. It follows that some diagonal entry u_{ss} of U is zero. If $u_{nn} = 0$, then the last row is 0, so that both $f(A)$ and $|A|$ are 0, as before. Otherwise, taking s as large as possible such that $u_{ss} = 0$, we have $u_{s+1\,s+1} = \cdots = u_{nn} = 1$. Then we can further row reduce U, using the operations Add$(s, r; -u_{sr})$ for

$r = s + 1, \ldots, n$ to make all entries of row s zero. Letting

$$E_{p+1} = A(s, s+1; -u_{s\,s+1}), \quad \ldots, \quad E_q = A(s, n; -u_{sn}),$$

where $q = p + n - s$, we find that the resulting matrix is $B = E_q \cdots E_{p+1} U$, and it has a row of zeros. But $f(B) = 0$ for any matrix B having a zero row (Prove!), so that

$$0 = f(E_q \cdots E_{p+1} U) = f(E_q \cdots E_1 A) = \cdots = |E_q| \cdots |E_1| f(A)$$

and $f(A)$ must be 0. Similarly,

$$0 = |E_q \cdots E_1 A| = |E_q| \cdots |E_1||A|$$

and $f(A) = 0 = |A|$. ∎

Theorem 6.4.1 is actually quite powerful. For example, we can use it to give yet a third proof of the multiplication property of determinants.

Theorem 6.4.2. For any two $n \times n$ matrices A and B, $|AB| = |A||B|$.

Proof: As in our first proof of the multiplication property, we easily reduce the proof to the case where B has nonzero determinant. Then, defining $f(A)$ as $\dfrac{|AB|}{|B|}$ for all A, we see that

$$f(EA) = \frac{|EAB|}{|B|} = \frac{|E||AB|}{|B|} = |E|f(A).$$

So f satisfies properties (1)–(4) of Theorem 6.4.1 and $f(A) = |A|$ for all A. This implies that $|A||B| = |AB|$. ∎

PROBLEMS

MORE THEORETICAL PROBLEMS

Easier Problems

1. The function $f(A) = \det(A')$ satisfies the following properties:
   (a) $f(I) = 1$;
   (b) $f(B) = -f(A)$ if B is obtained from A by interchanging two rows;
   (c) $f(B) = f(A)$ if B is obtained from A by adding a multiple of one row to another;
   (d) $f(B) = u(f(A))$ if B is obtained from A by multiplying some row by u.
   Then use this to show that $|A'| = |A|$ for all $n \times n$ matrices A.

CHAPTER 7

More on Systems of Linear Equations

7.1. LINEAR TRANSFORMATIONS FROM $F^{(n)}$ to $F^{(m)}$

In Chapter 2 we talked about systems of linear equations and how to solve them. Now we return to this topic and discuss it from a somewhat different perspective.

As in Chapter 2, we represent a system

$$
\begin{aligned}
a_{11}x_1 + \cdots + a_{1n}x &= y_1 \\
\vdots \qquad\qquad \vdots \quad &\vdots \\
a_{m1}x_1 + \cdots + a_{mn}x_n &= y_m
\end{aligned}
$$

of m linear equations in n variables by the matrix equation $Ax = y$, where A is the $m \times n$ matrix $\begin{bmatrix} a_{11} & \cdots & a_{1n} \\ \vdots & & \vdots \\ a_{m1} & \cdots & a_{mn} \end{bmatrix}$ and x and y are the column vectors $x = \begin{bmatrix} x_1 \\ \vdots \\ x_n \end{bmatrix}$ and $y = \begin{bmatrix} y_1 \\ \vdots \\ y_m \end{bmatrix}$. Now, as in the case $m = n$ in Chapter 3, we regard an $m \times n$ matrix A as a mapping from $F^{(n)}$ to $F^{(m)}$ sending $x \in F^{(n)}$ to $Ax \in F^{(m)}$, where Ax denotes the product of A and x and the system of equations above describes the mapping sending x to Ax in terms of coordinates. From the properties of matrix addition, multiplication, and scalar multiplication given in Theorem 6.1.1, *these mappings Ax are linear in the sense of*

Definition. A *linear mapping* from $F^{(n)}$ to $F^{(m)}$ is a function f from $F^{(n)}$ to $F^{(m)}$ such that $f(u + v) = f(u) + f(v)$ and $f(cv) = cf(v)$ for all $u, v \in F^{(n)}$, and $c \in F$.

Sec. 7.1]   Linear Transformations from $F^{(n)}$ to $F^{(m)}$ 283

The properties $f(u + v) = f(u) + f(v)$ and $f(cv) = cf(v)$ are the *linearity properties* of f.

EXAMPLE

The mapping $f\left(\begin{bmatrix} a \\ b \\ d \end{bmatrix}\right) = \begin{bmatrix} 3a + d \\ b - a \end{bmatrix}$ is a linear mapping. We can see this either by direct verification of the linearity properties for f or, as we now do, by finding a 2×3 matrix A such that $f\left(\begin{bmatrix} a \\ b \\ d \end{bmatrix}\right) = A\begin{bmatrix} a \\ b \\ d \end{bmatrix}$. If we let $A = \begin{bmatrix} 3 & 0 & 1 \\ -1 & 1 & 0 \end{bmatrix}$, then

$$A\begin{bmatrix} a \\ b \\ d \end{bmatrix} = \begin{bmatrix} 3 & 0 & 1 \\ -1 & 1 & 0 \end{bmatrix}\begin{bmatrix} a \\ b \\ d \end{bmatrix} = \begin{bmatrix} 3a + d \\ b - a \end{bmatrix} = f\left(\begin{bmatrix} a \\ b \\ d \end{bmatrix}\right).$$

So since Ax is a linear mapping and $Ax = f(x)$ for all x, f is a linear mapping.

We have the following theorem, whose proof is exactly the same as the proof of Theorem 3.2.2 for the $n \times n$ case.

Theorem 7.1.1. The linear mappings from $F^{(n)}$ to $F^{(m)}$ are just the $m \times n$ matrices.

In Chapter 2, given an $m \times n$ matrix A and vector $y \in F^{(m)}$, we asked: What is the set of all solutions to the equation $Ax = y$? From the point of view that A is a linear transformation, we are asking: What is the set of elements $x \in F^{(n)}$ that are mapped by A into y? Of course, this set of elements is nonempty if and only if y is the image Ax of some $x \in F^{(n)}$.

When $y = 0$, 0 is the image $A0$ of 0, so the set of elements $x \in F^{(n)}$ that are mapped into 0 is always nonempty. In fact, this set $N = \{x \in F^{(n)} | Ax = 0\}$ is a subspace of $F^{(n)}$ (Prove!) which we call the *nullspace* of A. This is of great importance since, as we are about to see, this implies that the set of solutions to $Ax = y$ is either empty or a *translation* $v + N$ of a subspace N of $F^{(n)}$.

Of course, the equation $Ax = y$ may have no solution x or it may have exactly one. Otherwise, there are at least two different solutions u and v, in which case the homogeneous equation $Ax = 0$ has infinitely many solutions $c(u - v)$, c being any scalar. Why? Since A is a linear transformation, we can use its linearity properties to compute

$$A(c(u - v)) = c(A(u - v)) = c(Au - Av) = c(y - y) = c(0) = 0.$$

So the equation $Ax = 0$ has infinitely many different solutions, in fact, one for each scalar value c. But then it follows that the equation $Ax = y$ also has infinitely many solutions $v + c(u - v)$, c being any scalar. Why? Again, simply compute

$$A(v + c(u - v)) = Av + A(c(u - v)) = Av + 0 = y + 0 = y.$$

These simple observations prove

Theorem 7.1.2. For any $m \times n$ matrix A and any $y \in F^{(n)}$, one of the following is true:

1. $Ax = y$ has no solution x;
2. $Ax = y$ has exactly one solution x;
3. $Ax = y$ has infinitely many solutions x.

More important than this theorem itself is the realization we get from its proof that any two solutions u and v to the equation $Ax = y$ give us a solution $u - v$ to the homogeneous equation $Ax = 0$: $A(u - v) = 0$. Why is this important? Letting N be the nullspace $\{w \in F^{(n)} \mid Aw = 0\}$, we have $u - v \in N$ and $u = v + w$ with $w \in N$ for any two solutions u, v of $Ax = y$. Conversely, if $Av = y$ and $u = v + w$ with $w \in N$, then $Au = y$. (Prove!) So, we get

Theorem 7.1.3. Let A be an $m \times n$ matrix and let $y \in F^{(m)}$. If $Av = y$, then the set of solutions x to the equation $Ax = y$ is $v + N = \{v + w \mid w \in N\}$, where N is the nullspace of A.

By this theorem, we can split up the job of finding the set $v + N$ of all the solutions to $Ax = 0$ to finding one v such that $Av = y$ and finding the nullspace N of A. So, we now are confronted with the question: What is the nullspace N of A, and how do we get one solution $x = v$ to the equation $Ax = y$, if it exists?

In Chapter 2, given an $m \times n$ matrix A and $y \in F^{(m)}$, we gave methods to determine whether the matrix equation $Ax = y$ has a solution x and, if so, to find the solutions:

1. Form the augmented matrix $[A, y]$ and reduce it to a row equivalent echelon matrix $[B, z]$.
2. Then $Ax = y$ has a solution if and only if $z_r = 0$ whenever row r of B is 0.
3. All solutions x can be found by solving $Bx = z$ by back substitution.

Now, we can get one solution v to $Bx = z$ and express the solutions to $Ax = y$ as $v + w$, where w is in the nullspace of B, so we have

Method to solve $Ax = y$ for x

1. Form the augmented matrix $[A, y]$ and reduce it to a row equivalent echelon matrix $[B, z]$.
2. If a solution exists, get one solution v to $Bv = z$ by setting all independent variables equal to zero and finding the corresponding solution v by back substitution.
3. Find the nullspace N of B, that is, the set of all solutions w to $Bw = 0$, by back substitution.
4. If a solution exists, the set of solutions x to $Ax = y$ is then $v + N = \{v + w \mid Bw = 0\}$.

EXAMPLE

Let's return to the problem in Section 2.2 of solving the matrix equation

$$\begin{bmatrix} 2 & 3 & 1 \\ -2 & 1 & 3 \end{bmatrix} \begin{bmatrix} p \\ q \\ r \end{bmatrix} = \begin{bmatrix} 6000 \\ 2000 \end{bmatrix}.$$ We reduce $\begin{bmatrix} 2 & 3 & 1 & 6000 \\ -2 & 1 & 3 & 2000 \end{bmatrix}$ successively to

$$\begin{bmatrix} 2 & 3 & 1 & 6000 \\ 0 & 4 & 4 & 8000 \end{bmatrix}, \quad \begin{bmatrix} 2 & 3 & 1 & 6000 \\ 0 & 1 & 1 & 2000 \end{bmatrix}, \quad \begin{bmatrix} 1 & \frac{3}{2} & \frac{1}{2} & 3000 \\ 0 & 1 & 1 & 2000 \end{bmatrix}.$$

To get one solution v of the equivalent matrix equation $\begin{bmatrix} 1 & \frac{3}{2} & \frac{1}{2} \\ 0 & 1 & 1 \end{bmatrix} \begin{bmatrix} p \\ q \\ r \end{bmatrix} = \begin{bmatrix} 3000 \\ 2000 \end{bmatrix}$, we set the independent variable r to the value $r = 0$ and solve for $q = 2000$ and $p = 0$, getting $v = \begin{bmatrix} 0 \\ 2000 \\ 0 \end{bmatrix}$. To get the elements of the nullspace of $\begin{bmatrix} 1 & \frac{3}{2} & \frac{1}{2} \\ 0 & 1 & 1 \end{bmatrix}$, we solve $\begin{bmatrix} 1 & \frac{3}{2} & \frac{1}{2} \\ 0 & 1 & 1 \end{bmatrix} \begin{bmatrix} p \\ q \\ r \end{bmatrix} = 0$ by back substitution, finding $q = -r$ and $p = r$ and getting $w = \begin{bmatrix} r \\ -r \\ r \end{bmatrix}$ as the general element. So the general solution is

$$v + w = \begin{bmatrix} 0 \\ 2000 \\ 0 \end{bmatrix} + \begin{bmatrix} r \\ -r \\ r \end{bmatrix} = \begin{bmatrix} r \\ 2000 - r \\ r \end{bmatrix},$$

which agrees with the solution we got in Section 2.2.

The principal players here are the subspace $AF^{(n)} = \{Ax \mid x \in F^{(n)}\}$ of images Ax ($x \in F^{(n)}$) of A and the nullspace $N = \{x \in F^{(n)} \mid Ax = 0\}$ of A. Since the image Ax of x under A is also the linear combination $x_1 v_1 + \cdots + x_n v_n$ of the columns $v_1, \ldots, v_n$ of A, $AF^{(n)}$ is also the span of the columns of A, that is, the *column space* of A. In terms of the nullspace and column space of A, the answer to the question "What is the set of all solutions of the equation $Ax = y$?" is:

1. $Ax = y$ has a solution x if and only if y is in the column space $AF^{(n)}$ of A.
2. For $y \in AF^{(n)}$, $Av = y$ implies that the set of solutions to $Ax = y$ is $v + N$, where N is the nullspace of A, and conversely.

We discuss the nullspace and column space of A in the next section, where an important relationship between them is determined.

PROBLEMS

NUMERICAL PROBLEMS

1. Show that the mapping $f\left(\begin{bmatrix} q \\ w \\ u \end{bmatrix}\right) = \begin{bmatrix} 2u - q - w \\ w - 2q - u \end{bmatrix}$ is a linear mapping from $F^{(3)}$ to $F^{(2)}$.

2. Describe the mapping f in Problem 1 as a 2×3 matrix.

3. Show that $\begin{bmatrix} 1 & 2 & 3 & 4 \\ 1 & 3 & 4 & 7 \\ 1 & 2 & 6 & 12 \end{bmatrix} \begin{bmatrix} a \\ s \\ d \\ c \end{bmatrix} = \begin{bmatrix} 0 \\ 0 \\ 3 \end{bmatrix}$ has infinitely many solutions $\begin{bmatrix} a \\ s \\ d \\ c \end{bmatrix}$.

4. Find a basis for the nullspace N of $\begin{bmatrix} 1 & 2 & 3 & 4 \\ 3 & 3 & 3 & 3 \\ 1 & 2 & 6 & 12 \end{bmatrix}$ and express the set of solutions to $\begin{bmatrix} 1 & 2 & 3 & 4 \\ 1 & 3 & 3 & 3 \\ 1 & 2 & 6 & 12 \end{bmatrix} \begin{bmatrix} a \\ b \\ c \\ d \end{bmatrix} = \begin{bmatrix} 3 \\ 3 \\ 8 \end{bmatrix}$ in the form $\begin{bmatrix} a \\ b \\ c \\ d \end{bmatrix} + N$ for some specific $\begin{bmatrix} a \\ b \\ c \\ d \end{bmatrix} \in F^{(4)}$.

5. If $F \in \mathbb{R}$, show that there is a unique solution $\begin{bmatrix} a \\ b \\ c \\ d \end{bmatrix}$ of shortest length to the equation in Problem 4.

MORE THEORETICAL PROBLEMS

Easier Problems

6. Let N be any subspace of $\mathbb{R}^{(n)}$. Show that N is the nullspace of some $n \times n$ matrix A.

Middle-Level Problems

7. Let N be any subspace of $\mathbb{R}^{(n)}$ and let v be any vector in $\mathbb{R}^{(n)}$. Then for any $m \geq n - \dim(N)$, show that there is an $m \times n$ matrix A such that $v + N$ is the set of solutions x to $Ax = y$ for some $y \in F^{(m)}$.

8. Show that if $A \in M_n(\mathbb{R})$ and Ax is the projection of y on the column space of A, then $A'Ax = A'y$.

7.2. THE NULLSPACE AND COLUMN SPACE OF AN $m \times n$ MATRIX

In Section 7.1 we found in discussing the set of solutions to an equation $Ax = y$ that the nullspace and column space of A played central roles. Here, you will recall, the column space of an $m \times n$ matrix A is the set $AF^{(n)} = \{Ax \mid x \in F^{(n)}\}$ of linear combinations Ax of the columns of A and the nullspace of A, denoted Nullspace (A), is $\{x \in F^{(n)} \mid Ax = 0\}$. The dimension of the column space of A is called the *rank* of A, denoted $r(A)$, whereas the dimension of Nullspace (A) is called the *nullity* of A, denoted $n(A)$.

Since $Ax = 0$ and $Bx = 0$ have the same solutions if A and B are row equivalent $m \times n$ matrices, we have

Theorem 7.2.1. The nullspaces of any two row equivalent $m \times n$ matrices are equal.

From this theorem it follows that the nullity of two row equivalent matrices are the same. Are their ranks the same, too? Yes! If $w_1, \ldots, w_r$ is a basis for the column space of an $m \times n$ matrix A and if U is an invertible $m \times m$ matrix, then $Uw_1, \ldots, Uw_r$ is a basis for the column space of UA (Prove!), so that A and UA have the same rank. This proves

Theorem 7.2.2. If A and B are row equivalent $m \times n$ matrices, then A and B have the same rank and nullity.

Using this theorem, we now proceed to show that the rank plus the nullity of any $m \times n$ matrix A is n. In Section 4.4 we showed this to be true for square matrices. For $m \times n$ matrices we really do not need to prove very much, since this result is simply an enlightened version of Theorem 2.4.1. Why is it an enlightened version of Theorem 2.4.1? The rank and nullity of A appear in disguise in Theorem 2.4.1 as the values $n - m'$ (the nullity of A) and m' (the rank of A), which are defined by taking m' to be the number of nonzero rows of any echelon matrix B which is row equivalent to A. What do we really need to prove? We just need to prove that m' is the rank of A (see Theorem 7.2.4) and $n - m'$ is the nullity of A (an easy exercise).

To keep our discussion self-contained, however, we do not use Theorem 2.4.1 in showing that the rank plus nullity is n. And we do not use the swift approach proof using inner products given in Chapter 4 in the case $m = n$, since it does not easily generalize. Instead, our approach is to replace the matrix A by a matrix row equivalent to it, which is a reduced echelon matrix in the sense of the following definition.

Definition. An $m \times n$ matrix A is a *reduced echelon matrix* if A is an echelon matrix such that the leading entry in any nonzero row is the only nonzero entry in its column.

Theorem 7.2.3. Any $m \times n$ matrix is row equivalent to a reduced echelon matrix.

Proof: We know from Chapter 2 that any $m \times n$ matrix A is row equivalent to an echelon matrix. But any echelon matrix can be reduced to a reduced echelon matrix using elementary row operations to get rid of nonzero entries above any given leading entry. ∎

The key to finding the rank of A is

Theorem 7.2.4. The rank $r(A)$ of an $m \times n$ matrix A is the number of nonzero rows in any echelon matrix B row equivalent to A.

Proof: The ranks of A and B are equal, by Theorem 7.2.2. If we further reduce B to a reduced echelon matrix, using elementary row operations, the rank is not changed and the number of nonzero rows is not changed. So we may as well assume that B itself is a *reduced* echelon matrix. Then each nonzero row gives us a leading entry 1, which in turn gives us the corresponding column in the $m \times m$ identity matrix. So if there are r nonzero rows, they correspond to the first r columns of the $m \times m$ identity matrix. In turn, these r columns of the $m \times m$ identity matrix form a basis for the column space of B, since all columns have all entries equal to 0 in row k for all $k > r$. This means that $r(A) = r$. ∎

Let's look at this in a specific example.

EXAMPLE

The rank of the echelon matrix $A = \begin{bmatrix} 1 & 3 & 0 & 0 \\ 0 & 0 & 1 & 0 \\ 0 & 0 & 0 & 1 \\ 0 & 0 & 0 & 0 \end{bmatrix}$ is 3 since A has three nonzero rows. The three nonzero rows correspond to the columns 1, 3, 4, which are the first three columns of the 4×4 identity matrix. On the other hand, the rank of the nonechelon matrix $B = \begin{bmatrix} 1 & 3 & 0 & 0 \\ 0 & 0 & 1 & 1 \\ 0 & 0 & 1 & 1 \\ 0 & 0 & 0 & 0 \end{bmatrix}$ is 2 even though it also has three nonzero rows.

The *row space* of an $m \times n$ matrix is the span of its rows over F. Since the row space does not change when an elementary row operation is performed (Prove!), the row spaces of any two row equivalent $m \times n$ matrices are the same. This is used in the proof of

Corollary 7.2.5. The row and column spaces of an $m \times n$ matrix A have the same dimension, namely $r(A)$.

Proof: The dimension of the row space of an $m \times n$ matrix A equals the dimension of the row space of any reduced echelon matrix row equivalent to it. So we may as well assume that the matrix A itself is a row-reduced echelon matrix. Then its nonzero rows are certainly linearly independent. Since the number of nonzero rows is $r(A)$, $r(A)$ is the dimension of the row space of A. ∎

Sec. 7.2]   The Nullspace and Column Space of an $m \times n$ Matrix

EXAMPLE

To find the rank (dimension of column space) of $\begin{bmatrix} 1 & 2 & 3 & 4 & 4 & 3 & 6 \\ 2 & 1 & 1 & 1 & 1 & 1 & 1 \\ 3 & 3 & 3 & 3 & 3 & 3 & 3 \end{bmatrix}$, calculate the dimension of the row space, which is 3 since the rows are linearly independent. (Verify!)

We now show that as for square matrices, the sum of the rank and nullity of an $m \times n$ matrix is n. Since the rank $r(A)$ and nullity $n(A)$ of an $m \times n$ matrix A do not change if an elementary row or column operation is applied to A (Prove!), we can use both elementary row and column operations in our proof. We use a column version of row equivalence, where we say that two $m \times n$ matrices A and B are *column equivalent* if there exists an invertible $n \times n$ matrix U such that $B = AU$.

Theorem 7.2.6.   Let A be an $m \times n$ matrix. Then the sum of the rank and nullity of A is n.

Proof: Since two row or column equivalent matrices have the same rank and nullity, and since A is row equivalent to a reduced echelon matrix, we can assume that A itself is a reduced echelon matrix. Then, using elementary column operations, we can further reduce A so that it has at most one nonzero entry in any row or column. Using column interchanges if necessary, we can even assume that the entries $a_{11}, \ldots, a_{rr}$ (where r is the rank of A) are nonzero, and therefore equal to 1, and all other entries of A are 0. But then

$$Ax = \begin{bmatrix} 1 & & & & & & 0 \\ & \ddots & & & & & \\ & & 1 & & & & \\ & & & 0 & & & \\ & & & & \ddots & & \\ 0 & & & & & & 0 \end{bmatrix} \begin{bmatrix} x_1 \\ \vdots \\ x_r \\ x_{r+1} \\ \vdots \\ x_n \end{bmatrix} = \begin{bmatrix} x_1 \\ \vdots \\ x_r \\ 0 \\ \vdots \\ 0 \end{bmatrix},$$

so $Ax = 0$ if and only if $x_1 = \cdots = x_r = 0$, which implies that the dimension of Nullspace (A) is $n - r$. ∎

EXAMPLE

To find the nullity of $\begin{bmatrix} 1 & 2 & 3 & 4 & 4 & 3 & 6 \\ 2 & 1 & 1 & 1 & 1 & 1 & 1 \\ 3 & 3 & 3 & 3 & 3 & 3 & 3 \end{bmatrix}$, find its rank r. The nullity is then $n - r$. We saw that the rank is 3 in the example above, so the nullity is $7 - 3 = 4$.

This theorem enables us to prove

Corollary 7.2.7. Let A be an $m \times n$ matrix A such that the equation $Ax = y$ has one and only one solution $x \in F^{(n)}$ for every $y \in F^{(m)}$. Then m equals n and A is invertible.

Proof: The nullity of A is 0, since $Ax = 0$ has only one solution. So the rank of A is $n - 0 = n$. Since $Ax = y$ has a solution for every $y \in F^{(m)}$, the column space of A is $F^{(m)}$. Since the dimension of $F^{(n)}$ is n, we get that $F^{(m)}$ has dimension equal to the rank of A, that is, n. So $m = n$. Since A has rank n, it follows that A is invertible. ∎

PROBLEMS

NUMERICAL PROBLEMS

1. Find the dimension of the column space of $\begin{bmatrix} 1 & 2 & 3 & 4 & 4 & 3 & 6 \\ 2 & 1 & 1 & 1 & 1 & 1 & 1 \\ 3 & 3 & 4 & 5 & 5 & 4 & 7 \end{bmatrix}$ by finding the dimension of its row space.

2. Find the nullity of $\begin{bmatrix} 1 & 2 & 3 & 4 & 4 & 3 & 6 \\ 2 & 1 & 1 & 1 & 1 & 1 & 1 \\ 3 & 3 & 4 & 5 & 5 & 4 & 7 \end{bmatrix}$ using Problem 1 and Theorem 7.2.4.

3. Find a reduced echelon matrix row equivalent to the matrix

$$\begin{bmatrix} 1 & 2 & 3 & 4 & 4 & 3 & 6 \\ 2 & 1 & 1 & 1 & 1 & 1 & 1 \\ 3 & 3 & 4 & 5 & 5 & 4 & 7 \end{bmatrix}$$

and use it to compute the nullspace of this matrix.

4. Find the nullspace of the transpose $\begin{bmatrix} 1 & 2 & 3 & 4 & 4 & 3 & 6 \\ 2 & 1 & 1 & 1 & 1 & 1 & 1 \\ 3 & 3 & 4 & 5 & 5 & 4 & 7 \end{bmatrix}$ of the matrix of Problem 3 by any method.

MORE THEORETICAL PROBLEMS

Easier Problems

5. Show that the rank and nullity of an $m \times n$ matrix A do not change if an elementary row or column operation is applied to A.

6. If A is an $m \times n$ matrix, show that the nullity of A equals the nullity of the transpose A' of A if and only if $m = n$.

7. If the set of solutions to $Ax = y$ is a subspace, show that $y = 0$.

Middle-Level Problems

8. Let A be a 3×2 matrix with real entries. Show that the equation $A'Ax = A'y$ has a solution x for all y in $\mathbb{R}^{(3)}$.

9. Show that if two reduced $m \times n$ echelon matrices are row equivalent, they are equal.
10. Show that if an $m \times n$ matrix is row equivalent to two reduced echelon matrices C and D, then $C = D$.

Harder Problems

11. Let A be an $m \times n$ matrix with real entries. Show that the equation $A'Ax = A'y$ has a solution x for all y in $\mathbb{R}^{(m)}$.

CHAPTER 8

Abstract Vector Spaces

8.1. INTRODUCTION, DEFINITIONS, AND EXAMPLES

Before getting down to the formal development of the mathematics involved, a few short words on what we'll be about. What will be done is from a kind of turned-about point of view. At first, most of the things that will come up will be things already encountered by the reader in $F^{(n)}$. It may seem a massive *deja-vu*. That's not surprising: it *is* a massive *deja-vu*, but with a twist.

Everything we did in Chapter 3 will make an appearance in due course, but in a new, abstract setting. In addition, some fundamental notions—notions that we have not seen to date, which are universal in all of mathematics—will show up.

Let F be $\mathbb{R}$ or $\mathbb{C}$, the real or complex numbers, respectively. For much of the development, whether F is $\mathbb{R}$ or $\mathbb{C}$ will have no relevance. As in our discussion of $F^{(n)}$, at some crucial point, where we shall require roots of polynomials, the use of $F = \mathbb{C}$ will become essential.

In Chapter 3 we saw that in $F^{(n)}$, the set of all column vectors $\begin{bmatrix} a_1 \\ a_2 \\ \vdots \\ a_n \end{bmatrix}$ over F, there are certain natural ways of combining elements—addition and multiplication by a scalar—which satisfied some nice, formal, computational rules. These properties, which are concrete and specific in $F^{(n)}$, serve to guide us to one of the most basic, abstract structures in all of mathematics, known as a *vector space*. The notion is closely modeled on the concrete $F^{(n)}$.

Definition. A set V is called a *vector space over F* if on V we have two operations, called *addition* and *multiplication by a scalar*, such that the following hold:

(A) *Rules for Addition*
1. If $v, w \in V$, then $v + w \in V$.
2. $v + w = w + v$ for all $v, w \in V$.
3. $v + (w + z) = (v + w) + z$ for all v, w, z in V.
4. There exists an element 0 in V such that $v + 0 = v$ for every $v \in V$.
5. Given $v \in V$ there exists a $w \in V$ such that $v + w = 0$.

(B) *Rules for Multiplication by a Scalar*
6. If $a \in F$ and $v \in V$, then $av \in V$.
7. $a(v + w) = av + aw$ for all $a \in F, v, w \in V$.
8. $(a + b)v = av + bv$ for all $a, b \in F, v \in V$.
9. $a(bv) = (ab)v$ for all $a, b \in F, v \in V$.
10. $1v = v$ for all $v \in V$, where 1 is the unit element of F.

These rules governing the operations in V seem very formal and formidable. They probably are. But what they really say is something quite simple: Go ahead and calculate in V as we have been used to doing in $F^{(n)}$. The rules just give us the enabling framework for carrying out such calculations.

Note that property 4 guarantees us that V has at least one element, hence V is *nonempty*.

Before doing anything with vector spaces it is essential that we see a large cross section of examples. In this way we can get the feeling of how wide and encompassing the concept really is.

In the examples to follow, we shall define the operations and verify, at most, that $v + w$ *and* av are in V for $v, w \in V$ and $a \in F$. The verification that each example is, indeed, an example of a vector space is quite straightforward, and is left to the reader.

EXAMPLES

1. Let $V = F^{(n)}$, or any subspace of $F^{(n)}$, with the operations of those of Chapter 3. Since these examples are the prime motivator for the general concept of vector space, it is not surprising that they are vector spaces over F. If $n = 1$, then $V = F^{(1)} = F$, showing how $V = F$ is itself regarded as a vector space over F.

2. As a variation of the theme of Example 1, let $F = \mathbb{R}$ and $V = \mathbb{C}^{(n)}$ with the usual addition and multiplication of elements in $\mathbb{C}^{(n)}$. It is immediate that $V = \mathbb{C}^{(n)}$ is a vector space over $F = \mathbb{R}$ with the usual product by a scalar between elements of F and of $\mathbb{C}$. If $n = 1$, then $V = \mathbb{C}^{(1)} = \mathbb{C}$, showing how $V = \mathbb{C}$ is regarded as a vector space over $\mathbb{R}$.

3. Let V consist of all the functions $a\cos(x) + b\sin(x)$, where $a, b \in \mathbb{R}$, with the usual operations used for functions. If

$$v = a\cos(x) + b\sin(x)$$

and

$$w = a_1\cos(x) + b_1\sin(x),$$

then
$$v + w = (a + a_1)\cos(x) + (b + b_1)\sin(x),$$

so $v + w$ is in V by the very criterion for membership in V. Similarly, $cv \in V$ if $c \in \mathbb{R}, v \in V$. So, V is a vector space over $\mathbb{R}$.

4. Let V consist of all real-valued solutions of the differential equation $\frac{d^2f(x)}{dx^2} + f(x) = 0$. From the theory of differential equations we know that V consists precisely of the elements $a\cos(x) + b\sin(x)$, with $a, b \in \mathbb{R}$. In other words, this example coincides with Example 3. For the sake of variety, we verify that $f + g$ is in V if f, g are in V by another method. Since $f, g \in V$, we have

$$\frac{d^2f(x)}{dx^2} + f(x) = 0,$$

$$\frac{d^2g(x)}{dx^2} + g(x) = 0,$$

whence

$$\frac{d^2}{dx^2}(f+g) = \frac{d^2f}{dx^2} + \frac{d^2g}{dx^2} = 0.$$

Thus $f + g$ is in V. Similarly, cf is in V if $c \in \mathbb{R}$ and $f \in V$.

5. Let V be the set of all real-valued solutions $f(x)$ of the differential equation $\frac{d^2f}{dx^2} + f(x) = 0$ such that $f(0) = 0$. It is clear that V is contained in the vector space of Example 4, and is itself a vector space over $\mathbb{R}$. So it is appropriate to call it a *subspace* (you can guess from Chapter 3 what this should mean) of Example 4. What is the form of the typical element of V?

6. Let V be the set of all polynomials $f(x) = a_0 x^n + \cdots + a_n$ where the a_i are in $\mathbb{C}$ and n is any nonnegative integer. We add and multiply polynomials by constants in the usual way. With these operations V becomes a vector space over $\mathbb{C}$.

7. Let W be the set of all polynomials over $\mathbb{C}$ of degree 4 or less. That is, $W = \{a_0 x^4 + a_1 x^3 + a_2 x^2 + a_3 x + a_4 \,|\, a_0, \ldots, a_4 \in \mathbb{C}\}$, with the operations those of Example 6. It is immediate that W is a vector space over $\mathbb{C}$ and is a subspace of Example 6.

8. Let V be the set of all sequences $\{a_r\}$, $a_r \in \mathbb{R}$, such that $\lim_{n \to \infty} a_n = 0$. Define addition and multiplication by scalars in V as follows:
   (a) $\{a_r\} + \{b_r\} = \{a_r + b_r\}$
   (b) $c\{a_r\} = \{ca_r\}$.

By the properties of limits, $\lim_{n \to \infty}(a_n + b_n) = \lim_{n \to \infty} a_n + \lim_{n \to \infty} b_n = 0$ if $\{a_i\}, \{b_i\}$ are in V, and $\lim_{n \to \infty} ca_n = c \lim_{n \to \infty} a_n = 0$. So V is a vector space over $\mathbb{R}$.

Note one little fact about V. Suppose we let

$$W = \{\{a_r\} \in V \,|\, a_r = 0 \text{ if } r \geq 3\}.$$

Then W is a *subspace* of V. If we write a sequence out as a ∞-tuple, then the typical element $\{a_r\}$ of W looks like $\{a_r\} = (a_1, a_2, 0, 0, \ldots, 0, \ldots)$ where a_1, a_2 are arbitrary in $\mathbb{R}$. In other words, W *resembles* $\mathbb{R}^{(2)}$ in some sense. What this sense is will be spelled out completely later. Of course, 2 is not holy in all this; for every $n \geq 1$ V contains, as a subspace something that resembles $\mathbb{R}^{(n)}$.

9. Let U be the subset of the vector space V of Example 8 consisting of all sequences $\{a_r\}$, where $a_s = 0$ for all $a \geq t$, for some t (depending on the sequence $\{a_r\}$). We leave it as an exercise that U is a subspace of V and is a vector space over $\mathbb{R}$. Note that, for instance, the sequence $\{a_r\}$, where $a_1, \ldots, a_4$ are arbitrary but $a_n = 0$ for $n > 4$, is in V.

10. Let V be the set, $M_n(F)$, of $n \times n$ matrices over F with the matrix addition and scalar multiplication used in Chapter 2. V is a vector space over F. If we view an $n \times n$ matrix (a_{rs}) as a strung-out n^2-tuple, then $M_n(F)$ resembles $F^{(n^2)}$ as a vector space. Thus if $\begin{bmatrix} a_{11} & a_{12} \\ a_{21} & a_{22} \end{bmatrix} \in M(F)$, we "identify" it with the column vector $\begin{bmatrix} a_{11} \\ a_{21} \\ a_{12} \\ a_{22} \end{bmatrix}$ in $F^{(4)}$ and we get the "resemblance" between $M_2(F)$ and $F^{(4)}$ mentioned above.

11. Let V be any vector space over F and let $v_1, \ldots, v_n$ be n given elements in V. An element $x = a_1 v_1 + \cdots + a_n v_n$, where the a_i are in F, is called a *linear combination* of $v_1, \ldots, v_n$ over F. Let $\langle v_1, \ldots, v_n \rangle$ denote the set of *all linear combinations of* $v_1, \ldots, v_n$ over F. We claim that $\langle v_1, \ldots, v_n \rangle$ is a vector space over F; it is also a subspace of V over F. We leave these to the reader. We call $\langle v_1, \ldots, v_n \rangle$ the *subspace of V spanned over F by* $v_1, \ldots, v_n$.

This is a very general construction of producing new vector spaces from old. We saw how important $\langle v_1, \ldots, v_n \rangle$ was in the study of $F^{(n)}$. It will be equally important here.

Let's look at an example for some given V of $\langle v_1, \ldots, v_n \rangle$. Let V be the set of all polynomials in x over $\mathbb{C}$ and let $v_1 = x$, $v_2 = 1 - x$, $v_3 = 1 + x + x^2$, $v_4 = x^3$. What is $\langle v_1, \ldots, v_n \rangle$? It consists of all $a_1 v_1 + \cdots + a_4 v_4$, where $a_1, \ldots, a_4$ are in $\mathbb{C}$. In other words, a typical element of $\langle v_1, \ldots, v_n \rangle$ looks like $a_1 x + a_2(1-x) + a_3(1 + x + x^2) + a_4 x^3 = (a_2 + a_3) + (a_1 - a_2 + a_3)x + a_3 x^2 + a_4 x^3$. We leave it to the reader to verify that *every* polynomial of degree 3 or less is so realizable. Thus, in this case, $\langle v_1, \ldots, v_4 \rangle$ consists of all polynomials over $\mathbb{C}$ of degree 3 or less. Put more tersely, we are saying that $\langle x, 1-x, 1+x+x^2, x^3 \rangle = \langle 1, x, x^2, x^3 \rangle = V$.

12. Let $V = M_n(F)$, viewed as a vector space over F. Let

$$W = \{(a_{rs}) \in M_n(F) \,|\, \text{tr}\,(a_{rs}) = 0\}.$$

Since $\operatorname{tr}(A + B) = \operatorname{tr}(A) + \operatorname{tr}(B)$ and $\operatorname{tr}(aA) = a\operatorname{tr}(A)$, we get that W is a vector space over F. Of course, W is a subspace of V.

13. Let $V = M_n(\mathbb{R})$ be viewed as a vector space over $\mathbb{R}$ and let $W = \{A \in M_n(\mathbb{R}) \mid A = A'\}$. Thus W is the set of symmetric matrices in $M_n(\mathbb{R})$. Because $(A + B)' = A' + B'$ and $(cA)' = cA'$, for $A, B \in M_n(\mathbb{R})$ and $c \in \mathbb{R}$, we see that W is a subspace of V over $\mathbb{R}$.

14. Let V be the set of all real-valued functions on the closed unit interval $[0, 1]$. For the operations in V we use the usual operations of functions. Then V is a vector space over $\mathbb{R}$. Let $W = \{f(x) \in V \mid f(\tfrac{1}{2}) = 0\}$. Thus if $f, g \in W$, then $f(\tfrac{1}{2}) = g(\tfrac{1}{2}) = 0$, hence $(f + g)(\tfrac{1}{2}) = f(\tfrac{1}{2}) + g(\tfrac{1}{2}) = 0$, and $af(\tfrac{1}{2}) = 0$ for $a \in \mathbb{R}$. So W is a vector space over $\mathbb{R}$ and is a subspace of V.

15. Let V be the set of all *continuous* real-valued functions on the closed unit interval. Since the sum of continuous functions is a continuous function and the product of a continuous function by a constant is continuous, we get that V is a vector space over $\mathbb{R}$. If W is the set of real-valued functions differentiable on the closed unit interval then $W \subset V$, and by the properties of differentiable functions, W is a vector space over $\mathbb{R}$ and a subspace of V.

The reader easily perceives that many of our examples come from the calculus and real variable theory. This is not just because these provide us with natural examples. We deliberately made the choice of these to stress to the reader that the concept of vector space is not just an algebraic concept. It turns up all over the place.

In Chapter 4, in discussing subspaces, we gave many diverse examples of subspaces of $F^{(n)}$. Each such, of course, provides us with an example of a vector space over F. We would suggest to readers that they go back to Chapter 4 and give those old examples a new glance.

8.2. SUBSPACES

With what we did in Chapter 4 on subspaces of $F^{(n)}$ there should be few surprises in store for us in the material to come. It might be advisable to look back and give oneself a quick review of what is done there.

Before getting to the heart of the matter, we must dispose of a few small, technical items. These merely tell us it is all right to go ahead and compute in a natural way.

To begin with, we assume that in a vector space V the sum of any two elements is again in V. By combining elements we immediately get from this that the sum of *any finite number of elements of V is again in V*. Also, the associative law

$$v + (w + z) = (v + w) + z$$

can be extended to any finite sum and *any* bracketing. So we can write such sums without the use of parentheses. To prove this is not hard but is tedious and noninstructive, so we omit it and go ahead using the general associative law without proof.

Possibly more to the point are some little computational details. We go through them in turn.

1. *If $v + w = v + z$, then $w = z$.* For there exists an element $u \in V$ such that $u + v = 0$; hence $u + (v + w) = u + (v + z)$. In other words,

$$w = 0 + w = (u + v) + w = u + (v + z) = (u + v) + z = 0 + z = z,$$

resulting in $w = z$. This little rule allows us to cancel in sums.

2. *If $0 \in F$ and $v \in V$, then $0v = 0$, the zero element of V.* Since

$$0v + 0w = (0 + 0)v = 0v = 0v + 0,$$

we get $0v + 0v = 0v + 0$, so that $0v = 0$ by (1) above.

3. $v + (-1)v = 0$. For $0 = (1 + (-1))v = v + (-1)v$. So $(-1)v$ acts as the negative for v in V. We denote it $(-1)v = -v$.

4. *If $a \neq 0$ is in F and $av = 0$, then $v = 0$.* For $0 = av$ gives us $0 = a^{-1}0 = a^{-1}(av) = (a^{-1}a)v = 1v = v$.

5. *If $a \neq 0$ is in F and $av + a_2v_2 + \cdots + a_kv_k = 0$, then $v = (-a_2/a)v_2 + \cdots + (-a_k/a)v_k$.*

We leave the proof of this to the reader. Again we stress that a result such as (5) is a license to carry on, in our usual way, in computing in V.

With these technical points out of the way we can get down to the business at hand.

Definition. Let V be a vector space over F. Then a nonempty subset W of V is called a *subspace* of V (over F) if, relative to the operations of V, W forms a vector space over F.

Two subspaces of V come immediately to mind. These are $W = \{0\}$ and $W = V$. We leave it to the reader to verify that these two are indeed subspaces of V. By a *proper subspace* W of V we shall mean a subspace of V such that $W \neq V$.

How do we recognize whether or not a given nonempty subset W of V is a subspace of V? Easily. All we have to do is to verify that $u, w \in W$ implies $u + w \in W$ and $au \in W$ for all $a \in F u \in W$. Why do we have to verify only these two properties? Because most of the other rules defining a vector space hold in W because they already hold in the larger set V. What we really need is that $0 \in W$ and, for every $w \in W$, $-w$ is in W. Now, since W is nonempty, let $z \in W$. Then $0z = 0$ is in W, hence $0 \in W$. Similarly, for w in W, $(-1)w = -w$ is in W. So W is a subspace of V.

We record this as

Lemma 8.2.1. *A nonempty subset W of V is a subspace of V if and only if:*

1. *$u, w \in W$ implies that $u + w \in W$;*
2. *$a \in F, w \in W$ implies that $aw \in W$.*

If we look back at the examples in Section 8.1, we note that we have singled out many of the examples as subspaces of some others. But just for the sake of practice we verify in a specific instance that something is a subspace.

Let V be the set of all real-valued functions defined on the closed unit interval. Then V is a vector space over $\mathbb{R}$. Let W be the set of all continuous functions in V. We assert that W is a subspace of V. Why? If $f, g \in W$, then each of them is continuous, but the sum of continous functions is continous. Hence $f + g$ is continous; and as such, it enjoys the condition for membership in W. In short, $f + g$ is in W. Similarly, $af \in W$ if $a \in F$ and $f \in W$.

Let U be the set of all differentiable functions in V. Because a differentiable function is automatically continuous, we get that $U \subset W$. Furthermore, if f, g are differentiable, so are $f + g$ and af, with $a \in F$, differentiable. Thus U is a subspace of V. Since U lies in W, U is also a subspace of W.

Of paramount importance to us is Example 11 of Section 8.1. It gives us a ready way of producing subspaces of a given vector space, at will. Recall that if $v_1, \ldots, v_n \in V$, then $\langle v_1, \ldots, v_n \rangle$ is the set of all linear combinations $a_1 v_1 + \cdots + a_n v_n$ of $v_1, \ldots, v_n$ over F. Let's verify that $\langle v_1, \ldots, v_n \rangle$ is really a subspace of V over F. If

$$u = a_1 v_1 + \cdots + a_n v_n \quad \text{and} \quad z = b_1 v_1 + \cdots + b_n v_n,$$

where the a_r and b_r are in F, then

$$u + z = (a_1 v_1 + \cdots + a_n v_n) + (b_1 v_1 + \cdots + b_n v_n)$$
$$= (a_1 + b_1) v_1 + \cdots + (a_n + b_n) v_n.$$

(Verify!) So $u + z$ is a linear combination of $v_1, \ldots, v_n$ over F, hence lies in $\langle v_1, \ldots, v_n \rangle$. Similarly, if $c \in F$, then

$$c(a_1 v_1 + \cdots + a_n v_n) = (ca_1) v_1 + \cdots + (ca_n) v_n,$$

so is a linear combination of $v_1, \ldots, v_n$ over F. As such it lies in $\langle v_1, \ldots, v_n \rangle$. We have checked out that $\langle v_1, \ldots, v_n \rangle$ is a subspace of V over F. We call it the subspace of V *spanned* by $v_1, \ldots, v_n$ over F.

In $F^{(3)}$, if we let $v_1 = \begin{bmatrix} 1 \\ 1 \\ 1 \end{bmatrix}$, $v_2 = \begin{bmatrix} 1 \\ 0 \\ 1 \end{bmatrix}$, $v_3 = \begin{bmatrix} 1 \\ 1 \\ 0 \end{bmatrix}$, $v_4 = \begin{bmatrix} 1 \\ 2 \\ 3 \end{bmatrix}$, then $\langle v_1, v_2, v_3, v_4 \rangle$ is the set of all $a_1 v_1 + a_2 v_2 + a_3 v_3 + a_4 v_4 = \begin{bmatrix} a_1 + a_2 + a_3 + a_4 \\ a_1 + a_3 + 2a_4 \\ a_1 + a_2 + 3a_4 \end{bmatrix}$. We leave it to the reader that *any* vector $\begin{bmatrix} b_1 \\ b_2 \\ b_3 \end{bmatrix}$ in $F^{(3)}$ is so realizable. Hence $\langle v_1, v_2, v_3, v_4 \rangle = F^{(3)}$.

This last example is quite illuminating. In it the vectors $v_1, \ldots, v_4$ spanned all of $V = F^{(3)}$ over F. This type of situation is the one that will be of greatest interest to us.

Definition. The vector space V over F is said to be *finite-dimensional over F* if $V = \langle v_1, \ldots, v_n \rangle$ for some vectors $v_1, \ldots, v_n$ in V.

In other words, V is finite-dimensional over F if there is a *finite* set $v_1, \ldots, v_n$ in V such that, given *any* $u \in V$, then $u = a_1 v_1 + \cdots + a_n v_n$ for some appropriate $a_1, \ldots, a_n$ in F.

Which of the examples in Section 8.1 are finite-dimensional over F? We give the answer as "yes" or "no" to each in turn. The reader should verify that these answers are the right ones:

Example 1: Yes.	Example 2: Yes.	Example 3: Yes.
Example 4: Yes.	Example 5: Yes.	Example 6: No.
Example 7: Yes.	Example 8: No.	Example 9: No.
Example 10: Yes.	Example 11: Yes.	Example 12: Yes.
Example 13: Yes.	Example 14: No.	Example 15: No.

By the very definition of finite-dimensionality we know that the span $W = \langle v_1, \ldots, v_n \rangle$ of elements $v_1, \ldots, v_n$ of a vector space V is not only a subspace of V but is also finite-dimensional over F.

For almost everything that we shall do in the rest of the book the vector space V will be finite-dimensional over F. Therefore, you may assume that V is finite-dimensional unless we indicate that it is not.

PROBLEMS

NUMERICAL PROBLEMS

1. Show that $W = \{0\}$ and $W = V$ are subspaces of V.
2. Verify that Examples 1 through 5, 7, and 10 through 13 are finite-dimensional over F and explain why Examples 6, 8, 9, 14, and 15 are not.
3. Show that $v_1 = \begin{bmatrix} 1 \\ 1 \\ 1 \end{bmatrix}$, $v_2 = \begin{bmatrix} 1 \\ 0 \\ 1 \end{bmatrix}$, $v_3 = \begin{bmatrix} 1 \\ 1 \\ 0 \end{bmatrix}$, $v_4 = \begin{bmatrix} 1 \\ 2 \\ 3 \end{bmatrix}$ span $F^{(3)}$ over F.
4. Let V be the vector space of polynomials in x having real coefficients. If $v_1 = x$, $v_2 = 2x$, $v_3 = x^2$, $v_4 = x - 6x^2$, find the span of v_1, v_2, v_3, v_4 over $\mathbb{R}$.
5. In Problem 4 show that $\langle v_1, v_2, v_3, v_4 \rangle = \langle v_1, v_4 \rangle$.

MORE THEORETICAL PROBLEMS

Easier Problems

6. If $v_1, \ldots, v_n$ span V over F, show that $v_1, \ldots, v_n, w_1, \ldots, w_m$ also span V over F, for any $w_1, w_2, \ldots, w_m$ in V.
7. If $v_1, \ldots, v_n$ span V over F, and if $a_1 v_1 + a_2 v_2 + \cdots + a_n v_n = 0$, where $a_1 \neq 0$, show that $v_2, \ldots, v_n$ already span V over F.
8. If $w_1, \ldots, w_k$ are in $\langle v_1, \ldots, v_n \rangle$, show that $\langle w_1, \ldots, w_k \rangle$ is contained in $\langle v_1, \ldots, v_n \rangle$.
9. Let V be a vector space over $\mathbb{C}$. As such we consider V also as a vector space over $\mathbb{R}$, by using as scalars only the elements of $\mathbb{R}$. Show that if V is finite-dimensional over $\mathbb{C}$, then V is finite-dimensional over $\mathbb{R}$.
10. In Problem 9 show that if V is finite-dimensional over $\mathbb{R}$, then V is also finite-dimensional over $\mathbb{C}$.

11. If U and W are subspaces of V, show that $U \cap W$ is a subspace of V.
12. If U and W are subspaces of V, let $U + W$ be defined by

$$U + W = \{u + w \mid u \in U, w \in W\}.$$

 Show that $U + W$ is a subspace of V over F.
13. If, in Problem 12, U and W are finite-dimensional over F, show that $U + W$ is finite-dimensional over F.
14. Let V and W be two vector spaces over F. Let $V \oplus W = \{(v, w) \mid v \in V, w \in W\}$, where we define $(v_1, w_1) + (v_2, w_2) = (v_1 + v_2, w_1 + w_2)$ and $a(v_1, w_1) = (av_1, aw_1)$ for all $v_1, v_2 \in V$ and $w_1, w_2 \in W$ and $a \in F$. Prove that $V \oplus W$ is a vector space over F. ($V \oplus W$ is called the *direct sum* of V and W.)
15. If V and W are finite-dimensional over F, show that $V \oplus W$ is also finite-dimensional over F.
16. Let $\tilde{V} = \{(v, 0) \mid v \in V\}$ and $\tilde{W} = \{(0, w) \mid w \in W\}$. Show that $\tilde{V}$ and $\tilde{W}$ are subspaces of $V \oplus W$ such that $V \oplus W = \tilde{V} + \tilde{W}$ and $\tilde{V} \cap \tilde{W} = \{0\}$.
17. If $V = \langle v_1, \ldots, v_m \rangle$ and $W = \langle w_1, \ldots, w_n \rangle$, show that

$$V \oplus W = \langle \tilde{v}_1, \ldots, \tilde{v}_m, \tilde{w}_1, \ldots, \tilde{w}_n \rangle,$$

 where $\tilde{v}_i = (v_i, 0)$ and $\tilde{w}_j = (0, w_j)$.
18. Show that in the vector space $V \oplus V$ over F, the set $V = \{(v, v)\}$ is a subspace of $V \oplus V$ over F.
19. If $v \in V$, let $Fv = \{av \mid a \in F\}$. Show that Fv is a subspace of V.
20. If $v_1, \ldots, v_n$ are in V, show that $\langle v_1, \ldots, v_n \rangle = Fv_1 + \cdots + Fv_n$ (see Problem 12).

Middle-Level Problems

21. Let V be a vector space over F and U, W subspaces of V. Define $f: U \oplus W \to V$ by $f(u, w) = u + w$. Show that f maps $U \oplus W$ onto $U + W$. Furthermore, show that if $X, Y \in U \oplus W$, then $f(X + Y) = f(X) + f(Y)$ and $f(aX) = af(X)$ for all $a \in F$.
22. In Problem 21, let $Z = \{(u, w) \mid f(u, w) = 0\}$. Show that Z is a subspace of $U \oplus W$ and is, in fact, $Z = \{(u, -u) \mid u \in U \cap W\}$.

Harder Problems

23. Let V be a vector space over F and suppose that V is the *set-theoretic union* of subspaces U and W. Prove that $V = U$ or $V = W$.

8.3. HOMOMORPHISMS AND ISOMORPHISMS

For the first time in a long time we are about to go into a territory that is totally new to us. What we shall talk about is the notion of a *homomorphism* of one vector space into another. This will be defined precisely below, but for the moment, suffice it to say that *a homomorphism is a mapping that preserves structure.*

The analog of this concept occurs in every part of mathematics. For instance, the analog in analysis might very well be that of a continuous function. In every part of algebra such decent mappings are defined by algebraic relations they satisfy.

A special kind of homomorphism is an *isomorphism*, which is really a homomorphism that is $1-1$ and onto. If there is an isomorphism from one vector space to another, the spaces are *isomorphic*. Isomorphic spaces are, in a very good sense, equal. Whatever is true in one space gets transferred, so it is also true in the other. For us, the importance will be that any *finite-dimensional* vector space V over F will turn out to be isomorphic to $F^{(n)}$ for some n. ⁷ s will allow us to transfer *everything* we did in $F^{(n)}$ to finite-dimensional vector spaces. In a sense, $F^{(n)}$ is the universal model of a finite-dimensional vector space, and in treating $F^{(n)}$ and the various concepts in $F^{(n)}$, we lose nothing in generality. These notions and results will be transferred to V by an isomorphism.

Of course, all this now seems vague. It will come more and more into focus as we proceed here and in the coming sections.

Definition. If V and W are vector spaces over F, then the mapping $\Phi: V \to W$ is said to be a *homomorphism* if

1. $\Phi(v_1 + v_2) = \Phi(v_1) + \Phi(v_2)$
2. $\phi(av) = a\phi(v)$

for all $v, v_1, v_2 \in V$ and all $a \in F$.

Note that in $\Phi(v_1 + v_2)$, the addition $v_1 + v_2$ takes place in V, whereas the addition $\Phi(v_1) + \Phi(v_2)$ takes place in W.

We did run into this notion when we talked about linear transformations of $F^{(n)}$ (and of $F^{(n)}$ into $F^{(m)}$) in Chapters 3 and 7. Here the context is much broader.

Decent concepts deserve decent examples. We shall try to present some now.

EXAMPLES

1. Let $V = F^{(4)}$ and $W = F^{(2)}$. Define $\Phi: V \to W$ by $\Phi \begin{bmatrix} a \\ b \\ c \\ d \end{bmatrix} = \begin{bmatrix} a + c - 3d \\ 4b - a \end{bmatrix}$.

We claim that Φ is a homomorphism of V into W. Given $\begin{bmatrix} a \\ b \\ c \\ d \end{bmatrix}$ and $\begin{bmatrix} a_1 \\ b_1 \\ c_1 \\ d_1 \end{bmatrix}$ in $F^{(4)}$, then

$$\Phi \begin{bmatrix} a \\ b \\ c \\ d \end{bmatrix} = \begin{bmatrix} a + c - 3d \\ 4b - a \end{bmatrix}, \Phi \begin{bmatrix} a_1 \\ b_1 \\ c_1 \\ d_1 \end{bmatrix} = \begin{bmatrix} a_1 + c_1 - 3d_1 \\ 4b_1 - a_1 \end{bmatrix},$$

while

$$\Phi\left(\begin{bmatrix} a \\ b \\ c \\ d \end{bmatrix} + \begin{bmatrix} a_1 \\ b_1 \\ c_1 \\ d_1 \end{bmatrix}\right) = \Phi\begin{bmatrix} a + a_1 \\ b + b_1 \\ c + c_1 \\ d + d_1 \end{bmatrix} = \begin{bmatrix} a + a_1 + c + c_1 & -3(d + d_1) \\ 4(b + b_1) & -(a + a_1) \end{bmatrix}$$

$$= \begin{bmatrix} a + c - 3d \\ 4b - a \end{bmatrix} + \begin{bmatrix} a_1 + c_1 - 3d_1 \\ 4b_1 - a_1 \end{bmatrix} = \Phi\left(\begin{bmatrix} a \\ b \\ c \\ d \end{bmatrix}\right) + \Phi\left(\begin{bmatrix} a_1 \\ b_1 \\ c_1 \\ d_1 \end{bmatrix}\right).$$

Similarly,

$$\Phi\left(a\begin{bmatrix} a \\ b \\ c \\ d \end{bmatrix}\right) = a\Phi\left(\begin{bmatrix} a \\ b \\ c \\ d \end{bmatrix}\right)$$

for all $a \in F$. So Φ is a homomorphism of $F^{(4)}$ into $F^{(2)}$. It is actually *onto* $F^{(2)}$. (Prove!)

2. Let $V = F^{(n)}$ and $W = F$ be viewed as vector spaces over F. Let $\Phi\begin{bmatrix} a_1 \\ \vdots \\ a_n \end{bmatrix} = a_1 + a_2 + \cdots + a_n$. We claim that Φ is a homomorphism of $F^{(n)}$ onto F.

First, why is it onto? That is easy, for given $a \in F$, then $a = \Phi\begin{bmatrix} a \\ 0 \\ \vdots \\ 0 \end{bmatrix}$, so Φ is onto. Why a homomorphism? By the definition,

$$\Phi\left(\begin{bmatrix} a_1 \\ \vdots \\ a_n \end{bmatrix} + \begin{bmatrix} b_1 \\ \vdots \\ b_n \end{bmatrix}\right) = \Phi\begin{bmatrix} a_1 + b_1 \\ \vdots \\ a_n + b_n \end{bmatrix} = (a_1 + b_1) + \cdots + (a_n + b_n)$$

$$= (a_1 + \cdots + a_n) + (b_1 + \cdots + b_n) = \Phi\begin{bmatrix} a_1 \\ \vdots \\ a_n \end{bmatrix} + \Phi\begin{bmatrix} b_1 \\ \vdots \\ b_n \end{bmatrix}.$$

Similarly,

$$\Phi\left(c\begin{bmatrix} a_1 \\ \vdots \\ a_n \end{bmatrix}\right) = c\Phi\begin{bmatrix} a_1 \\ \vdots \\ a_n \end{bmatrix}.$$

3. Let V be the set of all polynomials in x over $\mathbb{R}$. Define

$$F: V \to V \text{ by } \Phi(a_0 x^n + a_1 x^{n-1} + \cdots + a_n) = na_0 x^{n-1} + (n-1)a_1 x^{n-2} + \cdots + a_{n-1}.$$

So Φ is just the derivative of the polynomial in question. Either by a direct check, or by remembrances from the calculus, we have that Φ is a homomorphism of V into V. Is it onto?

4. Again let V be the set of polynomials over $\mathbb{R}$ and let $\Phi: V \to V$ be defined by

$$\Phi(p(x)) = \int_0^x p(t)\, dt.$$

Since we have that

$$\int_0^x (p(t) + q(t))\, dt = \int_0^x p(t)\, dt + \int_0^x q(t)\, dt$$

and

$$\int_0^x cp(t)\, dt = c \int_0^x p(t)\, dt \quad \text{for } c \in \mathbb{R}$$

we see that $\Phi(v_1 + v_2) = (v_1) + (v_2)$ and $(cv_1) = c\Phi(v_1)$ for all v_1, v_2 in V and all $c \in \mathbb{R}$. So Φ is a homomorphism of V into itself. Is it onto?

5. Let V be all infinite sequences over $\mathbb{R}$ where $\{a_i\} + \{b_i\} = \{a_i + b_i\}$ and $c\{a_i\} = \{ca_i\}$ for elements of V and element c in $\mathbb{R}$. Let $\Phi: V \to V$ be defined by

$$\Phi\{a_1, a_2, \ldots, a_n, \ldots\} = \{a_2, \ldots, a_n, \ldots\}.$$

So, for example, $\Phi\{1, 2, 1, 3, 0, \ldots\} = \{2, 1, 3, 0, \ldots\}$. It is immediate that Φ is a homomorphism of V onto itself. It is usually called a *left-shift operator*.

6. Let $V = F^{(5)}$ and $W = F^{(3)}$. Define $\Phi: V \to W$ by

$$\Phi \begin{bmatrix} a_1 \\ a_2 \\ \vdots \\ a_5 \end{bmatrix} = \begin{bmatrix} a_1 \\ a_2 \\ a_3 \end{bmatrix}.$$

It is immediate that Φ is a homomorphism of V onto W.

7. Let V and W be vector spaces over F and let $X = V \oplus W$. Define $\Phi: X \to V$ by

$$\Phi(v, w) = v.$$

We see here, too, that Φ is a homomorphism of X onto V. It is the *projection* of $V \oplus W$ onto the subspace V.

8. Let V be the complex numbers, viewed as a vector space over $\mathbb{R}$. If $z = a + bi$, define

$$\Phi(z) = a - bi = \bar{z}.$$

Then Φ is a homomorphism of V onto itself. It is called *complex conjugation*.

9. Let $V = M_n(F)$ and $W = F$ and let $\Phi: V \to W$ be defined by

$$\Phi(A) = \text{tr}(A).$$

Since $\text{tr}(A + B) = \text{tr}(A) + \text{tr}(B)$ and $\text{tr}(aA) = a\,\text{tr}(A)$, we get that Φ is a homomorphism of V onto W. It is the *trace function* on $M_n(F)$.

10. Let V be the set of polynomials in x of degree n or less over $\mathbb{R}$, and let W be the set of all polynomials in x over F. Define $\Phi: V \to W$ by

$$\Phi(p(x)) = x^2 p(x)$$

for all $p(x) \in V$. It is easy to see that Φ is a homomorphism of V into W, but not onto.

From the examples given, one sees that it is not too difficult to construct homomorphisms of vector spaces.

What are the important attributes of homomorphisms? For one thing they carry linear combinations into linear combinations.

Lemma 8.3.1. If Φ is a homomorphism of V into W, then

$$\Phi(a_1 v_1 + \cdots + a_k v_k) = a_1 \Phi(v_1) + \cdots + a_k \Phi(v_k)$$

for any $v_1, \ldots, v_k$ in V and $a_1, \ldots, a_k$ in F.

Proof: By the additive property of homomorphisms we have that

$$\Phi(a_1 v_1 + \cdots + a_k v_k) = \Phi(a_1 v_1) + \cdots + \Phi(a_k v_k).$$

Since $\Phi(a_i v_i) = a_i \Phi(v_i)$ this top relation becomes the one asserted in the lemma. ■

Another property is that homomorphisms carry subspaces into subspaces. This really is implicit in Lemma 8.3.1, but we state it as a lemma, leaving the proof to the reader.

Lemma 8.3.2. If V is a vector space, U a subspace of V and Φ a homomorphism of V into W, then $\Phi(U) = \{\Phi(u) \mid u \in U\}$ is a subspace of W.

Thus the image of V and any subspace of V under Φ is a subspace of W. Finally, another subspace crops up when dealing with homomorphisms.

Definition. If Φ is a homomorphism of V into W, then $\text{Ker}(\Phi)$, the *kernel* of Φ, is defined by $\text{Ker}(\Phi) = \{v \in V \mid \Phi(v) = 0\}$.

Does this remind you of the nullspace of a linear transformation? It should, because it is. For Φ is merely a linear transformation from V to W.

Lemma 8.3.3. $\text{Ker}(\Phi)$ is a subspace of V.

Proof: Suppose that $u, v \in \text{Ker}(\Phi)$ and $a \in F$. Then $\Phi(u + v) = \Phi(u) + \Phi(v) = 0$ since $\Phi(u) = \Phi(v) = 0$. Also, $\Phi(au) = a\Phi(u) = a0 = 0$. Thus $u + v$ and au are also in $\text{Ker}(\Phi)$, hence $\text{Ker}(\Phi)$ is a subspace of V. ∎

As we shall see later, we can have any subspace of V be the kernel of some homomorphism. Is there anything special that happens when $\text{Ker}(\Phi)$ is the particularly nice subspace consisting only of 0? Yes! Suppose that $\text{Ker}(\Phi) = \{0\}$ and that $\Phi(u) = \Phi(v)$. Then $0 = \Phi(u) - \Phi(v) = \Phi(u - v)$, hence $u - v$ is in $\text{Ker}(\Phi) = \{0\}$. Therefore, $u - v = 0$, so $u = v$. What this says is that Φ is then $1 - 1$. If Φ is $1 - 1$, it is trivial that $\text{Ker}(\Phi) = \{0\}$, hence

Lemma 8.3.4. The homomorphism $\Phi: V \to W$ is $1 - 1$ if and only if $\text{Ker}(\Phi) = \{0\}$.

A homomorphism that maps V *onto* W in a $1 - 1$ way is indeed something special. We give it a name.

Definition. The homomorphism $\Phi: V \to W$ is called an *isomorphism* if Φ is $1 - 1$ and onto W.

The importance of isomorphisms is that two vector spaces which are *isomorphic* (i.e., for which there is an isomorphism of one onto the other) are essentially the same. All that is different is the naming we give the elements. So if Φ is an isomorphism of V onto W, the renaming is given by calling $v \in V$ by the name $\Phi(v)$ in W. The fact that Φ preserves sums and products by scalars assures us that this renaming is consistent with the structure of V and W.

For practical purposes isomorphic vector spaces are the same. Anything true in one of these gets transferred to something true in the other. We shall see how the concept of isomorphism reduces the study of finite-dimensional vector spaces to that of $F^{(n)}$.

Definition. We shall denote that there is an isomorphism of V onto W by $V \simeq W$.

PROBLEMS

NUMERICAL PROBLEMS

1. In Example 1 show that the mapping Φ is onto $F^{(2)}$.
2. Verify that the mapping Φ in Example 3 is a homomorphism of V into itself.
3. Verify that the mapping Φ in Example 6 is a homomorphism of $F^{(5)}$ onto $F^{(3)}$.

4. Verify that the mapping $\Phi: V \oplus W \to V$ in Example 7 is a homomorphism.
5. Verify that the mapping Φ in Example 10 is a homomorphism of V into, but not onto, W.

Middle-Level Problems

6. In all Examples 1 through 10 given, find $\operatorname{Ker}(\Phi)$. That is, express the form of the general element in $\operatorname{Ker}(\Phi)$.
7. Prove that the identity mapping of a vector space V into itself is an isomorphism of V onto itself.
8. If $\Phi: V \to W$ is defined by $\Phi(v) = 0$ for all $v \in V$, show that Φ is a homomorphism.
9. Prove Lemma 8.3.2.
10. In Examples 1 through 10, determine which Φ are isomorphisms.

MORE THEORETICAL PROBLEMS

Easier Problems

11. Prove for U, V, W vector spaces over F that:
    (a) $V \simeq V$.
    (b) $U \simeq V$ implies that $V \simeq U$.
    (c) $U \simeq V, V \simeq W$ implies that $U \simeq W$.
12. Let V be the vector space of all polynomials in x over $\mathbb{R}$ of degree m or less. Prove that $V \simeq F^{(m+1)}$.
13. If $m \le n$ prove that there is a subspace V of $F^{(n)}$ such that $V \simeq F^{(m)}$.
14. If $\Phi: U \to V$ and $\psi: V \to W$ are homomorphisms of the vector spaces U, V, W over F, show that $\psi\Phi$ defined by $(\psi\Phi)(u) = \psi(\Phi(u))$ for every $u \in U$ is a homomorphism of U into W.
15. If $\Phi: V \to W$ is a homomorphism and if $a \in F$, define $\psi(v) = a\Phi(v)$ for every $v \in V$. Prove that ψ is a homomorphism of V into W.
16. Prove that $V \oplus W \simeq W \oplus V$.
17. Let $X = \{(v, v) | v \in V\}$. Show that $X \simeq V$.
18. If $\phi: M_n(F) \to F$ is defined by $\Phi(A) = \operatorname{tr}(A)$, find $\operatorname{Ker}(\Phi)$.
19. If Φ and ψ are homomorphisms of V into W, define $\Phi + \psi$ by $(\Phi + \psi)(v) = \Phi(v) + \psi(v)$ for every v in V. Prove that $\Phi + \psi$ is a homomorphism of V into W.
20. Combine the results of Problems 15 and 19 to show that the set of homomorphisms of V into W is a vector space over F.
21. Let $\Phi: V \oplus V \to V$ be defined by $\Phi(v_1, v_2) = v_1 + v_2$. Find $\operatorname{Ker}(\Phi)$ and show $V \oplus V = \{(v, v) | v \in V\} + \operatorname{Ker}(\Phi)$ (see Problem 14 of Section 8.2).

Middle-Level Problems

22. Suppose that U, W are subspaces of V. Define $\psi: U \oplus W \to V$ by $\psi(u, w) = u + w$. Show that $\operatorname{Ker}(\psi) \simeq U \cap W$.
23. If V is finite-dimensional over F and Φ is a homomorphism of V into W, show that $\phi(V) = \{\phi(v) | v \in V\}$ is finite-dimensional over F.

24. Let $V = F^{(n)}$ and $W = F^{(m)}$. Show that every homomorphism of V into W can be given by an $m \times n$ matrix over F.

25. Show that as vector spaces, $M_n(F) \simeq F^{(n^2)}$.

Very Hard Problems

26. Let V be a vector space over F. Show that V cannot be the *set-theoretic union* of a *finite* number of proper subspaces of V.

8.4. ISOMORPHISMS FROM V TO $F^{(n)}$

Recall that Φ is said to be an isomorphism of V *onto* W if Φ is a homomorphism of V *onto* W which is $1 - 1$. We also saw that Φ is $1 - 1$ if and only if $\text{Ker}(\Phi) = \{0\}$.

As we pointed out in Section 8.3, two vector spaces that are isomorphic are "essentially" equal. For one thing any result in one of these vector spaces is carried into an analogous result in the other one via the isomorphism.

This section has as its principal goal the theorem that if V is a finite-dimensional vector space over F, then $V \simeq F^{(n)}$ for some n. The dimension n of $F^{(n)}$ will become the dimension of V over F.

Every result proved in Chapters 3 and 4 for $F^{(n)}$ about linear independence, bases, dimension, and so on, becomes a theorem in V, established by use of the isomorphism of V onto $F^{(n)}$. We shall make a list of the principal results that transfer this way, referring back to the proofs given in $F^{(n)}$.

Suppose then that V is finite-dimensional over F. By the very definition of "finite-dimensional" we know that there is a finite set of elements, $v_1, \ldots, v_k$, in V that span V over F. Such a set of vectors is called a *generating set of V over F*. Since there is such a finite set, there is a finite set with the fewest number of elements that span V over F. We call such a smallest set, $u_1, \ldots, u_n$, a *minimal* generating set of V over F.

At this point we introduce concepts that we introduced and exploited heavily in Chapters 3 and 4, namely, linear independence and basis over F. As we shall soon see, a minimal generating set $u_1, \ldots, u_n$ of V over F will turn out to be a basis for V over F.

Definition. The elements $v_1, \ldots, v_m$ in V are said to be *linearly independent over F* if $a_1 v_1 + \cdots + a_m v_m = 0$ only if $a_1 = 0, \ldots, a_m = 0$. If $v_1, \ldots, v_m$ are *not* linearly independent over F, we say that they are *linearly dependent over F*.

Two small things should be pointed out.

To say that elements $v_1, \ldots, v_m$ of V are linearly independent is equivalent to saying that if v is in the span of $v_1, \ldots, v_m$, then v has a unique representation $v = a_1 v_1 + \cdots + a_m v_m$. That is, if v is also $v = b_1 v_1 + \cdots + b_m v_m$, then $a_1 = b_1$, $a_2 = b_2, \ldots, a_m = b_m$.

Why? Suppose first that $v_1, \ldots, v_m$ are linearly independent and that an element v in their span can be written as $v = a_1 v_1 + \cdots + a_m v_m$ and also as $v = b_1 v_1 + \cdots + b_m v_m$. Subtracting the first expression from the second, we get $0 = (b_1 - a_1) v_1 + \cdots + (b_m - a_m) v_m$ so that $(b_1 - a_1) = 0, \ldots, (b_m - a_m) = 0$ by the linear independence.

Thus $a_1 = b_1, \ldots, a_m = b_m$. Conversely, suppose that each element v in the span of $v_1, \ldots, v_m$ has a unique expression $v = a_1 v_1 + \cdots + a_m v_m$. Then the condition $a_1 v_1 + \cdots + a_m v_m = 0$ implies that $a_1 v_1 + \cdots + a_m v_m = 0 v_1 + \cdots + 0 v_m$, so that $a_1 = 0, \ldots, a_m = 0$ by the unicity of the representation.

If $v_1, \ldots, v_m$ are linearly independent over F, then none of $v_1, \ldots, v_m$ can be 0.

For suppose that $v_1 = 0$; then $av_1 + 0v_2 + \cdots + 0v_m = 0$ for *any* $a \in F$. By the paragraph above, we see that $v_1, \ldots, v_m$ then cannot be linearly independent over F.

In V, the vector space of all polynomials in x over F, the elements $1 + x$, $2 - x + x^2$, x^3, $\frac{1}{4}x^5$ can easily be shown to be linearly independent over F. (Do it!) On the other hand, the elements 5, $1 + x$, $2 + x$, $7 + x + x^2$ are linearly dependent over F, since $-\frac{1}{5}(5) + (-1)(1 + x) + 1(2 + x) + (0)(7 + x + x^2) = 0$.

Definition. A *basis* for V is a set $v_1, \ldots, v_n$ of vectors of V such that

1. The vectors $v_1, \ldots, v_n$ are linearly independent:
2. The vectors $v_1, \ldots, v_n$ span V.

A generating set $v_1, \ldots, v_n$ for a finite-dimensional vector space V is a basis if and only if it is linearly independent. If $v_1, \ldots, v_n$ is a basis for V over F, then each element $v \in V$ has a *unique* representation $v = a_1 v_1 + \cdots + a_n v_n$, where the $a_1, \ldots, a_n$ are from F. Using this, we now prove the

Theorem 8.4.1. *If $u_1, \ldots, u_n$ is a minimal generating set of a finite-dimensional vector space V over F, then $u_1, \ldots, u_n$ is a basis for V over F.*

Proof: Suppose to the contrary that $u_1, \ldots, u_n$ are linearly dependent, that is, $a_1 u_1 + \cdots + a_n u_n = 0$, where not all the a_i are 0. Since the numbering of $u_1, \ldots, u_n$ is not important, we may assume that $a_1 \neq 0$. Then $a_1 u_1 = -a_2 u_2 - \cdots - a_n u_n$, and since $a_1 \neq 0$, we

$$u_1 = (-a_2/a_1) u_2 - \cdots - (a_n/a_1) u_n = b_2 u_2 + \cdots + b_n u_n,$$

where the $b_r = (-a_r/a_1)$ are in F. Given $v \in V$, then, since $u_1, \ldots, u_n$ is a generating set of V over F, we have

$$v = c_1 u_1 + \cdots + c_n u_n = c_1(b_2 u_2 + \cdots + b_n u_n) + c_2 u_2 + \cdots + c_n u_n$$
$$= (c_1 b_2 + c_2) u_2 + \cdots + (c_1 b_n + c_n) u_n$$

and since the $c_r b_s + c_s$ are in F, we get that $u_2, \ldots, u_n$ already span V over F. Thus the set $u_2, \ldots, u_n$ has fewer elements than $u_1, \ldots, u_n$, and $u_1, \ldots, u_n$ has been assumed to be a *minimal* generating set. This is a contradiction. So we are forced to conclude that all of $a_1, \ldots, a_n$ are 0. In other words, $u_1, \ldots, u_n$ are linearly independent over F. Since they span V, they are a basis. ∎

Corollary 8.4.2. *Every finite-dimensional vector space has a basis.*

We are now in a position to prove the principal results of this section.

Theorem 8.4.3. If V is a vector space with basis $v_1, \ldots, v_n$ over F, then there is an isomorphism Φ from V to $F^{(n)}$ such that

$$\Phi(a_1 v_1 + \cdots + a_n v_n) = \begin{bmatrix} a_1 \\ \vdots \\ a_n \end{bmatrix}$$

for all $a_1, \ldots, a_n$ in F.

Proof: Let $v_1, \ldots, v_n$ be a basis for V over F. Given $u \in V$, u has a representation as $u = a_1 v_1 + \cdots + a_n v_n$. Define $\Phi: V \to F^{(n)}$ by

$$\Phi(u) = \Phi(a_1 v_1 + \cdots + a_n v_n) = \begin{bmatrix} a_1 \\ \vdots \\ a_n \end{bmatrix}.$$

Because the a_r are unique for u, this mapping Φ is well defined, that is, makes sense.

Clearly Φ maps V onto $F^{(n)}$, for given $\begin{bmatrix} c_1 \\ \vdots \\ c_n \end{bmatrix}$ in $F^{(n)}$, then $\begin{bmatrix} c_1 \\ \vdots \\ c_n \end{bmatrix} = \Phi(c_1 v_1 + \cdots + c_n v_n)$. We claim that Φ is a homomorphism of V onto $F^{(n)}$. Why? If u, w are in V, then

$$u = a_1 v_1 + \cdots + a_n v_n \text{ and } w = b_1 v_1 + \cdots + b_n v_n.$$

Thus

$$u + w = (a_1 v_1 + \cdots + a_n v_n) + (b_1 v_1 + \cdots + b_n v_n) = (a_1 + b_1) v_1 + \cdots + (a_n + b_n) v_n,$$

whence, by the definition of Φ,

$$\Phi(u + w) = \begin{bmatrix} a_1 + b_1 \\ \vdots \\ a_n + b_n \end{bmatrix} = \begin{bmatrix} a_1 \\ \vdots \\ a_n \end{bmatrix} + \begin{bmatrix} b_1 \\ \vdots \\ b_n \end{bmatrix}.$$

However, we recognize $\begin{bmatrix} a_1 \\ \vdots \\ a_n \end{bmatrix}$ as $\Phi(u)$ and $\begin{bmatrix} b_1 \\ \vdots \\ b_n \end{bmatrix}$ as $\Phi(w)$. Putting all these pieces together, we get that $\Phi(u + w) = \Phi(u) + \Phi(w)$. A very similar argument shows that $\Phi(au) = a\Phi(u)$ for $a \in F$, $u \in V$. In short, Φ is a homomorphism of V onto $F^{(n)}$.

To finish the proof, all we need to show now is that Φ is $1 - 1$. By Lemma 8.3.4 it is enough to show that $\text{Ker}(\Phi) = (0)$. What is $\text{Ker}(\Phi)$? If $z = c_1 v_1 + \cdots + c_n v_n$ is in $\text{Ker}(\Phi)$, then, by the very definition of kernel, $\Phi(z) = \begin{bmatrix} 0 \\ \vdots \\ 0 \end{bmatrix}$. But we know precisely what Φ does to z, by the definition of Φ, that is, we know that $\Phi(z) = \begin{bmatrix} c_1 \\ \vdots \\ c_n \end{bmatrix}$. Comparing these two evaluations of $\Phi(z)$, we get $\begin{bmatrix} c_1 \\ \vdots \\ c_n \end{bmatrix} = \begin{bmatrix} 0 \\ \vdots \\ 0 \end{bmatrix}$, that is, $c_1 = c_2 = \cdots = c_n = 0$. Thus Ker$(\Phi)$ consists only of 0. Thus Φ is an isomorphism of V onto $F^{(n)}$. This completes the proof of Theorem 8.4.3. ∎

Since every finite-dimensional vector space has a basis, by Corollary 8.4.2, we have the

Corollary 8.4.4. *Any finite-dimensional vector space over F is isomorphic to $F^{(n)}$ for some positive integer n.*

Theorem 8.4.3 opens a floodgate of results for us. By the isomorphism given in this theorem, we can carry over the whole corpus of results proved in Chapters 3 and 4 without the need to go into proof. Why? We can now use the following general *transfer principle*, thanks to Theorem 8.4.3 and Lemma 8.3.1: Any phenomena expressed in $F^{(n)}$ in terms of addition, scalar multiplication, or linear combinations are transferred by any isomorphism ϕ from V to $F^{(n)}$ to corresponding phenomena expressed in V in terms of addition, scalar multiplication, or linear combinations.

Let's be specific. In $F^{(n)}$, concepts of span, linear independence, basis, dimension, and so on, have been defined in such terms. In a finite-dimensional vector space V, we have defined span, linear independence along the same lines. Since $F^{(n)}$ has the standard basis $e_1, \ldots, e_n$ and any basis for $F^{(n)}$ has n elements by Corollary 3.6.4, $F^{(n)}$ has a basis and any two bases have the same number of elements. By our transfer principle, it follows that any finite-dimensional vector space has a basis and any two bases have the same number of elements.

We have now established the following theorems.

Theorem 8.4.5. *A finite-dimensional vector space V over F has a basis.*

Theorem 8.4.6. *Any two bases of a finite-dimensional vector space V over F have the same number of elements.*

We now can define dimension.

Definition. The *dimension* of a finite-dimensional vector space V over F is the number of elements in a basis. It is denoted by dim(V).

Of course, all results from Chapter 3 on $F^{(n)}$ that can be expressed in terms of linear transformations carry over by our transfer principle to finite-dimensional vector spaces. Here is a partial list corresponding to principal results:

1. $v_1, \ldots, v_n$ form a basis if and only if $v_1, \ldots, v_n$ is a minimal generating set.
2. $v_1, \ldots, v_m$ are linearly independent implies that $m \leq \dim(V)$.

3. $v_1, \ldots, v_n$ form a basis if and only if $v_1, \ldots, v_n$ is a maximal linearly independent subset of V.
4. Any linearly independent subset of V is contained in a basis of V.
5. Any linearly independent set of n elements of an n-dimensional vector space V is a basis.

We shall return to linear independence soon, because we need this concept even if V is not finite-dimensional over F.

As we shall see as we progress, using Theorem 8.4.3 we will be able to carry over the results on inner product spaces to the general context of vector spaces. Finally, and probably of greatest importance, everything we did for linear transformations of $F^{(n)}$—that is, for matrices—will go over neatly to finite-dimensional vector spaces. All these things in due course.

PROBLEMS

1. Make a list of all the results in Chapter 3 on $F^{(n)}$ as a vector space that can be carried over to a general finite-dimensional vector space by use of Theorem 8.4.3.
2. Prove the converse of Theorem 8.4.1—that any basis for V over F is a minimal generating set.

8.5. LINEAR INDEPENDENCE IN INFINITE-DIMENSIONAL VECTOR SPACES

We have described linear independence and dependence several times. Why, then, redo these things once again? The basic reason is that in all our talk about this we stayed in a finite-dimensional context. But these notions are important and meaningful even in the infinite-dimensional situation. This will be especially true when we pick up the subject of inner product spaces.

As usual, if V is any vector space over F, we say that the finite set $v_1, \ldots, v_n$ of elements in V is *linearly independent over F* if $a_1 v_1 + \cdots + a_n v_n = 0$, where $a_1, \ldots, a_n$ are in F, only if $a_1 = a_2 = \cdots = a_n = 0$. To avoid repetition of a phrase, we shall merely say "linearly independent," it being understood that this is over F. Of course, if $v_1, \ldots, v_n$ are not linearly independent, we say that they are *linearly dependent*.

But we can also speak about an infinite subset as a subset of linearly independent (or dependent) elements. We do so as follows:

Definition. If $S \subset V$ is any nonempty subset, we say that S is *linearly independent* if *any finite* subset of S consists of linearly independent elements.

For instance, if V is the vector space of polynomials in x over $\mathbb{R}$, then the set $S = \{1, x, x^2, \ldots, x^k, \ldots\}$ is a linearly independent set of elements. If $x^{i_1}, \ldots, x^{i_n}$ is any finite subset of S, where $0 \le i_1 < i_2 < \cdots < i_n$, and if $a_1 x^{i_1} + \cdots + a_n x^{i_n} = 0$, then, by invoking the definition of when a polynomial is identically zero, we get that $a_1, \ldots, a_n$ are all 0. Thus S is a linearly independent set.

It is easy to construct other examples. For instance, if V is the vector space of all real-valued differential functions, then the functions e^x, e^{2x}, ..., e^{nx}, ... are linearly independent over $\mathbb{R}$. Perhaps it would be worthwhile to verify this statement in this case, for it may provide us with a technique we could use elsewhere. Suppose then that $a_1 e^{m_1 x} + \cdots + a_k e^{m_k x} \equiv 0$, $a_k \neq 0$, where $1 \leq m_1 < m_2 < \cdots < m_k$ are integers. Suppose that this is the *shortest possible* such relation. We differentiate the expression to get $m_1 a_1 e^{m_1 x} + \cdots + m_k a_k e^{m_k x} \equiv 0$. Multiplying the first by m_k and subtracting the second we get $(m_k - m_1) a_1 e^{m_1 x} + \cdots + (m_k - m_{k-1}) a_{k-1} e^{m_{k-1} x} \equiv 0$. This is a *shorter* relation, so each $(m_k - m_r) a_r = 0$ for $1 \leq r \leq k-1$. But since $m_k > m_r$ for $r \leq k-1$, we end up with $a_r = 0$ for $r \leq k-1$. This leaves us with $a_k m^{kx} \equiv 0$, hence $a_k = 0$, a contradiction. Therefore, the functions $e^x, e^{2x}, \ldots, e^{mx}, \ldots$ are linearly independent over $\mathbb{R}$.

In the problem set there will be other examples. This is about all we wanted to add to the notion of linear independence, namely, to stress that the concept has significance even in the infinite-dimensional case.

PROBLEMS

NUMERICAL PROBLEMS

1. If V is the vector space of differentiable functions over $\mathbb{R}$, show that the following sets are linearly independent.
   (a) $\cos(x), \cos(2x), \ldots, \cos(nx), \ldots$.
   (b) $\sin(x), \sin(2x), \ldots, \sin(nx), \ldots$.
   (c) $\cos(x), \sin(x), \cos(2x), \sin(2x), \ldots, \cos(nx), \sin(nx), \ldots$.
   (d) $1, 2+x, 3+2x^2, 4+3x^3, \ldots, n+(n-1)x^{n-1}, \ldots$.

2. Which of the following sets are linearly independent, and which are not?
   (a) V is the vector space of polynomials over F, and the elements are
   $$1, 2+x, 3+2x, \ldots, n+(n-1)x, \ldots$$
   (b) V is the set of all real-valued functions on $[0,1]$ and the elements are
   $$\frac{1}{x+1}, \frac{1}{x+2}, \ldots, \frac{1}{x+n}, \ldots$$
   (c) V is the set of polynomials over $\mathbb{R}$ and the elements are
   $$1, x, 1+x+x^2, 1+x+x^2+x^3, \ldots, 1+x+\cdots+x^n, \ldots$$
   (d) V is the set of polynomials over F, and the elements are
   $$1, (x+2), (x+3)^2, \ldots, (x+n)^{n-1}, \ldots$$

MORE THEORETICAL PROBLEMS

If S is in V, we say that the *subspace of V spanned by S over F* is the set of *all finite linear combinations* of elements of S over F.

Easier Problems

3. Find the subspace of V spanned by the given elements:

  (a) V is the set of polynomials in x over F, and the elements are
  $$1, \ (x+2), \ \ldots, \ (x+n)^{n-1}, \ \ldots.$$

  (b) V is the set of polynomials in x over F, and the elements are
  $$1, \ x, \ x^2, \ \ldots, \ x^n, \ (x+1)^n, \ \ldots, \ (x+k)^n, \ \ldots.$$

Middle-Level Problems

4. Show that there is no *finite* subset of the vector space of polynomials in x over F which span all this vector space.

5. Let V be the vector space of all complex-valued functions, as a vector space over $\mathbb{C}$. Let e^{ix} denote the function $\cos(x) + i\sin(x)$, where $i^2 = -1$.

  (a) Show that $e^{\pm ix}, e^{\pm 2ix}, \ldots, e^{\pm inx}, \ldots$ are linearly independent over $\mathbb{C}$.

  (b) From Part (a), show that
  $$\cos(x), \ \sin(x), \ \ldots, \ \cos(nx), \ \sin(nx), \ \ldots$$
  are linearly independent over $\mathbb{C}$.

  (c) Show that
  $$e^{\pm ix}, \ \ldots, \ e^{\pm inx}, \ \ldots, \ \cos(x), \ \cos(2x), \ \ldots, \ \cos(nx), \ \ldots$$
  are *not* linearly independent over $\mathbb{C}$.

Harder Problems

6. Let V be the vector space of all continuous real-valued functions on the real line and let S be the set of all differentiable functions on the real line. Prove that S does not span V over $\mathbb{R}$ by showing that $f(x) = |x|$ is *not* in the span of S.

7. If $f_0(x), \ldots, f_n(x)$ are differentiable functions over $\mathbb{R}$, show that $f_0(x), \ldots, f_n(x)$ are linearly independent over $\mathbb{R}$ if

$$\begin{vmatrix} f_0(x) & \cdots & f_n(x) \\ f_0'(x) & \cdots & f_n'(x) \\ \cdots & \cdots & \cdots \\ f_0^{(n)}(x) & \cdots & f_n^{(n)}(x) \end{vmatrix}$$

is not identically 0. [This determinant is called the *Wronskian* of $f_0(x), \ldots, f_n(x)$.]

8. Use the result of Problem 7 to show that the set $e^x, e^{2x}, \ldots, e^{nx}$ is linearly independent.

9. Use the result of Problem 8 to show that the set $e^x, e^{2x}, \ldots, e^{nx}, \ldots$ is a linearly independent set.

8.6. INNER PRODUCT SPACES

Once again we return to a theme that was expounded on at some length in Chapter 3, the notion of an *inner product space*. Recall that in $\mathbb{C}^{(n)}$ we defined the inner product of the vectors $v = \begin{bmatrix} a_1 \\ \vdots \\ a_n \end{bmatrix}$, $w = \begin{bmatrix} b_1 \\ \vdots \\ b_n \end{bmatrix}$ by $(v, w) = \sum_{j=1}^{n} a_j \bar{b}_j$. We discussed many properties of this inner product and saw how useful a tool it was in deriving theorems about $F^{(n)}$ and about matrices.

In line with our general program we want to carry these ideas over to general vector spaces over $\mathbb{C}$ (or $\mathbb{R}$). In fact, for a large part we shall not even insist that the vector space be finite-dimensional. When we do focus on finite-dimensional inner product spaces we shall see that they are isomorphic to $F^{(n)}$ by an isomorphism that preserves inner products. This will become more precise when we actually produce this isomorphism.

It's easy to define what we shall be talking about.

Definition. A vector space V over $\mathbb{C}$ will be called an *inner product space* if there is defined a function from V to $\mathbb{C}$, denoted by (v, w), for pairs v, w in V, such that:

1. $(v, v) \geq 0$, and $(v, v) = 0$ only if $v = 0$.
2. $(v, w) = \overline{(w, v)}$, the complex conjugate of (v, w).
3. $(av, w) = a(v, w)$ for $a \in \mathbb{C}$.
4. $(v_1 + v_2, w) = (v_1, w) + (v_2, w)$.

Notice that (1)–(4) imply:

1. $(v, aw) = \bar{a}(v, w)$. For
$$(v, aw) = \overline{(aw, v)} = \overline{a(w, v)} \, [\text{by (3)}] = \bar{a}\overline{(w, v)} = \bar{a}(v, w) \, [\text{by (2)}].$$

2. $(v, w_1 + w_2) = (v, w_1) + (v, w_2)$. For
$$(v, w_1 + w_2) = \overline{(w_1 + w_2, v)} = \overline{(w_1, v) + (w_2, v)} \, [\text{by (4)}]$$
$$= \overline{(w_1, v)} + \overline{(w_2, v)} = (v, w_1) + (v, w_2) \, [\text{by (2)}].$$

3. If $(v, w) = 0$ for all $w \in V$, then $v = 0$. For in particular, (v, v) is then 0, hence by (1), $v = 0$.

It might be illuminating to see some examples of inner product spaces.

EXAMPLES

1. We already saw one example, namely, $V = \mathbb{C}^{(n)}$ and
$$(v, w) = \sum_{j=1}^{n} a_j \bar{b}_j, \quad \text{where } v = \begin{bmatrix} a_1 \\ \vdots \\ a_n \end{bmatrix}, \, w = \begin{bmatrix} b_1 \\ \vdots \\ b_n \end{bmatrix}.$$

2. This will be an example of an inner product space over $\mathbb{R}$. Let V be the set of all continuous functions on the closed interval $[-1, 1]$. For $f(x)$, $g(x)$ in V define

$$(f, g) = \int_{-1}^{1} f(x)g(x)\, dx.$$

For instance, if $f = x$, $g = x^3$, then

$$(x, x^3) = \int_{-1}^{1} x x^3\, dx = \int_{-1}^{1} x^4\, dx = \left[\frac{x^5}{5}\right]_{-1}^{1} = \tfrac{2}{5}.$$

What is (x, e^x)?

We verify the various properties of an inner product.

1. $(f, f) = \int_{-1}^{1} f(x)f(x)\, dx \geq 0$ since $[f(x)]^2 \geq 0$ and this integral is 0 if and only if $f(x) \equiv 0$.

2. $(f, g) = \int_{-1}^{1} f(x)g(x)\, dx = \int_{-1}^{1} g(x)f(x)\, dx = (g, f)$ [and $(g, f) = (f, g)$, since everything in sight is real].

3. $(f_1 + f_2, g) = \int_{-1}^{1} [f_1(x) + f_2(x)]g(x)\, dx = \int_{-1}^{1} [f_1(x)g(x) + f_2(x)g(x)]\, dx =$

$\int_{-1}^{1} f_1(x)g(x)\, dx + \int_{-1}^{1} f_2(x)g(x)\, dx = (f_1, g) + (f_2, g).$

4. $(af, g) = \int_{-1}^{1} af(x)g(x)\, dx = a \int_{-1}^{1} f(x)g(x)\, dx = a(f, g)$ for $a \in \mathbb{R}$.

So V is an inner product space over $\mathbb{R}$.

3. Let V be the set of differentiable functions on $[-\pi, \pi]$ and let W be the linear span of

$$\cos(x),\ \cos(2x),\ \ldots,\ \cos(nx),\ \ldots$$

over $\mathbb{R}$. We make V into an inner product space over $\mathbb{R}$ by defining

$$(f, g) = \int_{-\pi}^{\pi} f(x)g(x)\, dx.$$

As in Example 2, V is an inner product space relative to this inner product. We want to compute the inner product of the elements $\cos(mx)$ which span W. For m, n we have

$$(\cos(mx), \cos(nx)) = \int_{-\pi}^{\pi} \cos(mx)\cos(nx)\, dx.$$

If $m = n \neq 0$, we have

$$(\cos(mx), \cos(mx)) = \int_{-\pi}^{\pi} \cos^2(mx)\,dx = \pi,$$

and if $m \neq n$,

$$(\cos(mx), \cos(nx)) = \int_{-\pi}^{\pi} \cos(mx)\cos(nx)\,dx$$

$$= \left[\frac{1}{2(m+n)}\sin([m+n]x) + \frac{1}{2(m-n)}\sin([m-n]x)\right]_{-\pi}^{\pi} = 0.$$

Finally, if $m = n = 0$,

$$\int_{-\pi}^{\pi} \cos(mx)\cos(nx)\,dx = \int_{-\pi}^{\pi} dx = 2\pi.$$

The example is an interesting one. Relative to the inner product the functions

$$\cos(x),\ \cos(2x),\ \ldots,\ \cos(nx),\ \ldots$$

form an *orthogonal* set in V.

4. Let V be as in Example 3 and let

$$S = \{\cos(mx)\,|\,\text{all integers } m \geq 0\} \cup \{\sin(nx)\,|\,\text{all integers } n \geq 1\}.$$

As in Example 3, if we define the inner product

$$(f, g) = \int_{-\pi}^{\pi} f(x)g(x)\,dx.$$

Then S is a linearly independent set and $(f, g) = 0$ for $f \neq g$ in S. This remark is the foundation on what is known as *Fourier series* rests.

We now go on to some properties of inner product spaces. The first result is one we proved earlier, in Chapter 1. But we redo it here, in exactly the same way.

Lemma 8.6.1. *If $a > 0$, b, c are real numbers and $ax^2 + bx + c \geq 0$ for all real x, then $b^2 - 4ac \leq 0$.*

Proof: Since $ax^2 + bx + c \geq 0$ for all real x, it is certainly true for the real number $-b/2a$:

$$a\left(\frac{-b}{2a}\right)^2 + b\left(\frac{-b}{2a}\right) + c \geq 0.$$

Hence $b^2/4a - b^2/2a + c \geq 0$, that is, $c - b^2/4a \geq 0$. Because $a > 0$, this leads us to $b^2 \leq 4ac$. ∎

This technical result has a famous inequality as a consequence. It is known as the *Schwarz Inequality*.

Lemma 8.6.2. In an inner product space V, $|(v,w)|^2 \leq (v,v)(w,w)$ for all v, w in V.

Proof: If $v, w \in V$ and $x \in \mathbb{R}$ then, by Property 1 defining an inner product space, $(v + xw, v + xw) \geq 0$. Expanding this, we obtain $(v,v) + x(\overline{(v,w)} + (w,v)) + x^2(w,w) \geq 0$. We first settle the case in which (v,w) is real. Then $(v,w) = \overline{(w,v)} = (w,v)$. So the inequality above becomes $(w,w)x^2 + 2x(v,w) + (v,v) \geq 0$ for all real x. Invoking Lemma 8.6.1, with $a = (w,w)$, $b = 2(v,w)$, $c = (v,v)$ we get $4(v,w)^2 \leq 4(v,v)(w,w)$, so

$$|(v,w)|^2 \leq (v,v)(w,w),$$

the desired conclusion.

Suppose then that (v,w) is not real; then certainly $(v,w) \neq 0$. Let $z = v/(v,w)$, then $(z,w) = (v/(v,w), w) = \dfrac{1}{(v,w)}(v,w) = 1$. Thus (z,w) is real. By what we did above, $1 = |(z,w)|^2 \leq (z,z)(w,w)$. Now

$$(z,z) = \left(\frac{v}{(v,w)}, \frac{v}{(v,w)}\right) = \frac{1}{(v,w)}\frac{1}{\overline{(v,w)}}(v,v) = \frac{1}{|(v,w)|^2}(v,v)$$

since $(v,w)\overline{(v,w)} = |(v,w)|^2$. So we obtain that

$$1 = |(z,w)|^2 \leq \frac{1}{|(v,w)|^2}(v,v)(w,w),$$

from which we get

$$|(v,w)|^2 \leq (v,v)(w,w).$$

This finishes the proof. ∎

In $F^{(n)}$ we saw that this lemma implies that

$$\left(\left|\sum_{j=1}^{n} a_j \bar{b}_j\right|\right)^2 \leq \left(\sum_{j=1}^{n} |a_j|^2\right)\left(\sum_{j=1}^{n} |b_j|^2\right)$$

using the definition $(v,w) = \sum_{j=1}^{n} a_j \bar{b}_j$, where $v = \begin{bmatrix} a_1 \\ a_2 \\ \vdots \\ a_n \end{bmatrix}$ and $w = \begin{bmatrix} b_1 \\ b_2 \\ \vdots \\ b_n \end{bmatrix}$.

Perhaps of greater interest is the case of the real continuous functions, as a vector space over $\mathbb{R}$, with the inner product defined, say, by

$$(f,g) = \int_{-1}^{1} f(x)g(x)\,dx.$$

The Schwarz inequality then tell us that

$$\left| \int_{-1}^{1} f(x)g(x)\,dx \right|^2 \leq \left(\int_{-1}^{1} f^2(x)\,dx \right)\left(\int_{-1}^{1} g^2(x)\,dx \right).$$

Of course, we could define the inner product using other limits of integration. These integral inequalities are very important in analysis in carrying out estimates of integrals. For example, we could use this to estimate integrals where we cannot carry out the integration in closed form.

We apply it, however, to a case where we could do the integration. Suppose that we wanted some upper bound for $\int_{-\pi}^{\pi} x \sin(x)\,dx$. According to the Schwarz inequality

$$\left| \int_{-\pi}^{\pi} x \sin(x)\,dx \right|^2 \leq \left(\int_{-\pi}^{\pi} x^2\,dx \right)\left(\int_{-\pi}^{\pi} \sin^2(x)\,dx \right) = \left(\frac{2}{3}\pi^3 \right)(\pi) = \frac{2}{3}\pi^4.$$

We know that $\int x \sin(x)\,dx = \sin(x) - x\cos(x)$, hence $\int_{-\pi}^{\pi} x \sin(x)\,dx = 2\pi$. Hence $\left(\int_{-\pi}^{\pi} x \sin(x)\,dx \right)^2 = (2\pi)^2 = 4\pi^2$. So we see here that the upper bound $2\pi^4/3$ is crude, but not too crude.

In an inner product space we can introduce the concept of length.

Definition. If V is an inner product space, then the *length* of v in V, denoted by $\|v\|$, is defined by $\|v\| = \sqrt{(v,v)}$.

In $\mathbb{R}^{(n)}$ this coincides with our usual (and intuitive) notion of length. Note that $\|v\| = 0$ if and only if $v = 0$.

What properties does this length function enjoy? One of the basic ones is the *triangle inequality*.

Lemma 8.6.3. For $v, w \in V$, $\|v + w\| \leq \|v\| + \|w\|$.

Proof: By definition, $\|v + w\| = \sqrt{(v+w, v+w)}$. Now

$$(v+w, v+w) = (v,v) + (w,w) + (v,w) + (w,v).$$

By Lemma 8.6.2, $|(v,w)| = |(w,v)| \leq \sqrt{(v,v)(w,w)}$. So

$$\|v+w\|^2 = (v+w, v+w) = (v,v) + (w,w) + (v,w) + (w,v)$$
$$\leq (v,v) + (w,w) + 2\sqrt{(v,v)(w,w)}$$
$$= \|v\|^2 + \|w\|^2 + 2\|v\|\|w\|.$$

Sec. 8.6]   Inner Product Spaces

This gives us

$$\|v + w\|^2 \leq (\|v\|^2 + \|w\|^2 + 2\|v\|\|w\|) = (\|v\| + \|w\|)^2.$$

Taking square roots, we end up with $\|v + w\| \leq \|v\| + \|w\|$, the desired result. ∎

In working in $\mathbb{C}^{(n)}$ we came up with a scheme whereby from a set $v_1, \ldots, v_k$ of linearly independent vectors we created a set $w_1, \ldots, w_k$ of orthogonal vectors. We want to do the same thing in general, especially when V is infinite-dimensional over $\mathbb{C}$ (or $\mathbb{R}$). But first we better say what is meant by "orthogonal."

Definition. If v, w are in V, then v is said to be *orthogonal* to w if $(v, w) = 0$.

Since $(v, w) = \overline{(w, v)}$ we see immediately that if v is orthogonal to w, then w is orthogonal to v.

Let S be a set of nonzero orthogonal elements in V.

Lemma 8.6.4. A set S of nonzero orthogonal elements in V is a linearly independent set.

Proof: Suppose for some finite set of elements $s_1, \ldots, s_k$ in S that $a_1 s_1 + \cdots + a_k s_k = 0$. Thus

$$\begin{aligned} 0 &= (a_1 s_1 + \cdots + a_k s_k, s_j) \\ &= (a_1 s_1, s_j) + \cdots + (a_k s_k, s_j) \\ &= a_1(s_1, s_j) + \cdots + a_k(s_k, s_j). \end{aligned}$$

Because the elements of S are orthogonal, $(s_t, s_j) = 0$ for $t \neq j$. So the equation above reduces to $a_j(s_j, s_j) = 0$. The outcome of all this is that each $a_j = 0$. In other words, any finite subset of S consists of linearly independent elements. Hence S is a linearly independent set. ∎

Definition. A set S in V is said to be an *orthonormal* set if S is an orthogonal set and each element in S is of length 1.

Note that from an orthogonal set of nonzero elements we readily produce an orthonormal set. This is the content of

Lemma 8.6.5. If S is an orthogonal set of nonzero elements of V, then $\tilde{S} = \left\{ \dfrac{s}{\|s\|} \,\middle|\, s \in S \right\}$ is an orthonormal set.

Proof: Given $s \in S$, then $\left\| \dfrac{s}{\|s\|} \right\| = \dfrac{1}{\|s\|} \|s\| = 1$, so every element in $\tilde{S}$ has length 1. Also, if $s \neq t$ are in S, then $(s, t) = 0$, hence $\left(\dfrac{s}{\|s\|}, \dfrac{t}{\|t\|} \right) = \dfrac{1}{\|s\|} \dfrac{1}{\|t\|} (s, t) = 0$. In short, $\tilde{S}$ is an orthonormal set. ∎

PROBLEMS

NUMERICAL PROBLEMS

1. Compute the inner product (f, g) in V, the set of all real-valued continuous functions on $[-\pi, \pi]$, where $(f, g) = \int_{-\pi}^{\pi} f(x)g(x)\,dx$.

   (a) $f(x) = \cos(x)$, $\quad g(x) = x$.
   (b) $f(x) = e^x$, $\quad g(x) = \sin(x) + \cos(x)$.
   (c) $f(x) = \cos(4x)$, $\quad g(x) = \sin(x)$.
   (d) $f(x) = e^x$, $\quad g(x) = 1 - e^x$.

2. In V of Problem 1, find a vector w orthogonal to the given v.

   (a) $v = \cos(x) + \sin(2x) + \cos(3x)$.
   (b) $v = e^x$.
   (c) $v = 1 + x$.

3. Show that $(0, w) = (v, 0) = 0$ for all $v, w \in V$, where V is an inner product space.

4. In Problem 1 find the length of each of the given functions $g(x)$.

MORE THEORETICAL PROBLEMS

Easier Problems

5. If W is a subspace of V, let $W^\perp = \{v \in V \,|\, (v, w) = 0 \text{ for all } w \in W\}$.

   (a) Prove that $W^\perp$ is a subspace of V.
   (b) Show that $(W^\perp)^\perp \supset W$.

6. In Problem 5 show that $((W^\perp)^\perp)^\perp = W^\perp$.

7. If V is the vector space of polynomials with real coefficients with inner product given by $(f, g) = \int_{-1}^{1} f(x)g(x)\,dx$, find $W^\perp$ for

   (a) W is the linear span of $1, x, x^2$.
   (b) W is the linear span of $x + x^2, x^2 + x^3$.

8. Show that if W is a subspace of V, then $W \cap W^\perp = \{0\}$.

9. In V, an inner product space, define the *distance* between two elements v and w by $d(v, w) = \|v - w\|$. Prove:

   (a) $d(v, w) = 0$ if and only if $v = w$.
   (b) $d(v, w) = d(w, v)$.
   (c) $d(v, w) + d(w, z) \geq d(v, z)$ (triangle inequality).

10. If V is the inner product space of continuous real-valued functions on $[-\pi, \pi]$, find $d(v, w)$ for

    (a) $v = \cos(x)$, $\quad w = \sin(x)$.

(b) $v = e^x$, $\quad\quad\quad\quad w = e^{-x}$.

(c) $v = \cos(x) + \cos(2x)$, $\quad w = 3\sin(x) - 4\sin(2x)$.

Middle-Level Problems

11. If W is a subspace of V, let $W + W^\perp = \{u + v \mid u \in W, v \in W^\perp\}$. Show that if z is orthogonal to every element of $W + W^\perp$, then $z = 0$.

12. Let $V = M_n(\mathbb{R})$ and define for $A, B \in V$, the inner product of A and B by $(A, B) = \text{tr}(AB')$, where B' is the transpose of B. Show that this is an inner product on V.

13. Let E_{rs} be that matrix all of whose entries are 0 except the (r, s)-entry, which is 1. Prove that relative to the inner product of Problem 12, the E_{rs} form an orthonormal basis of V.

14. In Problem 12 show that $(AB, C) = (B, A'C)$.

Harder Problems

15. If V is an inner product space and $v \in V$, let $v^\perp = \{w \in V \mid (v, w) = 0\}$. Prove that $V = Fv + v^\perp$.

16. If V is an inner product space and W is a 2-dimensional subspace of V, show that $V = W + W^\perp$.

17. In Problem 12, if W is the subspace of all diagonal matrices, find $W^\perp$ and verify that $W + W^\perp = V = M_n(\mathbb{R})$.

8.7. MORE ON INNER PRODUCT SPACES

Once again we shall repeat something that was done earlier in $\mathbb{C}^{(n)}$ and $\mathbb{R}^{(n)}$. But here V will be an arbitrary inner product space over $\mathbb{R}$ or $\mathbb{C}$. The results we obtain are of importance not only for the finite-dimensional case but possibly more so for the infinite-dimensional one. One by-product of what we shall do is the fact that studying finite-dimensional inner product spaces is no more nor less than studying $F^{(n)}$ as an inner product space. This will be done by exhibiting a particular isomorphism. From this we shall be able to transfer everything we did in Chapters 3 and 4 for the special case of $F^{(n)}$ to arbitrary finite-dimensional vector spaces over F.

The principal tool we need is what was earlier called the *Gram–Schmidt process*. The outcome of this process is that for any set $S = \{s_r\}$ of linearly independent elements we can find an orthonormal set S_1 such that the linear span of S over F equals the linear span of S_1 over F.

Theorem 8.7.1 (Gram–Schmidt). *Let $S = \{s_r\}$ be a nonempty set of linearly independent elements in V. Then there is an orthonormal set S_1, contained in the linear span of S, such that the linear span of S_1 equals that of S.*

Proof: By Lemma 8.6.5 it is enough for us to produce an orthogonal set S_2 contained in the linear span of S, such that S_2 has the same linear span as does S. To modify S_2 to an orthonormal set—as is shown in Lemma 8.6.5—is easy.

Since S consists of linearly independent elements, we know, to begin with, that 0 is not in S. Thus $(s, s) \neq 0$ for every $s \in S$.

Suppose that the elements of S are enumerated as $s_1, s_2, \ldots, s_n, \ldots$. Step by step we shall construct nonzero elements $t_1, t_2, \ldots, t_n, \ldots$ such that t_j is a linear combination of $s_1, \ldots, s_j$, and such that $(t_j, t_k) = 0$ for $j \neq k$, and finally, such that the linear span of $S_1 = \{t_1, t_2, \ldots, t_n, \ldots\}$ equals that of S.

Where do we begin? We pick our first element t_1 in the easiest possible way, namely

$$t_1 = s_1.$$

How do we get t_2? Let $t_2 = as_1 + s_2 = at_1 + s_2$, where $a \in \mathbb{C}$ is to be determined. We want $(t_1, t_2) = 0$; that is, we want

$$0 = (as_1 + s_2, s_1) = a(s_1, s_1) + (s_2, s_1).$$

Since $(s_1, s_1) \neq 0$ we can solve for a as $a = -\dfrac{(s_2, s_1)}{(s_1, s_1)}$,

getting

$$t_2 = s_2 - \frac{(s_2, s_1)}{(s_1, s_1)} t_1.$$

Now we do have $(t_1, t_2) = 0$. Note, too, that since t_2 is a nonzero linear combination of s_1 and s_2, which are linearly independent, $t_2 \neq 0$. Also note that both s_1 and s_2 are expressible as linear combinations of t_1 and t_2. So the linear span of t_1 and t_2 equals that of s_1 and s_2.

We go on in this way, step by step. Suppose that we have constructed for $1 \leq j \leq k$ nonzero $t_1, t_2, \ldots, t_k$ which are orthogonal, where t_j is in the linear span of $s_1, \ldots, s_j$. We want to construct the next one, t_{k+1}. How? We want t_{k+1} to be a linear combination of $s_1, \ldots, s_{k+1}$, $t_{k+1} \neq 0$, and most important, $(t_{k+1}, t_j) = 0$ for $j \leq k$. Let

$$t_{k+1} = s_{k+1} - \frac{(s_{k+1}, t_1)}{(t_1, t_1)} t_1 - \frac{(s_{k+1}, t_2)}{(t_2, t_2)} t_2 - \cdots - \frac{(s_{k+1}, t_k)}{(t_k, t_k)} t_k.$$

Because $t_1, \ldots, t_k$ are linear combinations of the s_j, where $j \leq k < k + 1$, and since the s_i's are linearly independent, we see that $t_{k+1} \neq 0$. Also, t_{k+1} is in the linear span of $s_1, \ldots, s_{k+1}$. Finally, for $j \leq k$,

$$(t_{k+1}, t_j) = \left(s_{k+1} - \frac{(s_{k+1}, t_1)}{(t_1, t_1)} t_1 - \frac{(s_{k+1}, t_2)}{(t_2, t_2)} t_2 - \cdots \frac{(s_{k+1}, t_k)}{(t_k, t_k)} t_k, t_j \right)$$

$$= (s_{k+1}, t_j) - \frac{(s_{k+1}, t_1)}{(t_1, t_1)} (t_1, t_j) - \cdots - \frac{(s_{k+1}, t_j)}{(t_j, t_j)} (t_j, t_j) - \cdots - \frac{(s_{k+1}, t_k)}{(t_k, t_k)} (t_k, t_j).$$

Now the elements $t_1, \ldots, t_k$ had already been constructed in such a way that $(t_i, t_j) = 0$ for $i \neq j$. So the only nonzero contributions to our sum expression for

(t_{k+1}, t_j) come from (s_{k+1}, t_j) and $\dfrac{(s_{k+1}, t_j)}{(t_j, t_j)}(t_j, t_j)$. Thus this sum reduces to

$$(t_{k+1}, t_j) = (s_{k+1}, t_j) - \dfrac{(s_{k+1}, t_j)}{(t_j, t_j)}(t_j, t_j) = (s_{k+1}, t_j) - (s_{k+1}, t_j) = 0.$$

So we have gone one more step and have constructed t_{k+1} from $t_1, \ldots, t_k$ and s_{k+1}. Continue with the process. Note that from the form of the construction s_{k+1} is a linear combination of $t_1, \ldots, t_{k+1}$. This is true for all k. So we get that the linear span of the $\{t_1, t_2, \ldots, t_n, \ldots\}$ is equal to the linear span of $S = \{s_1, \ldots, s_n, \ldots\}$.

This finishes the proof of this very important construction. ∎

Corollary 8.7.2. If V is a finite-dimensional inner product space, then V has an orthonormal basis.

What the theorem says is that given a linearly independent set $\{s_1, \ldots s_n, \ldots\}$ we can modify it, using linear combinations of $s_1, \ldots, s_n, \ldots$ to achieve an orthonormal set that has the same span.

We have seen this theorem used in $F^{(n)}$, so it is important here to give examples in infinite-dimensional inner product spaces.

EXAMPLES

1. Let V be the inner product space of all polynomials in x with real coefficients, where we define $(f, g) = \int_{-1}^{1} f(x)g(x)\,dx$. How do we construct an orthonormal set

$$p_1(x), p_2(x), \ldots, p_n(x), \ldots$$

whose linear span is that of

$$s_1 = 1, s_2 = x, s_3 = x^2, \ldots, s_n = x^{n-1}, \ldots,$$

that is, all of V? The recipe for doing this is given in the proof of Theorem 8.7.1.

We first construct a suitable orthogonal set. Let $q_1(x) = 1$, $q_2(x) = s_2 + as_1 = x + a$, where $a = -\dfrac{(q_2, q_1)}{(q_1, q_1)} = -\int_{-1}^{1}(1)(x)\,dx \Big/ \int_{-1}^{1} 1\,dx = 0$. So $q_2(x) = x$.

We go on to $q_3(x)$. What do we require of $q_3(x)$? Only that it be of the form

$$q_3 = s_3 + aq_1 + bq_2$$

with $a = -\dfrac{(s_3, q_1)}{(q_1, q_1)}$, $b = -\dfrac{(s_3, q_2)}{(q_2, q_2)}$ (as in the proof). This ensures that $(q_3, q_1) = (q_3, q_2) = 0$. Since $(s_3, q_1) = \int_{-1}^{1}(x^2)(1)\,dx = [x^3/3]_{-1}^{1} = \tfrac{2}{3}$, $(q_1, q_1) = 2$ and

$$(s_3, q_2) = \int_{-1}^{1} (x^2)(x)\,dx = [x^4/4]_{-1}^{1} = 0, \text{ we get}$$

$$q_3(x) = x^2 - \frac{1}{3}.$$

We go on in this manner to $q_4(x)$, $q_5(x)$, and so on. The set we get consists of orthogonal elements. To make of them an orthonormal set, we merely must divide each $q_i(x)$ by its length

$$\sqrt{q_r(x)q_r(x)} = \sqrt{\left(\int_{-1}^{1} [q_r(x)]^2\,dx\right)}.$$

We leave the evaluation of these and of some of the other $q_r(x)$ to the problem set that follows.

This orthonormal set of polynomials constructed is a version of what are called the *Legendre polynomials*. They are of great importance in physics, chemistry, engineering, and mathematics.

2. Let V be the vector space of all sequences $\{a_1, a_2, \ldots, a_n, \ldots\}$ of real numbers, where after some point all the a_j are 0. We use as the inner product of two such sequences $\{a_r\}$ and $\{b_r\}$ the following: $(\{a_r\}, \{b_r\}) = \sum_r a_r b_r$. Since only a finite number of the a_i and b_i are nonzero, this seemingly infinite sum is really a finite sum, so makes sense.

Let

$$s_1 = \{1, 0, 0, 0, \ldots\},\ s_2 = \{1, 1, 0, 0, \ldots\}\ ,\cdots,\ s_n = \{\underbrace{1, 1, \ldots, 1, 1}_{(n\ 1\text{'s})}, 0, 0, \ldots\},\ \ldots.$$

These elements are linearly independent over $\mathbb{R}$. What is the orthonormal set $t_1, \ldots, t_n, \ldots$ that we get from these using the Gram–Schmidt construction? Again we merely follow the construction given in the proof, for this special case. So we start with $t_1 = s_1$, $t_2 = s_2 - \dfrac{(s_2, t_1)}{(t_1, t_1)} t_1, \ldots$. Now $(s_2, t_1) = 1$ and $(t_1, t_1) = 1$, hence $t_2 = s_2 - t_1 = s_2 - s_1 = \{0, 1, 0, 0, \ldots\}$. What about t_3? Again

$$t_3 = s_3 - \frac{(s_3, t_1)}{(t_1, t_1)} t_1 - \frac{(s_3, t_2)}{(t_2, t_2)} t_2.$$

Since $(s_3, t_1) = 1$, $(s_3, t_2) = 1$, we get that $t_3 = s_3 - t_1 - t_2 = \{0, 0, 1, 0, 0, \ldots\}$. How do you think the rest of it will go? We leave it to you to show that

$$t_n = \{0, 0, 0, \ldots, 1, 0, 0, \ldots\},$$

where the 1 occurs in the nth place.

In the finite-dimensional case of $F^{(n)}$ we exploited the result of the Gram–Schmidt construction to show that $F^{(n)} = W \oplus W^\perp$, where W is any subspace of $F^{(n)}$ and $W^\perp = \{v \in F^{(n)} \,|\, (v, w) = 0, \text{ all } w \in W\}$ is the *orthogonal complement* of W. We want to do something similar in general.

In the problem sets we were introduced to the *direct sum* of vector spaces V and W. This was defined as $V \oplus W = \{(v, w) \,|\, v \in V, w \in W\}$, where we added component-wise and $a(v, w) = (av, aw)$ for $a \in F$. We want to make a more formal introduction of this notion and some related ones.

Definition. Let V be a vector space over F and let U, W be subspaces of V. We say that V is the *direct sum* of U and W if every element z in V can be written in a *unique* way as $z = u + w$, where $u \in U$ and $w \in W$.

What does this have to do with the notion of $U \oplus W$ as ordered pairs described above? The answer is

Lemma 8.7.3. *If U, W are subspaces of V, then V is the direct sum of U and W if and only if*

1. $U + W = \{u + w \,|\, u \in U, w \in W\} = V$.
2. $U \cap W = \{0\}$.

Furthermore, in this case $V \simeq U \oplus W$, where $U \oplus W = \{(u, w) \,|\, u \in U, w \in W\}$.

Proof: If $V = U + W$, then every element v in V is of the form $v = u + w$. If $U \cap W = \{0\}$, we claim that this representation of v is unique, for suppose that $v = u + w = u_1 + w_1$, where $u, u_1 \in U$ and $w, w_1 \in W$. Thus $u - u_1 = w_1 - w$; but $u - u_1 \in U$ and $w_1 - w \in W$ and since $u - u_1 = w_1 - w$, we get $u - u_1 \in U \cap W = \{0\}$. Hence $u = u_1$. Similarly, $w = w_1$. So the representation of v in the desired form is indeed unique. Hence V is the direct sum of U and W.

To show that if V is the direct sum of U and W, then both $V = U + W$ and $U \cap W = \{0\}$, we leave as an exercise.

Now to the isomorphism of $U \oplus W$ and V. Define the mapping $\Phi: U \oplus W \to V$ by the rule

$$\Phi((u, w)) = u + w.$$

We leave to the reader the verification that Φ is a homomorphism of $U \oplus W$ into V. It is onto since $V = U + W$ (because V is the direct sum of U and W). To show that Φ is an isomorphism, we need but show that $\text{Ker}(\Phi) = \{(u, w) \in U \oplus W \,|\, \Phi((u, w)) = 0\}$ consists only of 0. If $\Phi((u, w)) = 0$, knowing that $\Phi((u, w)) = u + w$ leads us to $u + w = 0$, hence $u = -w \in U \cap W = \{0\}$. So $u = w = 0$. Thus $\text{Ker}(\Phi) = \{0\}$, and the lemma is proved. ∎

We are now able to prove the important analog of what we once did in $F^{(n)}$ for any inner product space.

Theorem 8.7.4. *If V is an inner product space and W is a finite-dimensional subspace* of V, then V is the direct sum of W and $W^\perp$. So $V \simeq W \oplus W^\perp$, the set of all ordered pairs (w, z), where $w \in W$, $z \in W^\perp$.

Proof: Since W is finite-dimensional over F, it has a finite basis over F. By Theorem 8.7.1 we know that W then has an orthonormal basis $w_1, \ldots, w_m$ over F. We claim that an element of z is in $W^\perp$ if and only if $(z, w_r) = 0$ for $r = 1, 2, \ldots, m$. Why? Clearly, if $z \in W^\perp$, then $(z, w) = 0$ for *all* $w \in W$, hence certainly $(z, w_r) = 0$. In the other direction, if $(z, w_r) = 0$ for $r = 1, 2, \ldots, m$, we claim that $(z, w) = 0$ for all $w \in W$. Because $w_1, w_2, \ldots, w_m$ is a basis of W over F, $w = a_1 w_1 + a_2 w_2 + \cdots + a_m w_m$. Thus

$$(z, w) = (z, a_1 w_1 + \cdots + a_m w_m) = (z, a_1 w_1) + \cdots + (z, a_m w_m)$$
$$= \bar{a}_1 (z, w_1) + \bar{a}_2 (z, w_2) + \cdots + \bar{a}_m (z, w_m) = 0.$$

Hence $z \in W^\perp$. Let $v \in V$; consider the element

$$z = v - (v, w_1) w_1 - \cdots - (v, w_m) w_m.$$

We assert that $(z, w_r) = 0$ for $r = 1, 2, \ldots, m$. Computing, we get

$$(z, w_r) = (v - (v, w_1) w_1 - \cdots - (v, w_m) w_m, w_r)$$
$$= (v, w_r) - (v, w_1)(w_1, w_r) - \cdots - (v, w_m)(w_m, w_r).$$

But the $w_1, \ldots, w_m$ are an orthonormal set, hence $(w_s, w_r) = 0$ if $s \neq r$ and $(w_r, w_r) = 1$. So the only nonzero contribution above is from the rth term; in other words,

$$(z, w_r) = (v, w_r) - (v, w_r)(w_r, w_r) = (v, w_r) - (v, w_r) = 0.$$

We have shown that $z = v - (v, w_1) w_1 - \cdots - (v, w_m) w_m$ is in $W^\perp$. But

$$v = z + [(v, w_i) w_1 + \cdots + (v, w_m) w_m]$$

and $[(v, w_i) w_1 + \cdots + (v, w_m) w_m]$ is in W. Thus $v \in W + W^\perp$. But $W \cap W^\perp = \{0\}$; hence, by Lemma 8.7.3, V is the direct sum of W and $W^\perp$. By Lemma 8.7.3 again, $V \simeq W \oplus W^\perp$. ∎

An important special situation in which Theorem 8.7.4 holds is that of V finite-dimensional over F, for in that case any subspace of V is also finite-dimensional over F. Thus

Theorem 8.7.5. *If V is a finite-dimensional inner product space, then for any subspace W of V, V is the direct sum of W and $W^\perp$.*

Because the direct sum of W and $W^\perp$ as subspaces of V is V, and since it is isomorphic to $W \oplus W^\perp$, we shall denote the fact that V is the direct sum of W and $W^\perp$ by $V = W \oplus W^\perp$. In these terms Theorem 8.7.5 reads as:

If V is a finite-dimensional inner product space, then $V = W \oplus W^\perp$ for any subspace W of V.

We already know that if V is a finite-dimensional inner product space, then V merely as a vector space is isomorphic to $F^{(n)}$, where $n = \dim(V)$. We might very well ask:

Is V isomorphic to $F^{(n)}$ as an inner product space?

What does this mean? It means precisely that there is an isomorphism Φ of V onto $F^{(n)}$ such that $(\Phi(u), \Phi(v)) = (u, v)$ for all $u, v \in V$. Here the inner product $(\Phi(u), \Phi(v))$ is that of $F^{(n)}$, while the inner product (u, v) is that of V. *In other words, Φ also preserves inner products.*

The answer is "yes," as is shown by

Theorem 8.7.6. *If V is a finite-dimensional inner product space, then V is isomorphic to $F^{(n)}$, where $n = \dim(V)$, as an inner product space.*

Proof: By Theorem 8.7.1, V has an orthonormal basis $v_1, \ldots, v_n$ where $n = \dim(V)$. Given $v \in V$, then $v = a_1 v_1 + \cdots + a_n v_n$, with $a_1, \ldots, a_n \in F$ uniquely determined by v. Define $\Phi: V \to F^{(n)}$ by $\Phi(v) = \begin{bmatrix} a_1 \\ \vdots \\ a_n \end{bmatrix}$. We saw in Theorem 8.4.3 that Φ is an isomorphism of V onto $F^{(n)}$.

Does Φ preserve the inner product? Let

$$v = a_1 v_1 + \cdots + a_n v_n \quad \text{and} \quad u = b_1 v_1 + \cdots + b_n v_n$$

for any u, v in V. Thus

$$(u, v) = (b_1 v_1 + \cdots + b_n v_n, a_1 v_1 + \cdots + a_n v_n) = \sum_{i=1}^{n} \sum_{j=1}^{n} b_i \bar{a}_j (v_i, v_j).$$

(Prove!) Now, since $v_1, \ldots, v_n$ are orthonormal, $(v_i, v_j) = 0$ if $i \neq j$ and $(v_i, v_i) = 1$. So the double sum for (u, v) reduces to $(u, v) = \sum_{j=1}^{n} b_i \bar{a}_i$. But $\sum_{j=1}^{n} b_i \bar{a}_i$ is precisely the inner product of the vectors $\Phi(u) = \begin{bmatrix} b_1 \\ \vdots \\ b_n \end{bmatrix}$ and $\Phi(v) = \begin{bmatrix} a_1 \\ \vdots \\ a_n \end{bmatrix}$ in $F^{(n)}$. In other words, $(\Phi(u), \Phi(v)) = (u, v)$. This completes the proof. ∎

In light of Theorem 8.7.6:

We can carry over everything we did in Chapters 3 and 4 for $F^{(n)}$ as an inner product space to all finite-dimensional inner product spaces over F.

For instance, we get from Theorem 4.3.7 that

Theorem 8.7.7. *If V is a finite-dimensional inner product space and W is a subspace of V, then $(W^\perp)^\perp = W$.*

Go back to Chapter 3 and see what else you can carry over to the general case by exploiting Theorem 8.7.6.

One could talk about vector spaces over F for systems F other than $\mathbb{R}$ or $\mathbb{C}$. However, we have restricted ourselves to working over $\mathbb{R}$ or $\mathbb{C}$. This allows us to show that any finite-dimensional vector space over $\mathbb{R}$ or $\mathbb{C}$ is, in fact, an inner product space.

Theorem 8.7.8. *If V is a finite-dimensional vector space over $\mathbb{R}$ or $\mathbb{C}$, then V is an inner product space.*

Proof: Because V is finite-dimensional over $\mathbb{R}$ (or $\mathbb{C}$) it has a basis $v_1, \ldots, v_n$ over $\mathbb{R}$ (or $\mathbb{C}$). We define an inner product on V by insisting that $(v_r, v_j) = 0$ for $i \ne j$ and $(v_r, v_r) = 1$. That is, we define $(u, v) = \sum_{j=1}^{n} a_j \bar{b}_j$, where $u = \sum_{r=1}^{n} a_r v_r$ and $v = \sum_{r=1}^{n} b_r v_r$.

We leave it to the reader to complete the proof that this is a legitimate inner product on V. ∎

Because of Theorem 8.7.8:

We can carry over to general finite-dimensional V over $F = \mathbb{R}$ or $\mathbb{C}$ everything we did in Chapter 2 for $F^{(n)}$ as a vector space over F.

A sample result of this kind is

If V is finite-dimensional over F and $u_1, \ldots, u_m$ in V are linearly independent over F, then we can fill $u_1, \ldots, u_m$ out to a basis of V over F. That is, we can find elements $w_1, \ldots, w_r$ in V such that $u_1, \ldots, u_m, w_1, \ldots, w_r$ is a basis of V over F (and so $\dim(V) = m + r$).

PROBLEMS

NUMERICAL PROBLEMS

1. In Example 1, find $q_4(x)$, $q_5(x)$, $q_6(x)$.
2. In Example 1, what is the length of $q_1(x)$, $q_2(x)$, $q_3(x)$?
3. Let V be the inner product space of Example 2. Let $s_1 = \{1, 0, 0, \ldots, 0, \ldots\}$. $s_2 = \{1, 2, 0, \ldots, 0, \ldots\}$, $\ldots$, $s_n = \{1, 2, \ldots, n, 0, \ldots, 0, \ldots\}$. Find an orthonormal set $t_1, \ldots, t_n, \ldots$ whose linear span is that of $s_1, s_2, \ldots, s_n, \ldots$.
4. If V is the space of Example 2 and $W = \{\{a, b, 0, \ldots, 0, \ldots\} \mid a, b \in F\}$, find $W^\perp$.
5. Make a list of all the results on inner product spaces that can be carried over to a general inner product space form $F^{(n)}$ by use of Theorem 8.7.5.
6. Let $V = M_3(\mathbb{R})$ with inner product given by $(A, B) = \text{tr}(AB')$, where $A, B \in M_3(\mathbb{R})$ and B' is the transpose of B. Find an orthonormal basis for V relative to this inner product.
7. If V is as in Problem 6 and $W = \left\{ \begin{bmatrix} a & b & c \\ 0 & 0 & 0 \\ 0 & 0 & 0 \end{bmatrix} \middle| a, b, c \in \mathbb{R} \right\}$, find $W^\perp$ and verify that $V = W \oplus W^\perp$.

8. Prove Theorem 8.7.7 using the results of Chapters 3 and 4 and Theorem 8.7.6.
9. Make a list of the results in Chapters 3 and 4 and hold for any finite-dimensional vector space over $\mathbb{R}$ or $\mathbb{C}$ using Theorems 8.7.6 and 8.7.8.
10. In the proof of Theorem 8.7.8, verify that $(u, v) = \sum_{j=1}^{n} a_j \bar{b}_j$.

MORE THEORETICAL PROBLEMS

Easier Problems

11. If $V = M_n(\mathbb{C})$, show that $(A, B) = \text{tr}(AB^*)$ for $A, B \in M_n(\mathbb{C})$ and where B^* is the Hermitian adjoint of B defines an inner product on V over $\mathbb{C}$.
12. In the V of Problem 11 find an orthonormal basis of V relative to the given inner product.
13. If in the V of Problem 11, W is the set of all diagonal matrices, find $W^\perp$ and verify that $V = W \oplus W^\perp$.
14. If W and U are finite-dimensional subspaces of V such that $W^\perp = U^\perp$, prove that $W = U$.

Middle-Level Problems

15. In Example 1 show that $q_n(x)$ is a polynomial of degree $n - 1$ over $\mathbb{R}$.
16. For $q_n(x)$ as in Problem 15 show:
    (a) If n is odd, then $q_n(x)$ is a polynomial in even powers of x.
    (b) If n is even, then $q_n(x)$ is a polynomial in odd powers of x.
    (*Hint*: Use induction.)
17. If V is finite-dimensional, show that for any subspace W of V, $\dim(V) = \dim(W) + \dim(W^\perp)$.

Very Hard Problems

18. If V is an infinite-dimensional inner product space and W is a finite-dimensional subspace of V, is $(W^\perp)^\perp = W$? Either prove or give a counterexample.
19. Give an example of an infinite-dimensional inner product space and an infinite-dimensional subspace $W \neq V$ such that $W^\perp = \{0\}$. (Hence $W \oplus W^\perp \neq V$.)

CHAPTER 9

Linear Transformations

9.1. INTRODUCTION

Earlier in the book we encountered linear transformations in the setting of $F^{(n)}$. The point of the material in this chapter is to consider linear transformations in the wider context of an arbitrary vector space. Although almost everything we do will be in the case of a *finite-dimensional* vector space V over F, in the definitions, early results, and examples we also consider the infinite-dimensional situation.

Our basic strategy will be to use the isomorphism established in Theorem 8.4.3 for a finite-dimensional vector space V over F, with $F^{(n)}$, where $n = \dim(V)$. This will also be exploited for inner product spaces. As we shall see, everything done for matrices goes over to general linear transformations on an arbitrary finite-dimensional vector space. In a nutshell, we shall see that such a linear transformation can be represented as a matrix and that the operations of addition and multiplication of linear transformations coincides with those of the associated matrices. Thus all the material developed for $n \times n$ matrices immediately can be transferred to the general case. Therefore, we often will merely sketch a proof or refer the reader back to the appropriate proof carried out in Chapters 3 and 4.

We can also speak about linear transformations—they are merely homomorphisms—of one vector space into another one. We shall touch on this lightly at the end of the chapter. Our main concern, however, will be linear transformations of a vector space V over F into itself, which is the more interesting situation where more can be said.

9.2. DEFINITION, EXAMPLES, AND SOME PRELIMINARY RESULTS

Let V be a vector space over F, where $F = \mathbb{R}$ or $F = \mathbb{C}$.

Definition. A *linear transformation* T of V into V over F is a mapping $T: V \to V$ such that

1. $T(v + w) = T(v) + T(w)$
2. $T(av) = aT(v)$

for all $v, w \in V$ and all $a \in F$.

Of course, properties (1) and (2) generalize to

$$T(a_1 v_1 + a_2 v_2 + \cdots + a_k v_k) = a_1 T(v_1) + a_2 T(v_2) + \cdots + a_k T(v_k)$$

for any finite positive integer k. (Prove!)

We consider several examples of linear transformations on a variety of vector spaces. We shall refer to them by their numbering from time to time later.

EXAMPLES

1. Recall first the basic example we studied in such detail in Chapter 2, namely, $V = F^{(n)}$ and $T = (a_{rs})$, where T acts on V by the rule

$$\begin{bmatrix} a_{11} & \cdots & a_{1n} \\ \vdots & & \vdots \\ a_{n1} & \cdots & a_{nn} \end{bmatrix} \begin{bmatrix} x_1 \\ \vdots \\ x_n \end{bmatrix} = \begin{bmatrix} y_1 \\ \vdots \\ y_n \end{bmatrix},$$

where $y_r = \sum_{s=1}^{n} a_{rs} x_s$. We saw that T acts on V as a transformation on V and furthermore, *every linear transformation* on V over F can be realized in this way.

2. Let V be the set of all polynomials in x over F. Define $T: V \to V$ by defining $T(1) = 1$, $T(x) = x + 1$, $T(x^2) = (x + 1)^2 = x^2 + 2x + 1, \ldots$, the general term being

$$T(x^n) = (x + 1)^n = \sum_{j=0}^{n} \binom{n}{j} x^j \left(\text{recall that } \binom{n}{j} = \frac{n!}{j!(n-j)!} \right);$$

and then defining, for any polynomial $p(x) = a_0 + a_1 x + a_2 x^2 + \cdots + a_k x^k$,

$$T(p(x)) = a_0 T(1) + a_1 T(x) + a_2 T(x^2) + \cdots + a_k T(x^k).$$

Then, for instance,

$$T(5 - x + x^2 + 6x^3) = 5T(1) - T(x) + T(x^2) + 6T(x^3)$$
$$= 5 - (x + 1) + (x + 1)^2 + 6(x + 1)^3$$
$$= 11 + 19x + 19x^2 + 6x^3.$$

This mapping T is a linear transformation on V because we have *forced* it to be one by the very definition of T. Specifying what T does to $1, x, x^2, \ldots, x^n, \ldots$, which generate V over F, and then *insisting* that T acts on

$$p(x) = a_0 + a_1 x + \cdots + a_k x^k$$

via

$$T(p(x)) = a_0 T(1) + a_1 T(x) + \cdots + a_k T(x^k)$$

guarantees that T will be a linear transformation on V.

If we pick a sequence of polynomials $p_0(x), p_1(x), \ldots, p_n(x), \ldots$ and define

$$T(x^j) = p_j(x)$$

and

$$T(a_0 + a_1 x + \cdots + a_k x^k) = a_0 p_0(x) + a_1 p_1(x) + \cdots + a_k p_k(x),$$

We will equally get a linear transformation on V over F.

So, for example, if we define $T(1) = x^5$, $T(x) = 6$, $T(x)^2 = x^7 + x$, and $T(x^k) = 0$ for $k > 2$, then T is a linear transformation on V when defined as above. Thus

$$T(a_0 + a_1 x + \cdots + a_k x^k) = a_0 T(1) + a_1 T(x) + a_2 T(x)^2 + \cdots + a_k T(x_k)$$
$$= a_0 x^5 + 6a_1 + a_2(x^7 + x).$$

3. Example 2 is quite illustrative. First notice that we never really used the fact that V was the set of polynomials over F. What we did was to exploit the fact that V is spanned by the set of all powers of x, and these powers of x are linearly independent over F. We can do something similar for any vector space that is spanned by some linearly independent set of elements. We make this more precise.

Let V be a vector space over F spanned by the linearly independent elements $v_1, \ldots, v_n, \ldots$. Let $w_1, w_2, \ldots, w_n, \ldots$ be *any* elements of V and define $T(v_j) = w_j$ for all j. Given any element u in V, u has a *unique* expansion of the form

$$u = a_1 v_1 + \cdots + a_k v_k,$$

where $a_1, \ldots, a_k$ are in F. Define

$$T(u) = a_1 T(v_1) + \cdots + a_k T(v_k) = a_1 w_1 + \cdots + a_k w_k.$$

By the linear independence of $v_1, \ldots, v_n, \ldots$ this mapping T is well defined. It is easy to check that it is a linear transformation on V.

Note how important it is that the elements $v_1, v_2, \ldots, v_n, \ldots$ are linearly independent. Take the example of the polynomials in x; the elements

$$1, \; x, \; 1 + x, \; x^2, \; \ldots$$

are *not* linearly independent. If we define

$$T(1) = 1, \; T(x) = x, \; T(1 + x) = x^3, \; \ldots$$

we see that T cannot be a linear transformation, for otherwise $T(1 + x)$ would equal $T(1) + T(x) = 1 + x$, yet we tried to insist that $T(1 + x) = x^3 \neq 1 + x$.

4. Let V be a finite-dimensional vector space over F and let $v_1, \ldots, v_n$ be a basis of V over F. If

$$u = a_1 v_1 + \cdots + a_n v_n,$$

define

$$T(u) = a_1 v_1 + a_3 v_3.$$

It is routine to verify that T is a linear transformation on V. If W is the subspace spanned by v_1 and v_3, we call T the *projection* of V onto W.

It is the first order of business that we establish some simple computational facts about linear transformations.

Lemma 9.2.1. If T is a linear transformation on V, then

1. $T(0) = 0$.
2. $T(-v) = -T(v) = (-1)T(v)$.

Proof: To see that $T(0) = 0$, note that $T(0 + 0) = T(0) + T(0)$. Yet $T(0 + 0) = T(0)$. Thus $T(0) + T(0) = T(0)$. There is an element $w \in V$ such that $T(0) + w = 0$, from the definition of a vector space.

So,

$$\begin{aligned} 0 &= T(0) + w \\ &= (T(0) + T(0)) + w \\ &= T(0) + (T(0) + w) \\ &= T(0) + 0 = T(0). \end{aligned}$$

We leave the proof of Part (2) to the reader. ∎

If V is any vector space over F, we can introduce two operations on the linear transformations on V. These are addition and multiplication by a scalar (i.e., by an element of F). We define both of these operations now.

Definition. If T_1, T_2 are two linear transformations of V into itself, then the *sum*, $T_1 + T_2$, of T_1 and T_2 is defined by $(T_1 + T_2)(v) = T_1(v) + T_2(v)$ for all $v \in V$.

Definition. If T is a linear transformation on V and $a \in F$, then aT, the *multiplication of T by the scalar a*, is defined by $(aT)(v) = aT(v)$ for all $v \in V$.

So, for instance, if V is the set of all polynomials in x over $\mathbb{R}$ and if T_1 is defined by

$$T_1(a_0 + a_1 x + a_2 x^2 + \cdots + a_n x^n) = 3a_0 + \tfrac{1}{2} a_1 x,$$

while T_2 is defined by

$$T_2(b_0 + b_1 x + b_2 x^2 + \cdots + b_n x^n) = b_3 x^3,$$

then

$$\begin{aligned}(T_1 + T_2)(c_0 + c_1 x + c_2 x^2 + \cdots + c_n x^n) &= T_1(c_0 + c_1 x + c_2 x^2 + \cdots + c_n x^n) \\ &\quad + T_2(c_0 + c_1 x + c_2 x^2 + \cdots + c_n xn) \\ &= 3c_0 + \tfrac{1}{2} c_1 x + c_3 x^3.\end{aligned}$$

Similarly, $15T_1$ acts on V by the action

$$(15T_1)(c_0 + c_1 x + c_2 x^2 + \cdots + c_n x^n) = 15(3c_0 + \tfrac{1}{2} c_1 x) = 45c_0 + \tfrac{15}{2} c_1 x.$$

One would hope that combining linear transformations and scalars in these ways always leads us again to linear transformations on V. There is no need to hope; we easily verify this to be the case.

Lemma 9.2.2. *If T_1, T_2 are linear transformations on V and if $a \in F$, then $T_1 + T_2$ and aT_1 are linear transformations on V.*

Proof: What must we show? To begin with, if T_1, T_2 are linear transformations on V and if $u, v \in V$, we need that

$$(T_1 + T_2)(u + v) = (T_1 + T_2)(u) + (T_1 + T_2)(v).$$

To see this we expand according to the definition of the sum

$$\begin{aligned}(T_1 + T_2)(u + v) = (T_1)(u + v) + (T_2)(u + v) &= T_1(u) + T_1(v) + T_2(u) + T_2(v) \\ \text{(since } T_1, T_2 \text{ are linear transformations)} &= T_1(u) + T_2(u) + T_1(v) + T_2(v) \\ &= (T_1 + T_2)(u) + (T_1 + T_2)(v)\end{aligned}$$

[again by the definition of $(T_1 + T_2)$], which is exactly what we wanted to show.

We also need to see that $(T_1 + T_2)(av) = a((T_1 + T_2)(v))$ for $v \in V$, $a \in F$. But we have

$$\begin{aligned}(T_1 + T_2)(av) &= T_1(av) + T_2(av) = aT_1(v) + aT_2(v) \\ &= a(T_1(v) + T_2(v)) = a((T_1 + T_2)(v)),\end{aligned}$$

as required.

So $T_1 + T_2$ is indeed a linear transformation on V. An analogous argument shows that bT_1 is a linear transformation on V for $b \in F$ and T_1 a linear transformation. ∎

Definition. If V is a vector space over F, then $L(V)$ is the set of all linear transformations on V over F.

What we showed was that if $T_1, T_2 \in L(V)$ and $a \in F$, then $T_1 + T_2$ and aT_1 are both in $L(V)$. We can now repeat the argument—especially after the next theorem—to show that if $T_1, \ldots, T_k \in L(V)$ and $a_1, \ldots, a_k \in F$, then $a_1 T_1 + \cdots + a_k T_k$ is again in $L(V)$.

In the two operations defined in $L(V)$ we have imposed an algebraic structure on $L(V)$. How does this structure behave?

We first point out one significant element present in $L(V)$. Let's define $O: V \to V$ by the rule

$$O(v) = 0$$

for all $v \in V$. Trivially, O is a linear transformation on V. Furthermore, if $v \in V$ and $T \in L(V)$, then $(T + O)(v) = T(v) + O(v) = T(v) + 0 = T(v)$, from which we deduce that

$$T + O = T.$$

Similarly, $O + T = T$. We shall usually write O merely as 0 and refer to it as the *zero element of* $L(V)$.

If $T \in L(V)$, consider the element

$$(-1)T = S,$$

which we know also lies in $L(V)$. For any $v \in V$,

$$(T + S)(v) = T(v) + S(v) = T(v) + (-1)T(v) = T(v + (-1)v) = T(0) = 0.$$

Hence

$$T + S = O = 0.$$

So $S = (-1)T$ acts as a negative for T in $L(V)$. We write it merely as $-T$.

With these preliminaries out of the way we have

Theorem 9.2.3. *$L(V)$ is itself a vector space over F.*

We leave the proof as a series of exercises. We must verify that each of the defining rules for a vector space holds in $L(V)$ under the operations we introduced.

PROBLEMS

NUMERICAL PROBLEMS

1. In Example 4 verify that the T defined is a linear transformation.
2. Show that $O: V \to V$ defined by the rule $O(v) = 0$ for all $v \in V$ is a linear transformation on V.

3. If V is the set of all polynomials in x over $\mathbb{R}$, show that $d: V \to V$, defined by $d(p(x)) = \dfrac{d}{dx} p(x)$, is a linear transformation on V.

4. If V is as in Problem 3 and $S: V \to V$ is defined by $S(p(x)) = \displaystyle\int_0^x p(t)\,dt$, show that S is a linear transformation on V.

5. Calculate $d + S$ and $-\frac{1}{2} S$ on
   (a) $p(x) = x^5 - \frac{1}{4}x + \pi$.
   (b) $p(x) = \frac{1}{6}x^6 - \frac{1}{7}x^7$.
   (c) $p(x) = 4\pi - 17x^{20}$.

6. In $L(V)$ show
   (a) $T_1 + T_2 = T_2 + T_1$
   (b) $(T_1 + T_2) + T_3 = T_1 + (T_2 + T_3)$
   (c) $(a + b)T_1 = aT_1 + bT_1$
   (d) $a(T_1 + T_2) = aT_1 + aT_2$
   for all $a, b \in F$ and all $T_1, T_2, T_3 \in L(V)$.

7. If V is the linear span over $\mathbb{R}$ of $\cos(x), \cos(2x), \cos(3x)$ in the set of all real-valued functions on $[0, 1]$, show that
   (a) The functions $\cos(x)$, $\cos(2x)$, $\cos(3x)$ are linearly independent over $\mathbb{R}$.
   (b) The mapping T on V defined by
   $$T(a\cos(x) + b\cos(2x) + c\cos(3x)) = c\cos(x) + a\cos(2x) + b\cos(3x)$$
   is a linear transformation.
   (c) T is $1-1$ and onto.

MORE THEORETICAL PROBLEMS

Easier Problems

8. Show that aT, where $a \in F$ and $T \in L(V)$, defined by $(aT)(v) = a(T(v))$ is a linear transformation on V.

9. If T is a linear transformation on V, show by mathematical induction that
$$T(a_1 v_1 + \cdots + a_k v_k) = a_1 T(v_1) + \cdots + a_k T(v_k)$$
for all $v_1, \ldots, v_k \in V$ and all $a_1, \ldots, a_k \in F$.

10. If W is a proper subspace of V and $T \in L(V)$ is such that $T(v) = 0$ for all $v \notin W$, prove that $T = 0$, the zero element of $L(V)$.

11. If $T_1, \ldots, T_k \in L(V)$ and $v \in V$, show that the set
$$\{(a_1 T_1 + \cdots + a_k T_k)(v) \mid a_1, \ldots, a_k \in F\}$$
is a finite-dimensional subspace of V.

12. Prove that if S is a finite-dimensional subspace of $L(V)$ and W is a finite-dimensional subspace of V, then
    (a) $S(W) = \{T(w) \mid T \in S, w \in W\}$ is a subspace of V.
    (b) $S(W)$ is finite-dimensional over F.
    (c) $\dim(S(W)) \leq (\dim(W))(\dim(S))$.

Middle-Level Problems

13. If $T \in L(V)$, show that
    (a) $U = \{v \in V \mid T(v) = 0\}$ is a subspace of V.
    (b) $T(V) = W = \{T(v) \mid v \in V\}$ is a subspace of V.
14. If W is a subspace of V, let $K = \{T \in L(V) \mid T(W) = 0\}$. Show that K is a subspace of $L(V)$.
15. Define the mapping $\Phi: L(V) \to V$ by $\Phi(T) = T(v_0)$, where $T \in L(V)$, v_0 is a *fixed* element of V. Prove that Φ is a homomorphism of $L(V)$ into V.
16. In Problem 15 show that the kernel of Φ cannot be merely $\{0\}$, if $\dim(V) > 1$.
17. Let W be a subspace of V and let $T \in L(V)$. If $U = \{v \in V \mid T(v) \in W\}$, show that U is a subspace of V. [Compare this with Part (a) of Problem 13.]

Harder Problems

18. If V is a finite-dimensional vector space over F and $T \in L(V)$ is such that T maps V onto V, prove that T is $1-1$.
19. If V is a finite-dimensional vector space over F and $T \in L(V)$ is $1-1$ on V, prove that T maps V onto V.

9.3. PRODUCTS OF LINEAR TRANSFORMATIONS

Given any set S and f, g mappings of S into S, we had earlier in the book defined the product or composition of f and g by the rule $(fg)(s) = f(g(s))$ for every $s \in S$. If T_1, T_2 are linear transformations on V over F, they are mappings of V into itself; hence their product, $T_1 T_2$, as mappings makes sense. *We shall use this product as an operation in $L(V)$*.

The first question that then comes to mind is: If T_1, T_2 are in $L(V)$, is $T_1 T_2$ also in $L(V)$?

The answer is "yes," as we see in

Lemma 9.3.1. *If $T_1, T_2 \in L(V)$, then $T_1 T_2 \in L(V)$.*

Proof: We must check out two things: For all $u, v \in V$, $a \in F$,

1. Is $(T_1 T_2)(u + v) = (T_1 T_2)(u) + (T_1 T_2)(v)$?
2. Is $(T_1 T_2)(av) = a((T_1 T_2)(v))$?

We look at these two in turn.

By definition,

$$(T_1T_2)(u+v) = T_1(T_2(u+v)) = T_1(T_2(u) + T_2(v)) \text{ [since } T_2 \in L(V)]$$
$$= T_1(T_2(u)) + T_1(T_2(v)) \text{ [since } T_1 \in L(V)]$$
$$= (T_1T_2)(u) + (T_1T_2)(v).$$

Thus (1) is established.

We leave the proof of (2) to the reader. ∎

Lemma 9.3.1 assures us that multiplying two elements of $L(V)$ throws us back into $L(V)$. Knowing this, we might ask what properties this product enjoys and how product, sum, and multiplication by a scalar interact.

There is a very special element, I, in $L(V)$ defined by

$$I(v) = v \quad \text{for all } v \in V.$$

It is easy to show that this I is indeed in $L(V)$. As the identity mapping on V we know from before that

$$TI = IT = T \quad \text{for all } T \in L(V).$$

We also know, just from the property of mappings, that

$$(T_1T_2)T_3 = T_1(T_2T_3).$$

If $T \in L(V)$ is *both* $1-1$ and onto V, there is a mapping S such that

$$ST = TS = I.$$

Is this mapping S also in $L(V)$? Let's check it out. Since $T \in L(V)$,

$$T(S(u) + S(v)) = (TS)(u) + (TS)(v) = I(u) + I(v) = u + v.$$

Therefore, multiplying this last relation on the left by S, we get

$$S(T(S(u) + S(v))) = S(u+v).$$

Since

$$S(T(S(u) + S(v))) = (ST)(S(u) + S(v)) = I(S(u) + S(v)) = S(u) + S(v),$$

we end up with the desired result that

$$S(u+v) = S(u) + S(v).$$

Similarly, we can show that

$$S(av) = aS(v)$$

for $a \in F$, $v \in V$. Consequently, $S \in L(V)$. If such an S exists for T—and it needn't for every T—we write it as T^{-1}, in accordance with what we did for mappings. So we have

Definition. If $S \in L(V)$ and $ST = TS = I$, we call S the *inverse* of T and denote it as T^{-1}.

We have shown

Lemma 9.3.2. If $T \in L(V)$ is $1-1$ and onto, then T^{-1} is also in $L(V)$.

The properties of the product by itself of linear transformations behaves exactly as the product of mappings, that is, decently. How do product and sum intertwine?

Lemma 9.3.3. If S, T, Q are in $L(V)$, then

1. $S(T + Q) = ST + SQ$;
2. $(T + Q)S = TS + QS$.

Proof: To prove each part we must show only that both sides of the alleged equalities agree on every element v in V. So, for any $v \in V$,

$$(S(T + Q))(v) = S(T(v) + Q(v)) = S(T(v)) + S(Q(v)) = (ST)(v) + (SQ)(v) = (ST + SQ)(v).$$

Therefore,

$$S(T + Q) = ST + SQ,$$

establishing Part (1). The proof of Part (2) is similar. ∎

We leave the proof of the next lemma to the reader. It is similar in spirit to the one just done.

Lemma 9.3.4. If $S, T \in L(V)$ and $a, b \in F$, then

1. $S(aT) = a(ST)$;
2. $(aS)(bT) = (ab)(ST)$.

For any $T \in L(V)$ we shall use the notation of exponents, that is, for $k \geq 0$, $T^0 = I$, $T^1 = T, \ldots, T^k = T(T^{k-1})$. The usual rules of exponents prevail, namely,

$$T^k T^m = T^{k+m} \quad \text{and} \quad (T^k)^m = T^{km}.$$

If T^{-1} exists in $L(V)$, we say that T is *invertible* in $L(V)$. In this case, we define T^{-n} for $n \geq 0$ by

$$T^{-n} = (T^{-1})^n.$$

Again, here, the usual rules of exponents hold.

PROBLEMS

NUMERICAL PROBLEMS

1. If V is the set of polynomials, of degree n or less, in x over $\mathbb{R}$ and $D: V \to V$ is defined by $D(p(x)) = \dfrac{d}{dx}(p(x))$, prove that $D^{n+1} = 0$.

2. If V is as in Problem 1 and $T: V \to V$ is defined by $T(p(x)) = p(x+1)$, show that T^{-1} exists in $L(v)$.

3. If V is the set of all polynomials in x over $\mathbb{R}$, let $D: V \to V$ by $D(p(x)) = \dfrac{d}{dx}(p(x))$; and $S: V \to V$ by $S(p(x)) = \displaystyle\int_0^x p(t)\, dt$. Prove:
   (a) $DS = I$.
   (b) $SD \neq I$.

4. If V is as in Problem 3 and $T: V \to V$ according to the rules $T(1) = 1$, $T(x) = x$, $T(x^2) = x^2$, and $T(x^k) = 0$ if $k > 2$, show that $T^2 = T$.

5. In Problem 4, if $S: V \to V$ according to $T(1) = x$, $T(x) = x^2$, $T(x^2) = 1$, $T(x^k) = x^k$ for $k > 2$, show that $T^3 = I$.

6. If V is the linear span over $\mathbb{R}$ in the set of all real-valued functions on $[0,1]$ of $\cos(x)$, $\sin(x)$, $\cos(2x)$, $\sin(2x)$, $\cos(3x)$, $\sin(3x)$, and $D: V \to V$ is defined by $D(v) = \dfrac{d}{dx}(v)$ for $v \in V$, show that
   (a) D maps V into V.
   (b) D maps V onto V.
   (c) D is a linear transformation on V.
   (d) D^{-1} exists on V.

7. In Problem 6 find the explicit form of D^{-1}.

MORE THEORETICAL PROBLEMS

Easier Problems

8. Let V be the set of all sequences of real numbers $(a_1, a_2, \ldots, a_n, \ldots)$. On V define S by $S((a_1, a_2, \ldots, a_n, \ldots)) = (0, a_1, a_2, \ldots, a_n, \ldots)$. Show that
   (a) S is a linear transformation on V.
   (b) S maps V into, but not onto, itself in a $1-1$ manner.
   (c) There exists an element $T \in L(V)$ such that $TS = I$.
   (d) For the T in Part (c), $ST \neq I$.
   (e) $(ST)^2 = ST$.
   (S is often called the *shift operator*.)

9. Prove Part (2) of Lemma 9.3.1.

10. In $L(V)$ prove that $(T + Q)S = TS + QS$.

11. Prove Lemma 9.3.4.

12. Prove that if $T \in L(V)$ and $a \in F$, then multiplication by the scalar a coincides with the product of the two linear transformations aI and T.

13. If $E \in L(V)$ satisfies $E^2 = E$ and E is invertible, what must E be?

14. If $F = \mathbb{C}$ and V is any vector space over $\mathbb{C}$, define $T: V \to V$ by $T(v) = av$ for all $v \in V$, where $a \in F$. Show that
  (a) T is a linear transformation on V.
  (b) If $a \neq 0$, then T is invertible in $L(V)$.
  (c) Find T^{-1} if $a \neq 0$.
  (d) If $a^n = 1$, then $T^n = I$.

15. If $T \in L(V)$ and $T^2 = I$, suppose, for some $0 \neq v \in V$, that $T(v) = av$, where $a \in F$. What are the possible values of a?

16. If $T^2 = I$ in $L(V)$ and $T \neq I$, find an element $v \neq 0$ in V and an element $w \neq 0$ in V such that $T(v) = v$ and $T(w) = -w$.

17. If $T \in L(V)$ satisfies $T^2 = T$ and $T \neq 0$, $T \neq I$, find an element $v \neq 0$ in V and an element $w \neq 0$ in V such that $T(v) = v$ and $T(w) = 0$.

Middle-Level Problems

18. If $T^2 = I$ in $L(V)$ and $T \neq I$, let $U = \{v \in V \mid T(v) = v\}$ and $W = \{v \in V \mid T(v) = -v\}$. Show that
  (a) $U \cap W = \{0\}$.
  (b) U and W are nonzero subspaces of V.
  (c) $V = U \oplus W$.

19. If $T^2 = T$ and $T \neq 0$, $T \neq I$, let $U = \{v \in V \mid T(v) = v\}$ and let $W = \{v \in V \mid T(v) = 0\}$. Prove:
  (a) U and W are nonzero subspaces of V.
  (b) $U \cap W = \{0\}$.
  (c) $V = U \oplus W$.
  [*Hint:* For Part (c), if $T^2 = T$, then $(I - T)^2 = I - T$.]

20. If $T \in L(V)$ is such that $T^2 - 2T + I = 0$, show that there is a $v \neq 0$ in V such that $T(v) = v$.

21. For T as in Problem 20, show that T is invertible in $L(V)$ and find T^{-1} explicitly.

Harder Problems

22. Consider the situation in Problems 3 and 8, where $S \in L(V)$ is such that $ST \neq I$ but $TS = I$ for some $T \in L(V)$. Prove that there is an *infinite* number of T's in $L(V)$ such that $TS = I$.

23. If $T \in L(V)$ satisfies $T^n + a_1 T^{n-1} + \cdots + a_{n-1} T + a_n I = 0$, where $a_1, \ldots, a_n$ are in F, prove:
  (a) T is invertible if $a_n \neq 0$.
  (b) If $a_n \neq 0$, find T^{-1} explicitly.

24. If $T \in L(V)$ is the T of Problem 20 and $v_0 \in V$, let

$$U = \{p(T)(v_0) \mid p(x) \text{ is any polynomial in } x \text{ over } F\}.$$

Prove:

(a) U is a subspace of V.

(b) U is finite-dimensional over F.

25. If $v_1,\ldots,v_n$ is a basis for V over F and if $T \in L(V)$ acts on V according to $T(v_1) = v_2$, $T(v_2) = v_3, \ldots, T(v_{n-1}) = v_n$, $T(v_n) = a_1 v_1 + \cdots + a_{n-1} v_{n-1}$, with $a_1, \ldots, a_{n-1} \in F$, show that there exist $b_1, \ldots, b_{n-1}$ in F such that

$$T^n + b_1 T^{n-1} + \cdots + b_{n-1} I = 0.$$

Can you describe the b_r explicitly?

9.4. LINEAR TRANSFORMATIONS AS MATRICES

In studying linear transformations on $V = F^{(n)}$, we showed that any linear transformation T in $L(V)$ can be realized as a matrix in $M_n(F)$. [And, conversely, every matrix in $M_n(F)$ defines a linear transformation on $F^{(n)}$.] Let's recall how this was done.

Let $v_1, \ldots, v_n$ be a basis of the n-dimensional vector space V over F. If $T \in L(V)$, to know what T does to any element it is enough to know what T does to each of the basis elements $v_1, \ldots, v_n$. This is true because, given $u \in V$, then $u = a_1 v_1 + \cdots + a_n v_n$, where $a_1, \ldots, a_n \in F$. Hence

$$T(u) = T(a_1 v_1 + \cdots + a_n v_n) = a_1 T(v_1) + \cdots + a_n T(v_n);$$

thus, knowing each of the $T(v_i)$ lets us know the value of $T(u)$.

So we only need to know $T(v_1), \ldots, T(v_n)$ in order to know T. Since $T(v_1)$, $T(v_2)$, ..., $T(v_n)$ are in V, and since $v_1, \ldots, v_n$ form a basis of V over F, we have that

$$T(v_s) = \sum_{r=1}^{n} t_{rs} v_r,$$

where the t_{rs} are in F. We associate with T the matrix

$$m(T) = (t_{rs}) = \begin{bmatrix} t_{11} & \cdots & t_{1n} \\ \vdots & & \vdots \\ t_{n1} & \cdots & t_{nn} \end{bmatrix}.$$

Notice that this matrix $m(T)$ depends heavily on the basis $v_1, \ldots, v_n$ used, as we take into account in

Definition. The *matrix of $T \in L(V)$ in the basis $v_1, \ldots, v_n$* is the matrix (t_{rs}) determined by the equations $Tv_s = \sum_{r=1}^{n} t_{rs} v_r$.

Thus, if we calculate $m(T)$ according to a basis $v_1, \ldots, v_n$ and you calculate $m(T)$ according to a basis $u_1, \ldots, u_n$, there is no earthly reason why we should get the same matrices. As we shall see, the matrices you get and we get are closely related but are not necessarily equal. In fact, we really should not write $m(T)$. Rather, *we should indicate the dependence on $v_1, \ldots, v_n$ by some device such as $m_{v_1, \ldots, v_n}(T)$*. But this is clumsy, so we take the easy way out and write it simply as $m(T)$.

Let's look at an example. Let V be the set of all polynomials in x over $\mathbb{R}$ of degree n or less. Thus $1, x, \ldots, x^n$ is a basis of V over $\mathbb{R}$. Let $D: V \to V$ be define by

$$D(p(x)) = \frac{d}{dx} p(x);$$

that is, $D(1) = 0$, $D(x) = 1$, $D(x^2) = 2x$, $\ldots$, $D(x^r) = rx^{r-1}$, $\ldots$, $D(x^n) = nx^{n-1}$. If we denote $v_1 = 1$, $v_2 = x$, $\ldots$, $v_k = x^{k-1}$, $\ldots$, $v_{n+1} = x^n$, then $D(v_k) = (k-1)v_{k-1}$. So the

matrix of D in this basis is $m(D) = \begin{bmatrix} 0 & 1 & 0 & \cdots & 0 \\ 0 & 0 & 2 & \cdots & 0 \\ \vdots & \vdots & \vdots & \ddots & \vdots \\ \vdots & \vdots & \vdots & & n \\ 0 & 0 & 0 & \cdots & 0 \end{bmatrix}$. For instance, if $n = 2$,

$m(D) = \begin{bmatrix} 0 & 1 & 0 \\ 0 & 0 & 2 \\ 0 & 0 & 0 \end{bmatrix}$.

What is $m(D)$ if we use the basis $1, 1+x, x^2, \ldots, x^n$ of V over F? If $u_1 = 1$, $u_2 = 1+x$, $u_3 = x^2$, $\ldots$, $u_{n+1} = x^n$, then, since $D(u_1) = 0$, $D(u_2) = 1 = u_1$, $D(u_3) = 2x = 2(u_2 - 1) = 2u_2 - 2u_1$, $D(u_4) = 3x^2 = 3u_3$, $\ldots$, the matrix $m(D)$ in this basis is

$$m(D) = \begin{bmatrix} 0 & 1 & -2 & 0 & \cdots & 0 \\ 0 & 0 & 2 & 0 & \cdots & 0 \\ 0 & 0 & 0 & 3 & & \vdots \\ \vdots & \vdots & \vdots & \vdots & \ddots & n \\ 0 & 0 & 0 & 0 & \cdots & 0 \end{bmatrix}.$$

For instance, if $n = 2$, $m(D) = \begin{bmatrix} 0 & 1 & -2 \\ 0 & 0 & 2 \\ 0 & 0 & 0 \end{bmatrix}$.

Let V be an n-dimensional vector space over F and let $v_1, \ldots, v_n$, $n = \dim(V)$, be a basis of V. *This basis $v_1, \ldots, v_n$ will be the one we use in the discussion to follow.*

How do the matrices $m(T)$, for $T \in L(V)$, behave with regard to the operations in $L(V)$? The proofs of the statements we are going to make are *precisely* the same as the proofs given in Section 3.9, where we considered linear transformations on $F^{(n)}$.

Theorem 9.4.1. *If $T_1, T_2 \in L(V)$, $a \in F$, then*

1. $m(T_1 + T_2) = m(T_1) + m(T_2)$;
2. $m(aT_1) = am(T_1)$;
3. $m(T_1 T_2) = m(T_1)m(T_2)$.

Furthermore, given any matrix A in $M_n(F)$, then $A = m(T)$ for some $T \in L(V)$. Finally, $m(T_1) = m(T_2)$ implies that $T_1 = T_2$. So the mapping $m: L(V) \to M_n(F)$ is $1-1$ and onto.

Corollary 9.4.2. $T \in L(V)$ is invertible if and only if $m(T) \in M_n(F)$ is invertible; and if T is invertible, then $m(T^{-1}) = m(T)^{-1}$.

Corollary 9.4.3. If $T \in L(V)$ and $S \in L(V)$ is invertible, then

$$m(S^{-1}TS) = m(S)^{-1}m(T)m(S).$$

What this theorem says is that m is a mechanism for carrying us from linear transformations to matrices in a way that preserves all properties. So we can argue back and forth readily using matrices and linear transformations interchangeably.

Corollary 9.4.4. If $T \in L(V)$, then the following conditions are equivalent:

1. The only solution to $Tx = 0$ is $x = 0$.
2. T is $1-1$.
3. T is onto.
4. T is invertible.

Proof: Letting A be the matrix $m(T)$ of T in a basis for V, we proved that conditions (1)–(4) with T replaced by A are equivalent. So by Theorem 9.4.1 and Corollary 9.4.2, conditions (1)–(4) for T are also equivalent. ∎

Corollary 9.4.5. If $T \in L(V)$ takes a basis $v_1, \ldots, v_n$ to a basis $Tv_1, \ldots, Tv_n$ of V, then T is invertible in $L(V)$.

Proof: Let $v_1, \ldots, v_n$ be a basis of V such that $Tv_1, \ldots, Tv_n$ is also a basis of V. Then, given $u \in V$, there are $a_1, \ldots, a_n \in F$ such that

$$u = a_1 Tv_1 + \cdots + a_n Tv_n = T(a_1 v_1 + \cdots + a_n v_n).$$

Thus T maps V onto V. Is T $1-1$? If $v = b_1 v_1 + \cdots + b_n v_n$ and $Tv = 0$, then $0 = T(b_1 v_1 + \cdots + b_n v_n) = b_1 Tv_1 + \cdots + b_n Tv_n$. So since $Tv_1, \ldots, Tv_n$ is a basis, the b_s are all 0 and $v = 0$. Thus T is invertible in $L(V)$ by Corollary 9.4.4. ∎

If $v_1, \ldots, v_n$ and $w_1, \ldots, w_n$ are two bases of V and T a linear transformation on V, we can compute, as above, the matrix $m_1(T)$ in the first basis and $m_2(T)$ in the second. How are these matrices related? If $m_1(T) = (t_{rs})$ and $m_2(T) = (\tau_{rs})$, then by the very definition of m_1 and m_2,

$$Tv_s = \sum_{r=1}^{n} t_{rs} v_r \tag{1}$$

$$Tw_s = \sum_{r=1}^{n} \tau_{rs} w_r \tag{2}$$

Sec. 9.4] Linear Transformations as Matrices 345

If S is the linear transformation defined by $w_s = Sv_s$ for $s = 1, 2, \ldots, n$, then since S takes a basis into a basis, it must be invertible.

Equation (2) becomes

$$TSv_s = \sum_{r=1}^{n} \tau_{rs} Sv_r = S\left(\sum_{r=1}^{n} \tau_{rs} v_r\right). \tag{3}$$

Hence

$$S^{-1}TSv_s = \sum_{r=1}^{n} \tau_{rs} v_r. \tag{4}$$

What (4) says is that the matrix of $S^{-1}TS$ in the basis $v_1, \ldots, v_n$ is precisely $(\tau_{rs}) = m_2(T)$. So $m_1(S^{-1}TS) = m_2(T)$. Using Corollary 9.4.3 we get $m_2(T) = m_1(S)^{-1} m_1(T) m_1(S)$.

We have proved

Theorem 9.4.6. *If $m_1(T)$ and $m_2(T)$ are the matrices of T in the bases $v_1, \ldots, v_n$ and $w_1, \ldots, w_n$, respectively, then $m_2(T) = m_1(S)^{-1} m_1(T) m_1(S)$, where S is defined by $w_s = Sv_s$ for $s = 1, 2, \ldots, n$.*

We illustrate the theorem with an example. Let V be the vector space of polynomials of degree 2 or less over F and let D be defined by $D(p(x)) = \dfrac{d}{dx} p(x)$. Then D is in $L(V)$. In the basis $v_1 = 1$, $v_2 = x$, $v_3 = x^2$ of V, we saw that

$$m_1(D) = \begin{bmatrix} 0 & 1 & 0 \\ 0 & 0 & 2 \\ 0 & 0 & 0 \end{bmatrix},$$

while in the basis $w_1 = 1$, $w_2 = 1 + x$, $w_3 = x^2$, we saw that

$$m_2(D) = \begin{bmatrix} 0 & 1 & -2 \\ 0 & 0 & 2 \\ 0 & 0 & 0 \end{bmatrix}.$$

What does $m_1(S)$ look like where $w_1 = Sv_1$, $w_2 = Sv_2$, $w_3 = Sv_3$? So $v_1 = w_1 = Sv_1$, $v_1 + v_2 = 1 + x = w_2 = Sv_2$, $v_3 = x^2 = w_3 = Sv_3$. Thus

$$m_1(S) = \begin{bmatrix} 1 & 1 & 0 \\ 0 & 1 & 0 \\ 0 & 0 & 1 \end{bmatrix}.$$

As is easily checked,

$$m_1(S)^{-1} = \begin{bmatrix} 1 & -1 & 0 \\ 0 & 1 & 0 \\ 0 & 0 & 1 \end{bmatrix}$$

and

$$\begin{bmatrix} 1 & -1 & 0 \\ 0 & 1 & 0 \\ 0 & 0 & 1 \end{bmatrix} \begin{bmatrix} 0 & 1 & 0 \\ 0 & 0 & 2 \\ 0 & 0 & 0 \end{bmatrix} \begin{bmatrix} 1 & 1 & 0 \\ 0 & 1 & 0 \\ 0 & 0 & 1 \end{bmatrix} = \begin{bmatrix} 0 & 1 & -2 \\ 0 & 0 & 2 \\ 0 & 0 & 0 \end{bmatrix} \begin{bmatrix} 1 & 1 & 0 \\ 0 & 1 & 0 \\ 0 & 0 & 1 \end{bmatrix}$$

$$= \begin{bmatrix} 0 & 1 & -2 \\ 0 & 0 & 2 \\ 0 & 0 & 0 \end{bmatrix} = m_2(D).$$

We saw earlier that in $M_n(F)$, $\text{tr}(C^{-1}AC) = \text{tr}(A)$ and $\det(C^{-1}AC) = \det(A)$. So, in light of Corollary 9.4.3, *no matter what bases* we use to represent $T \in L(V)$ as a matrix, then $m_2(T) = m_1(S)^{-1} m_1(T) m_1(S)$; hence $\text{tr}(m_2(T)) = \text{tr}(m_1(T))$ and $\det(m_1(T)) = \det(m_2(T))$. So we can make the meaningful

Definition. If $T \in L(V)$, then
1. $\text{tr}(T) = \text{tr}(m(T))$
2. $\det(T) = \det(m(T))$

for $m(T)$ the matrix of T in *any* basis of V.

We now can carry over an important result from matrices, the Cayley–Hamilton Theorem.

Theorem 9.4.7. If V is n-dimensional over F and $T \in L(V)$, then T satisfies $p_T(x) = \det(xI - T)$, where $m(T)$ is the matrix of T in any basis of V. So T satisfies a polynomial of degree n over F.

Proof: $m(xI - T) = xI - m(T)$; hence

$$\det(xI - T) = \det(m(xI - T)) = \det(xI - m(T)) = p_{m(T)}(m(T)) = 0$$

by the Cayley–Hamilton Theorem 5.10.2 for the matrix $m(T)$. ∎

We call $p_T(x) = \det(xI - T)$ the *characteristic polynomial* of T.

Thus if $a \in F$ is such that $\det(aI - T) = 0$, then $m(aI - T) = aI - m(T)$ cannot be invertible. Since m is $1-1$ and onto, $aI - T$ cannot be invertible. So there exists a vector $v \neq 0$ in V such that $(aI - T)v = 0$, that is, $Tv = av$. Furthermore, if $b \in F$ is *not* a root of $p_T(x)$, then $\det(bI - T) \neq 0$. This implies that $(bI - T)$ is invertible; hence there is *no* $v \neq 0$ such that $Tv = bv$. Thus

Theorem 9.4.8. If a is a root of $p_T(x) = \det(xI - T)$, then $Tv = av$ for some $v \neq 0$ in V, and if b is not a root of $p_T(x)$, there is no such $v \neq 0$ in V.

The roots of $p_T(x)$ are called the *characteristic roots* of T, and these are precisely the elements $a_1, \ldots, a_n$ in F such that $Tv_r = a_r v_r$ for some $v_r \neq 0$ in V. Also, v_r is called the *characteristic vector* associated with a_r.

With this said,

everything we did for characteristic roots and characteristic vectors of matrices carries over in its entirety to linear transformations on finite-dimensional vector spaces over F.

Readers should make a list of what strikes them as the most important results that can so be transferred.

A further word about this transfer:

Not only are the results on characteristic roots transferable to linear transformations, but all results proved for matrices have their exact analogs for linear transformations. The mechanism to do this is the one we used, namely, passing to the matrix of T in a given basis, and making use of the theorems proved about this.

PROBLEMS

NUMERICAL PROBLEMS

1. Find the matrix of D defined on the set of polynomials in x over F of degree at most 2 by $D(p(x)) = \dfrac{d}{dx} p(x)$ in the bases

   (a) $v_1 = 1$, $v_2 = 2 - x$, $v_3 = x^2 + 1$.

   (b) $v_1 = 1 + x$, $v_2 = x$, $v_3 = 1 + 2x + x^2$.

   (c) $v_1 = \dfrac{1-x}{2}$, $v_2 = \dfrac{1+x}{2}$, $v_3 = x^2$.

2. In Problem 1 find C such that $C^{-1} m_1(D) C = m_2(D)$, where $m_1(D)$ is the matrix of D in the basis of Part (a), and $m_2(D)$ is that in the basis of Part (c).

3. Do Problem 2 for the bases in Parts (b) and (c).

4. What are $\operatorname{tr}(D)$ and $\det(D)$ for the D in Problem 1?

5. Let V be all functions of the form $ae^x + be^{2x} + ce^{3x}$. If $D: V \to V$ is defined by $D(f(x)) = \dfrac{d}{dx} f(x)$, find:

   (a) The matrix of D in the basis

   $$v_1 = e^x, \quad v_2 = e^{2x}, \quad v_3 = e^{3x}.$$

   (b) The matrix of D in the basis

   $$v_1 = e^x + e^{2x}, \quad v_2 = e^{2x} + e^{3x}, \quad v_3 = e^x + e^{3x}.$$

   (c) Find a matrix C such that $C^{-1} m_1(D) C = m_2(D)$, where $m_1(D)$ is the matrix of D in the basis of Part (a) and $m_2(D)$ is that in the basis of Part (b).

   (d) Find $\det(m_1(D))$ and $\det(m_2(D))$, and $\operatorname{tr}(m_1(D))$ and $\operatorname{tr}(m_2(D))$.

6. In Problem 5, find the characteristic roots of D and their associated characteristic vectors.

7. Show that for D in Problem 5, $\det(D)$ is the product of the characteristic roots of D and $\text{tr}(D)$ is the sum of the characteristic roots of D.

8. Let V be the set of all polynomials in T of degree 3 or less over F. Define $T: V \to V$ by $T(f(x)) = \dfrac{1}{x} \displaystyle\int_0^x f(t)\,dt$ for $f(x) \neq 0$ and $T(0) = 0$. Is T a linear transformation on V? If not, why not?

9. Find the matrix of T defined on the V of Problem 8 by

$$T(1) = 1 + x, \quad T(x) = (1 + x)^2, \quad T(x^2) = (1 + x)^3, \quad T(x^3) = x;$$

and find $\det(T)$, $\text{tr}(T)$, and the characteristic polynomial $p_T(x)$.

MORE THEORETICAL PROBLEMS

Easier Problems

10. In V the set of polynomials in x over F of degree n or less, find $m(T)$, where T is defined in the basis $v_1 = 1$, $v_2 = x$, ..., $v_n = x^n$ by
    (a) $T(1) = x$, $T(x) = 1$, $T(x^k) = x^k$ for $1 < k \leq n$.
    (b) $T(1) = x$, $T(x) = x^2$, ..., $T(x^k) = x^{k+1}$ for $1 \leq k < n$, $T(x^n) = 1$.
    (c) $T(1) = x$, $T(x) = x^2$, $T(x^2) = x^3$, ..., $T(x^k) = x^k$ for $2 < k \leq n$.

11. For Parts (a), (b), and (c) of Problem 10, find $\det(T)$.

12. For Parts (a) and (c) of Problem 10, if $F = \mathbb{C}$, find the characteristic roots of T and the associated characteristic vectors.

13. Can $\begin{bmatrix} 1 & 2 & 5 \\ 0 & 1 & 6 \\ 1 & 0 & 1 \end{bmatrix}$ and $\begin{bmatrix} -1 & 0 & 1 \\ 0 & 4 & 2 \\ 0 & 1 & 3 \end{bmatrix}$ be the matrices of the same linear transformation in two different bases?

14. Show that $\begin{bmatrix} 1 & 2 & 3 \\ 0 & 1 & 2 \\ 0 & 0 & 1 \end{bmatrix}$ and $\begin{bmatrix} 1 & 2 & 5 \\ 0 & 1 & 2 \\ 0 & 0 & 1 \end{bmatrix}$ are the matrices of a linear transformation T on $F^{(3)}$ in two different bases.

Middle-Level Problems

15. If V is an n-dimensional vector space over $\mathbb{C}$ and $T \in L(V)$, show that we can find a basis $v_1, \ldots, v_n$ of V such that $T(v_r)$ is a linear combination of v_r and its predecessors $v_1, \ldots, v_{r-1}$ for $r = 1, 2, \ldots, n$ (see Theorem 4.7.1).

16. If $T \in L(V)$ is nilpotent, that is, $T^k = 0$ for some k, show that $T^n = 0$, where $n = \dim(V)$.

17. If $T \in L(V)$ is nilpotent, prove that 0 is the only characteristic root of T.

18. Prove that $m(T_1 + T_2) = m(T_1) + m(T_2)$ and $m(T_1 T_2) = m(T_1)m(T_2)$.

19. Prove that if V is n-dimensional over F, then m maps $L(V)$ onto $M_n(F)$ in a $1-1$ way.

20. If V is a vector space and W a subspace of V such that $T(W) \subset W$, show that we can find a basis of V such that $m(T) = \left[\begin{array}{c|c} A & * \\ \hline 0 & B \end{array}\right]$, where A is an $r \times r$ matrix $[r = \dim(W)]$ and B is an $(n-r) \times (n-r)$ matrix. (We don't care what $*$ is.)

21. If V is a vector space and U, W subspaces of V such that $V = U \oplus W$, show that if $T \in L(V)$ is such that $T(U) \subset U$ and $T(W) \subset W$, then there is a basis of V such that the matrix of T in that basis looks like $\left[\begin{array}{c|c} A & 0 \\ \hline 0 & B \end{array}\right]$, where A, B are $r \times r, (n-r) \times (n-r)$ matrices, respectively, where $r = \dim(U)$.

22. In Problem 20, what interpretation can you give the matrix A?

23. In Problem 21, what interpretation can you give the matrices A and B?

24. Prove that $T \in L(V)$ is invertible if and only if $\det(T) \neq 0$. What is $\det(T^{-1})$ in this case?

9.5. A DIFFERENT SLANT ON SECTION 9.4 (OPTIONAL)

We can look at the matters discussed in Section 9.4 in a totally different way, a way that is more abstract. Whereas what we did there depended heavily on the particular basis of V over F with which we worked, this more abstract approach is basis-free. Aside from not pinning us down to a particular basis, this basis-free approach is esthetically more satisfying.

But this new approach is not without its drawbacks. Seeing such an abstract argument for the first time might strike the reader as unnecessarily complicated, as overelaborated, and as downright hard. We pay a price, but it is worth it in the sense that this type of argument is the prototype of many arguments that occur in mathematics.

Let's recall that if V is an n-dimensional vector space over F, then, by Theorem 8.4.3, V is isomorphic to $F^{(n)}$. Let Φ denote the isomorphism of V onto $F^{(n)}$. We shall use Φ as the vehicle for transporting linear transformations on V to those on $F^{(n)}$.

Let $L(V)$ be the set of all linear transformations on V over F and consider $L(F^{(n)})$, the set of all linear transformations on $F^{(n)}$ over F. [Since the linear transformations of $F^{(n)}$ are just the $n \times n$ matrices, $L(F^{(n)})$ equals $M_n(F)$.] We *define* $\psi: L(V) \to L(F^{(n)})$ in

Definition. If $T \in L(V)$ and $v \in V$, then $\psi(T)(\Phi(v)) = \Phi(T(v))$.

We must first verify that $\psi(T)$ is a linear transformation on $F^{(n)}$ over F. First, since Φ maps V *onto* $F^{(n)}$, every element in $F^{(n)}$ appears in the form $\Phi(v)$ for some $v \in V$. Hence $\psi(T)$ is defined on *all* of $F^{(n)}$.

To show that ψ maps $L(V)$ into $L(F^{(n)})$, we must prove that $\psi(T)$ is a linear transformation on $F^{(n)}$ over F. We go about this now.

If $a \in F$ and $\bar{u}, \bar{v}$ are in $F^{(n)}$, then for some u, v in V, $a \in F$, $\bar{u} = \Phi(u)$, $\bar{v} = \Phi(v)$, hence

$$\psi(T)(a\bar{u} + \bar{v}) = \psi(T)(\Phi(au + v)) = \Phi(T(au + v))$$
$$= \Phi(aT(u) + T(v)) = a\Phi(T(u)) + \Phi(T(v))$$
$$= a\psi(T)(\Phi(u)) + \psi(T)(\Phi(v)) = a\psi(T)(\bar{u}) + \psi(T)(\bar{v}).$$

This shows that $\psi(T)$ is a linear transformation on $F^{(n)}$. Note that in the argument we used the fact that T is a linear transformation on V and that Φ is a vector space isomorphism of V onto $F^{(n)}$, several times.

So we know, now, that $\psi(L(V)) \subset L(F^{(n)})$. What further properties does the mapping ψ enjoy? We claim that ψ is not only a vector space isomorphism of $L(V)$ onto $L(F^{(n)})$ but, in addition, satisfies another nice condition, namely, that $\psi(ST) = \psi(S)\psi(T)$ for S, T in $L(V)$.

We check out all these claims now. First we need to see if

$$\psi(S + T) \rightleftharpoons \psi(S) + \psi(T).$$

What is $\psi(S+T)$? By definition, for all $v \in V$,

$$\psi(S+T)(\Phi(v)) = \Phi((S+T)(v)) = \Phi(S(v) + T(v)) = \Phi(S(v)) + \Phi(T(v))$$
$$= \psi(S)(\Phi(v)) + \psi(T)(\Phi(v)) = (\psi(S) + \psi((T))(\Phi(v)).$$

This gives us that

$$\psi(S) + \psi(T) = \psi(S + T).$$

A similar argument proves that $\psi(aS) = a\psi(S)$ for $a \in F$, $S \in L(V)$. Therefore, ψ is, at least, a vector space homomorphism of $L(V)$ into $L(F^{(n)})$.

If $S, T \in L(V)$, consider $\psi(ST)$. By definition, for every $v \in V$,

$$\psi((ST)(\Phi(v)) = \Phi((ST)(v)) = \Phi(S(T(v)))$$
$$= \psi(S)(\Phi(T(v)) = \psi(S)\psi(T)(\Phi(v)).$$

This implies that $\psi(ST) = \psi(S)\psi(T)$ for $S, T \in L(V)$. So, in addition to being a vector space homomorphism of $L(V)$ into $L(F^{(n)})$, ψ *also preserves the multiplication of elements in $L(V)$*.

To finish the job we must prove two other items, namely, that ψ is $1-1$ and that ψ maps $L(V)$ onto $L(F^{(n)})$.

For the $1-1$-ness, suppose that $\psi(S) = \psi(T)$, for $S, T \in L(V)$. We want to show that this forces $S = T$. If $v \in V$, then

$$\Phi(S(v)) = \psi(S)(\Phi(v)) = \psi(T)(\Phi(v)) = \Phi(T(v)).$$

Because Φ is an isomorphism of V onto $F^{(n)}$ we get from this last relation that $S(v) = T(v)$ for all $v \in V$. By the definition of the equality of two mappings, we then get that $S = T$. Therefore, ψ is $1-1$.

All we have left to do is to show that ψ maps $L(V)$ onto $L(F^{(n)})$. Let $\bar{T} \in L(F^{(n)})$; we want to find an element T is $L(V)$ such that $\psi(T) = \bar{T}$.

If $\bar{v} \in F^{(n)}$, then $\bar{v} = \Phi(v)$ for some $v \in V$, and $\bar{T}\bar{v} = \Phi(w)$ for some $w \in V$. These elements v and w are unique in V, since Φ is an isomorphism of V onto $F^{(n)}$. Define T by the rule $T(v) = w$. By the uniqueness of v and w, T is well defined. By definition, for every $v \in V$, $\psi(T)(\Phi(v)) = \Phi(T(v)) = \Phi(w) = \bar{T}(\Phi(v))$. Thus $\psi(T) = \bar{T}$ and ψ is onto.

This finishes the proof of everything we have claimed to hold for the mapping ψ of $L(V)$ onto $L(F^{(n)})$. The argument given allows us to transfer *any* statement about linear transformations on V to the *same* statement about linear transformations of $F^{(n)}$, making use of the mapping ψ—and most important, vice versa. So

everything we did in Chapters 3 and 4 for transformations on $F^{(n)}$ [and for matrices in $M_n(F) = L(F^{(n)})$] holds true for linear transformations on any n-dimensional vector space over F.

We amplify the remarks above with a simple example. We know that every element A in $M_2(F)$ satisfies a quadratic relation of the form $A^2 + aA + bI = 0$. In fact, we know what a and b are, namely,

$$a = -\text{tr}(A)$$

and

$$b = \det(A).$$

So every linear transformation T on $F^{(2)}$ satisfies a quadratic relation of the form $T^2 + uT + vI = 0$, where $u, v \in F$.

Suppose that V is a 2-dimensional vector space over F. By what we did above, there is an isomorphism ψ of $L(V)$ onto $L(F^{(2)})$ that preserves all the operations—addition, multiplication by a scalar, and the product in $L(V)$. Since $\psi(T) \in L(F^{(2)})$ for $T \in L(V)$, we know that $\psi(T)^2 + u\psi(T) + vI = 0$ for some $u, v \in F$. Since $\psi(T)^2 = \psi(T^2)$, and $u\psi(T) = \psi(uT)$, we get that

$$\psi(T^2) + \psi(uT) + \psi(vI) = 0.$$

However, ψ also preserves sums; thus we have that $\psi(T^2 + uT + vI) = 0$. Because ψ is $1-1$, this relation leads us to $T^2 + uT + vI = 0$. In other words, every element in $L(V)$ satisfies some quadratic relation over F.

Granted the example just done is a rather simple one, the technique employed in discussing it is typical of how the isomorphism ψ is exploited to prove all the results in $L(F^{(n)})$, that is, in $M_n(F)$, for a general vector space V which is n-dimensional over F.

9.6. HERMITIAN IDEAS

Since we are always working over $F = \mathbb{R}$ or $F = \mathbb{C}$, we saw that we can introduce an inner product structure on $F^{(n)}$. This depends heavily on the fact that $F = \mathbb{R}$ or $\mathbb{C}$. We could extend the notion of vector spaces over "number systems" F that are neither $\mathbb{R}$ nor $\mathbb{C}$. Depending on the nature of this "number system" the "vector space" might or might not be susceptible to the introduction of an inner product on it.

However, we are working only over $F = \mathbb{R}$ or $\mathbb{C}$, so we have no such worries. So given any finite-dimensional vector space V, we saw in Section 8.7 that we could make of V an inner product space. There we did it using the existence of orthonormal bases for V and for $F^{(n)}$.

We redo it here—to begin with—in a slightly more abstract way. It is healthy to see something from a variety of points of view. Given $F^{(n)}$ we had introduced on $F^{(n)}$ an inner product via $(v, w) = \sum_{r=1}^{n} a_r \bar{b}_r$, where $v = \begin{bmatrix} a_1 \\ \vdots \\ a_n \end{bmatrix}$ and $w = \begin{bmatrix} b_1 \\ \vdots \\ b_n \end{bmatrix}$. By Theorem 8.4.3 we know that there is an isomorphism Φ of V onto $F^{(n)}$, where $n = \dim(V)$. We exploit Φ to produce an inner product on V. How? Just define the function $[\cdot, \cdot]$ by the rule

$$[v, w] = (\Phi(v), \Phi(w)) \quad \text{for } v, w \in V.$$

Since $\Phi(v)$ and $\Phi(w)$ are in $F^{(n)}$ and $(\cdot, \cdot)$ is defined on $F^{(n)}$, the definition given above for $[\cdot, \cdot]$ makes sense.

It makes sense, but is it an inner product? The answer is "yes." We check out the rules that spell out an inner product, in turn.

1. If $v \in V$, then $[v, v] = (\Phi(v), \Phi(v)) \geq 0$. If $[v, v] = 0$, then $(\Phi(v), \Phi(v)) = 0$; hence $\Phi(v) = 0$ since $(\ ,\ )$ is an inner product on $F^{(n)}$. Because Φ is $1 - 1$ we end up with the desired result, $v = 0$.
2. Is $[v, w] = \overline{[w, v]}$ the complex conjugate of (v, w)? Well, $[v, w] = (\Phi(v), \Phi(w)) = \overline{(\Phi(w), \Phi(v))} = \overline{[w, v]}$.
3. Given $a \in F$, is $[av, w] = a[v, w]$? Again, $[av, w] = (\Phi(av), \Phi(w)) = (a\Phi(v), \Phi(w)) = a(\Phi(v), \Phi(w)) = a[v, w]$.
4. Finally, does $[u + v, w] = [u, w] + [v, w]$? Because $[u + v, w] = (\Phi(u+v), \Phi(w)) = (\Phi(u) + \Phi(v), \Phi(w)) = (\Phi(u), \Phi(w)) + (\Phi(v), \Phi(w)) = [u, w] + [v, w]$.

We have shown that $[\cdot, \cdot]$ is a legitimate inner product on V. Hence V is an inner product space. Thus

Theorem 9.6.1. *If V is finite-dimensional over $\mathbb{R}$ or $\mathbb{C}$, then V is an inner product space.*

Utilizing the isomorphisms Φ of V onto $F^{(n)}$, we propose to carry over to the abstract space V everything we did for $F^{(n)}$ as an inner product space. When this is done we shall define the Hermitian adjoint of any element in $L(V)$ and prove the analogs for Hermitian linear transformations of the results proved for Hermitian matrices.

Recall that the inner product on V is defined by $[v, w] = (\Phi(v), \Phi(w))$, where $(\cdot, \cdot)$ indicates the inner product in $F^{(n)}$. Let $e_1, \ldots, e_n$ be the canonical basis of $F^{(n)}$; so $e_1, \ldots, e_n$ is an orthonormal basis of $F^{(n)}$. Since Φ is an isomorphism of V onto $F^{(n)}$ there exist elements $v_1, \ldots, v_n$ in V such that $e_r = \Phi(v_r)$ for $r = 1, 2, \ldots, n$. Thus $[v_r, v_s] = (\Phi(v_r), \Phi(v_s)) = (e_r, e_s) = 1$ if $r = s$ and $= 0$ if $r \neq s$.

If we call $v, w \in V$ *orthogonal* if $[v, w] = 0$, then our basis $v_1, \ldots, v_n$ above consists of mutually orthogonal elements. Furthermore, if $\sqrt{[v, v]}$ is called the *length* of v, then each v_r is of length 1. So it is reasonable, using $F^{(n)}$ as a prototype, to call $v_1, \ldots, v_n$ an *orthonormal* basis of V.

What we did for V also holds for any subspace W of V. So W also has an orthonormal basis. This is

Theorem 9.6.2. Given a subspace W of V, then W has an orthonormal basis.

The analog of the *Gram–Schmidt process* discussed earlier also carries over. We feel it would be good for the reader to go back to where this was done in Section 4.3 and to carry over the proof there to the abstract space V.

So, according to the Gram–Schmidt process, given $u_1, \ldots, u_k$ in V, linearly independent over F, we can find an orthonormal bases $x_1, \ldots, x_k$ of $\langle u_1, \ldots, u_k \rangle$ (or cite Theorem 9.6.2). Let

$$S = \langle u_1, \ldots, u_k \rangle^\perp = \{ w \in V \mid [v, w] = 0 \text{ for all } v \in \langle u_1, \ldots, u_k \rangle \}.$$

Then $\langle u_1, \ldots, u_k \rangle^\perp$ is a subspace of V. If $v \in V$, then

$$s = v - [v, x_1] x_1 - [v, x_2] x_2 - \cdots - [v, x_k] x_k$$

has the property $[s, x_r] = 0$ for $r = 1, 2, \ldots, k$. Since s is orthogonal to every element of a basis of $\langle u_1, \ldots, u_k \rangle$, s must be orthogonal to every element of $\langle u_1, \ldots, u_k \rangle$. In short, $s \in S = \langle u_1, \ldots, u_k \rangle^\perp$. So $v = s + [v, x_1] x_1 + \cdots + [v, x_k] x_k$. Consequently, $v \in S + \langle u_1, \ldots, u_k \rangle$. Thus $S + \langle u_1, \ldots, u_k \rangle = V$. Also, $S \cap \langle u_1, \ldots, u_k \rangle = 0$ (Why?). Therefore, $V = S \oplus \langle u_1, \ldots, u_k \rangle$. (Verify all the statements made above!)

Given any subspace W of V, then W has a finite basis, so W is the linear span of a finite set of linearly independent elements. Thus, if $W^\perp = \{ s \in V \mid [w, s] = 0 \text{ for all } w \in W \}$, then by the above we have

Theorem 9.6.3. If W is a subspace of V, then $V = W \oplus W^\perp$.

Definition. $W^\perp$ is called the *orthogonal complement* of W.

If $U = (W^\perp)^\perp$, then for $u \in U$, $[u, s] = 0$ for all $s \in W^\perp$. Since $[w, s] = 0$ for $w \in W$, we know that $W \subset (W^\perp)^\perp = W^{\perp\perp}$. However, $V = W^\perp \oplus W^{\perp\perp}$; hence $\dim(V) = \dim(W^\perp) + \dim(W^{\perp\perp})$. (Prove!) But since $V = W \oplus W^\perp$, $\dim(V) = \dim(W) + \dim(W^\perp)$. The upshot of all this is that $\dim(W) = \dim(W^{\perp\perp})$. Because W is a subspace of $W^{\perp\perp}$ and of the same dimension as $W^{\perp\perp}$, we must have $W = W^{\perp\perp}$.

We have proved

Theorem 9.6.4. If W is a subspace of V, then $W = W^{\perp\perp}$.

We now turn our attention to linear transformations on V and how they interact with the inner product $[\cdot, \cdot]$ on V.

Let $v_1, \ldots, v_n$ be an orthonormal basis of V; hence $[v_r, v_s] = 0$ if $r \neq s$, $[v_r, v_r] = 1$, for $r, s = 1, 2, \ldots, n$. Let $T \in L(V)$ be a linear transformation on V. Thus for any

$s = 1, 2, \ldots, n$, $Tv_s = \sum_{r=1}^{n} t_{rs} v_r$, where the $t_{rs} \in F$. Define S by $Sv_s = \sum_{r=1}^{n} \sigma_{rs} v_r$ where $\sigma_{rs} = \bar{t}_{sr}$. We prove

Theorem 9.6.5. S is a linear transformation on V and for any $v, w \in V$, $[Tv, w] = [v, Sw]$.

Proof: It is enough to show that $[Tv_r, v_s] = [v_r, Sv_s]$ for all r, s. (Why?) But

$$[T(v_r), v_s] = \left[\sum_{k=1}^{n} t_{kr} v_k, v_s\right] = \sum_{k=1}^{n} t_{kr}[v_k, v_s] = t_{sr}$$

since $[v_k, v_s] = 0$ if $k \neq s$, and $[v_s, v_s] = 1$.

Now for the calculation of $[v_r, Sv_s]$. Since $Sv_s = \sum_{r=1}^{n} \sigma_{rs} v_r$ where $\sigma_{rs} = \bar{t}_{sr}$,

$$[v_r, Sv_s] = \left[v_r, \sum_{k=1}^{n} \sigma_{ks} v_k\right] = \sum_{k=1}^{n} \bar{\sigma}_{ks}[v_r, v_k] \text{ (Why?)} = \bar{\sigma}_{rs}$$

since $[v_r, v_k] = 0$ if $k \neq r$, and $= 1$ if $k = r$. So

$$[v_r, Sv_s] = \bar{\sigma}_{rs} = t_{sr} = [Tv_r, v_s].$$

This proves the theorem. ∎

Definition. We call S the *Hermitian adjoint* of T and denote it by T^*.

Note that if $m(T)$ is the matrix of T in an orthonormal basis of V, then the proof shows that $m(T^*) = m(T)^*$, where $m(T)^* = \overline{m(T)'}$, the Hermitian adjoint of the matrix $m(T)$. (Prove!)

We leave the proof of the next result to the reader.

Lemma 9.6.6. For $S, T \in L(V)$ and $a \in F$:

1. $T^{**} = (T^*)^* = T$;
2. $(aS + T)^* = \bar{a}S^* + T^*$;
3. $(ST)^* = T^*S^*$.

Definition. We call $T \in L(V)$ *Hermitian* if $T^* = T$. We call T *skew-Hermitian* if $T^* = -T$. We call T *unitary* if $TT^* = I = T^*T$.

Everything we proved for these kinds of matrices translates directly into a theorem for that kind of linear transformation. We give two sample results. The method used in proving them is that which is used in proving all such results.

Theorem 9.6.7. If T is Hermitian, then all its characteristic roots are real.

Proof: Let a be a characteristic root of T and $v \in V$ an associated characteristic vector. Thus $Tv = av$; hence $[Tv, v] = [av, v] = a[v, v]$. But $[Tv, v] = [v, T^*v] = [v, Tv] = [v, av] = \bar{a}[v, v]$. Therefore, $a[v, v] = \bar{a}[v, v]$, whence since $[v, v] \neq 0$, $a = \bar{a}$. Thus a is real. ∎

Theorem 9.6.8. $T \in L(V)$ is unitary if and only if $[Tv, Tw] = [v, w]$ for all $v, w \in V$.

Proof: Since T is unitary, $T^*T = I$. So $[v, w] = [Iv, w] = [T^*Tv, w] = [Tv, T^{**}w] = [Tv, Tw]$.

On the other hand, if $[Tv, Tw] = [v, w]$ for all $v, w \in V$, then $[T^*Tv, w] = [v, w]$; thus $[T^*Tv - v, w] = 0$ for all w. This forces $T^*Tv = v$ for all $v \in V$. (Why?) So $T^*T = I$. We leave the proof that $TT^* = I$ to the reader. ∎

The further properties of Hermitian, skew-Hermitian, and unitary linear transformations will be found in the exercises.

PROBLEMS

NUMERICAL PROBLEMS

1. For the following vector spaces V, find an isomorphism of V onto $F^{(n)}$ for appropriate n and use this isomorphism to define an inner product on V.
   (a) $V = $ all polynomials of degree 3 or less over F.
   (b) $V = $ all real-valued functions of the form
   $f(x) = a \cos(x) + b \sin(x)$.
   (c) $V = $ all real-valued functions of the form
   $f(x) = ae^x + be^{2x} + ce^{3x}$.

2. If V is the vector space of all polynomials of degree 4 or less and W is the subspace consisting of elements of the form $a + bx^2 + cx^4$, find via an isomorphism of V onto $F^{(5)}$ an inner product on V. Then find $W^\perp$.

3. For V as in Problem 2 find an orthonormal basis for $\langle v_1, v_2, v_3 \rangle$ where $v_1 = 1 + x$, $v_2 = x^2$, $v_3 = 1 + x + x^2 + x^3$.

4. For V as in Part (a) of Problem 1, define $(f(x), g(x)) = \int_0^1 f(x)\overline{g(x)}\,dx$. Prove this defines an inner product on V.

5. In Problem 4 find an orthonormal basis of V with respect to the inner product defined there.

6. For V as in Part (b) of Problem 1 define $(f(x), g(x)) = \int_0^\pi f(x)g(x)\,dx$.
   (a) Show that this defines an inner product on V.
   (b) Find an orthonormal basis of V with respect to this inner product.

7. For V as in Problem 6, let $T \in L(V)$ be define by

   $T(\cos(x)) = \sin(x)$, $T(\sin(x)) = \cos(x) + \sin(x)$.

   (a) Find the matrix of T in the orthonormal basis of Part (b) in Problem 6.
   (b) Express T^* as a linear transformation on V. [That is, what is $T^*(\cos(x))$, $T^*(\sin(x))$?]

8. If V is the vector space in Problem 5, using the orthonormal basis you found, find the form of T^* if $T(x^k) = x^{k+1}$ for $0 \leq k < 3$ and $T(x^3) = 1$.
9. If V, W are as in Problem 2, show that $W \oplus W^\perp = V$ and $W^{\perp\perp} = W$.

MORE THEORETICAL PROBLEMS

V is a vector space over $\mathbb{C}$.

Easier Problems

10. If $T \in L(V)$ and T is Hermitian, show that $iT = S$ is skew-Hermitian.
11. If S is skew-Hermitian, prove that $T = iS$ is Hermitian.
12. Prove that the characteristic roots of a skew-Hermitian T in $L(V)$ must be pure imaginaries.
13. Given $T \in L(V)$, show that $T = A + B$, where A is Hermitian, B is skew-Hermitian. Prove that A and B are unique.
14. Prove the following to be Hermitian:
    (a) TT^*
    (b) $T + T^*$
    (c) $TS + S^*T^*$
    for $S, T \in L(V)$.
15. If T is skew-Hermitian, prove that T^{2k} is Hermitian and T^{2k+1} is skew-Hermitian for all integers $k \geq 0$.
16. If T is unitary, prove that if a is a characteristic root of T, then $|a| = 1$.

Middle-Level Problems

17. Suppose that $W \subset V$ is a subspace of V such that $T(W) \subset W$, where $T^* = T$. Prove that $T(W^\perp) \subset W^\perp$.
18. If W and T are in Problem 17, show that $m(T) = \begin{bmatrix} A & 0 \\ 0 & B \end{bmatrix}$, where A is $r \times r$ and $r = \dim(W)$, for some suitable orthonormal basis of V.
19. Use Problem 18 and induction to prove that if $T^* = T$, then there exists an orthonormal basis of V such that $m(T)$ is a diagonal matrix in this basis.
20. If T is Hermitian and $T^k v = 0$, prove that $Tv = 0$.
21. If U in $L(V)$ is such that for some orthonormal basis $v_1, \ldots, v_n$ of V, the elements $w_1 = Uv_1, \ldots, w_n = Uv_n$ form an orthonormal basis of V, prove that U must be unitary.
22. If T is Hermitian, prove that both $\text{tr}(T)$ and $\det(T)$ are real.
23. In terms of the characteristic roots of T, where $T^* = T$, what are $\text{tr}(T)$ and $\det(T)$?
24. Call $T \in L(V)$ *normal* if $TT^* = T^*T$. Let $W = \{v \in V \mid Tv = 0\}$. Prove that $T^*(W) \subset W$ for T normal.
25. If $S, T \in L(V)$ and $W_a = \{v \in V \mid Tv = av\}$, prove that $S(W_a) \subset W_a$.

Harder Problems

26. If T, W are as in Problem 24, show that $T^*v = 0$ for $v \in W$ if $ST = TS$.

27. If T is normal and $T^k v = 0$, prove that $Tv = 0$. [*Hint*: Consider $(T^*T)^k v$.]
28. If T is normal, show that there is an orthonormal basis of V such that $m(T)$ is a diagonal matrix in this basis.
29. If T is normal, show
    (a) T is Hermitian if and only if its characteristic roots are all real.
    (b) T is skew-Hermitian if and only if its characteristic roots are pure imaginaries.
    (c) T is unitary if and only if each of its characteristic roots is of absolute value 1.
30. If T is skew-Hermitian, prove that $I - T$, $I + T$ are both invertible and $U = (I - T)(I + T)^{-1}$ is unitary.

9.7. QUOTIENT SPACES (OPTIONAL)

When we discussed homomorphisms of vector spaces we pointed out that this concept has its analog in almost every part of mathematics. Another of such a universal concept is that of the quotient space of a vector space V by a subspace W. It, too, has its analog almost everywhere in mathematics. As we shall see, the notion of quotient space is intimately related to that of homomorphism.

While the construction, at first glance, seems easy and straightforward, it is a very subtle one. This subtlety stems, in large part, from the fact that the *elements* of the quotient space V/W are *subsets* of V. In general, this can offer difficulties to someone seeing it for the first time.

Let V be a vector space over F and W a subspace of V. If $v \in V$, by $v + W$ we mean the subset $v + W = \{v + w \mid w \in W\}$.

Definition. If V is a vector space and W a subspace of V, then V/W, the *quotient space* of V by W, is the set of all $v + W$, where v is any element in V and where we define:

1. $(u + W) + (v + W) = (u + v) + W$
2. $a(v + W) = (av) + W$

for all $u, v \in V$ and all $a \in F$.

In a few moments, we shall show that V/W is a vector space over F. However, there are two slight difficulties that arise in the definition of V/W. These are

1. Is the addition we gave well defined?
2. Is the multiplication by a scalar we gave well defined?

What do these two questions even mean? Note that it is possible that $v + W = v_1 + W$ (as sets) with $v \neq v_1$. Yet we used a particular v to define the addition. For instance, in the $V = F^{(2)}$ and $W = \left\{ \begin{bmatrix} a \\ 0 \end{bmatrix} \middle| a \in F \right\}$ the subset $\begin{bmatrix} 0 \\ 1 \end{bmatrix} + W$ equals $\left\{ \begin{bmatrix} a \\ 1 \end{bmatrix} \middle| a \in F \right\}$. But $\begin{bmatrix} 0 \\ 1 \end{bmatrix} + W$ equally equals $\begin{bmatrix} 5 \\ 1 \end{bmatrix} + W = \left\{ \begin{bmatrix} 5 + a \\ 1 \end{bmatrix} \middle| a \in F \right\}$, since $5 + a$ gives every possible element of F as a runs over F.

We are obliged to show:

1. If $u + W = u_1 + W$ and $v + W = v_1 + W$, then in the definition given for addition in V/W, $(u + v) + W = (u + W) + (v + W) = (u_1 + W) + (v_1 + W) = (u_1 + v_1) + W$.
2. If $u + W = u_1 + W$ and $a \in F$, we must show that $(au) + W = a(u + W) = a(u_1 + W) = (au_1) + W$.

First note the simple fact that if $w_0 \in W$, then $w_0 + W = W$. This holds because, given any $w \in W$, then $w = w_0 + (w - w_0)$ and since W is a subspace of V, $w - w_0$ is in W. Thus $w \in w_0 + W$ for every $w \in W$, hence $W \subset w_0 + W$. Because $w_0 \in W$, $w_0 + W \subset W$. Therefore, $W = w_0 + W$.

If $u = u_1 + W$ and $v = v_1 + W$, then since $u = u + 0$ is in $u + W$, and since $u + W = u_1 + W$, we have that $u \in u_1 + W$. Thus $u = u_1 + w_0$ for some $w_0 \in W$. Similarly, $v = v_1 + w_1$ from some $w_1 \in W$. Therefore,

$$u + w = (u_1 + w_0) + (v_1 + w_1) = u_1 + v_1 + (w_0 + w_1)$$

and since $w_0 + w_1$ is in W, by the above we have that

$$(w_0 + w_1) + W = W.$$

Consequently,

$$u + v + W = u_1 + v_1 + (w_0 + w_1) + W = u_1 + v_1 + W.$$

With this we showed that the addition is indeed well defined. This clears us of obligation 1 above.

Note, too, that if $a \in F$, then $aW \subset W$. So if

$$u + W = u_1 + W, \text{ then } u = u_1 + w_0, w_0 \in W.$$

Therefore, $au = au_1 + aw_0$. Because $aw_0 \in W$ we get

$$au + W = au_1 + aw_0 + W = au_1 + W.$$

This clears us of obligation 2.

We have shown, at least, that the operations proposed to make of V/W a vector space, are well defined.

With that out of the way it is easy and routine to verify that V/W is a vector space. Notice that W acts as the 0-element for V/W and $-u + W$ acts as the negative of $u + W$. Although it is a tedious exercise to go through all the defining axioms for V/W, the reader should do it. It builds character, and patience.

We state the result as

Theorem 9.7.1. The quotient space $V/W = \{v + W | v \in V\}$ is a vector space over F relative to the operations

$$(u + W) + (v + W) = (u + v) + W$$

and
$$a(u + W) = au + W,$$
where $u, v \in V$ and $a \in F$.

Before proceeding to some examples of V/W—and because of the subtlety of the concept, we should see many examples—we dispose of one fact that ties in homomorphisms and quotient spaces.

Theorem 9.7.2. *If V is a vector space over F and W a subspace of V, then the mapping $\Phi: V \to V/W$ defined by $\Phi(v) = v + W$ is a homomorphism of V onto V/W. Moreover, $\text{Ker}(\Phi) = W$.*

Proof: To prove the result, why not do the obvious and define $\Phi: V \to V/W$ by $\Phi(v) = v + W$ for every $v \in V$? We check out the rules for a homomorphism.

1. $\Phi(u + v) = (u + v) + W = (u + W) + (v + W) = \Phi(v)$ from the definition of addition in V/W, for $u, v \in V$.
2. $\Phi(au) = (au) + W = a(u + W) = a\Phi(u)$, for $a \in F, u \in V$, by the definition of the multiplication by scalars in V/W.

Thus Φ is a homomorphism of V into V/W. It is onto, for, given $X \in V/W$, then $X = v + W$ for some $v \in V$, hence $X = v + W = \Phi(v)$.

Finally, what is the kernel of Φ? Remember that the 0-element of the space V/W is $0 + W = W$. Clearly, if $w \in W$, then $\Phi(w) = w + W = W$, hence $W \subset \text{Ker}(\Phi)$. On the other hand, if $u \in \text{Ker}(\Phi)$, then $\Phi(u) = W$, the 0-element of V/W. But by the definition of Φ, $\Phi(u) = u + W$. Thus $W = u + W$. This says that u, which is in $u + W$ (being of the form $u + 0$) and $0 \in W$), must be in $W = u + W$. Hence $\text{Ker}(\Phi) \subset W$. Therefore, we have $W = \text{Ker}(\Phi)$, as stated in the theorem.

We are now ready to look at some examples of V/W.

EXAMPLES

1. Let $V = F^{(n)}$ and let $W = \left\{ \begin{bmatrix} a \\ b \\ 0 \\ \vdots \\ 0 \end{bmatrix} \middle| a, b \in F \right\}$. W is a subspace of V. If

$$v = \begin{bmatrix} a_1 \\ \vdots \\ a_n \end{bmatrix} \in V, \text{ then } v + W = \begin{bmatrix} a_1 \\ \vdots \\ a_n \end{bmatrix} + W = \begin{bmatrix} 0 \\ 0 \\ a_3 \\ \vdots \\ a_n \end{bmatrix} + \begin{bmatrix} a_1 \\ a_2 \\ 0 \\ \vdots \\ 0 \end{bmatrix} + W.$$

But $\begin{bmatrix} a_1 \\ a_2 \\ 0 \\ \vdots \\ 0 \end{bmatrix} \in W$, hence $\begin{bmatrix} a_1 \\ a_2 \\ 0 \\ \vdots \\ 0 \end{bmatrix} + W = W$. So $v + W = \begin{bmatrix} 0 \\ 0 \\ a_3 \\ \vdots \\ a_n \end{bmatrix} + W$ is a typical element of V/W.

Note that if we map V/W onto $F^{(n-2)}$ by defining $\Phi\colon V/W \to F^{(n-2)}$ by the rule $\Phi(v + W) = \Phi\left(\begin{bmatrix} 0 \\ 0 \\ a_3 \\ \vdots \\ a_n \end{bmatrix}\right) = \begin{bmatrix} a_3 \\ a_4 \\ \vdots \\ a_n \end{bmatrix}$ that Φ is an isomorphism of V/W onto $F^{(n-2)}$. Thus, here, $V/W \simeq F^{(n-2)}$.

2. Let V be the set of all polynomials in x over F and let

$$W = \{x^3 p(x) \mid p(x) \in V\}.$$

W is a subspace of V. Given any polynomial $f(x)$ in V, then

$$f(x) = a_0 + a_1 x + a_2 x^2 + a_3 x^3 + \cdots + a_m x^m.$$

However,

$$g(x) = a_3 x^3 + \cdots + a_m x^m = x^3(a_3 + a_4 x + \cdots + a_m x^{m-3})$$

so is in W, hence $g(x) + W = W$. Now

$$f(x) = a_0 + a_1 x + a_2 x^2 + g(x),$$

whence

$$\begin{aligned} f(x) + W &= (a_0 + a_1 x + a_2 x^2 + g(x)) + W \\ &= (a_0 + a_1 x + a_2 x^2) + g(x) + W = a_0 + a_1 x + a_2 x^2 + W \\ &= (a_0 + W) + a_1(x + W) + a_2(x^2 + W) \end{aligned}$$

using the definition of the operations in V/W. If we let

$$\bar{v}_1 = 1 + W, \qquad \bar{v}_2 = x + W, \qquad \bar{v}_3 = x^2 + W,$$

then every element in V/W is of the form $c_1 \bar{v}_1 + c_2 \bar{v}_2 + c_3 \bar{v}_3$, where c_1, c_2, c_3 are in F. Furthermore, $\bar{v}_1, \bar{v}_2$, and $\bar{v}_3$ are linearly independent in V/W, for suppose that $b_1 \bar{v}_1 + b_2 \bar{v}_2 + b_3 \bar{v}_3 =$ zero element of $V/W = W$. This says that $b_1 + b_2 x + b_3 x^2 \in W$. However, every element in W is of the form $x^3 p(x)$, so is 0 or is of degree at least 3. The only way out is that $b_1 + b_2 x + b_3 x^2 = 0$, and so $b_1 = b_2 = b_3 = 0$. Hence, $\bar{v}_1, \bar{v}_2, \bar{v}_3$ are linearly independent over F.

Because V/W has $\bar{v}_1, \bar{v}_2, \bar{v}_3$ as a basis over F we know that V/W is of dimension 3 over F. Hence, by Theorem 8.4.3, $V/W \simeq F^{(3)}$.

3. Let $V = M_n(F)$ and let $W = \{A \in V \mid \operatorname{tr}(A) = 0\}$. Given $B \in V$, consider $A = B - \frac{1}{n}[\operatorname{tr}(B)I]$; if we take the trace of A we get

$$\operatorname{tr}(A) = \operatorname{tr}\left(B - \frac{1}{n}[\operatorname{tr}(B)I]\right)$$

$$= \operatorname{tr}(B) - \frac{1}{n}[\operatorname{tr}(B)][\operatorname{tr}(I)] = \operatorname{tr}(B) - \frac{n}{n}\operatorname{tr}(B) = 0.$$

Thus $A \in W$. So every element in V is of the form $B = A + aI$, where $a \in F$ and $\operatorname{tr}(A) = 0$ (so $A \in W$). In consequence of this, every element $B + W$ in V/W can be written as $B + W = (aI + A) + W = aI + W$ since $A \in W$. If we let $\bar{v} = I + W$, then every element in V/W is thus of the form $a\bar{v}$, where $a \in F$; therefore, V/W is 1-dimensional over F. Therefore, $V/W \simeq F$.

4. Let V be the vector space of all real-valued functions on the closed unit interval $[0, 1]$ over $\mathbb{R}$, and let $W = \{f(x) \in V \mid f(\frac{1}{2}) = 0\}$. Given $g(x) \in V$, let $h(x) = g(x) - g(\frac{1}{2})$; then $g(x) = g(\frac{1}{2}) + h(x)$ and $h(\frac{1}{2}) = g(\frac{1}{2}) - g(\frac{1}{2}) = 0$. Thus $h(x) \in W$. So $g(x) + W = g(\frac{1}{2}) + h(x) + W = g(\frac{1}{2}) + W$. But what is $g(\frac{1}{2})$? It is the evaluation of the function $g(x)$ at $x = \frac{1}{2}$, so is merely a real number, say $g(\frac{1}{2}) = a$. Then every element in V/W is of the form $a\bar{v}$, where $a \in \mathbb{R}$ and $\bar{v} = 1 + W$. This tells us that V/W is 1-dimensional over $\mathbb{R}$. So here $V/W \simeq \mathbb{R}$.

5. Let V be the set of all infinite-dimensional sequences $\{a_r\}$, where a_r is in F for every r. Let $W = \{a_1, a_2, 0, 0, 0, \ldots, 0, \ldots\}$. W is a subspace of V. Moreover, given any element $\{b_r\}$ in V, then

$$\{b_r\} = \{b_1, b_2, 0, 0, 0, \ldots, 0, \ldots\} + \{0, 0, b_3, b_4, \ldots, b_n, \ldots\}.$$

So

$$\{b_r\} + W = \{0, 0, b_3, b_4, \ldots, b_n, \ldots\} + \{b_1, b_2, 0, 0, 0, \ldots, 0, \ldots\} + W$$

and since $\{b_1, b_2, 0, 0, 0, \ldots, 0, \ldots\}$ is in W,

$$\{b_r\} + W = \{0, 0, b_3, b_4, 0, 0, 0, \ldots, 0, \ldots\} + W.$$

We leave as an exercise that $V/W \simeq V$.

Can you give an explicit isomorphism of V/W onto V?

6. Let V be any vector space, W a finite-dimensional subspace of V. By Theorem 8.7.4, $V = W \oplus W^\perp$. What does V/W look like? Given $v \in V$, then $v = w + z$, where $w \in W$, $z \in W^\perp$. So $v + W = w + z + W = z + W$ since $w \in W$. Therefore, every element of V/W is of the form $z + W$, where $z \in W^\perp$. Can such a $z + W = $ zero element of $V/W = W$? If so, $z + W = W$ forces $z \in W$ and since $z \in W^\perp$, and since $W \cap W^\perp = 0$, we get that $z = 0$. This tells us that $V/W \simeq W^\perp$. We leave the few details of proving this last remark as an exercise.

We shall refer back to these examples after the proof of the next theorem.

Theorem 9.7.3 (Homomorphism Theorem). If Φ is homomorphism of V onto W having kernel K, then $V/K \simeq W$.

Proof: Since Φ is a homomorphism of V onto W, every $w \in W$ is of the form $w = \Phi(v)$. On the other hand, every element of V/K is of the form $v + K$, where $v \in V$. What would be a natural mapping from W to V/K that could give us the desired isomorphism? Why not try the obvious one, namely, if $w = \Phi(v)$ is in W, let $\psi(w) = v + K$? First we must establish that ψ is well defined; that is, we must show that if $w = \Phi(u)$, then $u + K = \psi(w) = v + K$.

If $\Phi(u) = \Phi(v)$, then $\Phi(u - v) = \Phi(u) - \Phi(v) = 0$, whence $u - v \in K$. Therefore, $u - v + K = K$, which yields that $u + K = v + K$, the desired result. Hence ψ is well defined.

To show that ψ is an isomorphism of W onto V/K we must show that ψ is a homomorphism of W onto V/K having exactly the zero element of W as its kernel.

A typical element of V/K is $v + K$, and $v + K = \psi(w)$, where $w = \Phi(v)$. Thus ψ is certainly onto V/K. Also, if $\psi(w) =$ zero element of $V/K = K$, then $w = \Phi(v)$, $\psi(w) = v + K$. This implies that $v + K = K$, hence $v \in K$. But then $w = \Phi(v) = 0$, since K is the kernel of Φ. So if $\psi(w) = 0$, then $w = 0$.

All that is left to prove is that ψ is a homomorphism of W into V/K. Suppose, then, that $w_1 = \Phi(v_1)$, $w_2 = \Phi(v_2)$ are in W. Thus

$$w_1 + w_2 = \Phi(v_1) + \Phi(v_2) = \Phi(v_1 + v_2).$$

By the definition of ψ,

$$\psi(w_1 + w_2) = v_1 + v_2 + K = (v_1 + K) + (v_2 + K) = \psi(w_1) + \psi(w_2).$$

A similar argument shows that $\psi(aw) = a\psi(w)$ for $w \in W$ and $a \in F$.

We have shown that ψ is an isomorphism of W onto V/K. Hence $W \simeq K$. ∎

Let's see how Theorem 9.7.3 works for the examples of V/W given earlier.

EXAMPLES

1. Let $V = F^{(n)}$ and let $W = F^{(n-2)}$. Define $\Phi: V \to W$ by the rule
$\Phi\left(\begin{bmatrix} a_1 \\ a_2 \\ \vdots \\ a_n \end{bmatrix}\right) = \begin{bmatrix} a_3 \\ a_4 \\ \vdots \\ a_n \end{bmatrix}$. It is easy to see that Φ is a homomorphism of V onto W. What is $K = \text{Ker}(\Phi)$? If

$$\begin{bmatrix} 0 \\ 0 \\ \vdots \\ 0 \end{bmatrix} = \Phi\left(\begin{bmatrix} a_1 \\ a_2 \\ \vdots \\ a_n \end{bmatrix}\right) = \begin{bmatrix} a_3 \\ a_4 \\ \vdots \\ a_n \end{bmatrix},$$

Sec. 9.7] Quotient Spaces 363

we get that $a_3 = a_4 = \cdots = a_n = 0$. This gives us that $K = \text{Ker}(\Phi) = \left\{ \begin{bmatrix} a \\ b \\ 0 \\ \vdots \\ 0 \end{bmatrix} \right\}$. So $V/K \simeq W$ according to Theorem 9.7.3. This was verified in looking at the example.

2. Let V be the set of all polynomials in x over F. Define $\Phi: V \to F^{(3)}$ by defining $\Phi(a_0 + a_1 x + a_2 x^2 + \cdots + a_n x^n) = \begin{bmatrix} a_0 \\ a_1 \\ a_2 \end{bmatrix}$. Here, too, to verify that Φ is a homomorphism is straightforward. Also,

$$K = \text{Ker}(\Phi) = \{a_0 + a_1 x + \cdots \mid a_0 = a_1 = a_2 = 0\}.$$

So $f(x) = a_0 + a_1 x + a_2 x^2 + \cdots + a_n x^n$ is in K if and only if $a_0 = a_1 = a_2 = 0$. In that case $f(x) = x^3 p(x)$, hence $K = \{x^3 p(x) \mid p(x) \in V\}$. By Theorem 9.7.3, $V/K \simeq F^{(3)}$. This was explicitly shown in discussing the example.

3. $V = M_n(F)$. Define $\Phi: V \to F$ by $\Phi(A) = \text{tr}(A)$ for $A \in M_n(F)$. This mapping Φ is a homomorphism of V onto F and has kernel $K = \{A \in V \mid \text{tr}(A) = 0\}$. Therefore, $V/K \simeq F$ by Theorem 9.7.3. It also was shown explicitly in discussing the example.

4. $V = $ all real-valued functions on $[0, 1]$. Define $\Phi: V \to \mathbb{R}$ by $\Phi(f(x)) = f(\frac{1}{2})$; Φ is a homomorphism of V onto $\mathbb{R}$ with kernel $K = \{f(x) \mid f(\frac{1}{2}) = 0\}$. So $V/K \simeq \mathbb{R}$ by Theorem 9.7.2. This, too, was shown in the earlier example 4.

5. $V = $ all sequences $\{a_1, a_2, a_3, \ldots, a_n, \ldots\}$ of real numbers. Define $\Phi: V \to V$ by $\Phi(\{a_1, a_2, a_3, \ldots, a_n, \ldots\}) = \{a_3, a_4, a_5, \ldots, a_n, \ldots\}$. Φ is a homomorphism of V onto itself and has kernel $K = \{\{a_1, a_2, 0, 0, 0, \ldots, 0, \ldots\}\}$. So $V/K \simeq V$. (Prove!)

6. $V = W \oplus W^\perp$. Define $\Phi: V \to W$ by $\Phi(w, z) = w$ for $w \in W, z \in W^\perp$. Then $\text{Ker}(\Phi) = \{(0, z) \mid z \in W^\perp\}$. So $\text{Ker}(\Phi) \simeq W^\perp$ and $W \simeq V/W^\perp$.

The argument in example 6 shows that if U and W are vector spaces over F and $V = U \oplus W$, then the mapping $\Phi: V \to U$ defined by $\Phi((u, w)) = u$ is a homomorphism of V onto U with kernel $= \{(0, w) \mid w \in W\}$, and $\simeq W$ and $V/W \simeq U$.

We can say something about the dimension of V/W when V is finite-dimensional over F.

Theorem 9.7.4. If V is finite-dimensional over F and W is a subspace of V, then $\dim(V/W) = \dim(V) - \dim(W)$.

Proof: Let $w_1, \ldots, w_k$ be a basis for W. Then $k = \dim(W)$ and we can fill out $w_1, \ldots, w_k$ to a basis $w_1, \ldots, w_k, v_1, v_2, \ldots, v_r$ of V over F. Here $r + k = \dim(V)$.

Let $\bar{v}_1 = v_1 + W, \bar{v}_2 = v_2 + W, \ldots, \bar{v}_r = v_r + W$. We claim that $\bar{v}_1, \ldots, \bar{v}_r$ is a basis of V/W. Let $u \in V$; then $u = a_1 w_1 + \cdots + a_k w_k + b_1 v_1 + \cdots + b_r v_r$, where the a_j and b_j

are in F. So

$$u + W = a_1(w_1 + W) + \cdots + a_k(w_k + W) + b_1(v_1 + W) + \cdots + b_r(v_r + W).$$

Since $w_j \in W$, each $w_j + W = W$; hence

$$u + W = b_1(v_1 + W) + \cdots + b_r(v_r + W) = b_1\bar{v}_1 + \cdots + b_r\bar{v}_r.$$

Thus $\bar{v}_1, \ldots, \bar{v}_r$ span V/W over F. To show that $\dim(V/W) = r$ we must show that $\bar{v}_1, \ldots, \bar{v}_r$ is a basis of V/W over F, that is, to show that $\bar{v}_1, \ldots, \bar{v}_r$ are linearly independent over F.

If $c_1\bar{v}_1 + \cdots + c_r\bar{v}_r = 0$, then $c_1 v_1 + \cdots + c_r v_r$ is in W; therefore,

$$c_1 v_1 + \cdots + c_r v_r = d_1 w_1 + \cdots + d_k w_k.$$

But $w_1, \ldots, w_k, v_1, \ldots, v_r$ are linearly independent over F. Thus

$$c_1 = c_2 = \cdots = c_r = 0 \text{ (and } d_1 = d_2 = \cdots = d_k = 0).$$

So $\bar{v}_1, \ldots, \bar{v}_r$ are linearly independent over F. This finishes the proof. ∎

As a special case of this last theorem, if $V = U \oplus W$ and U, W are finite-dimensional over F, then $V/U \simeq W$. By Theorem 9.7.4, $\dim(W) = \dim(V/U) = \dim(V) - \dim(U)$, hence $\dim(V) = \dim(U) + \dim(W)$. We have proved

Theorem 9.7.5. *If U and W are finite-dimensional over F, then $\dim(U \oplus W) = \dim(U) + \dim(W)$.*

PROBLEMS

NUMERICAL PROBLEMS

1. If $V = F^{(n)}$ and $W = \left\{ \begin{bmatrix} a_1 \\ 0 \\ \vdots \\ 0 \end{bmatrix} \middle| a_1 \in F \right\}$, find the typical element of V/W.

2. For Problem 1 show that $V/W \simeq F^{(n-1)}$.

3. In V, the vector space of all polynomials in x over F, let

$$W = \{(x + 1)p(x) \mid p(x) \in V\}.$$

   (a) Find the form of the general element of V/W.
   (b) Show that $V/W \simeq F$.

4. Show that if $W = \{0\}$, then $V/W \simeq V$.

5. If $W = V$, show that V/W is a vector space having its zero element as its only element.

6. Let V be a vector space over F and let $X = V \oplus V$. If $W = \{(v, v) \mid v \in V\}$, show that W is a subspace of X and find the form of the general element X/V.

7. In $F = F^{(n)}$, let $W = \left\{ \begin{bmatrix} a_1 \\ a_2 \\ \vdots \\ a_n \end{bmatrix} \middle| a_1 + a_2 + \cdots + a_n = 0 \right\}$. Show that

(a) W is a subspace of V.
(b) $\dim(W) = n - 1$.

Find the form of the general element of V/W.

8. In Problem 7 prove that $V/W \simeq F$.

MORE THEORETICAL PROBLEMS

Easier Problems

9. If V is the vector space of polynomials in x over F and
$$W = \{(x^2 + 1)p(x) \mid p(x) \in V\},$$
show that $\dim(V/W) = 2$ by exhibiting a basis of V/W over F.

10. If V is a vector space and W is a subspace of V, define, for u, v in V, $u \sim v$ if $u - v$ is in W. Prove:

(a) $v \sim v$
(b) If $u \sim v$ then $v \sim u$
(c) If $u \sim v$ and $v \sim z$, then $u \sim z$

for all u, v, z in V.

11. In Problem 10 show that if $[v] = \{z \in V \mid z \sim v\}$, then for $u, v \in V$, either $[u] = [v]$ or $[u] \cap [v]$ is the empty set.

12. Let $T = \{[v] \mid v \in V\}$, where $[v]$ is defined in Problem 11. In T define $[u] + [v] = [u + v]$ and $a[u] = [au]$ for $u, v \in V$ and $a \in F$. Prove that these operations are well defined.

13. In Problem 12 prove that T is a vector space relative to the operations defined.

14. Prove that the mapping $\Phi: V \to T$ defined by $\Phi(v) = [v]$ is a homomorphism of V onto T having kernel W.

15. Show that $T = V/W$. (This is another way of defining V/W.)

Middle-Level Problems

16. If U and W are subspaces of V and if $U + W = \{u + w \mid u \in U, w \in W\}$, prove that $(U + W)/W \simeq U/(U \cap W)$.

17. Using the result of Problem 16, prove that if U and W are finite-dimensional subspaces of V, then
$$\dim(U + W) = \dim(U) + \dim(W) - \dim(U \cap W).$$

18. Let Φ be a homomorphism of V onto W. If W_0 is a subspace of W, let
$$V_0 = \{v \in V \mid \Phi(v) = W_0\}.$$

Prove:

(a) V_0 is a subspace of V.
(b) $V_0 \supset K$, where $K = \text{Ker}(\Phi)$.
(c) $V_0/K \simeq W_0$.

Harder Problems

19. If V is the vector space of polynomials in x over F and $W = \{p(x)f(x) \mid f(x) \in V\}$, show that W is a subspace of V and $V/W \simeq F^{(n)}$, where $n = \deg\{p(x)\}$.

20. If U and W are subspaces of V, show that
$$U + W \simeq \frac{U \oplus W}{K}$$
where $K = \{(u, -u) \mid u \in U \cap W\}$.

9.8. INVARIANT SUBSPACES (OPTIONAL)

Let V be a finite-dimensional vector space over F and W a subspace of V. Suppose that T in $L(V)$ is such that $T(W) \subset W$; we then say that W is *invariant* with respect to T.

Suppose then that W is invariant with respect to T. Thus given $w \in W$, we have that $Tw \in W$. So as a vector space in its own right, W has a mapping $\tilde{T}$ defined on it which is induced by T. In a word, $\tilde{T}w$ is defined by $\tilde{T}w = Tw$. Since T is a linear transformation on V, it is clear, or should be, that $\tilde{T}$ defines a linear transformation on W. That is, $\tilde{T} \in L(W)$. The following rules on $\sim$ are not hard and are left to the reader:

If W is invariant with respect to S and T in $L(V)$, then
$$\widetilde{(S+T)} = \tilde{S} + \tilde{T}.$$
$$\widetilde{(aS)} = a\tilde{S} \text{ for } a \in F.$$
$$\widetilde{(ST)} = \tilde{S}\tilde{T}.$$

Let $w_1, \ldots, w_k$ be a basis of W; here $k = \dim(W)$. We can fill this out to a basis of V, namely, $w_1, \ldots, w_k, v_{k+1}, \ldots, v_n$. If W is invariant with respect to T in $L(V)$, what does the matrix of T look like in this basis? Since $Tw_s \in W$ and $w_1, \ldots, w_k$ is a basis of W, $Tw_s = \sum_{r=1}^{k} t_{rs} w_r$ for appropriate t_{rs} in F. Since Tw_s involves only $w_1, \ldots, w_k$ (and so the coefficients of $v_{k+1}, \ldots, v_n$ in the expression for Tw_s are 0), we get that in the first k columns of $m(T)$, in this basis, all the entries below the kth one are 0. Hence the matrix of T in this basis looks like $m(T) = \begin{bmatrix} A & * \\ 0 & B \end{bmatrix}$, where A is $k \times k$ matrix and B is an $(n-k) \times (n-k)$ matrix, 0 is a $(n-k) \times k$ matrix consisting of all zeros, and $*$ is a $k \times (n-k)$ matrix in which we have little interest.

Can we identify A and B more closely and conceptually? For A it is easy, for as is not too hard to see, $A = m(\tilde{T})$, where $m(\tilde{T})$ is the matrix of $\tilde{T}$ in the basis $w_1, \ldots, w_k$. But what in the world is the matrix B? To answer this, we first make a slight, but relevant, digression.

Definition. If $W \subset V$ is a subspace invariant with respect to T, then $\hat{T}$ is defined on V/W by $\hat{T}(v + W) = T(v) + W$.

As it stands, the definition of $\hat{T}$ is really not a definition at all. We must show that $\hat{T}$ is well defined, that is, if $X = v_1 + W = v_2 + W$ is in V/W, then

$$\hat{T}(X) = T(v_1) + W = T(v_2) + W.$$

We go about showing this now.

Theorem 9.8.1. $\hat{T}$ is well defined and is a linear transformation on V/W.

Proof: Suppose that $X = v_1 + W = v_2 + W$ is in V/W. It is incumbent on us to show that $\hat{T}(X) = \hat{T}(v_1 + W) = \hat{T}(v_2 + W)$, that is, to show that $T(v_1) + W = T(v_2) + W$.

Because $v_1 + W = v_2 + W$, we know that $v_2 = v_1 + w$ for some $w \in W$. Hence $T(v_2 - v_1) = T(w) \in W$, which is to say, $T(v_2) - T(v_1)$ is in W. But then,

$$W = (T(v_2) - T(v_1)) + W,$$

hence we get the desired result $T(v_1) + W = T(v_2) + W$. Therefore, $\hat{T}$ is well defined.

Let $X = u + W$, $Y = v + W$ be in V/W and $a \in F$. Then

$$X + Y = (u + W) + (v + W) = (u + v) + W$$

by the definition of V/W. Thus

$$\hat{T}(X + Y) = T(u + v) + W = (T(u) + T(v)) + W$$
$$= (T(u) + W) + (T(v) + W) = \hat{T}(X) + \hat{T}(Y).$$

Similarly, $\hat{T}(aX) = a\hat{T}(X)$. Thus $\hat{T}$ is a linear transformation on V/W. ∎

Returning to our matrix $m(T) = \begin{bmatrix} A & * \\ 0 & B \end{bmatrix}$, we now claim

Theorem 9.8.2. B is the matrix, in the basis $v_{k+1} + W, \ldots, v_n + W$ of V/W, of the linear transformation $\hat{T}$.

Proof: First note that $v_{k+1} + W, \ldots, v_n + W$ is indeed a basis for V/W. This was implicitly done in the preceding section. We do it here explicitly.

Given $v \in V$, then

$$v = a_1 w_1 + \cdots + a_k w_k + b_1 v_{k+1} + \cdots + b_{n-k} v_n.$$

Thus

$$v + W = (a_1 w_1 + \cdots + a_k w_k + b_1 v_{k+1} + \cdots + b_{n-k} v_n) + W$$
$$= (a_1 w_1 + W) + \cdots + (a_k w_k + W) + (b_1 v_{k+1} + W) + \cdots + (b_{n-k} v_n + W)$$
$$= W + ((b_1 v_{k+1} + \cdots + b_{n-k} v_n) + W)$$

(since $a_r w_r + W = W$ because $a_r w_r \in W$)

$$= (b_1 v_{k+1} + W) + \cdots + (b_{n-k} v_n + W)$$
$$= b_1(v_{k+1} + W) + \cdots + b_{n-k}(v_n + W).$$

Therefore, the elements $v_1 + W, \ldots, v_n + W$ span V/W over F.

To show that they form a basis, we must show that they are linearly independent over F.

Suppose, then, that $c_1(v_{k+1} + W) + \cdots + c_{n-k}(v_n + W)$ is the zero element of V/W. Thus since the zero element of V/W is W, we get

$$c_1 v_{k+1} + \cdots + c_{n-k} v_n + W = W.$$

This tells us that

$$c_1 v_{k+1} + \cdots + c_{n-k} v_n$$

is an element of W, so must be a linear combination of $w_1, \ldots, w_k$. That is,

$$c_1 v_{k+1} + \cdots + c_{n-k} v_n = d_1 w_1 + \cdots + d_k w_k.$$

But $w_1, \ldots, w_k, v_{k+1}, \ldots, v_n$ are linearly independent over F. Hence

$$c_1 = c_2 = \cdots = c_{n-k} = 0. \quad (\text{Also, } d_1 = d_2 = \cdots = d_k = 0.)$$

Therefore, $v_{k+1} + W, \ldots, v_n + W$ are linearly independent over F. Hence they form a basis of V/W over F.

What does T look like in this basis? Given

$$v = a_1 w_1 + \cdots + a_k w_k + b_1 v_{k+1} + \cdots + b_{n-k} v_n,$$

then

$$T(v) = a_1 T(w_1) + \cdots + a_k T(w_k) + b_1 T(v_{k+1}) + \cdots + b_{n-k} T(v_n).$$

Because each $a_r T(w_r) \in W$, we get that $a_r T(w_r) + W = W$ and

$$T(v) + W = (b_1 T(v_{k+1}) + \cdots + b_{n-k} T(v_n)) + W + \sum_{r=1}^{k} (a_r T(w_r) + W)$$
$$= (b_1 T(v_{k+1}) + \cdots + b_{n-k} T(v_n)) + W.$$

This tells us that

$$\hat{T}(v + W) = T(v) + W$$
$$= \sum_{r=1}^{n-k} b_r T(v_{k+r}) + W.$$

If we compute $\hat{T}$ on the basis $v_{k+1} + W, \ldots, v_n + W$, we see from this last relation above that $m(T) = B$. We leave to the reader the few details needed to finish this. ∎

We illustrate this theorem with an old familiar friend, the derivative with respect to x on V, the set of polynomials of degree 6 or less over F. Let W be the set of polynomials over F of the form $a + bx^2 + cx^4 + dx^6$. Now $D(f(x)) = \dfrac{d}{dx} f(x)$ does not leave W invariant, but D^2 does, since

$$D^2(a + bx^2 + cx^4 + dx^6) = 2b + 12cx^2 + 30dx^4,$$

so is in W. $T = D^2$ defines a linear transformation on V that takes W into itself. A basis of W is $1, x^2, x^4, x^6$, and this can be filled out to a basis of V over F via

$$w_1 = 1, \quad w_2 = x^2, \quad w_3 = x^4, \quad w_4 = x^6,$$
$$v_5 = x, \quad v_6 = x^3, \quad v_7 = x^5.$$

Thus

$$T(w_1) = D^2(1) = 0,$$
$$T(w_2) = D^2(x^2) = 2 = 2w_1,$$
$$T(w_3) = D^2(x^4) = 12x^2 = 12w_2,$$
$$T(w_4) = D^2(x^6) = 30x^4 = 30w_3.$$

So the matrix of $\tilde{T}$ in the basis w_1, w_2, w_3, w_4 of W is

$$m(T) = \begin{bmatrix} 0 & 2 & 0 & 0 \\ 0 & 0 & 12 & 0 \\ 0 & 0 & 0 & 30 \\ 0 & 0 & 0 & 0 \end{bmatrix}.$$

Also

$$T(v_5) = D^2(x) = 0,$$
$$T(v_6) = D^2(x^3) = 6x = 6v_5,$$
$$T(v_7) = D^2(x^5) = 20x^3 = 20v_6.$$

So $\hat{T}$ on V/W looks like

$$\begin{bmatrix} 0 & 6 & 0 \\ 0 & 0 & 20 \\ 0 & 0 & 0 \end{bmatrix} = m(\hat{T}).$$

The matrix $m(T)$ on V using our basis above is

$$\begin{bmatrix} 0 & 2 & 0 & 0 & 0 & 0 & 0 \\ 0 & 0 & 12 & 0 & 0 & 0 & 0 \\ 0 & 0 & 0 & 30 & 0 & 0 & 0 \\ 0 & 0 & 0 & 0 & 0 & 0 & 0 \\ 0 & 0 & 0 & 0 & 0 & 6 & 0 \\ 0 & 0 & 0 & 0 & 0 & 0 & 20 \\ 0 & 0 & 0 & 0 & 0 & 0 & 0 \end{bmatrix} = \begin{bmatrix} m(\tilde{T}) & 0 \\ 0 & m(\hat{T}) \end{bmatrix}.$$

The fact that * here is 0 depends heavily on the specialty of the situation.

A special situation where the * is 0 is given us by

Theorem 9.8.3. If $V = U \oplus W$ where both $T(U) \subset U$ and $T(W) \subset W$ for some $T \in L(V)$, then in some basis of V, $m(T) = \begin{bmatrix} A & 0 \\ 0 & B \end{bmatrix}$, where A is the matrix of the linear transformation induced by T on U and B that of the linear transformation induced by T on W.

We leave the proof of this theorem to the reader.

PROBLEMS

NUMERICAL PROBLEMS

1. Let V be the vector space of polynomials in x over F, of degree 5 or less. Let $T: V \to V$ be defined by $T(1) = x^2 + x^4$, $T(x) = x + 1$, $T(x^2) = 1$, $T(x^3) = x^3 + x^2 + 1$, $T(x^4) = x^4$, $T(x^5) = 0$. If W is the linear span of $\{1, x^2, x^4\}$:
   (a) Show that $T(W) \subset W$.
   (b) Find the matrix of $\tilde{T}$ in a suitable basis of W.
   (c) Find the matrix of $\hat{T}$ in a suitable basis of V/W.
   (d) Find the matrix of T in a suitable basis of V.

2. Let V be the vector space of all real-valued functions of the form $a_1 e^x + a_2 e^{2x} + \cdots + a_{10} e^{10x}$. Let $T(f(x)) = \dfrac{d}{dx} f(x)$. If W is the subspace spanned by $e^{2x}, e^{4x}, \ldots, e^{10x}$, show that $T(W) \subset W$ and furthermore find
   (a) The matrix of $\tilde{T}$ in a suitable basis of W.
   (b) The matrix of $\hat{T}$ in a suitable basis of V/W.

3. Let V be the real-valued functions of the form

$$a_1 \sin(x) + a_2 \sin(2x) + a_3 \sin(3x) + b_1 \cos(x) + b_2 \cos(2x).$$

Defend $T: V \to V$ by $T(f(x)) = \dfrac{d^2}{dx^2} f(x)$. If W is the subspace of V spanned by $\cos(x)$ and $\cos(2x)$ over $\mathbb{R}$:
  (a) Show that $T(W) \subset W$.
  (b) Find the matrix of $\tilde{T}$ in a suitable basis of W.
  (c) Find the matrix of $\hat{T}$ in a suitable basis of V/W.
  (d) Combine (b) and (c) to get the matrix of T in some basis of V.

4. In Problem 3, define $(f(x), g(x)) = \displaystyle\int_{-\pi}^{\pi} f(x)g(x)\,dx$. Prove that $W^\perp$ is merely the linear span of $\sin(x)$, $\sin(2x)$, and $\sin(3x)$.

MORE THEORETICAL PROBLEMS

Easier Problems

5. Prove that if $S, T \in L(V)$ and $a \in F$, then
  (a) $\widetilde{(S+T)} = \tilde{S} + \tilde{T}$.
  (b) $\widetilde{(aS)} = a\tilde{S}$.
  (c) $\widetilde{(ST)} = \tilde{S}\tilde{T}$.
6. Show that A is indeed the matrix of $\tilde{T}$ in a certain basis of W.
7. Prove that $\hat{T}(aX) = a\hat{T}(X)$.
8. Finish the proof of Theorem 9.8.2 by showing that B is the matrix of $\hat{T}$ in the given basis of V/W.
9. Prove Theorem 9.8.3 completely.
10. Let $M = \{T \in L(V) \mid T(W) \subset W\}$. Show that
  (a) $S, T \in M$ implies that $S + T \in M$.
  (b) $S, T \in M$ implies that $ST \in M$.
  (c) If $\psi: M \to L(W)$ is defined by $\psi(T) = \tilde{T}$, then $\psi(S+T) = \psi(S) + \psi(T)$ and $\psi(ST) = \psi(S)\psi(T)$.
  (d) The kernel, K, of ψ defined by

$$K = \{T \in L(V) \mid \psi(T) = 0\}$$

satisfies $S, T \in K$, then $S + T \in K$ and $ST \in K$.
  (e) If $T \in K$ and $U \in M$, then both UT and TU are in K.
  (f) What the elements of K must look like.
11. Show that if $T(W) \subset W$, then the characteristic polynomial of T, $p_T(x)$, is given by $p_T(x) = p_{\tilde{T}}(x) p_{\hat{T}}(x)$, where $p_{\tilde{T}}(x)$, $p_{\hat{T}}(x)$ are the characteristic polynomials of $\tilde{T}$ and $\hat{T}$, respectively.
12. Show that the characteristic roots of T in Problem 11 are those of $p_{\tilde{T}}(x)$ and $p_{\hat{T}}(x)$.

Middle-Level Problems

13. If $E \in L(V)$ is such that $E^2 = E$, find the subspace W of elements w such that $Ew = w$ for all $w \in W$. Furthermore, show that there is a subspace U of V such that $Eu = 0$ for all $u \in U$ and that $V = W \oplus U$.
14. Find the matrix of E, using the result of Problem 13, in some basis of V.
15. What is the rank of E in Problem 13?

9.9. LINEAR TRANSFORMATIONS FROM ONE SPACE TO ANOTHER

Our main emphasis in this book has been on linear transformations of a given vector space into itself. But there is also some interest and need to speak about linear transformations from one vector space to another. In point of fact, we already have done so when we discussed homomorphisms of one vector space into another. In a somewhat camouflaged way we also discussed this in talking about $m \times n$ matrices where $m \neq n$. The unmasking of this camouflage will be done here, in the only theorem we shall prove in this section.

Let V and W be vector spaces of dimensions n and m, respectively, over F. We say that $T: V \to W$ is a *linear transformation* from V to W if

1. $T(v_1 + v_2) = T(v_1) + T(v_2)$.
2. $T(av_1) = aT(v_1)$

for all $v_1, v_2 \in V, a \in F$.

If you look back at the appropriate section, T is precisely what we called a homomorphism from V to W.

We prove, for the setup above

Theorem 9.9.1. Given a linear transformation from V to W, then T can be represented by an $m \times n$ matrix with entries from F. Furthermore, any such $m \times n$ matrix can be used to define a linear transformation from V to W.

Proof: Let $v_1, \ldots, v_n$ be a basis of V over F, and $w_1, \ldots, w_m$ a basis of W over F. Since $T(v_s) \in W$ for $s = 1, 2, \ldots, m$ and $w_1, \ldots, w_m$ is a basis of W, we have that $T(v_s) = \sum_{r=1}^{m} t_{rs} w_r$. Because any $v \in V$ is of the form $v = \sum_{s=1}^{n} a_s v_s$, with the $a_s \in F$, since T is a linear transformation from V to W, $T(v) = T\left(\sum_{s=1}^{n} a_s v_s\right) = \sum_{s=1}^{n} a_s T(v_s)$. Thus knowing $T(v_s)$ allows us to know exactly what $T(v)$ is for any $v \in V$. But all the information about $T(v_s)$ is contained in the $m \times n$ matrix $m(T) = \begin{bmatrix} t_{11} & t_{12} & \cdots & t_{1n} \\ t_{21} & t_{22} & \cdots & t_{2n} \\ \vdots & \vdots & & \vdots \\ t_{m1} & t_{m2} & \cdots & t_{mn} \end{bmatrix}$, for we can read off $T(v_s)$ from this matrix. So T is represented by this matrix $m(T)$. It is easy to verify that under addition and multiplication of linear transformations by scalars, $m(T_1 + T_2) = m(T_1) + m(T_2)$, and $m(aT_1) = am(T_1)$, for T_1, T_2 linear transformation froms V to W and $a \in F$. Of course, we are assuming that $T_1 + T_2$ is defined by

$$(T_1 + T_2)(v) = T_1(v) + T_2(v)$$

and

$$(aT_1)(v) = a(T_1(v)) \quad \text{for } v \in V, \quad a \in F.$$

Going the other way, given an $m \times n$ matrix

$$A = \begin{bmatrix} t_{11} & t_{12} & \cdots & t_{1n} \\ t_{21} & t_{22} & \cdots & t_{2n} \\ \vdots & \vdots & & \vdots \\ t_{m1} & t_{m2} & \cdots & t_{mn} \end{bmatrix},$$

with entries in F, define a linear transformation $T: V \to W$, making use of the bases $v_1, \ldots, v_n$ of V and $w_1, \ldots, w_m$ of W as follows:

$$T(v_s) = \sum_{r=1}^{m} t_{rs} w_r.$$

We leave it to the reader to prove that T defined this way leads us to a linear transformation of V to W.

Note that our constructions above made heavy use of particular bases of V and W. Using other bases we would get other matrix representations of T. These representations are closely related, but we shall not go into this relationship here. ∎

We give an example of the theorem. Let $V = F^{(2)}$ and $W = F^{(3)}$. Define $T: V \to W$ by the rule $T\begin{bmatrix} a_1 \\ a_2 \end{bmatrix} = \begin{bmatrix} a_1 + a_2 \\ a_1 - a_2 \\ a_2 \end{bmatrix}$. It is easy to see that T defines a linear transformation of V to W. Using the bases $v_1 = \begin{bmatrix} 1 \\ 0 \end{bmatrix}$, $v_2 = \begin{bmatrix} 0 \\ 1 \end{bmatrix}$ of V and $w_1 = \begin{bmatrix} 1 \\ 0 \\ 0 \end{bmatrix}$, $w_2 = \begin{bmatrix} 0 \\ 1 \\ 0 \end{bmatrix}$, $w_3 = \begin{bmatrix} 0 \\ 0 \\ 1 \end{bmatrix}$ of W, we have that

$$T(v_1) = T\begin{bmatrix} 1 \\ 0 \end{bmatrix} = \begin{bmatrix} 1 \\ 1 \\ 0 \end{bmatrix} = w_1 + w_2$$

and

$$T(v_2) = T\begin{bmatrix} 0 \\ 1 \\ 1 \end{bmatrix} = \begin{bmatrix} 1 \\ -1 \\ 1 \end{bmatrix} = w_1 - w_2 + w_3.$$

Thus the matrix in the given bases of T is given by $m(T) = \begin{bmatrix} 1 & 1 \\ 1 & -1 \\ 0 & 1 \end{bmatrix}$, a 2×3 matrix over F. What would $m(T)$ be if we used the basis $v_1 = \begin{bmatrix} 1 \\ -1 \end{bmatrix}$ and $v_2 = \begin{bmatrix} 1 \\ 1 \end{bmatrix}$ of V over F, and w_1, w_2, w_3 as above? Then

$$T(v_1) = \begin{bmatrix} 0 \\ 2 \\ -1 \end{bmatrix} = 2w_2 - w_3$$

and

$$T(v_2) = \begin{bmatrix} 2 \\ 0 \\ 1 \end{bmatrix} = 2w_1 + w_3.$$

Thus the matrix of T in this basis is $m(T) = \begin{bmatrix} 0 & 2 \\ 1 & 0 \\ -1 & 1 \end{bmatrix}$.

PROBLEMS

NUMERICAL PROBLEMS

1. If V is the vector space of all polynomials in x of degree 3 or less over F and $W = F^{(2)}$, let $T: V \to W$ be $T(a_0 + a_1 x + a_2 x^2 + a_3 x^3) = \begin{bmatrix} a_3 \\ 0 \end{bmatrix}$. Using as basis for V the powers $1, x, x^2, x^3$ of x, and as basis for, W $\begin{bmatrix} 1 \\ 0 \end{bmatrix}$ and $\begin{bmatrix} 0 \\ 1 \end{bmatrix}$, find the matrix of T.

2. In Problem 1, if we use $\begin{bmatrix} 1 \\ -1 \end{bmatrix}, \begin{bmatrix} 1 \\ 1 \end{bmatrix}$ as a basis of W, and the same basis of V, find the matrix of T.

3. Let V be the linear span of $\sin(x), \sin(2x), \sin(3x), \cos(x), \cos(2x)$ over $\mathbb{R}$, and let W be the linear span of $\sin(x), \sin(2x)$ over $\mathbb{R}$. Define $T: V \to W$ by $T(a \sin(x) + b \sin(2x) + c \sin(3x) + d \cos(x) + e \cos(2x)) = b \sin(x) - 5c \sin(2x)$. Using appropriate bases of V and W, find $m(T)$.

Sec. 9.9]   **Linear Transformations**   **375**

4. Let V be the real-valued functions of the form $a + be^x + ce^{-x} + de^{2x}$. Define $T: V \to F^{(3)}$ by the rule $T(a + be^x + ce^{-x} + de^{2x}) = \begin{bmatrix} b \\ -c \\ 2d \end{bmatrix}$. Find $m(T)$ in suitable bases of V and $F^{(3)}$.

5. Let V be as in Problem 4 and let W be the set of all real-valued functions of the form $g + ax + be^x + ce^{-x} + de^{2x}$. Consider $T: V \to W$ defined by

$$T(f(x)) = \int_0^x f(t)\,dt.$$

    (a) Prove that T is a linear transformation from V to W.
    (b) Find the matrix of T in appropriate bases of V and W.

MORE THEORETICAL PROBLEMS

6. Prove that $T_1 + T_2$ and aT_1 as defined are linear transformations from V to W.

7. If A is $m \times n$, the matrix (a_{rs}) in the canonical bases of $F^{(n)}$ and $F^{(m)}$ respectively,
    (a) Write down the linear transformation T defined by A using these canonical bases.
    (b) If $m(T)$ is the matrix of T in the canonical bases, prove that $m(T) = A$.

CHAPTER 10

The Jordan Canonical Form (Optional)

10.1. INTRODUCTION

The material that we consider in this chapter is probably the trickiest and most difficult that we consider in the book.

We saw earlier that if we had two bases of a finite-dimensional vector space V over F and T a linear transformation on V, then the matrices of T in these bases, A and B, were linked by a very nice relation, namely, $B = C^{-1}AC$, where C was the matrix of the change of basis.

This led us to declare two matrices A and B to be *similar* if, for some invertible C, $B = C^{-1}AC$. We denote similarity of A and B by writing $A \sim B$. We saw that this relation of similarity is very much like that of equality, namely

1. $A \sim A$.
2. $A \sim B$ implies $B \sim A$.
3. $A \sim B$ and $B \sim C$ implies $A \sim C$.

These three properties then led us to define the *similarity class* of A by

$$\mathrm{cl}(A) = \{B \in M_n(F) \mid B \sim A\},$$

and we found that *two such similarity classes either are equal or have no element in common.* The question becomes: Given the class $\mathrm{cl}(A)$, are there particular matrices in $\mathrm{cl}(A)$ that act like road signs which tell us if a given matrix B is in $\mathrm{cl}(A)$? These particular matrices are called canonical forms. There are several kinds of canonical forms; we choose to discuss one of them known as the *Jordan canonical form*.

Sec. 10.1] Introduction

Since we need the characteristic roots of all matrices, we work in $M_n(\mathbb{C})$ rather than in $M_n(\mathbb{R})$. To keep the notation simple, however, we will now use Latin rather than Greek letters to denote complex numbers.

We saw that if T is in $M_n(\mathbb{C})$, then for some invertible matrix C,

$$C^{-1}TC = \begin{bmatrix} a_1 & & * \\ & \ddots & \\ 0 & & a_n \end{bmatrix};$$

that is, $C^{-1}TC$ is upper triangular. Can we tighten the nature of $C^{-1}TC$ even further? The answer is "yes." We shall eventually see that *T is similar to a Jordan canonical matrix* in the sense of

Definition. A *Jordan canonical matrix* is an $n \times n$ matrix

$$\begin{bmatrix} a_1 & b_1 & & 0 \\ & \ddots & \ddots & \\ & & \ddots & b_{n-1} \\ 0 & & & a_n \end{bmatrix}$$

such that for each s, $1 \leq s \leq n$, either $b_s = 0$ or, on the other hand, $b_s = 1$ and $a_s = a_{s+1}$.

So a Jordan canonical matrix is an $n \times n$ matrix all of whose nonzero entries are contained on the diagonal and superdiagonal such that the diagonal entries are equal in blocks for which the superdiagonal entries are 1's. For example, the first such block of equal diagonal entries is $a_1 = a_2 = \cdots = a_k$, where $b_r = 1$ for $1 \leq r \leq k-1$ and $b_k = 0$ or $k = n$, so it is the $k \times k$ block

$$\begin{bmatrix} a & 1 & & 0 \\ & \ddots & \ddots & \\ & & \ddots & 1 \\ 0 & & & a \end{bmatrix},$$

where $a = a_r$ for $1 \leq r \leq k$.

EXAMPLE

The 3×3 Jordan canonical matrices are

$$\begin{bmatrix} a & 0 & 0 \\ 0 & a & 0 \\ 0 & 0 & a \end{bmatrix}, \begin{bmatrix} a & 1 & 0 \\ 0 & a & 0 \\ 0 & 0 & a \end{bmatrix}, \begin{bmatrix} a & 0 & 0 \\ 0 & a & 1 \\ 0 & 0 & a \end{bmatrix}, \begin{bmatrix} a & 1 & 0 \\ 0 & a & 1 \\ 0 & 0 & a \end{bmatrix} \text{ (any } a\text{)};$$

$$\begin{bmatrix} a & 0 & 0 \\ 0 & b & 0 \\ 0 & 0 & b \end{bmatrix}, \begin{bmatrix} b & 0 & 0 \\ 0 & a & 0 \\ 0 & 0 & b \end{bmatrix}, \begin{bmatrix} b & 0 & 0 \\ 0 & b & 0 \\ 0 & 0 & a \end{bmatrix}, \begin{bmatrix} a & 0 & 0 \\ 0 & b & 1 \\ 0 & 0 & b \end{bmatrix}, \begin{bmatrix} b & 1 & 0 \\ 0 & b & 0 \\ 0 & 0 & a \end{bmatrix} \text{ (any } a, b \text{ with } a \neq b\text{)};$$

$$\begin{bmatrix} a & 0 & 0 \\ 0 & b & 0 \\ 0 & 0 & c \end{bmatrix} \text{ (any } a, b, c \text{ with } a \neq b \neq c \neq a\text{).}$$

Let's consider the blocks $A_1, \ldots, A_k$ of equal diagonal entries of a Jordan canonical matrix A, and let $I_1, \ldots, I_k$ be the identity matrices of the same size. Then

$$A = \begin{bmatrix} A_1 & & 0 \\ & \ddots & \\ 0 & & A_k \end{bmatrix}$$

and

$$A_t = \begin{bmatrix} a_t & 1 & & 0 \\ & \ddots & \ddots & \\ & & \ddots & 1 \\ 0 & & & a_t \end{bmatrix} = a_t I_t + N_t,$$

where $a_t I_t = \begin{bmatrix} a_t & & 0 \\ & \ddots & \\ 0 & & a_t \end{bmatrix}$ and $N_t = \begin{bmatrix} 0 & 1 & & 0 \\ & \ddots & \ddots & \\ & & \ddots & 1 \\ 0 & & & 0 \end{bmatrix}$. We can interchange the positions of two different blocks A_r and A_s, and still stay in the same similarity class. For example, the Jordan canonical matrices $\begin{bmatrix} a & 1 & 0 \\ 0 & a & 0 \\ 0 & 0 & a \end{bmatrix}$ and $\begin{bmatrix} a & 0 & 0 \\ 0 & a & 1 \\ 0 & 0 & a \end{bmatrix}$ are similar. We leave it as an exercise for the reader to see how to find an invertible 3×3 matrix C such that $C^{-1} \begin{bmatrix} a & 1 & 0 \\ 0 & a & 0 \\ 0 & 0 & a \end{bmatrix} C = \begin{bmatrix} a & 0 & 0 \\ 0 & a & 1 \\ 0 & 0 & a \end{bmatrix}$.

When a Jordan canonical matrix is similar to T, it is called a *Jordan canonical form of* T. So what we shall eventually see is that every linear transformation T has a Jordan canonical form. For example, $\begin{bmatrix} 1 & 1 & 0 \\ 0 & 1 & 0 \\ 0 & 0 & 3 \end{bmatrix}$ is a Jordan canonical form of $\begin{bmatrix} 1 & 0 & 0 \\ 5 & 1 & 0 \\ 0 & 0 & 3 \end{bmatrix}$. (Prove!)

To get the idea of how this will come about, we picture, in a very sketchy way, what we shall do. We carry out a series of reductions to go from the general matrix T to very particular ones. The first step is to find an invertible $C \in M_n(\mathbb{C})$ such that

$$C^{-1}TC = \begin{bmatrix} A_1 & & 0 \\ & \ddots & \\ 0 & & A_k \end{bmatrix},$$

where each block A_t is an upper triangular matrix with only one distinct characteristic

Sec. 10.1] Introduction 379

root a_t ($1 \le t \le k$). This means that the matrix $N_t = A_t - a_t I$ is nilpotent, that is, $N_t^m = 0$ for some m. Then we need only find invertible matrices D_t (of the appropriate kind) such that the $D_t^{-1} A_t D_t$ have the desired form for all t, since

$$\begin{bmatrix} D_1^{-1} & & 0 \\ & \ddots & \\ 0 & & D_k^{-1} \end{bmatrix} \begin{bmatrix} A_1 & & 0 \\ & \ddots & \\ 0 & & A_k \end{bmatrix} \begin{bmatrix} D_1 & & 0 \\ & \ddots & \\ 0 & & D_k \end{bmatrix} = \begin{bmatrix} D_1^{-1} A_1 D_1 & & 0 \\ & \ddots & \\ 0 & & D_k^{-1} A_k D_k \end{bmatrix}$$

has the desired form. If $D_t^{-1} N_t D_t$ has the desired form, then so does $D_t^{-1} A_t D_t$, for the two are very closely related:

$$D_t^{-1} A_t D_t = D_t^{-1}(N_t + a_t I) D_t = D_t^{-1} N_t D_t + a_t I.$$

The heart of the argument then begins when we get down to this point.

As we said earlier, all this is not easy. But it has its payoffs. For one thing, we shall see how this Jordan canonical form allows us to solve a homogeneous system of linear differential equations in a very neat and efficient way.

PROBLEMS

NUMERICAL PROBLEMS

1. For $T = \begin{bmatrix} 1 & 2 \\ 0 & 3 \end{bmatrix}$, find an invertible 2×2 matrix C such that $C^{-1}TC$ is a Jordan canonical matrix.

2. In Problem 1, show that there is only one such Jordan canonical matrix, up to rearranging equal diagonal blocks.

3. For $T = \begin{bmatrix} 3 & 2 \\ 0 & 3 \end{bmatrix}$, show that T has a Jordan canonical form.

MORE THEORETICAL PROBLEMS

Easier Problem

4. For $T = \begin{bmatrix} 1 & 1 & 1 \\ 0 & 1 & 0 \\ 0 & 0 & 1 \end{bmatrix}$, show that $C^{-1}TC$ is never a diagonal matrix.

Middle-Level Problems

5. Show that if T is a matrix such that $(T - I)^e = 0$ for some positive integer e, then $C^{-1}TC$ is a diagonal matrix for some C only if $T = I$.

6. Find a Jordan canonical form for $\begin{bmatrix} 1 & 1 & 1 \\ 0 & 1 & 0 \\ 0 & 0 & 1 \end{bmatrix}$.

7. Compute all 3×3 matrices S that commute with $\begin{bmatrix} a & 1 & 0 \\ 0 & a & 1 \\ 0 & 0 & a \end{bmatrix}$.

8. Find an invertible 3×3 matrix C such that $C^{-1} \begin{bmatrix} a & 1 & 0 \\ 0 & a & 0 \\ 0 & 0 & a \end{bmatrix} C = \begin{bmatrix} a & 0 & 0 \\ 0 & a & 1 \\ 0 & 0 & a \end{bmatrix}$.

10.2. GENERALIZED NULLSPACES

We are heading toward the Jordan canonical form of a linear transformation T of a finite-dimensional vector space V over $\mathbb{C}$. To get there, we need a few preliminary results. In fact, the first of these—Theorem 10.2.1—is useful in a variety of places other than where we need it now, and a version of it occurs in other algebraic contexts.

Definition. The *generalized nullspace* of $T \in L(V)$ is the subspace

$$V_0(T) = \{v \in V \mid T^e v = 0 \text{ for some positive integer } e \text{ (depending on } v)\},$$

Similarly, if W is a subspace of V mapped into itself by T, then the *generalized nullspace* of T on W is the subspace

$$W_0(T) = \{v \in W \mid T^e v = 0 \text{ for some positive integer } e \text{ (depending on } v)\},$$

that is, the subspace $W_0(T) = W \cap V_0(T)$.

EXAMPLE

For $T = \begin{bmatrix} 1 & 1 & 1 \\ 0 & 1 & 0 \\ 0 & 0 & 2 \end{bmatrix}$, $W_0(T) = \{0\}$, $W_0(T - 1I)$ has the basis $\begin{bmatrix} 1 \\ 0 \\ 0 \end{bmatrix}$, $\begin{bmatrix} 0 \\ 1 \\ 0 \end{bmatrix}$ and $W_0(T - 2I)$ has the basis $\begin{bmatrix} 1 \\ 0 \\ 1 \end{bmatrix}$. We leave it to the reader as an exercise to verify this.

That $W_0(T)$ is a subspace is clear and left for verification by the reader. The reader should also try to show that if $T^e v = 0$, then $T^n v = 0$, where $n = \dim(V)$. So we do not have to chase all over the map for e; going to $e = n$ is enough.

We prove

Theorem 10.2.1. Given $T \in L(V)$ and a subspace W of V mapped into itself by T, then $W = W_0(T) \oplus W_*(T)$, where $W_*(T) = \bigcap_{e=1}^{\infty} T^e(W)$; that is, $W_*(T)$ is the intersection of $T^e(W)$ over all positive integers e.

Proof: If $n = \dim(W)$, we leave to the reader to prove that

$$T^n(W) = T^{n+1}(W) = \cdots = T^{n+e}(W)$$

for all $e > 1$. And since

$$T^n(W) \subseteq T^{n-1}(W) \subseteq \cdots \subseteq T(W),$$

we know that

$$W_*(T) = T^n(W).$$

Given $v \in W$, then since $T^n(W) = T^{2n}(W)$, $T^n(v) = T^{2n}(w)$ for some $w \in W$. Thus $T^n(v - T^n w) = 0$. Consequently, $v - T^n w \in W_0(T)$. But then

$$v = (v - T^n w) + T^n w,$$

so $v \in W_0(T) + W_*(T)$. In other words, $W_0(T) + W_*(T) = W$. What remains is to show that this sum is direct. This is equivalent to showing that $W_0(T) \cap W_*(T) = \{0\}$.

Remembering that $W_*(T) = T^n(W) = T^{n+1}(W)$, we see that

$$T(W_*(T)) = W_*(T),$$

so T maps $W_*(T)$ *onto* itself. Thus T must be $1 - 1$ on $W_*(T)$ by Theorem 3.8.1, and T^n is therefore also $1 - 1$ on $W_*(T)$.

Suppose that $w \in W_0(T) \cap W_*(T)$. Since $w \in W_0(T)$, $T^n w = 0$. Since $w \in W_*(T)$ and $T^n w = 0$, by the $1 - 1$-ness of T^n on $W_*(T)$, $w = 0$. Consequently, we have that $W_0(T) \cap W_*(T) = \{0\}$ and the sum $W_0(T) + W_*(T)$ is direct. This proves the theorem. ∎

Given $T \in L(V)$, $a \in F$ and a subspace W of V mapped into itself by T, we define

$$W_a(T) = \{w \in W \mid (T - aI)^e v = 0 \text{ for some positive integer } e\}.$$

This $W_a(T)$ is a subspace of W that we call the *generalized characteristic subspace* of T in W at a.

EXAMPLE

For

$$T = \begin{bmatrix} 1 & 1 & 1 \\ 0 & 1 & 0 \\ 0 & 0 & 2 \end{bmatrix},$$

we observed in the last example that

$$W_0(T) = \{0\}, W_0(T - I)$$

has the basis $\begin{bmatrix} 1 \\ 0 \\ 0 \end{bmatrix}, \begin{bmatrix} 0 \\ 1 \\ 0 \end{bmatrix}$ and $W_0(T - 2I)$ has the basis $\begin{bmatrix} 1 \\ 0 \\ 1 \end{bmatrix}$. These spaces are just the generalized characteristic spaces $W_0(T)$, $W_1(T)$, $W_2(T)$ of

$$T = \begin{bmatrix} 1 & 1 & 1 \\ 0 & 1 & 0 \\ 0 & 0 & 2 \end{bmatrix}.$$

Using Theorem 10.2.1, we now prove

Theorem 10.2.2. If $a_1, \ldots, a_p$ are the distinct characteristic roots of $T \in L(V)$, then

$$W = W_{a_1}(T) \oplus \cdots \oplus W_{a_p}(T)$$

for any subspace W of V mapped into itself by T.

Proof: We proceed by induction on m, where $m = \dim W$. If $m = 0$ or 1, the proof is trivial. Next, suppose that $m > 1$ and that the result has been proved for all subspaces W' of V mapped into themselves by T having dimension less than m. By Theorem 4.6.7 there is a characteristic vector $w \in W$ for T. Choose s such that $Tw = a_s w$, and renumber the a_r so that $s = 1$. By Theorem 10.2.1 we can decompose W as

$$W = W_0(T - a_1 I) \oplus W_*(T - a_1 I).$$

Since $W_0(T - a_1 I)$ contains w, $W_*(T - a_1 I)$ has dimension less than $\dim(W) = m$. Since $W_*(T - a_1 I) = (T - a_1 I)^n(W)$, the subspace $W_*(T - a_1 I)$ is mapped into itself by T. (Prove!) So, by induction it can be decomposed as

$$W_*(T - a_1 I) = W_*(T - a_1 I)_{a_1}(T) \oplus W_*(T - a_1 I)_{a_2}(T) \oplus \cdots \oplus W_*(T - a_1 I)_{a_p}(T).$$

Since $T - a_1 I$ is $1-1$ on $W_*(T - a_1 I)$, $W_*(T - a_1 I)_{a_1}(T) = \{0\}$, so that we actually have

$$W_*(T - a_1 I) = W_*(T - a_1 I)_{a_2}(T) \oplus \cdots \oplus W_*(T - a_1 I)_{a_p}(T).$$

This implies that

$$\begin{aligned} W &= W_0(T - a_1 I) \oplus W_*(T - a_1 I) \\ &= W_0(T - a_1 I) \oplus W_*(T - a_1 I)_{a_2}(T) \oplus \cdots \oplus W_*(T - a_1 I)_{a_p}(T) \\ &= W_{a_1}(T) \oplus W_*(T - a_1 I)_{a_2}(T) \oplus \cdots \oplus W_*(T - a_1 I)_{a_p}(T). \end{aligned}$$

But for $r \neq 1$, $W_*(T - a_1 I)_{a_r}(T) = ((T - a_1 I)^n(W)) \cap W_{a_r}(T) = W_{a_r}(T)$ since

$$(T - a_1 I)^n(W) \supset (T - a_1 I)^n(W_{a_r}(T)) = W_{a_r}(T). \quad \text{(Prove!)}$$

So, our decomposition is

$$W = W_{a_1}(T) \oplus W_{a_2}(T) \oplus \cdots \oplus W_{a_p}(T)$$

and the theorem is proved. ∎

Let $T \in L(V)$ and let W be a subspace of V mapped into itself by T. Then the minimal polynomial of the linear transformation of W that sends $w \in W$ to Tw is called *the minimal polynomial of T on W*. Using the minimal polynomial of T on W, we now can describe $W_0(T)$ and $W_*(T)$ in

Theorem 10.2.3. Let $T \in L(V)$ and let W be a subspace of V mapped into itself by T. Then

1. The minimal polynomial of T on $W_0(T)$ is X^e for some e;
2. The minimal polynomial $g(x)$ of T on $W_*(T)$ has nonzero constant term $g(0)$;
3. The minimal polynomial of T on W is given by $x^e g(x)$;
4. $W_0(T) = g(T)(w)$ and $W_*(T) = T^e(W)$.

Proof: Because $T^n(W_0(T)) = 0$ we have that the minimal polynomial of T on $W_0(T)$ is of the form x^e for some $e \leq n$, which proves (1).

If $g(x)$ is the minimal polynomial of T on $W_*(T)$, because T is invertible on $W_*(T)$, the constant coefficient, $g(0)$, of $g(x)$ is nonzero by Theorem 4.5.1. This proves (2).

We leave it to the reader to show that the minimal polynomial of T on W is $x^e g(x)$, proving (3).

Since $W = W_0(T) \oplus W_*(T)$, where $T^e(W_0(T)) = 0$ and $T^e(W_*(T)) = W_*(T)$, we have

$$\begin{aligned} T^e(W) &= T^e(W_0(T) \oplus W_*(T)) \\ &= T^e(W_0(T)) \oplus T^e(W_*(T)) = 0 \oplus T^e(W_*(T)) \\ &= W_*(T). \end{aligned}$$

Thus $W_*(T) = T^e(W)$, as claimed. Similarly,

$$\begin{aligned} g(T)(W) &= g(T)(W_0(T)) \oplus g(T)(W_*(T)) \\ &= g(T)(W_0(T)) \end{aligned}$$

since $g(T)(V_*(T)) = 0$. Because $g(0) \neq 0$, $g(T)$ can be written as

$$g(T) = T^k + a_1 T^{k-1} + \cdots + a_{k-1} T + g(0)I.$$

However, $(T^k + \cdots + a_{k-1}T)$ is nilpotent on $W_0(T)$ since T is nilpotent on $W_0(T)$. So $g(T)$ is of the form $g(T) = g(0)I + N$, where $g(0) \neq 0$ and N is nilpotent on $W_0(T)$. To get the inverse of $g(T)$ on $W_0(T)$ amounts to getting the inverse of $(1/g(0))g(T) = I - M$, where $M = (-1/g(0))N$. Since the inverse of $I - M$ on $W_0(T)$ is given by

$$(I + M + M^2 + \cdots + M^{n-1}),$$

(Prove!), the inverse of $g(T)$ on $W_0(T)$ is given by

$$g(T)^{-1} = (g(0)(I - M))^{-1}$$
$$= (1/g(0))(I - M)^{-1}$$
$$= (1/g(0))(I + M + M^2 + \cdots + M^{n-1}).$$

Hence

$$g(T)(W_0(T)) = W_0(T).$$

Thus $W_0(T) = g(T)(W_0(T)) = g(T)(W)$ from the above, and the theorem is proved. ∎

Theorem 10.2.3 has the following useful

Corollary 10.2.4. Let $T \in L(V)$, let W be a subspace of V mapped into itself by T, and let the minimal polynomial of T on W be $(x - a)^e g(x)$, where $g(a)$ is nonzero. Then $W_a(T) = g(T)W$ and $W_{a_2}(T) \oplus \cdots \oplus W_{a_p}(T) = (T - aI)^e(W)$, where $a_2, \ldots, a_p$ are the roots of $g(x)$.

EXAMPLE

For $T = \begin{bmatrix} 1 & 1 & 1 \\ 0 & 1 & 0 \\ 0 & 0 & 2 \end{bmatrix}$, the minimum polynomial is $(x - 1)^2(x - 2)$. Since

$$(T - 1I)^2 = \begin{bmatrix} 0 & 0 & 1 \\ 0 & 0 & 0 \\ 0 & 0 & 1 \end{bmatrix}$$

and

$$T - 2I = \begin{bmatrix} -1 & 1 & 1 \\ 0 & -1 & 0 \\ 0 & 0 & 0 \end{bmatrix},$$

the generalized characteristic spaces of $T = \begin{bmatrix} 1 & 1 & 1 \\ 0 & 1 & 0 \\ 0 & 0 & 2 \end{bmatrix}$ on $W = \mathbb{C}^{(3)}$ are

$$W_2(T) = \mathbb{C} \begin{bmatrix} 1 \\ 0 \\ 1 \end{bmatrix}$$

and

$$W_1(T) = \mathbb{C} \begin{bmatrix} 1 \\ -1 \\ 0 \end{bmatrix} + \mathbb{C} \begin{bmatrix} 1 \\ 0 \\ 0 \end{bmatrix}.$$

PROBLEMS

NUMERICAL PROBLEMS

1. For the matrix $T = \begin{bmatrix} 0 & 0 & -1 \\ 1 & 0 & 1 \\ 0 & 1 & 1 \end{bmatrix}$, do the following:

   (a) Find the characteristic roots of T.
   (b) Find the generalized characteristic subspaces of $\mathbb{C}^{(3)}$ with respect to T.
   (c) Find an invertible matrix C such that $C^{-1}TC = \begin{bmatrix} A & 0 \\ 0 & B \end{bmatrix}$, where B is an upper triangular 2×2 matrix.

2. Find the generalized characteristic subspaces for $T = \begin{bmatrix} 1 & 3 & 1 \\ 0 & 2 & 2 \\ 0 & 0 & 1 \end{bmatrix}$.

3. Find the generalized characteristic subspaces for $T = \begin{bmatrix} 2 & 3 & 1 \\ 0 & 2 & 2 \\ 0 & 0 & 1 \end{bmatrix}$.

4. For $T = \begin{bmatrix} 1 & 1 & 1 \\ 0 & 1 & 0 \\ 0 & 0 & 2 \end{bmatrix}$, verify that $W_0(T) = \{0\}$, $W_0(T-I)$ has the basis $\begin{bmatrix} 1 \\ 0 \\ 0 \end{bmatrix}, \begin{bmatrix} 0 \\ 1 \\ 0 \end{bmatrix}$ and $W_0(T-2I)$ has the basis $\begin{bmatrix} 1 \\ 0 \\ 1 \end{bmatrix}$.

MORE THEORETICAL PROBLEMS

5. Show that $W_0(T)$ and $W_*(T)$ are subspaces of W, for any $T \in L(V)$ and subspace W of V mapped into itself by T.

6. Show that $T(W_0(T)) \subseteq W_0(T)$ and $T(W_*(T)) \subseteq W_*(T)$, for any $T \in L(V)$ and subspace W of V mapped into itself by T.

7. If V is of dimension n and $T^e = 0$ for some $e \geq 1$, show that $T^n = 0$.

8. Show that the minimal polynomial $q(x)$ of T on a subspace W mapped into itself by T is $x^e g(x)$, as claimed in the proof of Theorem 10.2.3.

9. Let T be a linear transformation of $\mathbb{C}^{(n)}$ whose characteristic polynomial is $p_T(x) = (x - a_1)^{m_1} \cdots (x - a_k)^{m_k}$. Defining $p_t(x) = p_T(x)(x - a_t)^{-m_t}$, show that $V_{a_t}(T) = p_t(T)V$ for $1 \leq t \leq k$.

10. Show that if two linear transformations S and T of $\mathbb{C}^{(n)}$ commute, that is, $ST = TS$, then the generalized characteristic subspaces $V_{a_t}(T)$ ($1 \leq t \leq k$) of T are mapped into themselves by S.

10.3. THE JORDAN CANONICAL FORM

Given a matrix (or linear transformation) in $M_n(\mathbb{C})$, we discussed the notion of the similarity class, $\text{cl}(T)$, of T. This set $\text{cl}(T)$ was defined as the set of all matrices of the form $C^{-1}TC$, where $C \in M_n(\mathbb{C})$ is invertible.

We want to find a specific matrix in $\text{cl}(T)$, of a particularly nice form, which will act as an identifier of $\text{cl}(T)$. There are several possible such types of "identifier" matrices; they are called *canonical forms*. We have picked one of these, the *Jordan canonical form*, as the one whose properties we develop. Since for the Jordan form we need the characteristic roots of matrices, we work over $\mathbb{C}$ rather than over $\mathbb{R}$.

In Theorem 10.2.2 we saw that given T in $M_n(\mathbb{C})$, we can decompose the vector space $V\,(=\mathbb{C}^{(n)})$ as

$$V = V_{a_1}(T) \oplus \cdots \oplus V_{a_k}(T),$$

where $a_1, \ldots, a_k$ are the distinct characteristic roots of T and where

$$V_{a_r}(T) = \{v \in V \,|\, (T - a_r I)^n v = 0\}.$$

Since $T(V_{a_r}(T)) \subset V_{a_r}(T)$, T induces a linear transformation on $V_{a_r}(T)$ which has the property that $(T - a_r I)^n (V_{a_r}(T)) = 0$. So if we write

$$N_r = T - a_r I,$$

we have that in its action as a linear transformation on $V_{a_r}(T)$, $N_r^n = 0$. If we could find a basis of $V_{a_r}(T)$ for which N_r had a nice form, then since $T = N_r + a_r I$, the matrix of T would have a nice form in this basis. Then, by putting together these nice bases for $V_{a_1}(T), \ldots, V_{a_k}(T)$, we would get a nice basis for V from the point of view of T and its matrix in this basis.

All the words above can be reduced to the simple statement:

To settle the question of a canonical form (in our case, the Jordan canonical form) of a general matrix (or linear transformation) T, it is sufficient to do it merely for nilpotent matrices.

This explains our concern—and the *seeming* speciality—in the next theorem, which is about nilpotent matrices and linear transformations.

Theorem 10.3.1. Let $N \in L(V)$ be nilpotent. Then we can find a basis of V such that in this basis the matrix of N looks like

$$m(N) = \begin{bmatrix} N_1 & & & 0 \\ & N_2 & & \\ & & \ddots & \\ 0 & & & N_k \end{bmatrix},$$

Sec. 10.3]   The Jordan Canonical Form   387

where N_r is an $n_r \times n_r$ matrix and looks like

$$N_r = \begin{bmatrix} 0 & 1 & 0 & \cdots & 0 \\ 0 & 0 & 1 & \cdots & 0 \\ & & \ddots & \ddots & \\ & & & \ddots & 1 \\ 0 & 0 & 0 & \cdots & 0 \end{bmatrix}.$$

Proof: We go by induction on $\dim(V)$. If $\dim(V) = 1$, then $V = F$ and there is nothing to prove, for N must equal 0.

Suppose then that $\dim(V) = n$ and that the result is true for all subspaces of V of lower dimension. Since N is nilpotent there is an integer $m > 0$ such that $N^m = 0$ but $N^{m-1} \neq 0$. Thus $N^m(V) = 0$ but $N^{m-1}(V) \neq 0$. So we can find a w in V such that $N^{m-1}w \neq 0$.

We claim that $w, Nw, \ldots, N^{m-1}w$ are linearly independent. If

$$a_0 w + a_1 Nw + \cdots + a_{m-1} N^{m-1}w = 0,$$

multiplying this by N^{m-1} give us

$$a_0 N^{m-1}w = 0,$$

and since $N^{m-1}w \neq 0$, we end up with $a_0 = 0$. Now multiply the relation by N^{m-2}; this yields that $a_1 = 0$. Continuing in this way we get that $a_r = 0$ for all $r = 0, 1, 2, \ldots, m-1$. Thus $w, Nw, \ldots, N^{m-1}w$ are indeed linearly independent. Let X be the linear span of $w, Nw, \ldots, N^{m-1}w$. Then the matrix of N on X, using the basis $v_1 = N^{m-1}w, \ldots, v_{m-1} = Nw, v_m = w$, is the $m \times m$ matrix

$$\begin{bmatrix} 0 & 1 & 0 & \cdots & 0 \\ 0 & 0 & 1 & \cdots & 0 \\ & & \ddots & \ddots & \\ & & & \ddots & 1 \\ 0 & 0 & 0 & \cdots & 0 \end{bmatrix}.$$

Why? Because, for $r > 1$, $Nv_r = N(N^{m-r}w) = N^{m-r+1}w = v_{r-1}$, so that $Nv_1 = N(N^{m-1}w) = N^m w = 0$. Therefore, if $X = V$, we are done.

Suppose that $X \neq V$. Let Y be a subspace of V such that $N(Y) \subset Y$ and Y is of the *largest possible dimension* such that $X \cap Y = 0$. We will show that $V = X \oplus Y$. Once this is done we will know, by induction, that the theorem is correct on Y. Putting together the desired basis of Y with that above of X will give us the desired form for the matrix of N on all of V.

So *we have reduced the whole problem to showing that*

$$V = X \oplus Y.$$

Suppose then that $X \oplus Y \neq V$. So there is a $u \in V$ such that u is *not* in $X \oplus Y$. Because

$N^0 u = Iu = u$ is not in $X \oplus Y$ but $N^m u = 0 \in X \oplus Y$, there is a positive integer t, $1 \leq t \leq m$, such that $v = N^{t-1} u$ is not in $X \oplus Y$ but $Nv = N^t u \in X \oplus Y$.

Because $Nv \in X \oplus Y$ we can write Nv as $Nv = x + y$ with $x \in X$ and $y \in Y$. Thus

$$0 = N^{m-1}(Nv) = N^{m-1} x + N^{m-1} y,$$

which tells us that $N^{m-1} x = -N^{m-1} y$. But $N^{m-1} x \in X$ and $N^{m-1} y \in Y$, and since $N^{m-1} x = -N^{m-1} y$, we have that both $N^{m-1} x$ and $N^{m-1} y$ are in $X \cap Y$. But $X \cap Y = \{0\}$, which tells us that $N^{m-1} x = 0$ and $N^{m-1} y = 0$.

Since $x \in X$, we have

$$x = a_0 w + a_1 N w + \cdots + a_{m-1} N^{m-1} w$$

for some $a_0, \ldots, a_{m-1} \in F$. Thus

$$0 = N^{m-1} x = N^{m-1}(a_0 w + a_1 N w + \cdots + a_{m-1} N^{m-1} w),$$

which yields that

$$a_0 N^{m-1} w = 0$$

and so $a_0 = 0$. It follows that

$$\begin{aligned} x &= a_1 N w + \cdots + a_{m-1} N^{m-1} w \\ &= N(a_1 w + \cdots + a_{m-2} N^{m-2} w) \\ &= N x' \end{aligned}$$

for $x' \in X$. So

$$y = Nv - x = Nv - Nx' = N(v - x').$$

Thus N maps the element $v' = v - x'$ into Y and the subspace Y' generated by v' and Y is mapped into itself by N. Since v is not in $X \oplus Y$, v' is also not contained in $X \oplus Y$ and $\dim(Y') > \dim(Y)$. By the choice of Y we must have $X \cap Y' \neq \{0\}$, so $X \cap Y'$ contains some element $x'' = av' + y' \neq 0$, with $a \neq 0$ and $y' \in Y$. But then $av' = x'' - y' \in X \oplus Y$ and so $v' \in X \oplus Y$. Also, since $v' = v - x'$, we have $v = x' + v' \in X \oplus Y$, a contradiction. Thus $V = X \oplus Y$. With this the theorem is proved. ∎

With Theorem 10.3.1 established the hard work of proving the Jordan canonical form theorem is done.

From Theorem 10.3.1 we easily pass to

Theorem 10.3.2. *If $T \in L(V)$ and a_r is a characteristic root of T, then there is a basis of $V_{a_r}(T)$ such that the matrix of T in this basis is*

$$a_r I + \begin{bmatrix} N_1 & & & 0 \\ & N_2 & & \\ & & \ddots & \\ 0 & & & N_k \end{bmatrix} = \begin{bmatrix} a_r I_1 + N_1 & & & 0 \\ & a_r I_2 + N_2 & & \\ & & \ddots & \\ 0 & & & a_r I_k + N_k \end{bmatrix},$$

Sec. 10.3] The Jordan Canonical Form

where the matrices N_s are of the form described in Theorem 10.3.1 and the I_s are identity matrices of the same size as the N_s.

Proof: If $N = T - a_r I$, then we know that N is nilpotent on $V_{a_r}(T)$. So, by Theorem 10.3.1, there is a basis of $V_{a_r}(T)$ in which the matrix of $N = T - a_r I$ is

$$\begin{bmatrix} N_1 & & & 0 \\ & N_2 & & \\ & & \ddots & \\ 0 & & & N_k \end{bmatrix},$$

where the N_r are as described in Theorem 10.3.1. Thus the matrix of $T = a_r I + N$ in this basis is

$$a_r I + \begin{bmatrix} N_1 & & & 0 \\ & N_2 & & \\ & & \ddots & \\ 0 & & & N_k \end{bmatrix}.$$

Since

$$V = V_{a_1}(T) + \cdots + V_{a_k}(T)$$

and $T(V_{a_r}(T)) \subset V_{a_r}(T)$, by choosing bases of each $V_{a_r}(T)$ as in Theorem 10.3.2, we get the Jordan normal form for T, namely

Theorem 10.3.3. If $T \in L(V)$ and $a_1, \ldots, a_k$ are the distinct characteristic root of T, then there is a basis of V such that the matrix of T in this basis is

$$\begin{bmatrix} B_1 & & & 0 \\ & B_2 & & \\ & & \ddots & \\ 0 & & & B_k \end{bmatrix},$$

where

$$B_r = \begin{bmatrix} A_1^{(r)} & & 0 \\ & \ddots & \\ 0 & & A_k^{(r)} \end{bmatrix}$$

and the matrices $A_s^{(r)}$ each look like

$$A_s^{(r)} = \begin{bmatrix} a_r & 1 & 0 & \cdots & 0 \\ & & \ddots & \ddots & \vdots \\ & & & \ddots & 0 \\ & & & & 1 \\ 0 & & & & a_r \end{bmatrix}.$$

The matrices $A_s^{(r)}$ are called the *Jordan blocks* of T. The matrix

$$\begin{bmatrix} B_1 & & & 0 \\ & B_2 & & \\ & & \ddots & \\ 0 & & & B_k \end{bmatrix}$$

is called the *Jordan canonical form* of T.

The theorem just proved for linear transformations has a counterpart form in matrix terms. We leave the proof of this—the next theorem—to the reader.

Theorem 10.3.4. Let $A \in M_n(\mathbb{C})$. Then for some invertible matrix C in $M_n(\mathbb{C})$, we have that

$$C^{-1}AC = \begin{bmatrix} B_1 & & & 0 \\ & B_2 & & \\ & & \ddots & \\ 0 & & & B_k \end{bmatrix},$$

where $B_1, \ldots, B_k$ are as described in Theorem 10.3.3.

EXAMPLE

For $T = \begin{bmatrix} 0 & 0 & 1 \\ 1 & 0 & 1 \\ 0 & 1 & -1 \end{bmatrix}$, the characteristic roots are $1, -1, -1$ and the minimal polynomial is $(x-1)(x+1)^2$. (Prove!) By Corollary 10.2.4 we get a basis $\begin{bmatrix} 1 \\ 2 \\ 1 \end{bmatrix}$ for $V_1(T)$ by viewing it as the column space of $(T - 1I)^2$, and a basis $\begin{bmatrix} -1 \\ 0 \\ 1 \end{bmatrix}, \begin{bmatrix} -1 \\ 1 \\ 0 \end{bmatrix}$ for $V_{-1}(T)$ by viewing it as the column space of $(T - 1I)$. (Verify!) Letting C be the matrix of the change of basis $C = \begin{bmatrix} 1 & -1 & -1 \\ 2 & 0 & 1 \\ 1 & 1 & 0 \end{bmatrix}$, $C^{-1}TC$ is the Jordan canonical form $\begin{bmatrix} 1 & 0 & 0 \\ 0 & -1 & 1 \\ 0 & 0 & -1 \end{bmatrix}$ of T.

We could push further and prove a uniqueness theorem for the Jordan canonical form of T. This is even harder than the proof we gave for the existence of the Jordan canonical form. This section has been rather rough going. To do the uniqueness now would be too much, both for the readers and for us. At any rate that topic really belongs in discussing modules in a course on abstract algebra. We hope the readers in such a course will someday see the proof of uniqueness.

Except for permutations of the blocks in the Jordan form of T, this form is unique. With this understanding of uniqueness, the final theorem that can be proved is

Two matrices A and B (or two linear transformations) in $M_n(\mathbb{C})$ are similar if and only if they have the same Jordan canonical form.

As an example of this, suppose that A and B have, respectively, the following Jordan canonical forms:

$$\begin{bmatrix} 2 & 1 & 0 & 0 & 0 & 0 \\ 0 & 2 & 0 & 0 & 0 & 0 \\ 0 & 0 & -5 & 1 & 0 & 0 \\ 0 & 0 & 0 & -5 & 0 & 0 \\ 0 & 0 & 0 & 0 & -5 & 0 \\ 0 & 0 & 0 & 0 & 0 & 11 \end{bmatrix}, \begin{bmatrix} 2 & 1 & 0 & 0 & 0 & 0 \\ 0 & 2 & 0 & 0 & 0 & 0 \\ 0 & 0 & -5 & 1 & 0 & 0 \\ 0 & 0 & 0 & -5 & 1 & 0 \\ 0 & 0 & 0 & 0 & -5 & 0 \\ 0 & 0 & 0 & 0 & 0 & 11 \end{bmatrix}.$$

Since these Jordan canonical forms are different—they differ in the (4, 5) entry—we know that A and B cannot be similar.

PROBLEMS

NUMERICAL PROBLEMS

1. Find the Jordan canonical form of $\begin{bmatrix} 3 & 1 & 4 \\ 0 & 2 & 5 \\ 0 & 0 & 2 \end{bmatrix}$.

2. If $D^{-1}TD = \begin{bmatrix} 3 & 1 & 4 \\ 0 & 2 & 5 \\ 0 & 0 & 2 \end{bmatrix}$, where $D = \begin{bmatrix} 1 & 3 & 8 \\ 1 & 4 & 8 \\ 1 & 1 & 1 \end{bmatrix}$, find the Jordan canonical form for T.

3. Find the Jordan canonical form of $T = \begin{bmatrix} 4 & 1 & 2 \\ 0 & 4 & 3 \\ 0 & 0 & 2 \end{bmatrix}$.

MORE THEORETICAL PROBLEMS

Easier Problems

4. Find the characteristic and minimum polynomials for

$$\begin{bmatrix} a & 0 & 0 \\ 0 & b & 0 \\ 0 & 0 & c \end{bmatrix} \quad \text{and} \quad \begin{bmatrix} a & 0 & 0 \\ 0 & b & 1 \\ 0 & 0 & b \end{bmatrix} (a \neq b).$$

5. Find the characteristic and minimum polynomials for

$$\begin{bmatrix} a & 0 & 0 \\ 0 & a & 0 \\ 0 & 0 & a \end{bmatrix}, \begin{bmatrix} a & 1 & 0 \\ 0 & a & 0 \\ 0 & 0 & c \end{bmatrix}, \begin{bmatrix} a & 0 & 0 \\ 0 & a & 1 \\ 0 & 0 & a \end{bmatrix}, \begin{bmatrix} a & 1 & 0 \\ 0 & a & 1 \\ 0 & 0 & a \end{bmatrix}.$$

6. Prove for all $a \neq b$ that $\begin{bmatrix} a & 0 & 0 \\ 0 & b & 0 \\ 0 & 0 & b \end{bmatrix}$ and $\begin{bmatrix} a & 0 & 0 \\ 0 & b & 1 \\ 0 & 0 & b \end{bmatrix}$ are not similar.

7. Prove that $\begin{bmatrix} a & 0 & 0 \\ 0 & a & 0 \\ 0 & 0 & a \end{bmatrix}$ and $\begin{bmatrix} a & 0 & 0 \\ 0 & a & 1 \\ 0 & 0 & a \end{bmatrix}$ are not similar.

8. Prove that $\begin{bmatrix} a & 1 & 0 \\ 0 & a & 0 \\ 0 & 0 & b \end{bmatrix}$ and $\begin{bmatrix} b & 0 & 0 \\ 0 & a & 1 \\ 0 & 0 & a \end{bmatrix}$ are similar for all a and b.

Middle-Level Problems

9. Find the Jordan canonical form of $\begin{bmatrix} 1 & 2 & 0 & 0 \\ -2 & 0 & 0 & 0 \\ 0 & 0 & 0 & 4 \\ 0 & 0 & -4 & 0 \end{bmatrix}$.

10. Show that $\begin{bmatrix} 0 & 1 & 0 & 0 \\ 0 & 0 & 1 & 0 \\ 0 & 0 & 0 & 1 \\ 0 & 0 & 0 & 0 \end{bmatrix}$ and $\begin{bmatrix} 0 & 0 & 0 & 0 \\ 1 & 0 & 0 & 0 \\ 0 & 1 & 0 & 0 \\ 0 & 0 & 1 & 0 \end{bmatrix}$ are similar and have the same Jordan canonical form.

11. By a direct computation show that there is no invertible matrix C such that
$C^{-1}\begin{bmatrix} 1 & 1 & 0 \\ 0 & 1 & 1 \\ 0 & 0 & 1 \end{bmatrix}C = \begin{bmatrix} 1 & 1 & 0 \\ 0 & 1 & 0 \\ 0 & 0 & 1 \end{bmatrix}$; that is, show that $\begin{bmatrix} 1 & 1 & 0 \\ 0 & 1 & 1 \\ 0 & 0 & 1 \end{bmatrix}$ and $\begin{bmatrix} 1 & 1 & 0 \\ 0 & 1 & 0 \\ 0 & 0 & 1 \end{bmatrix}$ are not similar.

Harder Problems

12. If N_r is as described in Theorem 10.3.1 and N'_r is the transpose of N_r, prove that N_r and N'_r are similar.

13. If A and B are in $M_n(F)$, we declare A to be congruent to B, written $A \equiv B$, if there are invertible matrices P and Q in $M_n(F)$ such that $B = PAQ$. Prove:
    (a) $A \equiv A$
    (b) $A \equiv B$ implies that $B \equiv A$
    (c) $A \equiv B$ and $B \equiv C$ imply that $A \equiv C$
    for all A, B, C in $M_n(F)$.

14. In Problem 13, show that if A is invertible, then $A \equiv I$.

10.4. EXPONENTIALS

In so far as possible we have tried to keep the exposition of the material in the book self-contained. One exception has been our use of the Fundamental Theorem of Algebra, which we stated but did not prove. Another has been our use, at several places, of the calculus; this was based on the assumption that almost all the readers will have had some exposure to the calculus.

Now, once again, we come to a point where we diverge from this policy of self-containment. For use in the solution of differential equations, and for its own interest, we want to talk about power series in matrices. Most especially we would like to introduce the notion of the exponential, e^A, of a matrix A. We shall make some statements about these things that we do not verify with proofs.

Definition. The power series $a_0 I + a_1 A + \cdots + a_n A^n + \cdots$ in the matrix A [in $M_n(\mathbb{C})$ or $M_n(\mathbb{R})$] is said to *converge* to a matrix B if *in each entry* of "the matrix given by the power series" we have convergence.

For example, let's consider the matrix $A = \begin{bmatrix} 1 & 0 \\ 0 & 2 \end{bmatrix}$ and the power series

$$I + \left(\frac{1}{3}\right)A + \left(\frac{1}{3^2}\right)A^2 + \cdots + \left(\frac{1}{3^n}\right)A^n + \cdots,$$

which looks like

$$\begin{bmatrix} 1 & 0 \\ 0 & 1 \end{bmatrix} + \left(\frac{1}{3}\right)\begin{bmatrix} 1 & 0 \\ 0 & 2 \end{bmatrix} + \left(\frac{1}{3^2}\right)\begin{bmatrix} 1 & 0 \\ 0 & 2^2 \end{bmatrix} + \cdots + \left(\frac{1}{3^n}\right)\begin{bmatrix} 1 & 0 \\ 0 & 2^n \end{bmatrix} + \cdots.$$

Then we get convergence in the (1, 1) entry to

$$1 + \left(\frac{1}{3}\right) + \left(\frac{1}{3^2}\right) + \cdots + \left(\frac{1}{3^n}\right) + \cdots = \frac{3}{2},$$

in the (2, 1) and (1, 2) entries to 0, and in the (2, 2) entry to

$$1 + \left(\frac{2}{3}\right) + \left(\frac{2}{3}\right)^2 + \cdots + \left(\frac{2}{3}\right)^n + \cdots = 3.$$

So we declare that the power series

$$I + \left(\frac{1}{3}\right)A + \left(\frac{1}{3^2}\right)A^2 + \cdots + \left(\frac{1}{3^n}\right)A^n + \cdots$$

equals

$$\begin{bmatrix} \frac{3}{2} & 0 \\ 0 & 3 \end{bmatrix}.$$

Of course this example is a very easy one to compute. In general, things don't go so easily.

Definition. If $A \in M_n(F)$, $F = \mathbb{C}$ or $\mathbb{R}$, then the *exponential*, e^A, of A is defined by

$$e^A = I + A + \frac{A^2}{2!} + \frac{A^3}{3!} + \cdots + \frac{A^m}{m!} + \cdots.$$

The basic fact that we need is that the power series e^A converges, in the sense described above, for all matrices A in $M_n(F)$.

To prove this is not so hard, but would require a rather lengthy and distracting digression. So we wave our hands and make use of this fact without proof.

We know for real or complex numbers a and b that $e^{a+b} = e^a e^b$. This can be proved using the power series definition

$$e^x = 1 + x + \frac{x^2}{2!} + \frac{x^3}{3!} + \cdots + \frac{x^m}{m!} + \cdots$$

of e^x for real and complex numbers x.

For matrices, however, this is no longer true in general; that is, e^{A+B} is not, in general, equal to $e^A e^B$. Can you give an example where these are not equal? The reason for this is that to rearrange terms in the product $e^A e^B$ we need to know that A and B commute, that is, $AB = BA$. Then it is possible to rearrange the terms into the terms for e^{A+B}; that is, it is possible to prove:

If A and B commute, then $e^{A+B} = e^A e^B$.

In particular, if $A = aI$ is a scalar matrix, then

1. $e^{aI} = e^a I$.
2. Since $AB = BA$ for all B, $e^{aI+B} = e^{aI} e^B = e^a e^B$.

We illustrate what we have discussed with a simple example. Let $A = \begin{bmatrix} 1 & 1 \\ 0 & 1 \end{bmatrix}$. Then what is e^A? Note that $A = I + N$, where $N = \begin{bmatrix} 0 & 1 \\ 0 & 0 \end{bmatrix}$ and N satisfies $N^2 = 0$. Thus

$$A = I + N$$
$$A^2 = (I + N)^2 = I + 2N + N^2 = I + 2N,$$
$$A^3 = (I + N)^3 = I + 3N,$$
$$\vdots$$
$$A^k = (I + N)^k = I + kN.$$

Thus

$$e^A = I + I + \frac{I}{2!} + \frac{I}{3!} + \cdots + \frac{I}{k!} + \cdots$$
$$+ N + \frac{2N}{2!} + \frac{3N}{3!} + \cdots + \frac{kN}{k!} + \cdots$$
$$= I\left(1 + 1 + \frac{1}{2!} + \frac{1}{3!} + \cdots + \frac{1}{k!} + \cdots\right)$$
$$+ N\left(0 + 1 + \frac{2}{2!} + \frac{3}{3!} + \cdots + \frac{k}{k!} + \cdots\right)$$
$$= e(I + N)$$
$$= eA = e\begin{bmatrix} 1 & 1 \\ 0 & 1 \end{bmatrix}.$$

We can also compute e^A as

$$e^A = e^{I+N} = e^I e^N = e e^N = e(I + N) = e\begin{bmatrix} 1 & 1 \\ 0 & 1 \end{bmatrix},$$

ending up with the same answer.

There is one particular kind of matrix for which the power series really is a polynomial, namely a nilpotent matrix. If $N^k = 0$ for $N \in M_n(F)$, then $N^s = 0$ for all $s \geq k$, hence

$$e^N = I + N + \frac{N^2}{2!} + \frac{N^3}{3!} + \cdots + \frac{N^{k-1}}{(k-1)!},$$

a polynomial expression in N. We state this in

Theorem 10.4.1. If $N \in M_n(F)$ and $N^k = 0$ for some k, then

$$e^N = I + N + \frac{N^2}{2!} + \frac{N^3}{3!} + \cdots + \frac{N^{k-1}}{(k-1)!}.$$

A very simple example of this is afforded us by the $N = \begin{bmatrix} 0 & 1 \\ 0 & 0 \end{bmatrix}$ in the example above, where $N^2 = 0$ and e^N is just $I + N$.

In the next section we need to calculate the *matrix function e^{tT} of the scalar variable t* for any $n \times n$ matrix T. In our example above for $A = \begin{bmatrix} 1 & 1 \\ 0 & 1 \end{bmatrix}$, we computed e^A as $e\begin{bmatrix} 1 & 1 \\ 0 & 1 \end{bmatrix}$. Following in the footsteps of what we did there, we can compute e^{tA} as

$$e^{tA} = e^{tI+tN} = e^{tI}e^{tN} = e^t e^{tN} = e^t(I + tN) = e^t \begin{bmatrix} 1 & t \\ 0 & 1 \end{bmatrix}.$$

How do we compute e^{tT} in general? First, we find the Jordan canonical form $A = C^{-1}TC$ of T and express T as CAC^{-1}. Then we can calculate $T = CAC^{-1}$ by the formula

$$e^{CAC^{-1}} = Ce^A C^{-1}.$$

This formula for the *power series*

$$e^A = I + A + \frac{A^2}{2!} + \frac{A^3}{3!} + \cdots + \frac{A^m}{m!} + \cdots$$

is eminently reasonable, since we have already seen the formula

$$f(CAC^{-1}) = Cf(A)C^{-1}$$

for *polynomials*

$$f(A) = a_0 I + \cdots + a_k A^k.$$

In fact, the partial sums of the power series *are* polynomials, which means that it is possible to prove the formula for power series using it for polynomials. Having thus waved our hands at the proof, we now use the formula $e^{CAC^{-1}} = Ce^A C^{-1}$ to express $e^T = e^{CAC^{-1}}$ as $e^T = Ce^A C^{-1}$. Of course, we can also express e^{tT} as $e^{tT} = Ce^{tA}C^{-1}$.

To calculate e^{tA} for a Jordan canonical matrix $A = \begin{bmatrix} A_1 & & 0 \\ & \ddots & \\ 0 & & A_q \end{bmatrix}$, we first express e^{tA} as

$$e^{tA} = \begin{bmatrix} e^{tA_1} & & 0 \\ & \ddots & \\ 0 & & e^{tA_q} \end{bmatrix}.$$

Then we calculate e^{tA_r} by writing tA_r as

$$tA_r = \begin{bmatrix} ta_r & t & & 0 \\ & \ddots & \ddots & \\ & & \ddots & t \\ 0 & & & ta_r \end{bmatrix} = ta_r I_r + tN_r,$$

where

$$tN_r = \begin{bmatrix} 0 & t & & 0 \\ & \ddots & \ddots & \\ & & \ddots & t \\ 0 & & & 0 \end{bmatrix}.$$

Since the matrices $ta_r I_r$, tN_r commute and $ta_r I_r$ is a scalar, we get

$$e^{tA_r} = e^{ta_r I_r + tN_r} = e^{ta_r} e^{tN_r}.$$

Expressing e^{tN_r} as

$$e^{tN_r} = I + tN_r + \frac{t^2 N_r^2}{2!} + \frac{t^3 N_r^3}{3!} + \cdots + \frac{t^{k-1} N_r^{k-1}}{(k-1)!},$$

k being the number of columns of N_r, we get

$$e^{tN_r} = \begin{bmatrix} 1 & t & t^2/2! & t^3/3! & \cdots & t^{k-1}/(k-1)! \\ & 1 & t & t^2/2! & \cdots & t^{k-2}/(k-2)! \\ & & \cdot & \cdot & & \cdot \\ & & & \cdot & \cdot & \cdot \\ & & & & \cdot & t \\ 0 & & & & & 1 \end{bmatrix}.$$

So for $e^{tA_r} = e^{ta_r} e^{tN_r}$ we get

$$e^{tA_r} = e^{ta_r} \begin{bmatrix} 1 & t & t^2/2! & t^3/3! & \cdots & t^{k-1}/(k-1)! \\ & 1 & t & t^2/2! & \cdots & t^{k-2}/(k-2)! \\ & & \cdot & \cdot & & \cdot \\ & & & \cdot & \cdot & \cdot \\ & & & & \cdot & t \\ 0 & & & & & 1 \end{bmatrix}.$$

We summarize these findings in

Theorem 10.4.2. Let $T \in M_n(\mathbb{C})$ and let C be an invertible $n \times n$ matrix such that $A = C^{-1}TC$ is the Jordan canonical form of T. Then the matrix function e^{tT} is

$$e^{tT} = Ce^{tA}C^{-1}.$$

If $A = \begin{bmatrix} A_1 & & 0 \\ & \ddots & \\ 0 & & A_q \end{bmatrix}$, where $A_r = \begin{bmatrix} a_r & 1 & & 0 \\ & \ddots & \ddots & \\ & & \ddots & 1 \\ 0 & & & a_r \end{bmatrix}$ (a matrix whose size depends on r) for $1 \leq r \leq q$, then the matrix function e^{tA} is

$$e^{tA} = \begin{bmatrix} e^{tA_1} & & 0 \\ & \ddots & \\ 0 & & e^{tA_q} \end{bmatrix},$$

where

$$e^{tN_r} = \begin{bmatrix} 1 & t & t^2/2! & t^3/3! & \cdots & t^{k-1}/(k-1)! \\ & 1 & t & t^2/2! & \cdots & t^{k-2}/(k-2)! \\ & & \cdot & \cdot & & \cdot \\ & & & \cdot & \cdot & \cdot \\ & & & & \cdot & t \\ 0 & & & & & 1 \end{bmatrix}$$

for $1 \le r \le q$.

PROBLEMS

NUMERICAL PROBLEMS

1. Show that if $A = aI$, then $e^A = e^a I$.
2. If $N^2 = 0$, find the form of e^A, where $A = aI + N$.
3. If $N^3 = 0$, find the form of e^A, where $A = aI + N$.
4. Given $A = \begin{bmatrix} 1 & 1 \\ 0 & 0 \end{bmatrix}$ and $B = \begin{bmatrix} 0 & 0 \\ 1 & 1 \end{bmatrix}$ is $e^{A+B} = e^A e^B$?

MORE THEORETICAL PROBLEMS

Easier Problems

5. If $N^2 = 0$, find the form of e^{tA}, where $A = aI + N$.
6. If $N^3 = 0$, find the form of e^{tA}, where $A = aI + N$.

Middle-Level Problems

7. For real numbers a and b prove, using the power series definition of e^x, that $e^{a+b} = e^a e^b$.
8. Imitate the proof in Problem 7 to show that if $A, B \in M_n(F)$ and $AB = BA$, then $e^{A+B} = e^A e^B$.
9. Show that $e^A e^{-A} = I = e^{-A} e^A$, that is, e^A is invertible, for all A.
10. Define the power series

$$\sin A = A - \frac{A^3}{3!} + \frac{A^5}{5!} + \cdots$$

$$\cos A = I - \frac{A^2}{2!} + \frac{A^4}{4!} - \cdots$$

and prove that $(\sin A)^2 + (\cos A)^2 = I$.
11. Evaluate $\sin(e^{aI})$ and $\cos(e^{aI})$.

12. Show that the power series

$$1 + x + x^2 + \cdots + x^k + \cdots$$

does *not* converge when we replace x by $A = \begin{bmatrix} 1 & 1 \\ 1 & 1 \end{bmatrix}$ nor when we replace x by $I = \begin{bmatrix} 1 & 0 \\ 0 & 1 \end{bmatrix}$.

13. If $A = \begin{bmatrix} \frac{1}{2} & 0 & 0 \\ 0 & \frac{1}{3} & 0 \\ 0 & 0 & \frac{1}{4} \end{bmatrix}$, does the power series in Problem 12 converge for A? If so, to what matrix does it converge?

14. Evaluate e^A if $A = \begin{bmatrix} 0 & 1 \\ 1 & 0 \end{bmatrix}$.

15. Evaluate e^A if $A = \begin{bmatrix} 0 & -1 \\ 1 & 0 \end{bmatrix}$.

16. Show that $e^{iA} = \cos A + i \sin A$, where $\sin A$ and $\cos A$ are as defined in Problem 10.

17. If $A^2 = A$, evaluate e^A, $\cos A$, and $\sin A$.

18. If C is invertible and $A^2 = A$, show that $e^{C^{-1}AC} = C^{-1}e^A C$.

19. If C is invertible and $A^k = 0$, show that $e^{C^{-1}AC} = C^{-1}e^A C$.

20. The *derivative* of a matrix function $f(t) = \begin{bmatrix} a(t) & b(t) \\ c(t) & d(t) \end{bmatrix}$ is defined as $f'(t) = \begin{bmatrix} a'(t) & b'(t) \\ c'(t) & d'(t) \end{bmatrix}$. Using the definitions in Problem 10, show for $A = \begin{bmatrix} a & 0 \\ 0 & d \end{bmatrix}$ and $tA = \begin{bmatrix} ta & 0 \\ 0 & td \end{bmatrix}$ that the derivative of $\sin tA$ is $A \cos tA$. What is the derivative of $\cos tA$? Of e^{tA}?

Harder Problems

21. Find the derivative of e^{tA} for the matrices

(a) $A = \begin{bmatrix} a & 0 \\ 0 & b \end{bmatrix}$.

(b) $A = \begin{bmatrix} a & 1 \\ 0 & a \end{bmatrix}$.

(c) $A = \begin{bmatrix} a & 1 & 0 \\ 0 & a & 0 \\ 0 & 0 & b \end{bmatrix}$.

(d) $A = \begin{bmatrix} a & 1 & 0 \\ 0 & a & 1 \\ 0 & 0 & a \end{bmatrix}$.

(e) $A = \begin{bmatrix} a & 1 & 0 & 0 \\ 0 & a & 1 & 0 \\ 0 & 0 & a & 1 \\ 0 & 0 & 0 & a \end{bmatrix}$.

22. For each A in Problem 21, prove that the derivative of e^{tA} equals Ae^{tA}.

10.5. SOLVING HOMOGENEOUS SYSTEMS OF LINEAR DIFFERENTIAL EQUATIONS

Using the Jordan canonical form and exponential of an $n \times n$ matrix $T = (t_{rs})$, we can solve the homogeneous system

$$\begin{aligned} x_1'(t) &= t_{11}x_1(t) + \cdots + t_{1n}x_n(t), & x_1(0) &= x_{01} \\ &\vdots & &\vdots \\ x_n'(t) &= t_{n1}x_1(t) + \cdots + t_{nn}x_n(t), & x_n(0) &= x_{0n} \end{aligned}$$

of linear differential equations in a very neat and efficient way. We represent the system by a *matrix differential equation*

$$x'(t) = Tx(t), \quad x(0) = x_0.$$

Here T is the $n \times n$ coefficient matrix (t_{rs}), $x(t)$ is the vector function

$$x(t) = \begin{bmatrix} x_1(t) \\ \vdots \\ x_n(t) \end{bmatrix},$$

$x'(t)$ is its derivative

$$x'(t) = \begin{bmatrix} x_1'(t) \\ \vdots \\ x_n'(t) \end{bmatrix},$$

and x_0 is the vector of initial conditions x_{0r}, that is,

$$x_0 = \begin{bmatrix} x_{01} \\ \vdots \\ x_{0n} \end{bmatrix}.$$

Sec. 10.5] Solving Homogeneous Systems 401

Differentiation of vector functions satisfies the properties

1. The derivative of $ax(t) = \begin{bmatrix} ax_1(t) \\ \vdots \\ ax_n(t) \end{bmatrix}$ is $ax'(t)$ (where a is a constant).

2. The derivative of $x(t) + y(t) = \begin{bmatrix} x_1(t) + y_1(t) \\ \vdots \\ x_n(t) + y_n(t) \end{bmatrix}$ is $x'(t) + y'(t)$.

3. The derivative of $f(t)y(t) = \begin{bmatrix} f(t)y_1(t) \\ \vdots \\ f(t)y_n(t) \end{bmatrix}$ is

$$f(t)y'(t) + f'(t)y(t) = \begin{bmatrix} f(t)y'_1(t) \\ \vdots \\ f(t)y'_n(t) \end{bmatrix} + \begin{bmatrix} f'(t)y_1(t) \\ \vdots \\ f'(t)y_n(t) \end{bmatrix}.$$

4. If $T \in M_n(\mathbb{C})$, then the derivative of $e^{tT}x_0$ is $Te^{tT}x_0$.

We leave proving properties 1–3 as an exercise for the reader. Property 4 is easy to prove *if we assume that the series $e^{tT}x_0$ can be differentiated term by term*. We could prove this assumption here, but its proof, which requires background on power series, would take us far afield. So we omit its proof and assume it instead. Given this assumption, how do we prove property 4? Easily. Letting $f(t) = e^{tT}x_0$, we differentiate the series

$$f(t) = x_0 + \left(\frac{tT}{1!}\right)x_0 + \left(\frac{t^2T^2}{2!}\right)x_0 + \cdots + \left(\frac{t^kT^k}{k!}\right)x_0 + \cdots$$

term by term, getting its derivative

$$f'(t) = 0 + Tx_0 + \left(\frac{tT^2}{1!}\right)x_0 + \cdots + \left(\frac{t^{k-1}T^k}{(k-1)!}\right)x_0 + \cdots$$

$$= T\left(x_0 + \left(\frac{tT}{1!}\right)x_0 + \cdots + \left(\frac{t^{k-1}T^{k-1}}{(k-1)!}\right)x_0 + \cdots\right)$$

$$= Te^{tT}x_0.$$

Since $x(t) = e^{tT}x_0$ has derivative $x'(t) = Te^{tT}x_0 = Tx(t)$ and initial value $x(0) = e^{0I}x_0 = x_0$ by property (4), we have

Theorem 10.5.1. For any $T \in M_n(\mathbb{C})$ and $x_0 \in \mathbb{C}^{(n)}$, $x(t) = e^{tT}x_0$ is a solution to the matrix differential equation $x'(t) = Tx(t)$ such that $x(0) = x_0$.

How do we calculate the solution $x(t) = e^{tT}x_0$ given in Theorem 10.5.1? By

Theorem 10.4.2, it is just $x(t) = Ce^{tA}C^{-1}x_0$, where $A = C^{-1}TC$ is the Jordan canonical form of T. So all that is left for us to do is compute e^{tA}. By Theorem 10.4.2, we can do this by looking at the Jordan blocks in $A = \begin{bmatrix} A_1 & & 0 \\ & \ddots & \\ 0 & & A_q \end{bmatrix}$ and

$$A_r = \begin{bmatrix} a_r & 1 & & 0 \\ & \ddots & \ddots & \\ & & \ddots & 1 \\ 0 & & & a_r \end{bmatrix} \text{ (size depending on } r\text{)}$$

for $1 \leq r \leq q$, thereby getting

$$e^{tA} = \begin{bmatrix} e^{tA_1} & & 0 \\ & \ddots & \\ 0 & & e^{tA_q} \end{bmatrix},$$

where

$$e^{tA_r} = e^{ta_r}\begin{bmatrix} 1 & t & t^2/2! & t^3/3! & \cdots & t^{k-1}/(k-1)! \\ & 1 & t & t^2/2! & \cdots & t^{k-2}/(k-2)! \\ & & \ddots & \ddots & & \vdots \\ & & & \ddots & \ddots & \vdots \\ & & & & \ddots & t \\ 0 & & & & & 1 \end{bmatrix}$$

for $1 \leq r \leq q$.

EXAMPLE

Suppose that predators A and their prey B, for example two kinds of fish, have populations $P_A(t)$, $P_B(t)$ at time t with growth rates

$$P'_A(t) = 0P_A(t) + 1P_B(t)$$
$$P'_B(t) = -2P_A(t) + 3P_B(t).$$

These growth rates reflect how much the predators A depend on their current population and the availability $P_B(t)$ of the prey B for their population growth; and how much the growth in population of the prey B is impeded by the presence $P_A(t)$ of the predators A. Assuming that at time $t = 0$ we have initial populations $P_A(0) = 30,000$ and $P_B(0) = 40,000$, then how do we find the population functions $P_A(t)$ and $P_B(t)$?

We represent the system

$$P'_A(t) = 0P_A(t) + 1P_B(t)$$
$$P'_B(t) = -2P_A(t) + 3P_B(t)$$

of growth rate equations by the differential matrix equation

$$\begin{bmatrix} P'_A(t) \\ P'_B(t) \end{bmatrix} = \begin{bmatrix} 0 & 1 \\ -2 & 3 \end{bmatrix} \begin{bmatrix} P_A(t) \\ P_B(t) \end{bmatrix}.$$

The initial value $P(0)$ is then $\begin{bmatrix} P_A(t) \\ P_B(0) \end{bmatrix} = \begin{bmatrix} 30{,}000 \\ 40{,}000 \end{bmatrix}$. To get the Jordan canonical form of $T = \begin{bmatrix} 0 & 1 \\ -2 & 3 \end{bmatrix}$, we find the characteristic roots of T, namely the roots 1 and 2 of $\begin{vmatrix} -x & 1 \\ -2 & 3-x \end{vmatrix} = x^2 - 3x + 2$, and observe that $(1I - T)(2I - T) = 0$ by the Cayley–Hamilton Theorem. Then $v_1 = (1I - T)e_1 = \begin{bmatrix} 1 \\ 2 \end{bmatrix}$ and $v_2 = (2I - T)e_1 = \begin{bmatrix} 1 \\ 1 \end{bmatrix}$ form a nice basis for $\mathbb{R}^{(2)}$ with respect to T, since $Tv_1 = 2v_1$ and $Tv_2 = 1v_2$. Letting $C = [v_1, v_2] = \begin{bmatrix} 1 & 1 \\ 2 & 1 \end{bmatrix}$, the matrix $C^{-1}TC = \begin{bmatrix} 2 & 0 \\ 0 & 1 \end{bmatrix}$ of T in the basis v_1, v_2 is the Jordan canonical form of T. It follows that our solution is

$$\begin{bmatrix} P_A(t) \\ P_B(t) \end{bmatrix} = Ce^{t\begin{bmatrix} 2 & 0 \\ 0 & 1 \end{bmatrix}} C^{-1} \begin{bmatrix} 30{,}000 \\ 40{,}000 \end{bmatrix} = C \begin{bmatrix} e^{2t} & 0 \\ 0 & e^{t} \end{bmatrix} C^{-1} \begin{bmatrix} 30{,}000 \\ 40{,}000 \end{bmatrix}$$

$$= C \begin{bmatrix} e^{2t} & 0 \\ 0 & e^{t} \end{bmatrix} \begin{bmatrix} 10{,}000 \\ 20{,}000 \end{bmatrix} = \begin{bmatrix} 1 & 1 \\ 2 & 1 \end{bmatrix} \begin{bmatrix} 10{,}000 e^{2t} \\ 20{,}000 e^{1t} \end{bmatrix}$$

$$= \begin{bmatrix} 10{,}000 e^{2t} + 20{,}000 e^{1t} \\ 20{,}000 e^{2t} + 20{,}000 e^{1t} \end{bmatrix}.$$

In the example above, the characteristic roots are real. Even when this is not the case, we can use this method to get real solutions. How do we do this? Since the entries of the coefficient matrix and initial values are real,

The real part of the complex solution is a real solution.

To see why, just split up the complex solution as a sum of its real and complex parts, put them into the matrix differential equation, and compare the real and complex parts of both sides of the resulting equation.

EXAMPLE

How do we solve the differential equation

$$\begin{bmatrix} x'_1(t) \\ x'_2(t) \end{bmatrix} = \begin{bmatrix} 0 & 1 \\ -2 & 2 \end{bmatrix} \begin{bmatrix} x_1(t) \\ x_2(t) \end{bmatrix}$$

with initial values

$$\begin{bmatrix} x_{01} \\ x_{02} \end{bmatrix} = \begin{bmatrix} x_1(0) \\ x_2(0) \end{bmatrix} = \begin{bmatrix} 2 \\ 4 \end{bmatrix}?$$

From the characteristic polynomial $x^2 - 2x + 2$ of $T = \begin{bmatrix} 0 & 1 \\ -2 & 2 \end{bmatrix}$, we get the characteristic roots $1+i$, $1-i$ of T. The corresponding vectors

$$v_1 = ((1+i)I - T)e_1 = \begin{bmatrix} 1+i \\ 2 \end{bmatrix}$$

and $v_2 = ((1-i)I - T)e_1 = \begin{bmatrix} 1-i \\ 2 \end{bmatrix}$ form a nice basis for $\mathbb{C}^{(2)}$ with respect to T, since

$$\begin{bmatrix} 0 & 1 \\ -2 & 2 \end{bmatrix} \begin{bmatrix} 1+i \\ 2 \end{bmatrix} = \begin{bmatrix} 2 \\ 2-2i \end{bmatrix} = (1-i) \begin{bmatrix} 1+i \\ 2 \end{bmatrix}$$

$$\begin{bmatrix} 0 & 1 \\ -2 & 2 \end{bmatrix} \begin{bmatrix} 1-i \\ 2 \end{bmatrix} = \begin{bmatrix} 2 \\ 2+2i \end{bmatrix} = (1+i) \begin{bmatrix} 1-i \\ 2 \end{bmatrix}.$$

So letting $C = [v_1, v_2] = \begin{bmatrix} 1+i & 1-i \\ 2 & 2 \end{bmatrix}$, the matrix $C^{-1}TC = \begin{bmatrix} 1-i & 0 \\ 0 & 1+i \end{bmatrix}$ of T in v_1, v_2 is the Jordan canonical form of T. As in the example above,

$$\begin{bmatrix} x_1(t) \\ x_2(t) \end{bmatrix} = Ce^{t\begin{bmatrix} 1-i & 0 \\ 0 & 1+i \end{bmatrix}} C^{-1} \begin{bmatrix} 2 \\ 4 \end{bmatrix} = Ce^{t\begin{bmatrix} 1-i & 0 \\ 0 & 1+i \end{bmatrix}} \begin{bmatrix} 1 \\ 1 \end{bmatrix}$$

$$= C \begin{bmatrix} e^{(1-i)t} & 0 \\ 0 & e^{(1+i)t} \end{bmatrix} \begin{bmatrix} 1 \\ 1 \end{bmatrix} = C \begin{bmatrix} e^{(1-i)t} \\ e^{(1+i)t} \end{bmatrix}$$

$$= \begin{bmatrix} 1+i & 1-i \\ 2 & 2 \end{bmatrix} \begin{bmatrix} e^{(1-i)t} \\ e^{(1+i)t} \end{bmatrix}$$

or

$$\begin{bmatrix} x_1(t) \\ x_2(t) \end{bmatrix} = \begin{bmatrix} 1+i & 1-i \\ 2 & 2 \end{bmatrix} \begin{bmatrix} e^{(1-i)t} \\ e^{(1+i)t} \end{bmatrix},$$

which represents the system

$$x_1(t) = (i+1)e^{(1-i)t} + (1-i)e^{(1+i)t}$$
$$x_2(t) = \quad 2e^{(1-i)t} + \quad 2e^{(1+i)t}.$$

Using our newly acquired principle that the real part of the complex solution is a real solution, we get the real parts of the functions

$$x_1(t) = (i + 1)e^{(1-i)t} + (1 - i)e^{(1+i)t}$$
$$x_2(t) = 2e^{(1+i)t} + 2e^{(1+i)t}.$$

Recall that for t real, $e^{it} = \cos t + i \sin t$. Using this, we find that the real parts are

$$x_{1r}(t) = 2e^t(\cos t + \sin t)$$
$$x_{2r}(t) = 4e^t \cos t.$$

They give us a solution

$$x_r(t) = \begin{bmatrix} x_{1r}(t) \\ x_{2r}(t) \end{bmatrix} = \begin{bmatrix} 2e^t(\cos t + \sin t) \\ 4e^t \cos t \end{bmatrix}$$

to $x'(t) = Tx(t)$ such that $x_r(0) = \begin{bmatrix} x_{1r}(0) \\ x_{2r}(0) \end{bmatrix}$ is the real part of $x(0) = \begin{bmatrix} 2 \\ 4 \end{bmatrix}$.
(Verify!)

By what we have done so far, we have established the *existence* part of

Theorem 10.5.2. Let $T \in M_n(\mathbb{C})$ and $x_0 \in \mathbb{C}^{(n)}$. Then the differential equation $x'(t) = Tx(t)$ has one and only one solution x such that $x(0) = x_0$, namely $x = e^{tT}x_0$. Moreover:

1. This solution can be calculated as $e^{tT} = Ce^{tA}C^{-1}x_0$, where $A = C^{-1}TC$ is the Jordan canonical form of T.
2. If $T \in M_n(\mathbb{R})$ and $x_0 \in \mathbb{R}^{(n)}$, then the real part of $x(t)$ is a real solution to $x'(t) = Tx(t)$, which equals x_0 when $t = 0$.

Proof: By Theorem 10.5.1, $x = e^{tT}x_0$ is a solution to the equation $x'(t) = Tx(t)$ such that $x(0) = Ce^{0A}C^{-1}x_0 = x_0$; and by the discussion following Theorem 10.5.1, it can be calculated as $e^t = Ce^{tA}C^{-1}x_0$. For $x_0 \in \mathbb{R}^{(n)}$, it is also clear that the real part x_{real} of x is a real solution to $x' = Tx$ such that $x_{\text{real}}(0) = x_0$. It now remains only to show that *this solution is the only one*. Letting $u(t)$ denote any other such solution, we define $v = u - e^{tT}x_0$. Then

$$v(0) = u(0) - e^{0T}x_0 = x_0 - x_0 = 0.$$

To prove that $u = e^{tT}x_0$, we now simply use the equation $v(0) = 0$ to show that $v = 0$. Suppose, to the contrary, that v is nonzero. Choosing $d > 0$ so that T^d is linearly dependent on $T^0, T^1, \ldots, T^{d-1}$, the vector function $T^d v$ is a linear combination of $v, Tv, \ldots, T^{d-1}v$. It follows that the span V of $v, Tv, \ldots, T^{d-1}v$ over $\mathbb{C}$ is mapped into itself by T. Let w be a characteristic vector of T in V, so that $Tw = aw$ for some scalar a. Since $w \in V$, and since $v' = Tv$ and $v(0) = 0$, w satisfies the conditions $w' = Tw$ and $w(0) = 0$. So, $w' = aw$, from which it follows easily that w is of the form $w = \begin{bmatrix} w_1 e^{at} \\ \vdots \\ w_n e^{at} \end{bmatrix}$ (Prove!). Since $0 = w(0) = \begin{bmatrix} w_1 \\ \vdots \\ w_n \end{bmatrix}$, it follows that $w = 0$. But this contradicts our choice of w as a characteristic (and, therefore, nonzero) vector. So our hypothesis that v be nonzero leads to a contradiction. We conclude that $v = 0$ and $u = e^{tT} x_0$. ∎

The function $e^{(t-t_0)T} x_0$ is a solution $x(t)$ to the equation $x'(t) = Tx(t)$ such that $x'(t_0) = x_0$. It is also the *unique* solution to $x'(t) = Tx(t)$ such that $x(0) = (e^{-t_0 T}) x_0$. If $y'(t) = Ty(t)$ and $y'(t_0) = x_0$, then we know from Theorem 10.5.2 that $y(t) = e^{tT} y_0$ where $y_0 = y(0)$. But then we have $x_0 = y(t_0) = e^{t_0 T} y_0$, so that $y_0 = (e^{-t_0 T}) x_0$. Since $x(t)$ is the unique solution to $x'(t) = Tx(t)$ such that $x(0) = (e^{-t_0 T}) x_0$, it follows that the functions $x(t)$ and $y(t)$ are equal, which proves

Theorem 10.5.3. For any $T \in M_n(\mathbb{C})$, initial vector $x_0 \in \mathbb{C}^{(n)}$, and initial scalar t_0, the differential equation $x'(t) = Tx(t)$ has a unique solution $x(t)$ such that $x(t_0) = x_0$, namely, $x(t) = e^{(t-t_0)T} x_0$.

If we are given solutions $f_r(t) = e^{tT} v_r$ to $x' = Tx$, when can we find a solution x' to $x' = Tx$ such that $x(t_0) = x_0$ in terms of the f_r? If the vectors $v_1, \ldots, v_n \in \mathbb{C}^{(n)}$ are linearly independent, the vectors $e^{tT} v_1, \ldots, e^{tT} v_n$ are linearly independent for all t, since e^{tT} is invertible. In other words, the *Wronskian*

$$\det [e^{tT} v_1, \ldots, e^{tT} v_n] = \det (e^{tT} [v_1, \ldots, v_n]) = \det e^{tT} \det [v_1, \ldots, v_n]$$

of the set of solutions $e^{tT} v_1, \ldots, e^{tT} v_n$ is nonzero for some t if and only if it is nonzero for all t if and only if the v_r are linearly independent. So if we have solutions $f_1(t), \ldots, f_n(t)$ that we know to be linearly independent for some value of t, their Wronskian $|f_1(t), \ldots, f_n(t)|$ is nonzero for all t. This enables us to find a vector function $x(t)$ expressed in the form

$$x(t) = c_1 f_1(t) + \cdots + c_n f_n(t)$$

such that $x'(t) = Tx(t)$ and $x(t_0) = x_0$. How? Simply solve the matrix equation

$$[f_1(t_0), \ldots, f_n(t_0)] \begin{bmatrix} c_1 \\ \vdots \\ c_n \end{bmatrix} = x_0 \quad \text{(e.g., by row reduction).}$$

PROBLEMS

NUMERICAL PROBLEMS

1. Verify that the real part $\begin{bmatrix} x_{1r}(t) \\ x_{2r}(t) \end{bmatrix}$ of our solution $\begin{bmatrix} x_1(t) \\ x_2(t) \end{bmatrix}$ to

$$\begin{bmatrix} x_1'(t) \\ x_2'(t) \end{bmatrix} = \begin{bmatrix} 0 & 1 \\ -2 & 2 \end{bmatrix} \begin{bmatrix} x_1(t) \\ x_2(t) \end{bmatrix}$$

in the second example satisfies

$$\begin{bmatrix} x_{1r}'(t) \\ x_{2r}'(t) \end{bmatrix} = \begin{bmatrix} 0 & 1 \\ -2 & 2 \end{bmatrix} \begin{bmatrix} x_{1r}(t) \\ x_{2r}(t) \end{bmatrix}.$$

2. Find a solution to the equation $x'(t) = Tx(t)$, where $T = \begin{bmatrix} 1 & 1 & 2 \\ 0 & 1 & 2 \\ 0 & 0 & 2 \end{bmatrix}$.

MORE THEORETICAL PROBLEM

Easier Problems

3. Prove that the real part $\begin{bmatrix} x_{1r}'(t) \\ x_{2r}'(t) \end{bmatrix}$ of any solution $\begin{bmatrix} x_1'(t) \\ x_2'(t) \end{bmatrix}$ to a differential equation $\begin{bmatrix} x_1'(t) \\ x_2'(t) \end{bmatrix} = \begin{bmatrix} a & b \\ c & d \end{bmatrix} \begin{bmatrix} x_1(t) \\ x_2(t) \end{bmatrix}$ with real coefficient matrix $\begin{bmatrix} a & b \\ c & d \end{bmatrix}$ satisfies

$$\begin{bmatrix} x_{1r}'(t) \\ x_{2r}'(t) \end{bmatrix} = \begin{bmatrix} a & b \\ c & d \end{bmatrix} \begin{bmatrix} x_{1r}(t) \\ x_{2r}(t) \end{bmatrix}.$$

CHAPTER

11

Applications (Optional)

11.1. FIBONACCI NUMBERS

The ancient Greeks attributed a mystical and aesthetic significance to what is called the *golden section*. The golden section is the division of a line segment into two parts such that the smaller one, a, is to the larger one, b, as b is to the total length of the segment. In mathematical terms this translates into $\frac{a}{b} = \frac{b}{a+b}$, hence $a(a+b) = b^2$, and so $a^2 + ab - b^2 = 0$. In particular, if we normalize things by putting $b = 1$, we get $a^2 + a - 1 = 0$. Hence, by the quadratic formula,

$$a = \frac{-1 \pm \sqrt{1+4}}{2} = \frac{-1 \pm \sqrt{5}}{2}.$$

The particular value, $a = \frac{\sqrt{5}-1}{2}$, is called the *golden mean*. This number also comes up in certain properties of regular pentagons.

To the Greeks, constructions where the ratio of the sides of a rectangle was the golden mean were considered to have a high aesthetic value. This fascination with the golden mean persists to this day. For example, in the book *A Maggot* by the popular writer John Fowles, he expounds on the golden mean, the Fibonacci numbers, and a hint that there is a relation between them. He even cites the Fibonacci numbers in relation to the rise of the rabbit population, which we discuss below.

We introduce the sequence of numbers known as the Fibonacci numbers. They derive their name from the Italian mathematician Leonardo Fibonacci, who lived in Pisa and is often known as Leonardo di Pisa. He lived from 1180 to 1250. To quote

Boyer in his *A History of Mathematics*: "Leonardo of Pisa was without doubt the most original and most capable mathematician of the medieval Christian world."

Again quoting Boyer, Fibonacci in his book *Liber Abaci* posed the following problem: "How many pairs of rabbits will be produced in a year, beginning with a single pair, if in every month each pair bears a new pair which becomes productive from the second month on?"

As is easily seen, this problem gives rise to the sequence of numbers

$$a_0 = 0, \quad a_1 = 1, \quad a_2 = 1, \quad a_3 = 2, \quad a_4 = 3, \quad a_5 = 5, \ldots$$

where the $(n + 1)$st number, a_{n+1}, is determined by $a_{n+1} = a_n + a_{n-1}$ for $n > 1$. In other words, the $(n + 1)$st member of the sequence is the sum of its two immediate predecessors. This sequence of numbers is called the *Fibonacci sequence*, and its terms are known as the *Fibonacci numbers*.

Can we find a nice simple formula that expresses a_n as a function of n? Indeed we can, as we shall demonstrate below. Interestingly enough, the answer we get involves the golden mean, which was mentioned above.

There are many approaches to the problem of finding a formula for a_n. The approach we take is by means of the 2×2 matrices, not that this is necessarily the shortest or best method of solving the problem. What it does do is to give us a nice illustration of how one can exploit characteristic roots and their associated characteristic vectors. Also, this method lends itself to easy generalizations.

In $\mathbb{C}^{(2)}$ we write down a sequence of vectors built up from the Fibonacci numbers as follows:

$$v_0 = \begin{bmatrix} a_0 \\ a_1 \end{bmatrix} = \begin{bmatrix} 0 \\ 1 \end{bmatrix}, \quad v_1 = \begin{bmatrix} a_1 \\ a_2 \end{bmatrix} = \begin{bmatrix} 1 \\ 1 \end{bmatrix}, \quad v_2 = \begin{bmatrix} a_2 \\ a_3 \end{bmatrix} = \begin{bmatrix} 1 \\ 2 \end{bmatrix}, \quad v_3 = \begin{bmatrix} a_3 \\ a_4 \end{bmatrix} = \begin{bmatrix} 2 \\ 3 \end{bmatrix}, \quad \ldots,$$

and generally, $v_n = \begin{bmatrix} a_n \\ a_{n+1} \end{bmatrix}$ for all positive integers n.

Consider the matrix $A = \begin{bmatrix} 0 & 1 \\ 1 & 1 \end{bmatrix}$. How does A behave vis-à-vis the sequence of vectors $v_0, v_1, v_2, \ldots, v_n, \ldots$? Note that

$$Av_0 = \begin{bmatrix} 0 & 1 \\ 1 & 1 \end{bmatrix}\begin{bmatrix} 0 \\ 1 \end{bmatrix} = \begin{bmatrix} 1 \\ 1 \end{bmatrix} = v_1, \quad Av_1 = \begin{bmatrix} 0 & 1 \\ 1 & 1 \end{bmatrix}\begin{bmatrix} 1 \\ 1 \end{bmatrix} = \begin{bmatrix} 1 \\ 2 \end{bmatrix} = v_2, \quad \ldots,$$

$$Av_n = \begin{bmatrix} 0 & 1 \\ 1 & 1 \end{bmatrix}\begin{bmatrix} a_n \\ a_{n+1} \end{bmatrix} = \begin{bmatrix} a_{n+1} \\ a_n + a_{n+1} \end{bmatrix} = \begin{bmatrix} a_{n+1} \\ a_{n+2} \end{bmatrix} = v_{n+1}.$$

Thus A carries each term of the sequence into its successor in the sequence. Going back to the beginning,

$$v_1 = Av_0, \quad v_2 = Av_1 = A(Av_0) = A^2 v_0, \quad v_3 = Av_2 = A(A^2 v_0) = A^3 v_0,$$

and

$$v_{n+1} = Av_n = A(A^n v_0) = A^{n+1} v_0.$$

If we knew exactly what $A^n v_0 = v_n$ equals, then, by reading off the first component, we would have the desired formula for a_n.

We ask: What are the characteristic roots of $A = \begin{bmatrix} 0 & 1 \\ 1 & 1 \end{bmatrix}$? Since its characteristic polynomial is

$$p_A(x) = \det(xI - A) = \det\begin{bmatrix} x & -1 \\ -1 & x-1 \end{bmatrix} = x^2 - x - 1,$$

from the quadratic formula the roots of $p_A(x)$, that is, the characteristic roots of A, are $\dfrac{1+\sqrt{5}}{2}$ and $\dfrac{1-\sqrt{5}}{2}$. Note that the second of these is just the negative of the golden mean.

We want to find explicitly the characteristic vectors w_1 and w_2 associated with $\dfrac{1+\sqrt{5}}{2}$ and $\dfrac{1-\sqrt{5}}{2}$, respectively. Because $\left(\dfrac{1+\sqrt{5}}{2}\right)\left(\dfrac{-1+\sqrt{5}}{2}\right) = 1$ we see that if $w_1 = \begin{bmatrix} (-1+\sqrt{5})/2 \\ 1 \end{bmatrix}$, then

$$Aw_1 = \begin{bmatrix} 0 & 1 \\ 1 & 1 \end{bmatrix} \begin{bmatrix} (-1+\sqrt{5})/2 \\ 1 \end{bmatrix} = \begin{bmatrix} 1 \\ 1 + (-1+\sqrt{5})/2 \end{bmatrix}$$

$$= \begin{bmatrix} 1 \\ (1+\sqrt{5})/2 \end{bmatrix} = (1+\sqrt{5})/2 \begin{bmatrix} (-1+\sqrt{5})/2 \\ 1 \end{bmatrix}.$$

Thus w_1 is a characteristic vector associated with the characteristic root $(1+\sqrt{5})/2$. A similar computation shows that $w_2 = \begin{bmatrix} (-1-\sqrt{5})/2 \\ 1 \end{bmatrix}$ is a characteristic vector associated with $[1-\sqrt{5})/2$, that is, $Aw_2 = ((1-\sqrt{5})/2)w_2$.

From $Aw_1 = ((1+\sqrt{5})/2)w_1$ we easily get, by successively multiplying by A, that

$$A^2 w_1 = A(Aw_1) = A\left(\frac{(1+\sqrt{5})}{2}\right)w_1 = \left(\frac{(1+\sqrt{5})}{2}\right)Aw_1$$

$$= \left(\frac{1+\sqrt{5}}{2}\right)^2 w_1, \ldots, A^n w_1 = \left(\frac{1+\sqrt{5}}{2}\right)^n w_1.$$

Similarly, $A^n w_2 = ([1-\sqrt{5}]/2)^n w_2$.

Can v_0 be expressed as a combination of w_1 and w_2? We write down this expression, which you can verify or find for yourself.

$$v_0 = \left(\frac{1+\sqrt{5}}{2\sqrt{5}}\right) w_1 - \left(\frac{1-\sqrt{5}}{2\sqrt{5}}\right) w_2.$$

Therefore,

$$v_n = A^n v_0 = A^n \left(\left(\frac{1+\sqrt{5}}{2\sqrt{5}} \right) w_1 - \left(\frac{1-\sqrt{5}}{2\sqrt{5}} \right) w_2 \right)$$

$$= \left(\frac{1+\sqrt{5}}{2\sqrt{5}} \right) \left(\frac{1+\sqrt{5}}{2} \right)^n w_1 - \left(\frac{1-\sqrt{5}}{2\sqrt{5}} \right) \left(\frac{1-\sqrt{5}}{2} \right)^n w_2$$

$$= \frac{1}{\sqrt{5}} \begin{bmatrix} \left(\frac{1+\sqrt{5}}{2}\right)^n - \left(\frac{1-\sqrt{5}}{2}\right)^n \\ \left(\frac{1+\sqrt{5}}{2}\right)^{n+1} - \left(\frac{1-\sqrt{5}}{2}\right)^{n+1} \end{bmatrix} = \begin{bmatrix} a_n \\ a_{n+1} \end{bmatrix}.$$

Thus $a_n = \frac{1}{\sqrt{5}} \left(\left(\frac{1+\sqrt{5}}{2} \right)^n - \left(\frac{1-\sqrt{5}}{2} \right)^n \right)$. (Verify all the arithmetic!) We have found the desired formula for a_n.

Note one little consequence of this formula for a_n. By its construction, a_n is an *integer*; therefore, the strange combination $\frac{1}{\sqrt{5}} \left(\left(\frac{1+\sqrt{5}}{2} \right)^n - \left(\frac{1-\sqrt{5}}{2} \right)^n \right)$ is an *integer*. Can you find a direct proof of this without resorting to the Fibonacci numbers?

There are several natural variants, of the Fibonacci numbers that one could introduce. To begin with, we need not begin the proceedings with

$$a_0 = 0, \quad a_1 = 1, \quad a_2 = 1, \quad \ldots;$$

we could start with

$$a_0 = a, \quad a_1 = b, \quad a_2 = a + b, \quad \ldots, \quad a_{n+1} = a_n + a_{n-1}.$$

The same matrix $\begin{bmatrix} 0 & 1 \\ 1 & 1 \end{bmatrix}$, the same characteristic roots, and the same characteristic vectors w_1 and w_2 arise. The only change we have to make is to express $\begin{bmatrix} a \\ b \end{bmatrix}$ as a combination of w_1 and w_2. The rest of the argument runs as above.

A second variant might be: Assume that

$$a_0 = 0, \quad a_1 = 1, \quad a_2 = c, \quad \ldots, \quad a_{n+1} = ca_n + da_{n-1}, \ldots \rightarrow \ldots$$

where c and d are fixed integers. The change from the argument above would be that we use the matrix $B = \begin{bmatrix} 0 & 1 \\ d & c \end{bmatrix}$; note that

$$B \begin{bmatrix} a_{n-1} \\ a_n \end{bmatrix} = \begin{bmatrix} 0 & 1 \\ d & c \end{bmatrix} \begin{bmatrix} a_{n-1} \\ a_n \end{bmatrix} = \begin{bmatrix} a_n \\ da_{n-1} + ca_n \end{bmatrix} = \begin{bmatrix} a_n \\ a_{n+1} \end{bmatrix}.$$

If the characteristic roots of B are distinct, to find the formula for a_n we must find the characteristic roots and associated vectors of B, express the first vector $\begin{bmatrix} 0 \\ 1 \end{bmatrix}$ as a combination of these characteristic vectors, and proceed as before.

We illustrate this with a simple example. Suppose that

$$a_0 = 0, \quad a_1 = 1, \quad a_2 = 1, \quad a_3 = a_2 + 2a_1 = 3, \quad \ldots, \quad a_n = a_{n-1} + 2a_{n-2}$$

for $n \geq 2$. Thus our matrix B is merely $\begin{bmatrix} 0 & 1 \\ 2 & 1 \end{bmatrix}$. The characteristic roots of B are 2 and -1, and if $w_1 = \begin{bmatrix} 1 \\ 2 \end{bmatrix}$, $w_2 = \begin{bmatrix} -1 \\ 1 \end{bmatrix}$, then $Bw_1 = 2w_1$, $Bw_2 = -w_2$. Also, $\begin{bmatrix} 0 \\ 1 \end{bmatrix} = \tfrac{1}{3} w_1 + \tfrac{1}{3} w_2$, hence

$$v_n = B^n v_0 = \frac{1}{3} B^n w_1 + \frac{1}{3} B^n w_2 = \frac{2^n}{3} w_1 + \frac{(-1)^n}{3} w_2 = \begin{bmatrix} \dfrac{2^n}{3} + \dfrac{(-1)^{n+1}}{3} \\ \dfrac{2^{n+1}}{3} + \dfrac{(-1)^{n+2}}{3} \end{bmatrix}.$$

Thus $a_n = \dfrac{2^n + (-1)^{n+1}}{3}$ for all $n \geq 0$.

The last variant we want to consider is one in which we don't assume that a_{n+1} is merely a fixed combination of its two predecessors. True, the Fibonacci numbers use only the two predecessors, but that doesn't make "two" sacrosanct. We might assume that $a_0, a_1, \ldots, a_{k-1}$ are given integers and for $n \geq k$,

$$a_n = c_1 a_{n-1} + c_2 a_{n-2} + \cdots + c_k a_{n-k},$$

where $c_1, \ldots, c_k$ are fixed integers. The solutions would involve the k-tuples $\begin{bmatrix} a_n \\ a_{n+1} \\ \vdots \\ a_{n+k} \end{bmatrix}$ and the matrix

$$C = \begin{bmatrix} 0 & 1 & 0 & 0 & 0 & 0 & \cdots & 0 & 0 \\ 0 & 0 & 1 & 0 & 0 & 0 & \cdots & 0 & 0 \\ 0 & 0 & 0 & 1 & 0 & 0 & \cdots & 0 & 0 \\ \vdots & \vdots & \vdots & \vdots & \vdots & \vdots & & \vdots & \vdots \\ 0 & 0 & 0 & 0 & 0 & 0 & \cdots & 0 & 1 \\ c_k & c_{k-1} & c_{k-2} & c_{k-3} & c_{k-4} & c_{k-5} & \cdots & c_2 & c_1 \end{bmatrix}.$$

This would entail the roots of the characteristic polynomial

$$p_C(x) = x^k - c_1 x^{k-1} - c_2 x^{k-2} - \cdots - c_{k-1} x - c_k.$$

Sec. 11.1] Fibonacci Numbers 413

In general, we cannot find the roots of $p_C(x)$ exactly. However, if we could find the roots, and if they were distinct, we could find the associated characteristic vectors $w_1, \ldots, w_k$ and express $\begin{bmatrix} a_0 \\ a_1 \\ \vdots \\ a_{k-1} \end{bmatrix}$ as a combination of $w_1, \ldots, w_k$. We would then proceed as before, noting that

$$C^n \begin{bmatrix} a_0 \\ a_1 \\ \vdots \\ a_{k-1} \end{bmatrix} = \begin{bmatrix} a_n \\ a_{n+1} \\ \vdots \\ a_{n+k-1} \end{bmatrix}.$$

PROBLEMS

NUMERICAL PROBLEMS

1. Find a_n if
   (a) $a_0 = 0$, $a_1 = 4$, $a_2 = 4$, ..., $a_n = a_{n-1} + a_{n-2}$ for $n \geq 2$.
   (b) $a_0 = 1$, $a_1 = 1$, ..., $a_n = 4a_{n-1} + 3a_{n-2}$ for $n \geq 2$.
   (c) $a_0 = 1$, $a_1 = 3$, ..., $a_n = 4a_{n-1} + 3a_{n-2}$ for $n \geq 2$.
   (d) $a_0 = 1$, $a_1 = 1$, ..., $a_n = 2(a_{n-1} + a_{n-2})$ for $n \geq 2$.

MORE THEORETICAL PROBLEMS

2. For $v_0 = \begin{bmatrix} 0 \\ 1 \end{bmatrix}$ we found that $a_n = A^n v_0$. Show that for $\tilde{v}_0 = \begin{bmatrix} 1 \\ 0 \end{bmatrix}$ that $\tilde{a}_n = A^n \tilde{v}_0$ and $\tilde{a}_n = a_{n-1}$.

3. For $v_0 = \begin{bmatrix} x \\ y \end{bmatrix}$ and $a_n = A^n v_0$, use the result of Problem 2 to show that $a_n = ya_n + xa_{n-1}$.

4. Show directly that $\dfrac{1}{\sqrt{5}}(([1 + \sqrt{5}]/2)^n - ([1 - \sqrt{5}]/2)^n)$ is an integer and is positive.

5. Show that $(1 + \sqrt{5})^n + (1 - \sqrt{5})^n$ is an integer.

6. If w_1 and w_2 are as in the discussion of the Fibonacci numbers, show that $w_1 - w_2 = \sqrt{5} \begin{bmatrix} 1 \\ 0 \end{bmatrix}$ and therefore that

$$\begin{bmatrix} 0 \\ 1 \end{bmatrix} = \begin{bmatrix} 0 & 1 \\ 1 & 1 \end{bmatrix} \begin{bmatrix} 1 \\ 0 \end{bmatrix} = \begin{bmatrix} 0 & 1 \\ 1 & 1 \end{bmatrix} \left(\frac{1}{\sqrt{5}} (w_1 - w_2) \right)$$

$$= \frac{1 + \sqrt{5}}{2\sqrt{5}} w_1 - \frac{1 - \sqrt{5}}{2\sqrt{5}} w_2.$$

11.2. EQUATIONS OF CURVES

Determinants provide us with a tool to determine the equation of a curve of a given kind that passes through enough given points. We can immediately obtain the equation of the curve in determinant form. To get it more explictly would require the expansion of the determinant. Of course, if the determinant is of high order, this is no mean job.

Be that as it may, we do get an effective method of determining curves through points this way. We illustrate what we mean with a few examples.

We all know that the equation of a straight line is of the form $ax + by + c = 0$, where a, b, c are constants. Furthermore, we know that a line is determined by two distinct points on it. Given $(x_1, y_1) \neq (x_2, y_2)$, what is the equation of the straight line through them? Suppose that the equation of the line is $ax + by + c = 0$. Thus since (x_1, y_1) and (x_2, y_2) are on the line, we have

$$ax + by + c = 0$$
$$ax_1 + by_1 + c = 0$$
$$ax_2 + by_2 + c = 0.$$

In order that this system of equations have a nontrivial solution for (a, b, c), we must have $\begin{vmatrix} x & y & 1 \\ x_1 & y_1 & 1 \\ x_2 & y_2 & 1 \end{vmatrix} = 0$. (Why?) Hence, expanding yields

$$x(y_1 - y_2) - y(x_1 - x_2) + (x_1 y_2 - x_2 y_1) = 0.$$

This is the equation of the straight line we are looking for. Of course, it is an easy enough matter to do this without determinants. But doing it with determinants illustrates what we are talking about.

Consider another situation, that of a circle. The equation of the general circle is given by $a(x^2 + y^2) + bx + cy + d = 0$. (If $a = 0$, the circle degenerates to a straight line.) Given three points, they determine a unique circle (if the points are collinear, this circle degenerates to a line). What is the equation of this circle? If the points are (x_1, y_1), (x_2, y_2), (x_3, y_3) we have the four equations

$$a(x^2 + y^2) + bx + cy + d = 0$$
$$a(x_1^2 + y_1^2) + bx_1 + cy_1 + d = 0$$
$$a(x_2^2 + y_2^2) + bx_2 + cy_2 + d = 0$$
$$a(x_3^2 + y_3^2) + bx_3 + cy_3 + d = 0.$$

Since not all of a, b, c, d are zero, we must have the determinant of their coefficients equal to 0. Thus

$$\begin{vmatrix} x^2 + y^2 & x & y & 1 \\ x_1^2 + y_1^2 & x_1 & y_1 & 1 \\ x_2^2 + y_2^2 & x_2 & y_2 & 1 \\ x_3^2 + y_3^2 & x_3 & y_3 & 1 \end{vmatrix} = 0$$

is the equation of the sought-after circle.

Consider a specific case the three points are $(1, -1)$, $(1, 2)$, $(2, 0)$. These are not collinear. The circle through them is given by

$$\begin{vmatrix} x^2 + y^2 & x & y & 1 \\ 1 + (-1)^2 & 1 & -1 & 1 \\ 1^2 + 2^2 & 1 & 2 & 1 \\ 2^2 + 0^2 & 2 & 0 & 1 \end{vmatrix} = 0.$$

Expanding and simplifying this, we get the equation

$$x^2 + y^2 - x - y - 2 = 0,$$

that is,

$$(x - \tfrac{1}{2})^2 + (y - \tfrac{1}{2})^2 = \tfrac{5}{2}.$$

This is a circle, center at $(\tfrac{1}{2}, \tfrac{1}{2})$ and radius $= \sqrt{(5/2)}$.

We could go on with specific examples; however, we leave these to the exercises.

PROBLEMS

1. The equation of a conic having its axes of symmetry on the x- and y-axes is of the form $ax^2 + by^2 + cx + dy + e = 0$. Find the equation of the conic through the given points.

   (a) $(1, 2)$, $(2, 1)$, $(3, 2)$, $(2, 3)$.
   (b) $(0, 0)$, $(5, -4)$, $(1, -2)$, $(1, 1)$.
   (c) $(5, 6)$, $(0, 1)$, $(1, 0)$, $(0, 0)$.
   (d) $(-1, -1)$, $(2, 0)$, $(0, \tfrac{1}{2})$, $(-\tfrac{1}{2}, \tfrac{1}{3})$.

2. Identity the conics found in Problem 1.

3. Prove from the determinant form of a circle why, if the three given points are collinear, the equation of the circle reduces to that of a straight line.

4. Find the equations of the curve of the form

$$a \sin(x) + b \sin(2y) + c$$

passing through $(\pi, \pi/2)$ and $(\pi/2, \pi/4)$.

5. Find the equation of the curve $ae^x + be^{-y} + ce^{2x} + de^{-2y}$ passing through $(1, 1)$, $(2, 2)$ and $(2, -1)$.

11.3. MARKOV PROCESSES

A country holding regular elections is always governed by one of three parties, A, B, and C. At the time of any election, the probabilities for transition from one party to another are as follows:

A to B: 49%,	A to C: 1%,	A to A: 50% (the remaining probability)
B to A: 49%,	B to C: 1%,	B to B: 50% (the remaining probability)
C to A: 5%,	C to B: 5%,	C to C: 90% (the remaining probability)

Given these *transition probabilities*, we ask:

1. If party B was chosen in the first election, what is the probability that the party chosen in the fifth election is party C?

2. If $s_1^{(k)}$ is the probability of choosing party A in the kth election, $s_2^{(k)}$ that of choosing party B and $s_3^{(k)}$ that of choosing party C, what are the probabilities $s_1^{(k+p)}$ of choosing party A in the $(k+p)$th election, $s_2^{(k+p)}$ that of choosing party B, and $s_3^{(k+p)}$ that of choosing party C?

3. Is there a *steady state* for this transition; that is, are there nonnegative numbers s_1, s_2, s_3 which add up to 1 such that if s_1 is the probability of having chosen party A in the last election, s_2 that of having chosen party B, and s_3 that of having chosen party C, then the probabilities of choosing parties A, B and C in the next election are still s_1, s_2, and s_3, respectively?

To answer these questions, we let $S^{(k)}$ be the vector $\begin{bmatrix} s_1^{(k)} \\ s_2^{(k)} \\ s_3^{(k)} \end{bmatrix}$, where $s_1^{(k)}$ is the probability of choosing party A in the kth election, $s_2^{(k)}$ that of choosing party B, and $s_3^{(k)}$ that of choosing party C. So the entries $s_1^{(k)}$, $s_2^{(k)}$, $s_3^{(k)}$ of $S^{(k)}$ are nonnegative, which add up to 1. We call $S^{(k)}$ the *state* after the kth election. If the current party is A, then $S^{(1)} = \begin{bmatrix} 1 \\ 0 \\ 0 \end{bmatrix}$ and $S^{(2)} = \begin{bmatrix} .5 \\ .49 \\ .01 \end{bmatrix}$ since the probabilities for transition to parties A, B, and C from party A are 50%, 49%, and 1%. Similarly, if the current party is B, then $S^{(1)} = \begin{bmatrix} 0 \\ 1 \\ 0 \end{bmatrix}$ and $S^{(2)} = \begin{bmatrix} .49 \\ .50 \\ .01 \end{bmatrix}$; and if the current party is C, then $S^{(1)} = \begin{bmatrix} 0 \\ 0 \\ 1 \end{bmatrix}$ and $S^{(2)} = \begin{bmatrix} .05 \\ .05 \\ .90 \end{bmatrix}$. So the transition from $S^{(1)}$ to $S^{(2)}$ results from multiplying $S^{(1)}$ by the transition matrix $T = \begin{bmatrix} .50 & .49 & .05 \\ .49 & .50 & .05 \\ .01 & .01 & .90 \end{bmatrix}$. Assuming that *transition is linear*, that is,

Sec. 11.3] Markov Processes 417

$$S^{(k)} = s_1^{(k)} \begin{bmatrix} 1 \\ 0 \\ 0 \end{bmatrix} + s_2^{(k)} \begin{bmatrix} 0 \\ 1 \\ 0 \end{bmatrix} + s_3^{(k)} \begin{bmatrix} 0 \\ 0 \\ 1 \end{bmatrix} \text{ implies that}$$

$$S^{(k+1)} = s_1^{(k)} \begin{bmatrix} .50 \\ .49 \\ .01 \end{bmatrix} + s_2^{(k)} \begin{bmatrix} .49 \\ .50 \\ .01 \end{bmatrix} + s_3^{(k)} \begin{bmatrix} .05 \\ .05 \\ .90 \end{bmatrix},$$

we get

$$S^{(k+1)} = s_1^{(k)} T \begin{bmatrix} 1 \\ 0 \\ 0 \end{bmatrix} + s_2^{(k)} T \begin{bmatrix} 0 \\ 1 \\ 0 \end{bmatrix} + s_3^{(k)} T \begin{bmatrix} 0 \\ 0 \\ 1 \end{bmatrix} = TS^{(k)},$$

that is, $S^{(k+1)} = TS^{(k)}$. From this, it is apparent that $S^{(k+p)} = T^p S^{(k)}$ for any positive integer p, and that $S^{(k)} = T^{k-1} S^{(1)}$, which provides us with answers to our first two questions:

1. If party B was chosen in the first election, the probability that the party chosen in the 5th election is party C is the third entry of the second column
$$\begin{bmatrix} .50 & .49 & .05 \\ .49 & .50 & .05 \\ .01 & .01 & .90 \end{bmatrix}^4 \begin{bmatrix} 0 \\ 1 \\ 0 \end{bmatrix} \text{ of } \begin{bmatrix} .50 & .49 & .05 \\ .49 & .50 & .05 \\ .01 & .01 & .90 \end{bmatrix}^4.$$

2. If $S^{(k)}$ is the state after the kth election, then the state after the $(k+p)$th election is $T^p S^{(k)}$.

For example, if the chances for choosing parties A, B, and C, respectively, in the twelfth election are judged to be 30%, 45%, 25%, then the state after the fifteenth election is $\begin{bmatrix} .50 & .49 & .05 \\ .49 & .50 & .05 \\ .01 & .01 & .90 \end{bmatrix}^3 \begin{bmatrix} .30 \\ .45 \\ .25 \end{bmatrix}$, that is, the first, second, and third entries of this vector give the probabilities of choosing parties A, B, and C, respectively, in the fifteenth election.

Let's now look for a steady state $S = \begin{bmatrix} s_1 \\ s_2 \\ s_3 \end{bmatrix}$, that is, a vector S whose entries are nonnegative and add up to 1 such that $TS = S$. To find S it is necessary that the determinant of $T - 1I = \begin{bmatrix} -.50 & .49 & .05 \\ .49 & -.50 & .05 \\ .01 & .01 & -.10 \end{bmatrix}$ be 0, which it is since the rows add up to zero. Now S must be in the nullspace of $T - 1I$. Since the second row is a linear combination of the first and third, the $T - 1I$ row reduces to the following matrices:

$$\begin{bmatrix} .01 & .01 & -.10 \\ -.50 & .49 & .05 \\ .00 & .00 & .00 \end{bmatrix}, \begin{bmatrix} 1 & 1 & -10 \\ 0 & 99 & -495 \\ 0 & 0 & 0 \end{bmatrix}, \begin{bmatrix} 1 & 1 & -10 \\ 0 & 1 & -5 \\ 0 & 0 & 0 \end{bmatrix}, \begin{bmatrix} 1 & 0 & -5 \\ 0 & 1 & -5 \\ 0 & 0 & 0 \end{bmatrix}.$$

Since $\begin{bmatrix} 5 \\ 5 \\ 1 \end{bmatrix}$ is a basis for the nullspace of $\begin{bmatrix} 1 & 0 & -5 \\ 0 & 1 & -5 \\ 0 & 0 & 0 \end{bmatrix}$, the answer to our second question is: There is one and only one steady state $S = \begin{bmatrix} \frac{5}{11} \\ \frac{5}{11} \\ \frac{1}{11} \end{bmatrix}$.

Since 1 is a characteristic not for T, we can find the other two characteristic values for T and use them to get a simple formula for $S^{(k)} = T^{k-1}S^{(1)}$. To find them we go to the column space $(T - 1I)\mathbb{R}^{(3)}$, which has as basis the second two columns of $\begin{bmatrix} -.50 & .49 & .05 \\ .49 & -.50 & .05 \\ .01 & .01 & -.10 \end{bmatrix}$. To get the coefficients a, b, c, d, in the equations

$$\begin{bmatrix} .50 & .49 & .05 \\ .49 & .50 & .05 \\ .01 & .01 & .90 \end{bmatrix} \begin{bmatrix} .49 \\ -.50 \\ .01 \end{bmatrix} = a \begin{bmatrix} .49 \\ -.50 \\ .01 \end{bmatrix} + c \begin{bmatrix} .05 \\ .05 \\ -.10 \end{bmatrix}$$

$$\begin{bmatrix} .50 & .49 & .05 \\ .49 & .50 & .05 \\ .01 & .01 & .90 \end{bmatrix} \begin{bmatrix} .05 \\ .05 \\ -.10 \end{bmatrix} = b \begin{bmatrix} .49 \\ -.50 \\ .01 \end{bmatrix} + d \begin{bmatrix} .05 \\ .05 \\ -.10 \end{bmatrix},$$

we calculate

$$\begin{bmatrix} .50 & .49 & .05 \\ .49 & .50 & .05 \\ .01 & .01 & .90 \end{bmatrix} \begin{bmatrix} .49 \\ -.50 \\ .01 \end{bmatrix} = \begin{bmatrix} .0005 \\ -.0094 \\ .0089 \end{bmatrix},$$

$$\begin{bmatrix} .50 & .49 & .05 \\ .49 & .50 & .05 \\ .01 & .01 & .90 \end{bmatrix} \begin{bmatrix} .05 \\ .05 \\ -.10 \end{bmatrix} = \begin{bmatrix} .0445 \\ .0445 \\ -.0890 \end{bmatrix}$$

and solve the equations

$$\begin{bmatrix} .49 & .05 \\ -.50 & .05 \\ .01 & -.10 \end{bmatrix} \begin{bmatrix} a \\ c \end{bmatrix} = \begin{bmatrix} .0005 \\ -.0094 \\ .0089 \end{bmatrix},$$

$$\begin{bmatrix} .49 & .05 \\ -.50 & .05 \\ .01 & -.10 \end{bmatrix} \begin{bmatrix} b \\ d \end{bmatrix} = \begin{bmatrix} .0445 \\ .0445 \\ -.0890 \end{bmatrix},$$

Sec. 11.3] Markov Processes 419

getting $\begin{bmatrix} a \\ c \end{bmatrix} = \begin{bmatrix} .01 \\ -.088 \end{bmatrix}$, $\begin{bmatrix} b \\ d \end{bmatrix} = \begin{bmatrix} 0 \\ .89 \end{bmatrix}$. Since the matrix $\begin{bmatrix} a & b \\ c & d \end{bmatrix}$ is $\begin{bmatrix} .01 & 0 \\ -.088 & .89 \end{bmatrix}$, the characteristic roots in addition to 1 are .89 and .01. For the characteristic root 1, we already have the characteristic vector $\begin{bmatrix} \frac{5}{11} \\ \frac{5}{11} \\ \frac{1}{11} \end{bmatrix}$ or $\begin{bmatrix} 5 \\ 5 \\ 1 \end{bmatrix}$; and for .89 we have the characteristic vector $\begin{bmatrix} .05 \\ .05 \\ -.10 \end{bmatrix}$ or $\begin{bmatrix} 1 \\ 1 \\ -2 \end{bmatrix}$. For .01, we reduce

$$T - .01I = \begin{bmatrix} .49 & .49 & .05 \\ .49 & .49 & .05 \\ .01 & .01 & .89 \end{bmatrix}$$

to $\begin{bmatrix} 01 & 01 & 89 \\ 49 & 49 & 05 \\ 00 & 00 & 00 \end{bmatrix}$ and then to $\begin{bmatrix} 01 & 01 & 89 \\ 00 & 00 & 01 \\ 00 & 00 & 00 \end{bmatrix}$, and find the characteristic vector $\begin{bmatrix} 1 \\ -1 \\ 0 \end{bmatrix}$ in its nullspace. Letting $C = \begin{bmatrix} 5 & 1 & 1 \\ 5 & 1 & -1 \\ 1 & -2 & 0 \end{bmatrix}$, we have

$$C^{-1}TC = \begin{bmatrix} 1 & 0 & 0 \\ 0 & .89 & 0 \\ 0 & 0 & .01 \end{bmatrix} \quad \text{or} \quad T = C \begin{bmatrix} 1 & 0 & 0 \\ 0 & .89 & 0 \\ 0 & 0 & .01 \end{bmatrix} C^{-1}.$$

So we have the formula

$$S^{(k)} = T^{k-1}S^{(1)} = C \begin{bmatrix} 1 & 0 & 0 \\ 0 & (.89)^{k-1} & 0 \\ 0 & 0 & (.01)^{k-1} \end{bmatrix} C^{-1}S^{(1)}.$$

One calculates that

$$C^{-1} = (\tfrac{1}{22}) \begin{bmatrix} 2 & 2 & 2 \\ 1 & 1 & -10 \\ 11 & -11 & 0 \end{bmatrix},$$

so that

$$S^{(k)} = (\tfrac{1}{22}) \begin{bmatrix} 5 & 1 & 1 \\ 5 & 1 & -1 \\ 1 & -2 & 0 \end{bmatrix} \begin{bmatrix} 1 & 0 & 0 \\ 0 & (.89)^{k-1} & 0 \\ 0 & 0 & (.01)^{k-1} \end{bmatrix} \begin{bmatrix} 2 & 2 & 2 \\ 1 & 1 & -10 \\ 11 & -11 & 0 \end{bmatrix} S^{(1)}.$$

From the formula for $S^{(k)}$, we see that the limit $S^{(\infty)}$ of $S^{(k)}$ as k goes to ∞ exists and is

$$S^{(\infty)} = (\tfrac{1}{22}) \begin{bmatrix} 5 & 1 & 1 \\ 5 & 1 & -1 \\ 1 & -2 & 0 \end{bmatrix} \begin{bmatrix} 1 & 0 & 0 \\ 0 & 0 & 0 \\ 0 & 0 & 0 \end{bmatrix} \begin{bmatrix} 2 & 2 & 2 \\ 1 & 1 & -10 \\ 11 & -11 & 0 \end{bmatrix} = \begin{bmatrix} \tfrac{5}{11} & \tfrac{5}{11} & \tfrac{5}{11} \\ \tfrac{5}{11} & \tfrac{5}{11} & \tfrac{5}{11} \\ \tfrac{1}{11} & \tfrac{1}{11} & \tfrac{1}{11} \end{bmatrix} S^{(1)}.$$

Since the sum of the entries of $S^{(1)}$ is 1, it follows that $S^{(\infty)}$ is $\begin{bmatrix} \tfrac{5}{11} \\ \tfrac{5}{11} \\ \tfrac{1}{11} \end{bmatrix}$ for any initial state $S^{(1)}$. Since this vector is the steady state S, this means that if k is large, then S_k is close to the steady state $\begin{bmatrix} \tfrac{5}{11} \\ \tfrac{5}{11} \\ \tfrac{1}{11} \end{bmatrix}$, regardless of the results of the first election.

The process of passing from $S^{(k)}$ to $S^{(k+1)}$ ($k = 1, 2, 3, \ldots$) described above is an example of a Markov process. What is a Markov process in general? A *transition matrix* is a matrix $T \in M_n(\mathbb{R})$ such that each entry of T is nonnegative and the entries of each column add up to 1. A *state* is any element $S \in \mathbb{R}^{(n)}$ whose entries are all nonnegative and add up to 1. Any transition matrix $T \in M_n(\mathbb{R})$ has the property that TS is a state for any state $S \in \mathbb{R}^{(n)}$. (Prove!) A *Markov process* is any process involving transition from one of several mutually exclusive conditions $C_1, \ldots, C_n$ to another such that for some transition matrix T, the following is true:

If at any stage in the process, the probabilities of the conditions C_r are the entries S_r of the state S, then the probability of condition C_r at the next stage in the process is entry r of TS, for all r.

For any such Markov process, T is called the *transition matrix* of the process and its entries are called the *transition probabilities*. The transition probabilities can be described conveniently by giving a *transition diagram*, which in our example with

$$T = \begin{bmatrix} .50 & .49 & .05 \\ .49 & .50 & .05 \\ .01 & .01 & .90 \end{bmatrix} \text{ is}$$

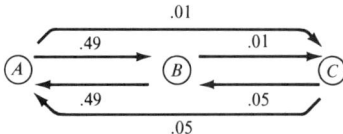

Note that the probabilities for remaining at any of the nodes A, B, C are not given, since they can be determined from the others.

Let's now discuss an arbitrary Markov process along the same lines as our example above. Letting the transition matrix be T, the state $S^{(k)}$ resulting from $S^{(1)}$ after

Sec. 11.3]   Markov Processes   421

$k - 1$ transitions is certainly $S^{(k)} = T^{k-1}S^{(1)}$. If a state S has the property that $TS = S$, it is called a *steady state*. In our example, there was exactly one steady state. Why?

The reason is that the transition matrix $T = \begin{bmatrix} .50 & .49 & .05 \\ .49 & .50 & .05 \\ .01 & .01 & .90 \end{bmatrix}$ is regular in the following sense.

Definition. A transition matrix T is *regular* if all entries of T^k are positive for some positive integer k.

Theorem 11.3.1. Let T be the transition matrix of a Markov process. Then there exists a steady state S. If T is regular, there is only one steady state.

Proof: Since the sum of the elements of any column of T is 1, the sum of the rows of $T - 1I$ is 0 and the rows of $T - 1I$ are linearly dependent. So 1 is a root of the characteristic polynomial $\det(T - xI)$ of T. Any characteristic vector s for T corresponding to 1 has the property that $TS = S$. Let's write $S = S_+ + S_-$ where:

If $s_r < 0$, then entry r of S_+ is 0 and entry r of S_- is s_r; and
If $s_r \geq 0$, then entry r of S_- is 0 and entry r of S_+ is s_r.

Since $S_+ + S_- = S = TS = TS_+ + TS_-$, and since T is a transition matrix, it follows that $TS_+ = S_+$ and $TS_- = S_-$. Since S is nonzero, one of the vectors $S_+, -S_-$ is nonzero and has nonzero sum of entries m. Replacing S by this vector multiplied by $1/m$, the new S is a steady state.

Suppose next that T is regular and choose k so that the entries of T^k are all positive. To show that T has only one steady state, we take steady states S' and S'' for T and proceed to show that they are equal. Let $S = S'' - S'$. Suppose first that $S \neq 0$, and write $S = S_+ + S_-$ as before. Since the sum of all entries is 1 for S' and S'', the sum of all entries is 0 for S, so that it is nonzero for S_+ and for S_-. Let S''' be S_+ divided by the sum of its entries, so that S''' is a state. Since S_- is nonzero, some entry s_r''' of S''' is zero. Since $TS = S$, we also have $TS_+ = S_+$, as before, so that $TS''' = S'''$. But then we also have $S''' = T^k S'''$. The entry s_r''' of $T^k S'''$ is the sum $t_{r1}^{(k)} s_1''' + \cdots + t_{rn}^{(k)} s_n'''$, where the $t_{rj}^{(k)}$ are the entries of row r of T^k and are, therefore, positive by our choice of k. Since s_r''' is 0 and the s_j''' are nonnegative, it follows that $s_j''' = 0$ for all j so that $S''' = 0$, a contradiction. So the assumption that S is nonzero leads to a contradiction, and $S = 0$. But then $0 = S = S'' - S'$ and $S' = S''$. This completes the proof of the theorem. ∎

We could now go on to prove that if T is a regular transition matrix of a Markov process, the limit $S^{(k)}$ as k goes to infinity always exists and equals the steady state S, regardless of the initial state $S^{(k)}$. From this, taking $S^{(k)}$ to be column j of the identity matrix, it follows that the limit of column j of T^k as k goes to infinity always exists and equals the steady state S. In other words, the limit of the matrix T^k exists and its columns all equal the steady state S. However, we've already seen an instance

Applications

of this in detail in our example where the limit of $\begin{bmatrix} .50 & .49 & .05 \\ .49 & .50 & .05 \\ .01 & .01 & .90 \end{bmatrix}^k$ turned out to be

$\begin{bmatrix} \frac{5}{11} & \frac{5}{11} & \frac{5}{11} \\ \frac{5}{11} & \frac{5}{11} & \frac{5}{11} \\ \frac{1}{11} & \frac{1}{11} & \frac{1}{11} \end{bmatrix}$. So we omit the proof and go on, instead, to other things.

PROBLEMS

NUMERICAL PROBLEMS

1. Find a steady state for $T = \begin{bmatrix} .2 & .3 \\ .8 & .7 \end{bmatrix}$, and find an expression for the state $S^{(304)}$ if $S^{(1)}$ is $\begin{bmatrix} .5 \\ .5 \end{bmatrix}$.

2. Suppose that companies R, S, and T all compete for the same market, which 10 years ago was held completely by company S. Suppose that at the end of each of these 10 years, each company lost a percentage of its market to the other companies as follows:

    R to S: 49% R to T: 1%
    S to R: 49% S to T: 1%
    T to R: 5% T to S: 5%

    Describe the process whereby the distribution of the market changes as a Markov process and give an expression for the percentage of the market that will be held by company S at the end of 2 more years. Find a steady state for the distribution of the market.

3. Suppose that candidates U, V, and W are running for mayor and that at the end of each week as the election approaches, each candidate loses a percentage of his or her share of the vote to the other candidates as follows:

    U to V: 5% U to W: 5%
    V to W: 49% V to U: 1%
    W to V: 49% W to U: 1%

    (a) Describe the process whereby the distribution of the vote changes as a Markov process.
    (b) Assuming that at the end of week 2, U has 80% of the vote, V has 10% of the vote, and W has 10% of the vote, give an expression for the percentage of the voters who, at the end of week 7, plan to vote for U.
    (c) Assuming that the the number of voters who favor the different candidates is the same for weeks 5 and 6, what are the percentages of voters who favor the different candidates at the end of week 6?
    (d) Assuming that the number of voters who favor the different candidates is

Sec. 11.4] Incidence Models

the same for weeks 5 and 6, and assuming that the total number of voters is constant, was it the same for weeks 4 and 5 as well? Explain.

MORE THEORETICAL PROBLEMS

Easier Problem

4. In Problem 2, give a formula for the percentage of the market that will be held by company S at the end of p more years and determine the limit of this percentage as p goes to infinity.

11.4. INCIDENCE MODELS

Models used to determine *local prices* of goods transported among various cities, models used to determine *electrical potentials* at nodes of a network of electrical currents, and models for determining the displacements at nodes of a mechanical structure under stress all have one thing in common. When these models are stripped of the trappings that go with the particular model, what is left is an *incidence diagram*, such as the one shown here.

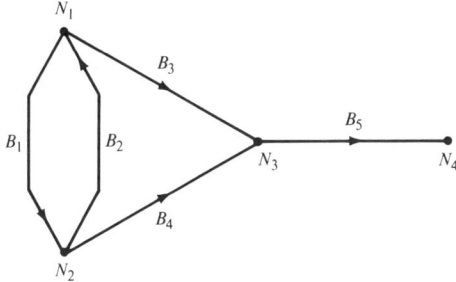

What, exactly, is an incidence diagram? It is a diagram with a certain number m of *nodes* and a certain number n of oriented *branches*, each of which begins and ends at different nodes, such that every node is the *beginning* or *ending* of some branch. How is an incidence diagram represented in a mathematical model? Corresponding to each incidence diagram is an *incidence matrix*, that is, a matrix such that each column has one entry whose value is 1, one entry whose value is -1, and 0 for all other entries. To form this matrix, let there be one row for each node N_r and one column for each branch B_s. If node N_r is the beginning of branch B_s, let the (r, s) entry of the matrix be -1. If N_r is the ending of branch B_s, let the (r, s) entry of the matrix by 1. Otherwise, let the (r, s) entry be 0, indicating that N_r is neither the beginning nor ending of branch B_s. For example, the matrix of the foregoing incidence diagram is

$$\begin{bmatrix} -1 & 1 & -1 & 0 & 0 \\ 1 & -1 & 0 & -1 & 0 \\ 0 & 0 & 1 & 1 & -1 \\ 0 & 0 & 0 & 0 & 1 \end{bmatrix}.$$ Conversely, given an $m \times n$ incidence matrix T, there

is a corresponding incidence diagram with m nodes $N_1, \ldots, N_m$ corresponding to the rows of T and n oriented branches $B_1, \ldots, B_n$ corresponding to the columns of T. If row r has -1 in column s, then node N_r is the *beginning* of branch B_s. On the other hand, if row r has 1 in column s, then N_r is the *end* of branch B_s. For example,

$$T = \begin{bmatrix} -1 & 1 & -1 & 0 & 0 \\ 1 & -1 & 0 & -1 & 0 \\ 0 & 0 & 1 & 1 & -1 \\ 0 & 0 & 0 & 0 & 1 \end{bmatrix}$$ is an incidence matrix whose incidence diagram is the one illustrated here.

We may as well just consider *connected* incidence diagrams, where every node can be reached from the first node. These can be studied inductively, since we can always remove *some* node and still have a connected diagram. How do we find such a node? Think of the nodes as lead weights or sinkers and the branches as fishline interconnecting them. Lift the first node high enough that all nodes hang from it without touching bottom. Then slowly lower it until one node just touches bottom. That is the node that we can remove without disconnecting the diagram. This proves

Lemma 11.4.1. If **D** is a connected incidence diagram with $m > 2$ nodes, then for some node N_r of **D**, removal of N_r and all branches of **D** that begin or end at N_r leaves a connected incidence diagram $\mathbf{D}^{\#}$ with one node fewer.

Theorem 11.4.2. Let T be an incidence matrix of a connected incidence diagram with m nodes. Then the rank of T is $m - 1$.

Proof: If S and T are incidence matrices for the same incidence diagram with respect to a different order for the nodes and for the branches, S can be obtained from T by row and column interchanges. So S and T have the same rank. We now show by mathematical induction that any connected incidence diagram with m nodes has an incidence matrix T of rank $m - 1$. If $m = 2$, this is certainly true, so we suppose that $m > 2$ and that it is true for diagrams with fewer than m nodes. By Lemma 11.4.1 our connected incidence diagram **D** has a node that can be removed without disconnecting the diagram; that is, the diagram $\mathbf{D}^{\#}$ of $m - 1$ nodes and $n - k$ branches, for some $k > 0$, that we get by removing this node and all branches that begin or end with it is a connected incidence diagram. By induction, $\mathbf{D}^{\#}$ has an incidence matrix $T^{\#}$ of rank $m - 2$. Now we put back the removed node and the k removed branches, and add to $T^{\#}$ k columns and a row with 0 as value for entries $1, \ldots, n - k$ and -1 and 1 for the other entries, to get an incidence matrix T for **D**. Since the rank of $T^{\#}$ is $m - 2$, the rank of T is certainly m or $m - 1$. But the rank of T is not m, since the sum of its rows is 0, which implies that its rows are linearly dependent. So the rank of T is $m - 1$. ∎

For example, our theorem tells us that the rank of the incidence matrix
$$\begin{bmatrix} -1 & 1 & -1 & 0 & 0 \\ 1 & -1 & 0 & -1 & 0 \\ 0 & 0 & 1 & 1 & -1 \\ 0 & 0 & 0 & 0 & 1 \end{bmatrix}$$ is 3, which we can see directly by row reducing it to the echelon matrix $\begin{bmatrix} 1 & -1 & 1 & 0 & 0 \\ 0 & 0 & 1 & 1 & 0 \\ 0 & 0 & 0 & 0 & 1 \\ 0 & 0 & 0 & 0 & 0 \end{bmatrix}$, which has three nonzero rows.

EXAMPLE

In a transportation model, the nodes N_r of our incidence diagram represent cities and the branches B_s transportation routes between the cities. Suppose that beer is produced and consumed in the cities N_r and transported along the routes B_s so that:

1. For each city N_r, if we add the rates F_s at which beer (measured in cans of beer per hour) is transported on routes B_s heading into city N_r and then subtract the rates F_s for routes B_s heading out of city N_r, we get the difference G_r between the rate of production and the rate of consumption of beer in city N_r. So for city N_2, since route B_1 heads in, whereas routes B_2 and B_4 head out, $F_1 - F_2 - F_4$ equals G_2.
2. If we let P_r denote the price of beer (per can) in city N_r and E_s the price at the end of the route B_s minus the price at the beginning of the route B_s, there is a positive constant value R_s such that $R_s F_s = E_s$. This constant R_s reflects distance and other *resistance to flow* along route B_s that will increase the price difference E_s. For $s = 3$, the price at the beginning of B_3 is P_1 and the price at the end of B_3 is P_3, so that $R_3 F_3 = E_3$, where $E_3 = P_3 - P_1$.

In terms of the incidence matrix, we now have

Kirchhoff's First Law: $TF = G$

Kirchoff's Second Law: $T'P = E$

Ohm's Law: $RF = E$, where $R = \begin{bmatrix} R_1 & & 0 \\ & \ddots & \\ 0 & & R_n \end{bmatrix}$.

For example, for the incidence matrix above, the equation $T'P = E$ is

$$\begin{bmatrix} -1 & 1 & 0 & 0 \\ 1 & -1 & 0 & 0 \\ -1 & 0 & 1 & 0 \\ 0 & -1 & 1 & 0 \\ 0 & 0 & -1 & 1 \end{bmatrix} \begin{bmatrix} P_1 \\ P_2 \\ P_3 \\ P_4 \end{bmatrix} = \begin{bmatrix} P_2 - P_1 \\ P_1 - P_2 \\ P_3 - P_1 \\ P_3 - P_2 \\ P_4 - P_3 \end{bmatrix} = \begin{bmatrix} E_1 \\ E_2 \\ E_3 \\ E_4 \\ E_5 \end{bmatrix},$$

and the equation $RF = E$ is

$$\begin{bmatrix} R_1 & & & & 0 \\ & R_2 & & & \\ & & R_3 & & \\ & & & R_4 & \\ 0 & & & & R_5 \end{bmatrix} \begin{bmatrix} F_1 \\ F_2 \\ F_3 \\ F_4 \\ F_5 \end{bmatrix} = \begin{bmatrix} R_1 F_1 \\ R_2 F_2 \\ R_3 F_3 \\ R_4 F_4 \\ R_5 F_5 \end{bmatrix} = \begin{bmatrix} E_1 \\ E_2 \\ E_3 \\ E_4 \\ E_5 \end{bmatrix}.$$

Since the sum of the rows of T is 0, the sum of the entries of G is 0, which means that *the total production equals the total consumption of beer*. For instance, for the incidence diagram above, the equation $TF = G$ is

$$\begin{bmatrix} -1 & 1 & -1 & 0 & 0 \\ 1 & -1 & 0 & -1 & 0 \\ 0 & 0 & 1 & 1 & -1 \\ 0 & 0 & 0 & 0 & 1 \end{bmatrix} \begin{bmatrix} F_1 \\ F_2 \\ F_3 \\ F_4 \\ F_5 \end{bmatrix} = \begin{bmatrix} F_2 - F_1 - F_3 \\ F_1 - F_2 - F_4 \\ F_3 + F_4 - F_5 \\ F_5 \end{bmatrix} = \begin{bmatrix} G_1 \\ G_2 \\ G_3 \\ G_4 \end{bmatrix},$$

and the local differences between the rates of production and consumption add up to

$$G_1 + G_2 + G_3 + G_4 = F_2 - F_1 - F_3 + F_1 - F_2 - F_4 \\ + F_3 + F_4 - F_5 + F_5 = 0.$$

Assuming that this is so, we ask:

Given T, R and G, what is the price vector P?

Substituting $T'P$ for E in $RF = E$, we find that $F = R^{-1}T'P$, so that

$$(TR^{-1}T')P = TF = G.$$

Can we now solve $(TR^{-1}T')P = G$ for P? The answer is "yes," as we show in Theorem 11.4.5. First, however, we need

Theorem 11.4.3. Let T be an $m \times n$ matrix with real entries and let D be a diagonal $n \times n$ matrix with positive entries. Then the nullspace of T' equals the nullspace of TDT'.

Proof: If $T'v = 0$, certainly $TDT'v = 0$. Suppose, conversely, that $TDT'v = 0$. It follows that $v'TDT'v = 0$. Let $(T'v)_r$ denote entry r of $T'v$. The only nonzero entries of D are the positive numbers $d_{11}, \ldots, d_{nn}$, so that

$$0 = v'TDT'v = d_{11}(T'v)_1^2 + \cdots + d_{nn}(T'v)_n^2.$$

But this implies that all of the $(T'v)_r$ are 0, so that $T'v = 0$. So the nullspace of T' equals the nullspace of TDT'. ∎

Corollary 11.4.4. Let T be an $m \times n$ matrix with real entries and let D be a diagonal $n \times n$ matrix with positive entries. Then

1. If T has rank m, then TDT' is invertible.
2. If T has rank $m - 1$ and v is a nonzero solution to $T'v = 0$, then $TDT'x = y$ has a solution x for every y orthogonal to v; and every other solution is of the form $x + cv$ for some $c \in \mathbb{R}$.

Proof: If T has rank m, the nullspace of T' is 0; so, by the theorem, the nullspace of the square matrix TDT' is 0 and it is invertible. This proves (1). Suppose next that T has rank $m - 1$ and v is a nonzero solution to $T'v = 0$. Then the nullspace of T' is $\mathbb{R}v$ and, by the theorem, the nullspace of TDT' is also $\mathbb{R}v$. It follows that the rank of the $m \times m$ matrix TDT' is $m - 1$. Since $v'TDT'$ is 0, $TDT'\mathbb{R}^{(m)}$ is contained in $v^\perp$. Since the rank of TDT' is $m - 1$ and the dimension of $v^\perp$ is also $m - 1$, it follows that $TDT'\mathbb{R}^{(m)}$ equals $v^\perp$. This proves that $TDT'x = y$ has a solution x for every y orthogonal to v; and every other solution is of the form $x + cv$ for some $c \in \mathbb{R}$, since $\mathbb{R}v$ is the nullspace of TDT'. This proves (2). ∎

Theorem 11.4.5. Suppose that a transportation model is defined by a connected incidence diagram with $m \times n$ incidence matrix T, resistance matrix $R = \begin{bmatrix} R_1 & & 0 \\ & \ddots & \\ 0 & & R_n \end{bmatrix}$

with positive diagonal entries, and a production and consumption vector $G \in \mathbb{R}^{(m)}$ whose entries add up to 0 (total production equals total consumption). Then for any arbitrarily selected price p at one specified node N_r, there is one and only one solution P (distribution of prices) to $(TR^{-1}T')P = G$ such that the price P_r assigned by P for goods in the city N_r is p.

Proof: By Theorem 11.4.2 the rank of T is $m - 1$, so that we can use Corollary 11.4.4. The sum of the rows of T is 0, so that $T'v = 0$, where $v \in \mathbb{R}^{(m)}$ has all entries equal to 1. Since we are assuming that the sum of the entries of G is 0 (total production equals total consumption), G is orthogonal to v, so that $(TR^{-1}T')P = G$ has a solution $P = Q$ by the corollary. Moreover, every other solution P is of the form $P = Q + cv$ for $c \in \mathbb{R}$. Solving $Q_r + c1 = p$ for c, we get a unique solution $P = Q + cv$ such that $P_r = Q_r + c1 = p$. ∎

EXAMPLE

Let's now look at an *electrical network model*. This is really a transportation model in disguise, where the goods being transported in routes are electrical charges flowing as current in network elements, the production and consumption come from power sources and incoming and exiting currents and the price distribution resulting from transportation constraints is the distribution of potentials (electrical pressures) resulting from resistance to current flow. So it is not surprising that there are important similarities. In the model, each node N_r of the incidence diagram represents a connection point and each branch B_s represents a network element consisting of a *resistance* R_s in series with a *source potential* (pressure or voltage from a source of electrical power) e_s. When these network elements are connected together, *currents* I_s begin to flow in the branches B_s. The network may *receive* positive or negative *external currents* from another network connected to it, namely the currents G_1, G_2, G_3, G_4 indicated in the diagram at the connection points N_1, N_2, N_3, N_4. As a result of the flow of all currents, *potentials* (electrical pressures) P_r are created at each node N_r. The potential at the end of a branch B_s minus the potential at the beginning of B_s is the *potential difference* E_s for the branch B_s. For example, $E_2 = P_1 - P_2$ and $E_3 = P_3 - P_1$. We assume that the following conditions involving the incidence matrix T and the vectors R, G, I, P, E are satisfied:

1. For each connection point N_r, if we add the currents I_s flowing in branches B_s heading into N_r and then subtract the currents I_s in branches B_s heading out of N_r, we get the external current G_r. So for N_2, since route B_1 heads in, whereas routes B_2 and B_4 head out, $I_1 - I_2 - I_4$ equals G_2.

2. For any branch B_s, we have $R_s I_s = E_s$. If $s = 3$, for instance, the potential at the beginning of B_3 is P_1 and the potential at the end of B_3 is P_3, so that $R_3 I_3 = E_3$, where $E_3 = P_3 - P_1$.

In terms of the incidence matrix $T = \begin{bmatrix} -1 & 1 & -1 & 0 & 0 \\ 1 & -1 & 0 & -1 & 0 \\ 0 & 0 & 1 & 1 & -1 \\ 0 & 0 & 0 & 0 & 1 \end{bmatrix}$, we now have:

Kirchhoff's First Law: $TI = G$

Kirchoff's Second Law: $T'P = E + e$

Ohm's Law: $RI = E$, where $R = \begin{bmatrix} R_1 & & 0 \\ & \ddots & \\ 0 & & R_5 \end{bmatrix}$.

Since the sum of the rows of T is 0, the sum of the entries of G is 0, which means that *the sum of all external currents is 0*. So, regarding our network as a node in a bigger one, the external currents at this node become the internal currents at this node in the bigger network and their sum is 0 (Kirchhoff's First Law at nodes without external current). Assuming that this is so, we ask:

Given T, R, and G, what is the potential vector P?

Substituting $T'P - e$ for E in $RI = E$ and then solving for I, we have $I = R^{-1}T'P - R^{-1}e$, so that $G = TI = TR^{-1}T'P - TR^{-1}e$ and $(TR^{-1}T')P = G + TR^{-1}e$. We can now solve $(TR^{-1}T')P = G + TR^{-1}e$ for P, by

Theorem 11.4.6. Suppose that an electrical network model is defined by a connected incidence diagram with $m \times n$ incidence matrix T, a resistance matrix

Sec. 11.4] Incidence Models

$$R = \begin{bmatrix} R_1 & & 0 \\ & \ddots & \\ 0 & & R_n \end{bmatrix}$$ with positive diagonal entries, an arbitrary distribution $e \in \mathbb{R}^{(n)}$ of source voltage, and an external current vector $G \in \mathbb{R}^{(m)}$ whose entries add up to 0 (total external current is 0). Then for any arbitrarily selected potential p at one specified node N_r, there is one and only one solution P (distribution of potential) to $(TR^{-1}T')P = G + TR^{-1}e$ such that the potential P_r assigned by P to node N_r is p.

Proof: The sum of the entries of G are 0, by assumption, and the sum of the entries of $TR^{-1}e$ are zero since the sum of the rows of T is 0. So the sum of the entries of the right-hand-side vector $G + TR^{-1}e$ is zero. The remainder of the proof follows in the footsteps of the proof of Theorem 11.4.5. ∎

When we are given a network with specified T, R, G and e, why is finding the node potentials P_r so important? *The distribution P of node potentials is the key to the network*. Given P, everything else is determined:

1. The distribution of potential differences in the branches is given by $E = T'P - e$.
2. Once we have E, from (1), the distribution of network currents in the branches is given by $I = R^{-1}E$.

PROBLEMS

NUMERICAL PROBLEMS

1. Draw the incidence diagram for $T = \begin{bmatrix} 1 & -1 & 0 & 1 \\ -1 & 0 & 1 & 0 \\ 0 & 1 & -1 & -1 \end{bmatrix}$.

2. For the electrical network model defined by

    $$T = \begin{bmatrix} 1 & -1 & 0 & 1 \\ -1 & 0 & 1 & 0 \\ 0 & 1 & -1 & -1 \end{bmatrix},$$

    $R_1 = 2$, $R_2 = 1$, $R_3 = 3$, $R_4 = 1$,

    $e_1 = -5$, $e_2 = e_3 = e_4 = 0$,

    $G_1 = 1$, $G_2 = 1$, $G_3 = -2$,

    is there a solution P (distribution of potential) such that $P_3 = 872$? If so, show how to find it in terms of T, R, E, and G.

MORE THEORETICAL PROBLEMS

Easier Problems

3. Show that if **D** is an incidence diagram with m nodes that is not *connected* (some node cannot be reached from the first node), then the corresponding incidence matrix has rank less than $m - 1$.

11.5. DIFFERENTIAL EQUATIONS

Finding the set of all solutions to the homogenenous differential equation $\frac{d^n v}{dt^n} + b_{n-1}\frac{d^{n-1} v}{dt^{n-1}} + \cdots + b_0 v = 0$ is really the same as finding the nullspace of a corresponding linear transformation. Why? The mapping T that maps a function v of a real variable t to its derivative $Tv = \frac{dv}{dt}$ is linear on the vector space F_n of n times differentiable complex-valued functions v of t. $\Bigl($Here, as in Section 10.5, the derivative of the function $v(t) = a(t) + b(t)i$ is defined as $\frac{dv}{dt} = \frac{da}{dt} + \frac{db}{dt}i.\Bigr)$ So if we let $f(x)$ denote the polynomial $f(x) = x^n + b_{n-1}x^{n-1} + \cdots + b_0$, $f(T)$ is linear on F_n and we can express the set of solutions to $\frac{d^n v}{dt^n} + b_{n-1}\frac{d^{n-1} v}{dt^{n-1}} + \cdots + b_0 v = 0$ as the nullspace $W = \{v \in F_n \mid f(T)v = 0\}$ of $f(T)$.

To determine the nullspace W of $f(T)$, note that $TW \subseteq W$, take any $v \in W$ and form the subspace V of W spanned over $\mathbb{C}$ by the functions $v, Tv, \ldots, T^{n-1}v$. Since $f(T)v = 0$, $T^n v$ is a linear combination of $v, Tv, \ldots, T^{n-1}v$, so is contained in V. It follows that T maps $v, Tv, \ldots, T^{n-1}v$ into V, that is, $TV \subseteq V$.

Now we can regard T as a linear transformation of the finite-dimensional vector space V. Since $f(T)V = 0$, V is the direct sum $V = V_{a_1}(T) \oplus \cdots \oplus V_{a_k}(T)$, where $f(x) = (x - a_1)^{m_1} \cdots (x - a_k)^{m_k}$ and $V_{a_1}(T), \ldots, V_{a_k}(T)$ are the generalized characteristic subspaces introduced in Section 10.2. Writing v as $v = v_1 + \cdots + v_k$ with $v_r \in V_{a_r}(T)$ for all r, we have $(T - a_r I)^{m_r} v_r = 0$.

Let's fix r and denote $a_r = a$, $v_r = w$ and $m_r = m$. Since $T(e^{at}u) = ae^{at}u + e^{at}Tu$ for any differentiable u, we have $(T - aI)(e^{at}u) = e^{at}Tu$. Applying $(T - aI)$ in this fashion $m - 1$ more times, we also have $(T - aI)^m(e^{at}u) = e^{at}T^m u$. Replacing u by $e^{-at}w$, we have $(T - aI)^m(e^{at}e^{-at}w) = e^{at}T^m(e^{-at}w)$, that is, $(T - aI)^m w = e^{at}T^m(e^{-at}w)$. So our condition $(T - aI)^m w = 0$ is equivalent to the condition $T^m(e^{-at}w) = 0$, which in turn is equivalent to the condition that $e^{-at}w$ is a polynomial $p(t)$ of degree less than m. But then $w = p(t)e^{at}$; that is, w is a linear combination of the functions

$$e^{at}, te^{at}, \ldots, t^{m-1}e^{at}.$$

Conversely, it is easy to check that the functions $e^{at}, te^{at}, \ldots, t^{m-1}e^{at}$ are solutions to the differential equation $(T - aI)^m = 0$. Since $(x - a)^m = (x - a_r)^{m_r}$ is a factor of $f(x)$, they are also solutions to the differential equation $f(T)v = 0$.

Since each solution $v \in W$ is a sum of functions v_r each of which is a linear combination of $e^{a_r t}, te^{a_r t}, \ldots, t^{m_r - 1}e^{a_r t}$, we conclude that

$$\{t^{n_r} e^{a_r t} \mid 1 \le r \le k, \ 0 \le n_r \le m_r - 1\}$$

spans W. So W is finite-dimensional. Regarding T as a linear transformation of W,

Sec. 11.5] Differential Equations 431

we have $f(T)W = 0$ and $W = W_{a_1}(T) \oplus \cdots \oplus W_{a_k}(T)$, where $W_{a_1}(T), \ldots, W_{a_k}(T)$ are the generalized characteristic subspaces introduced in Section 10.2. Since the functions

$$e^{a_r t}, te^{a_r t}, \ldots, t^{m_r - 1} e^{a_r t}$$

are linearly independent elements of $W_{a_r}(T)$ for each r (Prove!), the set $\{t^{n_r} e^{a_r t} \mid 1 \le r \le k,\ 0 \le n_r \le m_r - 1\}$ is linearly independent and we have

Theorem 11.5.1. The space W of n times differentiable complex-valued solutions v to
$$\frac{d^n v}{dt^n} + b_{n-1} \frac{d^{n-1} v}{dt^{n-1}} + \cdots + b_0 v = 0 \text{ is } n\text{-dimensional over } \mathbb{C}. \text{ A basis for } W \text{ is}$$

$$\{t^{n_r} e^{a_r t} \mid 1 \le r \le k,\ 0 \le n_r \le m_r - 1\},$$

where the polynomial $x^n + a_{n-1} x^{n-1} + \cdots + a_0$ factors as $(x - a_1)^{m_1} \cdots (x - a_k)^{m_k}$.

Can we get a solution f to $\dfrac{d^n v}{dt^n} + b_{n-1} \dfrac{d^{n-1} v}{dt^{n-1}} + \cdots + b_0 v = 0$ satisfying initial conditions $f^{(r)}(t_0) = v_r (0 \le r \le n - 1)$ at a specified value t_0 of t? Letting the basis for the solution space W be $f_1, \ldots, f_n$, we want $f = c_1 f_1 + \cdots + c_n f_n$ to satisfy the system of equations

$$c_1 f_1(t_0) + \cdots + c_n f_n(t_0) = y_0$$
$$\vdots \qquad \qquad \vdots \qquad \qquad \vdots$$
$$c_1 f_1^{(n-1)}(t_0) + \cdots + c_n f_n^{(n-1)}(t_0) = y_{n-1},$$

where $f_s^{(r)}$ is the rth derivative of f_s. Of course, we can solve this system of equations for the coefficients c_r if we know that the determinant

$$\begin{vmatrix} f_1(t_0) & \cdots & f_n(t_0) \\ f_1^{(1)}(t_0) & \cdots & f_n^{(1)}(t_0) \\ \vdots & & \vdots \\ f_1^{(n)}(t_0) & \cdots & f_n^{(n)}(t_0) \end{vmatrix},$$

called the *Wronskian* of $f_1, \ldots, f_n$, is nonzero at t_0. Since we have the basis $\{t^{n_r} e^{a_r t} \mid 1 \le r \le k,\ 0 \le n_r \le m_r - 1\}$, we could, *in principle*, compute the Wronskian for this basis, which we would find to be nonzero at all t_0. In fact, it is an interesting and easy exercise to do so in two extreme cases: when all the m_r are 1, and when $k = 1$.

Instead of following such an approach, which has the drawback that we have to compute the entries $f^{(r)}(t_0)$ and then solve for the coefficients c_r, we exhibit the desired solution explicitly in terms of the exponential function e^{tT}. For this we consider the system

$$u_1' = u_2$$
$$u_2' = u_3$$
$$\vdots$$
$$u_n' = -b_0 u_1 - \cdots - b_{n-1} u_n$$

of n linear differential equations in n unknowns represented by the matrix equation $u' = Tu$, where T is

$$\begin{bmatrix} 0 & & & & -a_0 \\ 1 & \cdot & & & \\ 0 & \cdot & \cdot & & \vdots \\ \vdots & & \cdot & \cdot & \\ 0 & \cdots & 0 & 1 & -a_{n-1} \end{bmatrix},$$

the *companion matrix* to the polynomial $b_0 + \cdots + b_{n-1}x^{n-1} + x^n$ with the same coefficients as the differential equation. How does this system of linear differential equations relate to the differential equation

$$\frac{d^n v}{dt^n} + b_{n-1}\frac{d^{n-1}v}{dt^{n-1}} + \cdots + b_0 v = 0?$$

The condition $u' = Tu$ implies that $v = u_1$ satisfies the conditions

$$u_1 = v^{(0)}, \quad u_2 = v^{(1)}, \quad \ldots, \quad u_n = v^{(n-1)}.$$

But then

$$v^{(n)} = u'_n = -b_0 u_1 - \cdots - b_{n-1} u_n,$$

that is, $v^{(n)} = -b_0 v^{(0)} - \cdots - b_{n-1} v^{(n-1)}$. So Theorem 10.5.3 now gives us

Theorem 11.5.2. For any t_0 and $v_0, \ldots, v_{n-1}$, the differential equation

$$\frac{d^n v}{dt^n} + b_{n-1}\frac{d^{n-1}v}{dt^{n-1}} + \cdots + b_0 v = 0$$

has one and only one solution v such that $v^{(r)}(t_0) = v_r$ for $0 \le r \le n-1$, namely,

$$v = u_1 \text{ where } u = e^{(t-t_0)T}\begin{bmatrix} v_0 \\ \vdots \\ v_{n-1} \end{bmatrix}.$$

PROBLEMS

NUMERICAL PROBLEMS

1. Find a basis over $\mathbb{C}$ for the solution space of $\dfrac{d^3 v}{dt^3} - 2\dfrac{d^2 v}{dt^2} + \dfrac{dv}{dt} = 0$.

2. In Problem 1, find a real solution v such that $v(0) = 3$, $v'(0) = 3$, $v''(0) = 4$,
3. where $v'(t)$ denotes the derivative of $v(t)$.

MORE THEORETICAL PROBLEMS

Easier Problems

4. Use the formula $T(e^{at}u) = ae^{at}u + e^{at}Tu$ to determine the matrix of T on the space W introduced in the discussion preceding Theorem 11.5.1 with respect to the basis $\{t^{n_r}e^{a_r t} \mid 1 \le r \le k,\ 0 \le n_r \le m_r - 1\}$.

5. Show that $\{(t^{n_r}e^{a_r t}/n_r!) \mid 1 \le r \le k,\ 0 \le n_r \le m_r - 1\}$ is a Jordan basis for W with respect to T and describe the Jordan canonical form of T.

6. Show that if the coefficients c_r are real, then the real part v_{real} of any solution v to the differential equation

$$\frac{d^n v}{dt^n} + c_{n-1}\frac{d^{n-1} v}{dt^{n-1}} + \cdots + c_0 v = 0$$

is also a solution.

CHAPTER

12

Least Squares Methods (Optional)

12.1. APPROXIMATE SOLUTIONS OF SYSTEMS OF LINEAR EQUATIONS

Since the system of m linear equations in n variables with matrix equation $Ax = y$ has a solution x if and only if y is in the column space of A, many systems of linear equations have no solutions. What shall we do with an equation $Ax = y$ such as

$$\begin{bmatrix} 6 & 3 \\ 2 & 1 \end{bmatrix} \begin{bmatrix} x_1 \\ x_2 \end{bmatrix} = \begin{bmatrix} 2 \\ 2 \end{bmatrix}$$

which has no solution? We can *always* find an *approximate solution* x by replacing y by the vector y' in the column space of A nearest to y and solving $Ax = y'$ instead.
In the case of the equation $\begin{bmatrix} 6 & 3 \\ 2 & 1 \end{bmatrix} \begin{bmatrix} x_1 \\ x_2 \end{bmatrix} = \begin{bmatrix} 2 \\ 2 \end{bmatrix}$, where A and y are $\begin{bmatrix} 6 & 3 \\ 2 & 1 \end{bmatrix}$ and $\begin{bmatrix} 2 \\ 2 \end{bmatrix}$, the column space of A is the span

$$W = \mathbb{R} \begin{bmatrix} 3 \\ 1 \end{bmatrix}$$

of $\begin{bmatrix} 3 \\ 1 \end{bmatrix}$. So y' is the multiple $y' = t \begin{bmatrix} 3 \\ 1 \end{bmatrix}$ of $\begin{bmatrix} 3 \\ 1 \end{bmatrix}$ such that the length

$$\|y - y'\| = \left\| \begin{bmatrix} 2 \\ 2 \end{bmatrix} - t \begin{bmatrix} 3 \\ 1 \end{bmatrix} \right\|$$

Sec. 12.1]   Solutions of Systems of Linear Equations   435

is as small as possible. We can represent y, y', and W pictorially as follows:

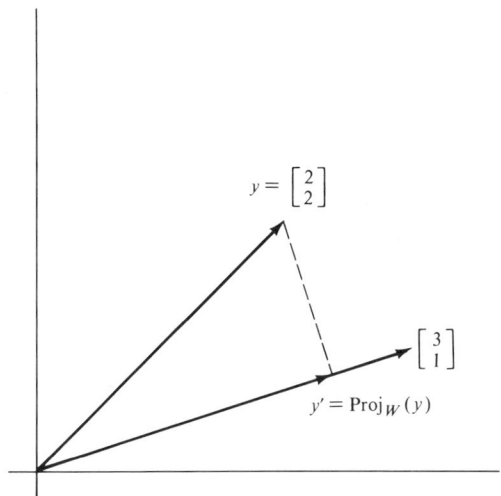

To get an explicit expression for the vector y' in the column space W of A nearest to y, we need

Definition. Let W be a subspace of $\mathbb{R}^{(n)}$ and write $y \in \mathbb{R}^{(n)} = W \oplus W^\perp$ as $y_1 + y_2$, where $y_1 \in W$ and $y_2 \in W^\perp$. Then y_1 is called the *projection* of y on W and is denoted $y_1 = \text{Proj}_W(y)$.

For the equation $\begin{bmatrix} 6 & 3 \\ 2 & 1 \end{bmatrix} \begin{bmatrix} x_1 \\ x_2 \end{bmatrix} = \begin{bmatrix} 2 \\ 2 \end{bmatrix}$, we take W to be the span $\mathbb{R} \begin{bmatrix} 3 \\ 1 \end{bmatrix}$ of $\begin{bmatrix} 3 \\ 1 \end{bmatrix}$. Then $W^\perp$ is the span of $\begin{bmatrix} -1 \\ 3 \end{bmatrix}$. So we get that $\begin{bmatrix} 2 \\ 2 \end{bmatrix} = (\tfrac{4}{5})\begin{bmatrix} 3 \\ 1 \end{bmatrix} + (\tfrac{2}{5})\begin{bmatrix} -1 \\ 3 \end{bmatrix}$, the first term of which is $\text{Proj}_W \begin{bmatrix} 2 \\ 2 \end{bmatrix} = (\tfrac{4}{5})\begin{bmatrix} 3 \\ 1 \end{bmatrix}$. To minimize the length

$$\|y - y'\| = \left\| \begin{bmatrix} 2 \\ 2 \end{bmatrix} - t \begin{bmatrix} 3 \\ 1 \end{bmatrix} \right\|$$

$$= \left\| (\tfrac{4}{5})\begin{bmatrix} 3 \\ 1 \end{bmatrix} + (\tfrac{2}{5})\begin{bmatrix} -1 \\ 3 \end{bmatrix} - t\begin{bmatrix} 3 \\ 1 \end{bmatrix} \right\|$$

$$= \left\| ((\tfrac{4}{5}) - t)\begin{bmatrix} 3 \\ 1 \end{bmatrix} + (\tfrac{2}{5})\begin{bmatrix} -1 \\ 3 \end{bmatrix} \right\|,$$

since $\begin{bmatrix} 3 \\ 1 \end{bmatrix}$ and $\begin{bmatrix} -1 \\ 3 \end{bmatrix}$ are orthogonal we must take $t = \tfrac{4}{5}$ to eliminate the term involving $\begin{bmatrix} 3 \\ 1 \end{bmatrix}$. (Prove!) So the element y' of W nearest $\begin{bmatrix} 2 \\ 2 \end{bmatrix}$ is

$$y' = \text{Proj}_W \begin{bmatrix} 2 \\ 2 \end{bmatrix} = (\tfrac{4}{5}) \begin{bmatrix} 3 \\ 1 \end{bmatrix}.$$

In fact, we have

Theorem 12.1.1. Let W be a subspace of $\mathbb{R}^{(n)}$. Then the element y' of W nearest to $y \in \mathbb{R}^{(n)}$ is $y' = \text{Proj}_W(y)$.

Proof: Write $y = y_1 + y_2$ with $y_1 \in W$ and $y_2 \in W^\perp$, so $y_1 = \text{Proj}_W(y)$. For $w \in W$, the squared distance $\|y - w\|^2$ is

$$\|(y_1 - w) + y_2\|^2 = ((y_1 - w) + y_2, (y_1 - w) + y_2)$$
$$= (y_1 - w, y_1 - w) + (y_2, y_2) = \|y_1 - w\|^2 + \|y_2\|^2,$$

which is minimal if and only if $w = y_1 = \text{Proj}_W(y)$. ∎

Given a vector y and subspace W, the method of going from y to the vector $y' = \text{Proj}_W(y)$ in W nearest to y is sometimes called the *method of least squares* since the sum of squares of the entries of $y - y'$ is thereby minimized.

How do we compute $y' = \text{Proj}_W(y)$? Let's first consider the case where W is the span $\mathbb{R}w$ of a single nonzero vector w. Then $V = W \oplus W^\perp$ and we have the simple formula

$$\text{Proj}_W(y) = \frac{(y, w)}{(w, w)} w$$

since

1. $\dfrac{(y, w)}{(w, w)} w$ is in W.

2. $y - \dfrac{(y, w)}{(w, w)} w$ is in $W^\perp$. (Prove!)

3. $y = \dfrac{(y, w)}{(w, w)} w + \left(y - \dfrac{(y, w)}{(w, w)} w \right)$.

So if W is spanned by a single element w, we get a formula for $\text{Proj}_W(y)$ involving only y and w. In the case where W is the span $W = \mathbb{R} \begin{bmatrix} 3 \\ 1 \end{bmatrix}$ of $\begin{bmatrix} 3 \\ 1 \end{bmatrix}$, our formula gives us

$$\text{Proj}_W \begin{bmatrix} 2 \\ 2 \end{bmatrix} = \frac{2 \cdot 3 + 2 \cdot 1}{3 \cdot 3 + 1 \cdot 1} \begin{bmatrix} 3 \\ 1 \end{bmatrix} = (\tfrac{4}{5}) \begin{bmatrix} 3 \\ 1 \end{bmatrix},$$

which is what we got when we computed it directly.

Sec. 12.1] Solutions of Systems of Linear Equations 437

What happens for an arbitrary subspace W? If we are given an orthogonal basis for W, we get a formula for $\text{Proj}_W(y)$ which involves only y and that orthogonal basis, as we now show in

Theorem 12.1.2. Let W be a subspace of $\mathbb{R}^{(n)}$ with orthogonal basis $w_1, \ldots, w_k$ and let $y \in \mathbb{R}^{(n)}$. Then $\text{Proj}_W(y) = \dfrac{(y, w_1)}{(w_1, w_1)} w_1 + \cdots + \dfrac{(y, w_k)}{(w_k, w_k)} w_k$.

Proof: Let $y_1 = \dfrac{(y, w_1)}{(w_1, w_1)} w_1 + \cdots + \dfrac{(y, w_k)}{(w_k, w_k)} w_k$. Then we have $(y - y_1, w_j) = (y, w_j) - \dfrac{(y, w_j)}{(w_j, w_j)}(w_j, w_j) = 0$ for $1 \leq j \leq k$. So $y - y_1 \in W^\perp$ and $y_1 = \text{Proj}_W(y)$. ∎

A nice geometric interpretation of this theorem is that

the projection of y on the span W of mutually orthogonal vectors $w_1, \ldots, w_k$ equals the sum of the projections of y on the lines $\mathbb{R}w_1, \ldots, \mathbb{R}w_k$.

For example, if $n = 3$ and $k = 2$, we can represent the projection $y' = \text{Proj}_W(y)$ of a vector y onto the span W of nonzero orthogonal vectors w_1, w_2 pictorially as follows, where y_1 and y_2 are the projections of y on $\mathbb{R}w_1$ and $\mathbb{R}w_2$:

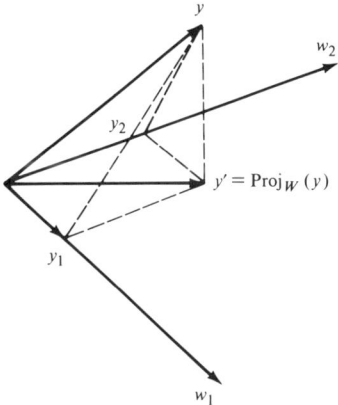

If W is a subspace of $\mathbb{R}^{(n)}$ with orthonormal basis $w_1, \ldots, w_k$ and $y \in \mathbb{R}^{(n)}$. Then the formula in our theorem simplifies to $\text{Proj}_W(y) = (y, w_1)w_1 + \cdots + (y, w_k)w_k$.

EXAMPLE

Let's find the element of W nearest $\begin{bmatrix} 1 \\ 2 \\ 3 \end{bmatrix}$, where W is the span of vectors

$$\begin{bmatrix} 1 \\ 2 \\ -1 \end{bmatrix}, \begin{bmatrix} -1 \\ 4 \\ 1 \end{bmatrix}.$$

Using the Gram–Schmidt process, we get an orthonormal basis

$$1/\sqrt{6}\begin{bmatrix} 1 \\ 2 \\ -1 \end{bmatrix}, \ 1/\sqrt{3}\begin{bmatrix} -1 \\ 1 \\ 1 \end{bmatrix}$$

for W. Then the projection of $\begin{bmatrix} 1 \\ 2 \\ 3 \end{bmatrix}$ on W is

$$\text{Proj}_W \begin{bmatrix} 1 \\ 2 \\ 3 \end{bmatrix} = \frac{1}{\sqrt{6}}(1+4-3)\frac{1}{\sqrt{6}}\begin{bmatrix} 1 \\ 2 \\ -1 \end{bmatrix} + \frac{1}{\sqrt{3}}(-1+2+3)\frac{1}{\sqrt{3}}\begin{bmatrix} -1 \\ 1 \\ 1 \end{bmatrix}$$

$$= (\tfrac{1}{3})\begin{bmatrix} 1 \\ 2 \\ -1 \end{bmatrix} + (\tfrac{4}{3})\begin{bmatrix} -1 \\ 1 \\ 1 \end{bmatrix} = \begin{bmatrix} -1 \\ 2 \\ 1 \end{bmatrix}$$

So, we have

$$\begin{bmatrix} 1 \\ 2 \\ 3 \end{bmatrix} = \begin{bmatrix} -1 \\ 2 \\ 1 \end{bmatrix} + \begin{bmatrix} 2 \\ 0 \\ 2 \end{bmatrix}$$

in $W \oplus W^{\perp}$ and the element of W closest to $\begin{bmatrix} 1 \\ 2 \\ 3 \end{bmatrix}$ in W is $\text{Proj}_W \begin{bmatrix} 1 \\ 2 \\ 3 \end{bmatrix} = \begin{bmatrix} -1 \\ 2 \\ 1 \end{bmatrix}$.

Now we have a way to compute $\text{Proj}_W(y)$. First, find an orthonormal basis for W (e.g., by the Gram-Schmidt orthogonalization process) and then use the formula for $\text{Proj}_W(y)$ given in Theorem 12.1.2.

EXAMPLE

Suppose that we have a supply of 5000 units of S, 4000 units of T, and 2000 units of U, materials used in manufacturing products P and Q. We ask: If each unit of P uses 2 units of S, 0 units of T and 0 units of U, and each unit of Q uses 3 units of S, 4 units of T and 1 unit of U, how many units p and q of P and Q should we make if we want to use up the entire supply? The system is represented by the equation

$$\begin{bmatrix} 2 & 3 \\ 0 & 4 \\ 0 & 1 \end{bmatrix} \begin{bmatrix} p \\ q \end{bmatrix} = \begin{bmatrix} 5000 \\ 4000 \\ 2000 \end{bmatrix}.$$

Since the vector $\begin{bmatrix} 5000 \\ 4000 \\ 2000 \end{bmatrix}$ is not a linear combination $p\begin{bmatrix} 2 \\ 0 \\ 0 \end{bmatrix} + q\begin{bmatrix} 3 \\ 4 \\ 1 \end{bmatrix}$ of the columns of $\begin{bmatrix} 2 & 3 \\ 0 & 4 \\ 0 & 1 \end{bmatrix}$, *there is no exact solution* $\begin{bmatrix} p \\ q \end{bmatrix}$. So we get an approximate solution $\begin{bmatrix} p \\ q \end{bmatrix}$ by finding those values of p and q for which the distance from $p\begin{bmatrix} 2 \\ 0 \\ 0 \end{bmatrix} + q\begin{bmatrix} 3 \\ 4 \\ 1 \end{bmatrix}$ to $\begin{bmatrix} 5000 \\ 4000 \\ 2000 \end{bmatrix}$ is as small as possible. To do this, we first find the vector in the space W of linear combinations $p\begin{bmatrix} 2 \\ 0 \\ 0 \end{bmatrix} + q\begin{bmatrix} 3 \\ 4 \\ 1 \end{bmatrix}$ that is closest to $\begin{bmatrix} 5000 \\ 4000 \\ 2000 \end{bmatrix}$. We've just seen that this vector is the projection of $\begin{bmatrix} 5000 \\ 4000 \\ 2000 \end{bmatrix}$ on W. Computing this by the formula

$$\text{Proj}_W(v) = (v, w_1)w_1 + (v, w_2)w_2$$

of Theorem 12.1.2, where w_1, w_2 is the orthonormal basis

$$w_1 = \begin{bmatrix} 1 \\ 0 \\ 0 \end{bmatrix}, \quad w_2 = \frac{1}{c}\begin{bmatrix} 0 \\ 4 \\ 1 \end{bmatrix} \quad (c = \sqrt{17})$$

of W, we get

$$\text{Proj}_W\left(\begin{bmatrix} 5000 \\ 4000 \\ 2000 \end{bmatrix}\right) = \left(\begin{bmatrix} 5000 \\ 4000 \\ 2000 \end{bmatrix}, \begin{bmatrix} 1 \\ 0 \\ 0 \end{bmatrix}\right)\begin{bmatrix} 1 \\ 0 \\ 0 \end{bmatrix} + \left(\begin{bmatrix} 5000 \\ 4000 \\ 2000 \end{bmatrix}, \frac{1}{c}\begin{bmatrix} 0 \\ 4 \\ 1 \end{bmatrix}\right)\frac{1}{c}\begin{bmatrix} 0 \\ 4 \\ 1 \end{bmatrix}$$

$$= (5000)\begin{bmatrix} 1 \\ 0 \\ 0 \end{bmatrix} + \left(\frac{1}{c}16000 + \frac{1}{c}2000\right)\frac{1}{c}\begin{bmatrix} 0 \\ 4 \\ 1 \end{bmatrix}$$

$$= (5000)\begin{bmatrix} 1 \\ 0 \\ 0 \end{bmatrix} + \frac{18{,}000}{17}\begin{bmatrix} 0 \\ 4 \\ 1 \end{bmatrix}.$$

To get p and q amounts to expressing $\text{Proj}_W\left(\begin{bmatrix} 5000 \\ 4000 \\ 2000 \end{bmatrix}\right)$ as a linear combination,

$$\text{Proj}_W\left(\begin{bmatrix} 5000 \\ 4000 \\ 2000 \end{bmatrix}\right) = (5000)\begin{bmatrix} 1 \\ 0 \\ 0 \end{bmatrix} + \frac{18000}{17}\begin{bmatrix} 0 \\ 4 \\ 1 \end{bmatrix}$$

$$= \left(2500 - \left(\frac{3}{2}\right)\left(\frac{18000}{17}\right)\right)\begin{bmatrix} 2 \\ 0 \\ 0 \end{bmatrix} + \frac{18000}{17}\begin{bmatrix} 3 \\ 4 \\ 1 \end{bmatrix}$$

of $\begin{bmatrix} 2 \\ 0 \\ 0 \end{bmatrix}$ and $\begin{bmatrix} 3 \\ 4 \\ 1 \end{bmatrix}$. So we get $p = 2500 - \left(\frac{3}{2}\right)\left(\frac{18000}{17}\right) = 911.76$ and $q = \frac{18000}{17} = 1058.82$ and our approximate solution is $\begin{bmatrix} p \\ q \end{bmatrix} = \begin{bmatrix} 911.76 \\ 1058.82 \end{bmatrix}$. By making 911.76 units of P and 1058.82 units of Q, the vector representing supplies used is

$$911.76\begin{bmatrix} 2 \\ 0 \\ 0 \end{bmatrix} + 1058.82\begin{bmatrix} 3 \\ 4 \\ 1 \end{bmatrix} = \text{Proj}_W\left(\begin{bmatrix} 5000 \\ 4000 \\ 2000 \end{bmatrix}\right)$$

$$= (5000)\begin{bmatrix} 1 \\ 0 \\ 0 \end{bmatrix} + \frac{18000}{17}\begin{bmatrix} 0 \\ 4 \\ 1 \end{bmatrix}$$

$$= \begin{bmatrix} 5000 \\ 4235.29 \\ 1058.82 \end{bmatrix}.$$

So we use exactly 5000 units of S as desired, 4235.29 units of T (we need 235.29 units of T), and 1058.82 units of U (we have 941.18 units of U left over).

In the example, we found an approximate solution in the following sense.

Definition. For any $m \times n$ matrix A with real entries, an *approximate solution* x to an equation $Ax = y$ is a solution to the equation $Ax = \text{Proj}_{A(\mathbb{R}^{(n)})}(y)$.

In the example, W is the column space $A(\mathbb{R}^{(2)})$ of A and we found

$$\text{Proj}_{A(\mathbb{R}^{(n)})}\left(\begin{bmatrix} 5000 \\ 4000 \\ 2000 \end{bmatrix}\right)$$

by using the formula $\text{Proj}_{A(\mathbb{R}^{(2)})}(v) = (v, w_1)w_1 + (v, w_2)w_2$ of Theorem 12.1.2, where w_1, w_2 was an orthonormal basis for $A(\mathbb{R}^{(2)})$. We were then able to solve the equation $\begin{bmatrix} 2 & 3 \\ 0 & 4 \\ 0 & 1 \end{bmatrix} x = \text{Proj}_{A(\mathbb{R}^{(2)})}\left(\begin{bmatrix} 5000 \\ 4000 \\ 2000 \end{bmatrix}\right)$. Until we have a better method, we can use

Sec. 12.1]   Solutions of Systems of Linear Equations   441

this same method for any $m \times n$ matrix A with real entries:

1. Compute the projection $\text{Proj}_{A(\mathbb{R}^{(n)})}(y)$ of y onto the column space of A.
2. Find one solution $x = v$ to the equation $Ax = \text{Proj}_{A(\mathbb{R}^{(n)})}(y)$.
3. The set of approximate solutions x to $Ax = y$ is then the set $v + \text{Nullspace}(A)$ of solutions to the equation $Ax = \text{Proj}_{A(\mathbb{R}^{(n)})}(y)$.

How can we find the *shortest* approximate solution x to $Ax = y$? We use

Corollary 12.1.3. Let W be a subspace of $\mathbb{R}^{(n)}$ and let $y \in \mathbb{R}^{(n)}$. Then the element of $y + W = \{y + w \mid w \in W\}$ of shortest length is $y - \text{Proj}_W(y)$.

Proof: The length $\|y + w\|$ of $y + w$ is the distance from y to $-w$. To minimize this distance over all $w \in W$, we simply take $-w = \text{Proj}_W(y)$, by Theorem 12.1.1. Then $y + w = y - \text{Proj}_W(y)$. ∎

By this corollary, we get the shortest element of $v + \text{Nullspace}(A)$ by replacing v by $v - \text{Proj}_N(v) = \text{Proj}_{N^\perp}(v)$, where $N = \text{Nullspace}(A)$. Since $N^\perp$ is the column space of the transpose A' of A (Prove!), $\text{Proj}_{N^\perp}(v)$ is just the projection $\text{Proj}_{A'(\mathbb{R}^{(m)})}(v)$ of v on the column space of A'. So we can find the shortest approximate solution x to $Ax = y$ as follows:

1. Find one approximate solution v to the equation $Av = y$ by any method (e.g., by the one given above).
2. Replace the approximate solution v by $x = \text{Proj}_{A'(\mathbb{R}^{(m)})}(v)$.

EXAMPLE

Let's find the shortest approximate solution to

$$\begin{bmatrix} 2 & 3 & 5 \\ 0 & 4 & 4 \\ 0 & 1 & 1 \end{bmatrix} x = \begin{bmatrix} 5000 \\ 4000 \\ 2000 \end{bmatrix}.$$

Since the column space of $A = \begin{bmatrix} 2 & 3 & 5 \\ 0 & 4 & 4 \\ 0 & 1 & 1 \end{bmatrix}$ is the same as the column space

W of the matrix $\begin{bmatrix} 2 & 3 \\ 0 & 4 \\ 0 & 1 \end{bmatrix}$ in our earlier example, the approximate solution

$\begin{bmatrix} p \\ q \end{bmatrix} = \begin{bmatrix} 911.76 \\ 1058.82 \end{bmatrix}$ to the equation

$$\begin{bmatrix} 2 & 3 \\ 0 & 4 \\ 0 & 1 \end{bmatrix} \begin{bmatrix} p \\ q \end{bmatrix} = \begin{bmatrix} 5000 \\ 4000 \\ 2000 \end{bmatrix},$$

which satisfies the equation

$$\begin{bmatrix} 2 & 3 \\ 0 & 4 \\ 0 & 1 \end{bmatrix} \begin{bmatrix} p \\ q \end{bmatrix} = \text{Proj}_{A(\mathbb{R}^{(3)})} \left(\begin{bmatrix} 5000 \\ 4000 \\ 2000 \end{bmatrix} \right),$$

leads to the approximate solution

$$v = \begin{bmatrix} p \\ q \\ 0 \end{bmatrix} = \begin{bmatrix} 911.76 \\ 1058.82 \\ 0 \end{bmatrix}$$

to the equation $\begin{bmatrix} 2 & 3 & 5 \\ 0 & 4 & 4 \\ 0 & 1 & 1 \end{bmatrix} v = \begin{bmatrix} 5000 \\ 4000 \\ 2000 \end{bmatrix}$, since it satisfies the equation

$$\begin{bmatrix} 2 & 3 & 5 \\ 0 & 4 & 4 \\ 0 & 1 & 1 \end{bmatrix} x = \text{Proj}_{A(\mathbb{R}^{(3)})} \left(\begin{bmatrix} 5000 \\ 4000 \\ 2000 \end{bmatrix} \right).$$

So to get the shortest approximate solution, we replace v by $v - \text{Proj}_N(v)$, where $v = \begin{bmatrix} 911.76 \\ 1058.82 \\ 0 \end{bmatrix}$ and $N = \text{Nullspace} \left(\begin{bmatrix} 2 & 3 & 5 \\ 0 & 4 & 4 \\ 0 & 1 & 1 \end{bmatrix} \right)$. Since N is the set of vectors rw, where $w = \begin{bmatrix} -1 \\ -1 \\ 1 \end{bmatrix}$ (Verify!), we replace v by

$$v - \text{Proj}_N(v) = v - \frac{(v, w)}{(w, w)} w = \begin{bmatrix} 911.76 \\ 1058.82 \\ 0 \end{bmatrix} - \frac{-911.76 - 1058.82}{3} \begin{bmatrix} -1 \\ -1 \\ 1 \end{bmatrix}$$

$$= \begin{bmatrix} 911.76 \\ 1058.82 \\ 0 \end{bmatrix} - \begin{bmatrix} 656.86 \\ 656.86 \\ -656.86 \end{bmatrix} = \begin{bmatrix} 254.90 \\ 401.96 \\ 656.86 \end{bmatrix}.$$

The approximate solution $\begin{bmatrix} 911.76 \\ 1058.82 \\ 0 \end{bmatrix}$ has length 1397.29. Our new approximate solution $\begin{bmatrix} 254.90 \\ 401.96 \\ 656.86 \end{bmatrix}$ has length 811.18.

We've seen that there is a shortest approximate solution v to the equation $Ax = y$

and that it satisfies the two conditions

1. $Av = \text{Proj}_{A(\mathbb{R}^{(n)})}(y)$;
2. v is in the column space of A'.

Suppose that both u and v satisfy these two conditions. Then the vector $w = v - u$ is in the nullspace N of A, since

$$Aw = Av - Au = \text{Proj}_{A(\mathbb{R}^{(n)})}(y) - \text{Proj}_{A(\mathbb{R}^{(n)})}(y) = 0.$$

Moreover, $w = v - u$ is orthogonal to N, since u and v are in the column space of A'. Since w is in N, w is orthogonal to w, that is, w is 0. But then $u = v$. This proves

Theorem 12.1.4. Let A be an $m \times n$ matrix with real entries and let $y \in \mathbb{R}^{(m)}$. Then $Ax = y$ has one and only one shortest approximate solution x. Necessary and sufficient conditions that x be the shortest approximate solution to $Ax = y$ are that

1. $Ax = \text{Proj}_{A(\mathbb{R}^{(n)})}(y)$; and
2. x is in the column space of A'.

PROBLEMS

NUMERICAL PROBLEMS

1. Find all approximate solutions to $\begin{bmatrix} 2 & 1 & 5 \\ 0 & 2 & 2 \\ 0 & 0 & 0 \end{bmatrix} x = \begin{bmatrix} 2 \\ 0 \\ 2 \end{bmatrix}$.

2. Find the shortest approximate solution to $\begin{bmatrix} 2 & 1 & 5 \\ 0 & 2 & 2 \\ 0 & 0 & 0 \end{bmatrix} x = \begin{bmatrix} 2 \\ 0 \\ 2 \end{bmatrix}$.

3. Find the shortest approximate solution to $\begin{bmatrix} 1 & 1 \\ 0 & 2 \\ 1 & 1 \end{bmatrix} x = \begin{bmatrix} 2 \\ 1 \\ 0 \end{bmatrix}$.

MORE THEORETICAL PROBLEMS

Easier Problems

4. If $y' = \text{Proj}_W(y)$, show that $y - y' = \text{Proj}_{W^\perp}(y)$.
5. Using Problem 4, show that if W is the plane in $\mathbb{R}^{(3)}$ orthogonal to a nonzero vector $N \in \mathbb{R}^{(3)}$, then $\text{Proj}_W(y) = y - \dfrac{(y, N)}{(N, N)} N$.

12.2. THE APPROXIMATE INVERSE OF AN $m \times n$ MATRIX

By Corollary 7.2.7, any $m \times n$ matrix A such that the equation $Ax = y$ has one and only one solution $x \in \mathbb{R}^{(n)}$ for every $y \in \mathbb{R}^{(m)}$ is an invertible $n \times n$ matrix. What, then, should we be able to say now that we know from the preceding section that for *any* $m \times n$ matrix A with real entries the equation $Ax = y$ has one and only shortest approximate solution $x \in \mathbb{R}^{(n)}$ for every $y \in \mathbb{R}^{(m)}$? *We denote the shortest approximate solution to $Ax = y$ by $A^-(y)$ for all y in $\mathbb{R}^{(m)}$. By Theorem 12.1.4, this means that $A^-(y)$ is the unique vector $v \in \mathbb{R}^{(n)}$ such that*

1. $Av = \operatorname{Proj}_{A(\mathbb{R}^{(n)})}(y)$;
2. v is in the column space of A'.

We claim that the mapping A^- from $\mathbb{R}^{(m)}$ to $\mathbb{R}^{(n)}$ is a linear mapping. Letting $y, z \in \mathbb{R}^{(m)}$ and $c \in \mathbb{R}$, we have

1. $A(A^-(y) + A^-(z)) = AA^-(y) + AA^-(z)$
    $= \operatorname{Proj}_{A(\mathbb{R}^{(n)})}(y) + \operatorname{Proj}_{A(\mathbb{R}^{(n)})}(z)$
    $= \operatorname{Proj}_{A(\mathbb{R}^{(n)})}(y + z)$ (Prove!);
2. $A^-(y) + A^-(z)$ is in the column space of A', since $A^-(y)$ and $A^-(z)$ are.

It follows that $A^-(y) + A^-(z) = A^-(y + z)$. Similarly, we have

1. $A(cA^-(y)) = cAA^-(y) = c\operatorname{Proj}_{A(\mathbb{R}^{(n)})}(y) = \operatorname{Proj}_{A(\mathbb{R}^{(n)})}(cy)$ (Prove!);
2. $cA^-(y)$ is in the column space of A', since $A^-(y)$ is.

From this it follows that $A^-(cy) = cA^-(y)$. So the mapping A^- is linear. We have now proved

Theorem 12.2.1. The mapping A^- from $\mathbb{R}^{(m)}$ to $\mathbb{R}^{(n)}$ is linear.

By Theorems 12.2.1 and 7.1.1 we get that A^- is an $n \times m$ matrix.

Definition. For any $m \times n$ matrix A with real entries, we call the $n \times m$ matrix A^- the *approximate inverse* of A, since A^-y is the shortest approximate solution x to $Ax = y$ for all y.

The approximate inverse is also called the *pseudoinverse*. How do we find the $n \times m$ matrix A^-? Its columns are $A^-(e_1), \ldots, A^-(e_m)$, which we can compute as the shortest approximate solutions to the equations $Ax = e_1, \ldots, Ax = e_m$, where $e_1, \ldots, e_m$ is the standard basis for $\mathbb{R}^{(m)}$.

EXAMPLE

Let's find the approximate inverse A^- of the 3×2 matrix $A = \begin{bmatrix} 2 & 3 \\ 0 & 4 \\ 0 & 1 \end{bmatrix}$

and use it to find the shortest approximate solution

$$A^- \begin{bmatrix} 5000 \\ 4000 \\ 2000 \end{bmatrix}$$

Sec. 12.2]   The Approximate Inverse of an $m \times n$ Matrix   445

of

$$\begin{bmatrix} 2 & 3 \\ 0 & 4 \\ 0 & 1 \end{bmatrix} \begin{bmatrix} p \\ q \end{bmatrix} = \begin{bmatrix} 5000 \\ 4000 \\ 2000 \end{bmatrix}.$$

The projections of $\begin{bmatrix} 1 \\ 0 \\ 0 \end{bmatrix}, \begin{bmatrix} 0 \\ 1 \\ 0 \end{bmatrix}, \begin{bmatrix} 0 \\ 0 \\ 1 \end{bmatrix}$ on the column space of $\begin{bmatrix} 2 & 3 \\ 0 & 4 \\ 0 & 1 \end{bmatrix}$ are

$$\begin{bmatrix} 1 \\ 0 \\ 0 \end{bmatrix}, \begin{bmatrix} 0 \\ \frac{16}{17} \\ \frac{4}{17} \end{bmatrix}, \begin{bmatrix} 0 \\ \frac{4}{17} \\ \frac{1}{17} \end{bmatrix}$$

and the solutions to the equations

$$\begin{bmatrix} 2 & 3 \\ 0 & 4 \\ 0 & 1 \end{bmatrix} x = \begin{bmatrix} 1 \\ 0 \\ 0 \end{bmatrix}, \begin{bmatrix} 2 & 3 \\ 0 & 4 \\ 0 & 1 \end{bmatrix} x = \begin{bmatrix} 0 \\ \frac{16}{17} \\ \frac{4}{17} \end{bmatrix}, \begin{bmatrix} 2 & 3 \\ 0 & 4 \\ 0 & 1 \end{bmatrix} x = \begin{bmatrix} 0 \\ \frac{4}{17} \\ \frac{1}{17} \end{bmatrix}$$

are

$$\begin{bmatrix} \frac{1}{2} \\ 0 \end{bmatrix}, \begin{bmatrix} -\frac{6}{17} \\ \frac{4}{17} \end{bmatrix}, \begin{bmatrix} -\frac{3}{34} \\ \frac{1}{17} \end{bmatrix}.$$

Since column space of the transpose of $\begin{bmatrix} 2 & 3 \\ 0 & 4 \\ 0 & 1 \end{bmatrix}$ is $\mathbb{R}^{(2)}$, the projections of $\begin{bmatrix} \frac{1}{2} \\ 0 \end{bmatrix}$,

$\begin{bmatrix} -\frac{6}{17} \\ \frac{4}{17} \end{bmatrix}, \begin{bmatrix} -\frac{3}{34} \\ \frac{1}{17} \end{bmatrix}$ on the column space of the transpose of $\begin{bmatrix} 2 & 3 \\ 0 & 4 \\ 0 & 1 \end{bmatrix}$ are just

$\begin{bmatrix} \frac{1}{2} \\ 0 \end{bmatrix}, \begin{bmatrix} -\frac{6}{17} \\ \frac{4}{17} \end{bmatrix}, \begin{bmatrix} -\frac{3}{34} \\ \frac{1}{17} \end{bmatrix}$ themselves. So the approximate inverse of $\begin{bmatrix} 2 & 3 \\ 0 & 4 \\ 0 & 1 \end{bmatrix}$ is

$$\begin{bmatrix} \frac{1}{2} & -\frac{6}{17} & -\frac{3}{34} \\ 0 & \frac{4}{17} & \frac{1}{17} \end{bmatrix}$$

and the shortest approximate solution of $\begin{bmatrix} 2 & 3 \\ 0 & 4 \\ 0 & 1 \end{bmatrix} \begin{bmatrix} p \\ q \end{bmatrix} = \begin{bmatrix} 5000 \\ 4000 \\ 2000 \end{bmatrix}$ is

$$\begin{bmatrix} p \\ q \end{bmatrix} = \begin{bmatrix} \frac{1}{2} & -\frac{6}{17} & -\frac{3}{34} \\ 0 & \frac{4}{17} & \frac{1}{17} \end{bmatrix} \begin{bmatrix} 5000 \\ 4000 \\ 2000 \end{bmatrix} = \begin{bmatrix} 911.76 \\ 1058.82 \end{bmatrix},$$

the value that we got in the second example of Section 12.1.

We close this section by interrelating the matrices A, A^-, $\text{Proj}_{A(\mathbb{R}^{(n)})}$ [projection from $\mathbb{R}^{(m)}$ to $A(\mathbb{R}^{(n)})$, viewed as an $n \times m$ matrix] and $\text{Proj}_{A'(\mathbb{R}^{(m)})}$ [projection from $\mathbb{R}^{(n)}$ to $A'(\mathbb{R}^{(m)})$, viewed as an $m \times n$ matrix].

Theorem 12.2.2. $AA^- = \text{Proj}_{A(\mathbb{R}^{(n)})}$ and $A^-A = \text{Proj}_{A'(\mathbb{R}^{(m)})}$.

Proof: Since A^- maps y to the shortest x such that $Ax = \text{Proj}_{A(\mathbb{R}^{(n)})} y$, AA^- maps y to $\text{Proj}_{A(\mathbb{R}^{(n)})} y$. And since A maps x to y whereupon A^- maps y to $\text{Proj}_{A'(\mathbb{R}^{(m)})} x$, A^-A maps x to $\text{Proj}_{A'(\mathbb{R}^{(m)})} x$. ∎

PROBLEMS

NUMERICAL PROBLEMS

1. Find the approximate inverse of $A = \begin{bmatrix} 1 \\ 0 \\ 1 \end{bmatrix}$ and use it to find approximate solutions to the equations $Ax = \begin{bmatrix} 1 \\ 1 \\ 1 \end{bmatrix}$ and $Ax = \begin{bmatrix} 1 \\ 2 \\ 1 \end{bmatrix}$.

2. Find the approximate inverse of $A = \begin{bmatrix} 1 & 1 \\ 0 & 2 \\ 1 & 1 \end{bmatrix}$ and use it to find an approximate solution to $Ax = \begin{bmatrix} 2 \\ 1 \\ 1 \end{bmatrix}$.

3. Find the approximate inverse of $A = \begin{bmatrix} 1 & 0 & 1 \\ 1 & 2 & 1 \end{bmatrix}$ and use it to find an approximate solution to $Ax = \begin{bmatrix} 1 \\ 3 \end{bmatrix}$.

MORE THEORETICAL PROBLEMS

4. Prove that $\text{Proj}_W(y)$ is linear in y, that is $\text{Proj}_W(y + z) = \text{Proj}_W(y) + \text{Proj}_W(z)$ and $\text{Proj}_W(cy) = c\text{Proj}_W(y)$.

12.3. SOLVING A MATRIX EQUATION USING ITS NORMAL EQUATION

Up to now, we have found approximate solutions to $Ax = y$ and computed the approximate inverse of A directly from the definitions. Are there better methods? Fortunately, we can find the approximate solutions x to $Ax = y$ by finding the solutions x to the corresponding *normal equation* $A'Ax = A'y$. In most applications, finding the solutions to $A'Ax = A'y$ is easier than finding the approximate solutions to

$Ax = y$ directly. One reason for this is that $A'A$ is an $n \times n$ matrix when A is an $m \times n$ matrix. So if $n < m$, which is true in most applications, $A'A$ being $n \times n$ is of smaller size than A, which is $m \times n$. Another is that $A'A$ is symmetric, so that it is similar to a diagonal matrix.

Why can we find the approximate solutions x to $Ax = y$ by finding the solutions x to the corresponding normal equation $A'Ax = A'y$? The condition $Ax = \text{Proj}_{A\mathbb{R}^{(n)}}(y)$ on the element Ax of the column space of A is that $y - Ax$ be orthogonal to the column space of A, that is, that $A'(y - Ax) = 0$. But this is just the condition that x be a solution of the equation $A'Ax = A'y$. To require further that x be the shortest approximate solution to $Ax = y$ is, by Theorem 12.1.4, equivalent to requiring that x be in the column space of A'. This proves

Theorem 12.3.1. Let $y \in \mathbb{R}^{(m)}$. Then

1. The approximate solutions x to $Ax = y$ are just the solutions x to the corresponding normal equation $A'Ax = A'y$; and
2. $x = A^- y$ if and only if $A'Ax = A'y$ and x is in the column space of A'.

Before we discuss the general case further, let's first look at the special case where the columns of A are linearly independent. We need

Theorem 12.3.2. If the columns of A are linearly independent, then the matrix $A'A$ is invertible.

Proof: Since $A'A$ is a square matrix, it is invertible if and only if its nullspace is 0. So letting x be any vector such that $A'Ax = 0$, it suffices to show that $x = 0$. Multiplying by x', we have $0 = x'A'Ax = (Ax)'(Ax)$. This implies that the length of Ax is 0, so that Ax is 0. Since the columns of A are linearly independent, it follows that x is 0. (Prove!). ∎

EXAMPLE

Since the columns of $A = \begin{bmatrix} 2 & 3 \\ 0 & 4 \\ 0 & 1 \end{bmatrix}$ are linearly independent, the matrix $A'A = \begin{bmatrix} 2 & 0 & 0 \\ 3 & 4 & 1 \end{bmatrix} \begin{bmatrix} 2 & 3 \\ 0 & 4 \\ 0 & 1 \end{bmatrix} = \begin{bmatrix} 4 & 6 \\ 6 & 26 \end{bmatrix}$ is invertible, with inverse $(\frac{1}{34}) \begin{bmatrix} 13 & -3 \\ -3 & 2 \end{bmatrix}$.

If the columns of A are linearly independent, we know from Theorem 12.3.2 that $A'A$ is invertible. Since, by Theorem 12.3.1, x is an approximate solution of $Ax = y$ if and only if x is a solution of $A'Ax = A'y$, it follows that x is an approximate solution of $Ax = y$ if and only if $x = (A'A)^{-1}A'y$, which proves

Corollary 12.3.3. Suppose that the columns of A are linearly independent. Then for any y in the $\mathbb{R}^{(m)}$, there is one and only one approximate solution x to $Ax = y$, namely $x = (A'A)^{-1}A'y$.

Theorem 12.3.4. If the columns of A are linearly independent, then the approximate inverse of A is $A^- = (A'A)^{-1}A'$.

EXAMPLE

Since the columns of $A = \begin{bmatrix} 2 & 3 \\ 0 & 4 \\ 0 & 1 \end{bmatrix}$ are linearly independent, the approximate inverse of A is

$$A^- = \left(\begin{bmatrix} 2 & 0 & 0 \\ 3 & 4 & 1 \end{bmatrix}\begin{bmatrix} 2 & 3 \\ 0 & 4 \\ 0 & 1 \end{bmatrix}\right)^{-1}\begin{bmatrix} 2 & 0 & 0 \\ 3 & 4 & 1 \end{bmatrix} = \begin{bmatrix} 4 & 6 \\ 6 & 26 \end{bmatrix}^{-1}\begin{bmatrix} 2 & 0 & 0 \\ 3 & 4 & 1 \end{bmatrix}$$

$$= (\tfrac{1}{34})\begin{bmatrix} 13 & -3 \\ -3 & 2 \end{bmatrix}\begin{bmatrix} 2 & 0 & 0 \\ 3 & 4 & 1 \end{bmatrix} = \begin{bmatrix} \tfrac{1}{2} & -\tfrac{6}{17} & -\tfrac{3}{34} \\ 0 & \tfrac{4}{17} & \tfrac{1}{17} \end{bmatrix},$$

which agrees with our calculation of A^- in the corresponding example of Section 12.2.

It is very useful to have the explicit formula $A^- = (A'A)^{-1}A'$ for the approximate inverse of A in the case that the columns of A are linearly independent. Is there such a formula in general? Yes! The shortest approximate solution x to $Ax = y$ is the shortest solution (and therefore shortest approximate solution) x to the normal equation $(A'A)x = A'y$. This x is just $x = (A'A)^-(A'y)$, where $(A'A)^-$ is the approximate inverse of $A'A$. This proves

Theorem 12.3.5. The approximate inverse of any $m \times n$ matrix A with real entries is $A^- = (A'A)^- A'$.

Of course, to make this theorem work for us, we need to find $(A'A)^-$. This is now easier, since $A'A$ is a *symmetric* matrix and, in most applications, $A'A$ is much smaller than A. One method to find the approximate inverse of a symmetric matrix B, such as $A'A$, is to compute the approximate inverse of its matrix $Q'BQ$ in a different orthonormal basis Q and use

Theorem 12.3.6. Let A be a real $m \times n$ matrix, let P be a real unitary $m \times m$ matrix, and let Q be a real unitary $n \times n$ matrix. Then $(PAQ')^- = QA^-P'$.

Proof: The following conditions on $x \in \mathbb{R}^{(n)}$ and $y \in \mathbb{R}^{(m)}$ are equivalent, since each one is equivalent to the next:

1. $Qx = (PAQ')^- Py$;
2. Qx is the shortest approximate solution to $PAQ'(Qx) = Py$;
3. Qx is the shortest solution to $(PAQ')'PAQ'(Qx) = (PAQ')'Py$;
4. Qx is the shortest solution to $QA'P'PAQ'Qx = QA'P'Py$;
5. x is the shortest solution to $QA'Ax = QA'y$;

Sec. 12.3]   Solving a Matrix Equation   449

6. x is the shortest solution to $A'Ax = A'y$;
7. x is the shortest approximate solution to $Ax = y$;
8. $x = A^- y$;
9. $Qx = QA^- P'Py$.

Since $(PAQ')^-$ and $QA^- P'$ have the same effect on all vectors Py, they are equal. ∎

We can now compute the approximate inverse of a real symmetric $n \times n$ matrix A by taking a real unitary matrix Q such that $Q'AQ$ is a diagonal matrix D, by Section 4.6, and using

Theorem 12.3.7. Let A be a symmetric $n \times n$ matrix with real entries and let Q be a unitary $n \times n$ matrix with real entries such that $Q'AQ = D$ is a diagonal matrix. Then the approximate inverse of A is $A^- = QD^-Q'$, where D^- is the diagonal matrix whose (r, r) entry is d_{rr}^-, where d_{rr}^- is d_{rr}^{-1} if $d_{rr} \neq 0$ and 0 if $d_{rr} = 0$ for $1 \leq r \leq n$.

Proof: Using Theorem 12.3.6, we have $A^- = (QDQ')^- = QD^-Q'$, where D^- is the approximate inverse of D. So it remains only to show that the approximate inverse D^- of a diagonal matrix D is the diagonal matrix whose (r, r) entry is d_{rr}^{-1} if d_{rr} is nonzero and 0 if $d_{rr} = 0$ for $1 \leq r \leq n$. Why is this so? The shortest solution $\begin{bmatrix} x_1 \\ \vdots \\ x_n \end{bmatrix}$ to the equation

$$\begin{bmatrix} d_{11} & & 0 \\ & \ddots & \\ 0 & & d_{nn} \end{bmatrix} \begin{bmatrix} x_1 \\ \vdots \\ x_n \end{bmatrix} = \begin{bmatrix} d_{11}x_1 \\ \vdots \\ d_{nn}x_n \end{bmatrix} = \text{Proj}_{D\mathbb{R}^{(n)}} \begin{bmatrix} y_1 \\ \vdots \\ y_n \end{bmatrix}$$

is

$$\begin{bmatrix} x_1 \\ \vdots \\ x_n \end{bmatrix} = \begin{bmatrix} d_{11}^- & & 0 \\ & \ddots & \\ 0 & & d_{nn}^- \end{bmatrix} \begin{bmatrix} y_1 \\ \vdots \\ y_n \end{bmatrix} = \begin{bmatrix} d_{11}^- y_1 \\ \vdots \\ d_{nn}^- y_n \end{bmatrix},$$

since $\text{Proj}_{D\mathbb{R}^{(n)}} \begin{bmatrix} y_1 \\ \vdots \\ y_n \end{bmatrix}$ is obtained from $\begin{bmatrix} y_1 \\ \vdots \\ y_n \end{bmatrix}$ by replacing y_r by 0 if entry d_{rr} of D is 0 and since we want x_r to be 0 if d_{rr} is 0 to make x as short as possible. ∎

Since the approximate inverse of any $m \times n$ matrix A with real entries is $A^- = (A'A)^- A'$, our theorem proves

Corollary 12.3.8. Let A be an $m \times n$ matrix with real entries and let Q be a unitary matrix with real entries such that $Q'A'AQ = D$ is a diagonal matrix. Then $A^- = QD^-Q'A'$.

PROBLEMS

NUMERICAL PROBLEMS

1. Using the formula $A^- = (A'A)^{-1}A'$, find the approximate inverse of $A = \begin{bmatrix} 1 \\ 0 \\ 1 \end{bmatrix}$ and compare with your answer to Problem 1 of Section 12.2.

2. Using the formula $A^- = (A'A)^{-1}A'$, find the approximate inverse of $A = \begin{bmatrix} 1 & 1 \\ 0 & 2 \\ 1 & 1 \end{bmatrix}$ and compare with your answer to Problem 2 of Section 12.2.

3. Using the formula $A^- = (A'A)^-A'$, find the approximate inverse of $A = \begin{bmatrix} 1 & 2 \\ 0 & 0 \\ 1 & 2 \end{bmatrix}$. Then calculate the projection $AA^- = A(A'A)^-A'$ onto the column space of A. Finally, use it to find the approximate solutions to the equations $Ax = \begin{bmatrix} 1 \\ 1 \\ 1 \end{bmatrix}$ and $Ax = \begin{bmatrix} 2 \\ 1 \\ 1 \end{bmatrix}$.

4. Verify that the matrix $\begin{bmatrix} 1 & 0 & 0 \\ 0 & \frac{16}{17} & \frac{4}{17} \\ 0 & \frac{4}{17} & \frac{1}{17} \end{bmatrix}$, which we calculated as the projection onto the column space of a matrix, does actually coincide with its square.

MORE THEORETICAL PROBLEMS

Easier Problems

5. Viewing real numbers d as 1×1 matrices, show directly that the approximate inverse of a real number d is d^-, where d^- is $1/d$ if d is nonzero and d^- is 0 if d is 0.

Middle-Level Problems

6. Show that $U \operatorname{Proj}_{A(\mathbb{R}^{(n)})}(y) = \operatorname{Proj}_{UA(\mathbb{R}^{(n)})}(Uy)$ for any real $m \times n$ matrix A and real $m \times m$ unitary matrix U.

12.4. FINDING FUNCTIONS THAT APPROXIMATE DATA

In an experiment having input variable x and an output variable y, we get output values $y_0, \ldots, y_m$ corresponding to input values $x_0, \ldots, x_m$ from data generated in the experiment. We then seek to find a useful *functional approximation* to the mapping $y_r = f(x_r)$ $(0 \leq r \leq m)$, that is, for instance, we want to find a function such as

Sec. 12.4] Finding Functions that Approximate Data 451

$y = ax^2 + bx + c$ such that y_r and $ax_r^2 + bx_r + c$ are equal or nearly equal for $0 \le r \le m$. If we seek a functional approximation *of order n*, that is, a function of the form

$$y = p(x) = c_0 + c_1 x + \cdots + c_n x^n,$$

how do we choose the coefficients c_r? We write down the equations

$$c_0 + c_1 x_0 + \cdots + c_n x_0^n = y_0$$
$$\vdots \qquad \qquad \vdots$$
$$c_0 + c_1 x_m + \cdots + c_n x_m^n = y_m,$$

viewing the x_r^s as the entries of the coefficient matrix A and the c_r as the unknowns. Then we find the approximate solution $\begin{bmatrix} c_0 \\ \vdots \\ c_n \end{bmatrix}$ to

$$\begin{bmatrix} 1 & x_0 & \cdots & x_0^n \\ \vdots & \vdots & & \vdots \\ 1 & x_m & \cdots & x_m^n \end{bmatrix} \begin{bmatrix} c_0 \\ \vdots \\ c_n \end{bmatrix} = \begin{bmatrix} y_0 \\ \vdots \\ y_m \end{bmatrix},$$

for example by calculating the approximate inverse A^- of $A = \begin{bmatrix} 1 & x_0 & \cdots & x_0^n \\ \vdots & \vdots & & \vdots \\ 1 & x_m & \cdots & x_m^n \end{bmatrix}$

and letting $\begin{bmatrix} c_0 \\ \vdots \\ c_n \end{bmatrix} = A^- \begin{bmatrix} y_0 \\ \vdots \\ y_m \end{bmatrix}.$

The functional approximation $y = p(x) = c_0 + c_1 x + \cdots + c_n x^n$ for the mapping $y_r = f(x_r)$ that we obtain in this way is the polynomial $p(x)$ of degree n for which the sum of squares $(y_0 - p(x_0))^2 + \cdots + (y_m - p(x_m))^2$ is as small as possible. This method of finding a functional approximation is often called *the method of least squares*.

EXAMPLE

In a time-study experiment that we conduct to find a functional relationship between the duration x of the coffee break (in minutes) and the value y (in thousands of dollars) of the work performed the same day by a group of employees, coffee breaks of $x_0 = 10$, $x_1 = 15$, $x_2 = 21$, $x_3 = 5$ minutes duration and values $y_0 = 10$, $y_1 = 14$, $y_2 = 13$, $y_3 = 10$ of thousands of dollars' worth of work performed were observed on the four successive days of the experiment. We decide to analyze the data two ways, namely, to use it to get a first-order approximation, and then to use it to get a second-order approximation:

1. To get the first-order approximation, we first get the general approximate solution to $\begin{bmatrix} 1 & x_0 \\ \vdots & \vdots \\ 1 & x_m \end{bmatrix} \begin{bmatrix} c_0 \\ c_1 \end{bmatrix} = \begin{bmatrix} y_0 \\ \vdots \\ y_m \end{bmatrix}$. This is $\begin{bmatrix} c_0 \\ c_1 \end{bmatrix} = A^- \begin{bmatrix} y_0 \\ \vdots \\ y_m \end{bmatrix} = (A'A)^{-1} A' \begin{bmatrix} y_0 \\ \vdots \\ y_m \end{bmatrix},$

where $A = \begin{bmatrix} 1 & x_0 \\ \vdots & \vdots \\ 1 & x_m \end{bmatrix}$, that is,

$$\begin{bmatrix} c_0 \\ c_1 \end{bmatrix} = \left(\begin{bmatrix} 1 & \cdots & 1 \\ x_0 & \cdots & x_m \end{bmatrix} \begin{bmatrix} 1 & x_0 \\ \vdots & \vdots \\ 1 & x_m \end{bmatrix} \right)^{-1} \begin{bmatrix} 1 & \cdots & 1 \\ x_0 & \cdots & x_m \end{bmatrix} \begin{bmatrix} y_0 \\ \vdots \\ y_m \end{bmatrix}$$

$$= \begin{bmatrix} m+1 & \mathbf{1} \cdot x \\ x \cdot \mathbf{1} & x \cdot x \end{bmatrix}^{-1} \begin{bmatrix} \mathbf{1} \cdot y \\ x \cdot y \end{bmatrix}$$

or

$$\begin{bmatrix} c_0 \\ c_1 \end{bmatrix} = \begin{bmatrix} m+1 & \mathbf{1} \cdot x \\ x \cdot \mathbf{1} & x \cdot x \end{bmatrix}^{-1} \begin{bmatrix} \mathbf{1} \cdot y \\ x \cdot y \end{bmatrix},$$

where $\mathbf{1} = \begin{bmatrix} 1 \\ \vdots \\ 1 \end{bmatrix}$ and $u \cdot v$ denotes (u, v). In the time study, m is 3 and we have

$$x \cdot \mathbf{1} = 10 + 15 + 21 + 5 = 51$$
$$x \cdot x = 100 + 225 + 441 + 25 = 791$$
$$\mathbf{1} \cdot y = 10 + 14 + 13 + 10 = 47$$
$$x \cdot y = 100 + 210 + 273 + 50 = 633.$$

So

$$\begin{bmatrix} c_0 \\ c_1 \end{bmatrix} = \begin{bmatrix} m+1 & \mathbf{1} \cdot x \\ x \cdot \mathbf{1} & x \cdot x \end{bmatrix}^{-1} \begin{bmatrix} \mathbf{1} \cdot y \\ x \cdot y \end{bmatrix} = \begin{bmatrix} c_0 \\ c_1 \end{bmatrix} = \begin{bmatrix} 4 & 51 \\ 51 & 791 \end{bmatrix}^{-1} \begin{bmatrix} 47 \\ 633 \end{bmatrix}$$

$$= \left(\frac{1}{563}\right) \begin{bmatrix} 791 & -51 \\ -51 & 4 \end{bmatrix} \begin{bmatrix} 47 \\ 633 \end{bmatrix} = \begin{bmatrix} 8.69 \\ .24 \end{bmatrix}.$$

So the linear function $y = 8.69 + .24x$ is the approximation of first order. Let's see how well it approximates:

x	10	15	21	5
actual y	10	14	13	10
approximating $y = 8.69 + .24x$	11.57	12.29	13.73	9.89

2. For the second-order approximation, the general solution to

$$\begin{bmatrix} 1 & x_0 & x_0^2 \\ \vdots & \vdots & \vdots \\ 1 & x_m & x_m^2 \end{bmatrix} \begin{bmatrix} c_0 \\ c_1 \\ c_2 \end{bmatrix} = \begin{bmatrix} y_0 \\ \vdots \\ y_m \end{bmatrix}$$

Sec. 12.4]   Finding Functions that Approximate Data   453

is $\begin{bmatrix} c_0 \\ c_1 \\ c_2 \end{bmatrix} = A^- \begin{bmatrix} y_0 \\ \vdots \\ y_m \end{bmatrix} = (A'A)^{-1}A' \begin{bmatrix} y_0 \\ \vdots \\ y_m \end{bmatrix}$, where $A = \begin{bmatrix} 1 & x_0 & x_0^2 \\ \vdots & \vdots & \vdots \\ 1 & x_m & x_m^2 \end{bmatrix}$,

that is,

$$\begin{bmatrix} c_0 \\ c_1 \\ c_2 \end{bmatrix} = \left(\begin{bmatrix} 1 & \cdots & 1 \\ x_0 & \cdots & x_m \\ x_0^2 & \cdots & x_m^2 \end{bmatrix} \begin{bmatrix} 1 & x_0 & x_0^2 \\ \vdots & \vdots & \vdots \\ 1 & x_m & x_m^2 \end{bmatrix} \right)^{-1} \begin{bmatrix} 1 & \cdots & 1 \\ x_1 & \cdots & x_m \\ x_1^2 & \cdots & x_m^2 \end{bmatrix} \begin{bmatrix} y_0 \\ \vdots \\ y_m \end{bmatrix}$$

$$= \begin{bmatrix} m+1 & \mathbf{1} \cdot x & \mathbf{1} \cdot x^2 \\ x \cdot \mathbf{1} & x \cdot x & x \cdot x^2 \\ x^2 \cdot \mathbf{1} & x^2 \cdot x & x^2 \cdot x^2 \end{bmatrix}^{-1} \begin{bmatrix} \mathbf{1} \cdot y \\ x \cdot y \\ x^2 \cdot y \end{bmatrix}$$

or

$$\begin{bmatrix} c_0 \\ c_1 \\ c_2 \end{bmatrix} = \begin{bmatrix} m+1 & \mathbf{1} \cdot x & \mathbf{1} \cdot x^2 \\ x \cdot \mathbf{1} & x \cdot x & x \cdot x^2 \\ x^2 \cdot \mathbf{1} & x^2 \cdot x & x^2 \cdot x^2 \end{bmatrix}^{-1} \begin{bmatrix} \mathbf{1} \cdot y \\ x \cdot y \\ x^2 \cdot y \end{bmatrix},$$

where $\mathbf{1} = \begin{bmatrix} 1 \\ \vdots \\ 1 \end{bmatrix}$. So

$$\begin{bmatrix} c_0 \\ c_1 \\ c_2 \end{bmatrix} = \begin{bmatrix} 4 & 51 & 791 \\ 51 & 791 & 13761 \\ 791 & 13761 & 255231 \end{bmatrix}^{-1} \begin{bmatrix} 47 \\ 633 \\ 10133 \end{bmatrix}.$$

For these c_0, c_1, c_2, the approximation of second order is the quadratic polynomial $y = c_0 + c_1 x + c_2 x^2$.

In order to use the formula $A^- = (A'A)^{-1}A'$ for the approximate inverse of $A = \begin{bmatrix} 1 & x_0 & \cdots & x_0^n \\ \vdots & \vdots & & \vdots \\ 1 & x_m & \cdots & x_m^n \end{bmatrix}$ in finding the approximating function of order n for the mapping $y_r = f(x_r)$ ($1 \leq r \leq m$), we need to know that the columns of A are linearly independent. Assuming that the x_r are all different, this is true if $m \geq n$. Why? For $m \geq n$, A has an invertible $n \times n$ submatrix by the following theorem.

Theorem 12.4.1. The matrix $A = \begin{bmatrix} 1 & x_0 & \cdots & x_0^n \\ \vdots & \vdots & & \vdots \\ 1 & x_n & \cdots & x_n^n \end{bmatrix}$ is invertible if and only the $x_0, \ldots, x_n$ are all different.

Proof: If two of the x_r are the same, then A has two identical rows, so it is not invertible. Suppose, conversely, that A is not invertible. Then the system of equations

$$c_0 + c_1 x_0 + \cdots + c_n x_0^n = 0$$
$$\vdots \qquad \vdots$$
$$c_0 + c_1 x_n + \cdots + c_n x_n^n = 0$$

has a nonzero solution $\begin{bmatrix} c_0 \\ \vdots \\ c_n \end{bmatrix}$; that is, there is a nonzero polynomial $p(x) = c_0 + c_1 x + \cdots + c_n x^n$ of degree n which vanishes at all of the $n + 1$ numbers $x_0, \ldots, x_n$. Since a polynomial $p(x)$ of degree n has at most n roots, and since $x_0, \ldots, x_n$ are $n + 1$ in number, two of them are equal. ∎

PROBLEMS

NUMERICAL PROBLEMS

1. Find an approximating function of order 1 for the function that maps 2 to 4, 3 to 5, and 4 to 6.

2. Find an approximating function of order 1 for the function that maps 2 to 5, 3 to 10, and 4 to 17.

3. Give an expression for an approximating function of order 2 for the function that maps 2 to 5, 3 to 10, and 4 to 17.

MORE THEORETICAL PROBLEMS

Easier Problems

4. Using the formula $A^- = (A'A)^{-1} A'$ for the approximate inverse of A, give the formula for $A^- \begin{bmatrix} y_0 \\ \vdots \\ y_m \end{bmatrix}$ for $A = \begin{bmatrix} 1 & x_0 & \cdots & x_0^n \\ \vdots & \vdots & & \vdots \\ 1 & x_m & \cdots & x_m^n \end{bmatrix}$ that generalizes the formula

$$A^- \begin{bmatrix} y_0 \\ \vdots \\ y_m \end{bmatrix} = \begin{bmatrix} m+1 & \mathbf{1} \cdot x & \mathbf{1} \cdot x^2 \\ x \cdot \mathbf{1} & x \cdot x & x \cdot x^2 \\ x^2 \cdot \mathbf{1} & x^2 \cdot x & x^2 \cdot x^2 \end{bmatrix}^{-1} \begin{bmatrix} \mathbf{1} \cdot y \\ x \cdot y \\ x^2 \cdot y \end{bmatrix}$$

for $A = \begin{bmatrix} 1 & x_0 & x_0^2 \\ \vdots & \vdots & \vdots \\ 1 & x_m & x_m^2 \end{bmatrix}$, and show that $x^r \cdot x^s = \mathbf{1} \cdot x^{r+s}$ for all r, s.

5. Show that if $f(x)$ is a polynomial of degree n, $x_0, \ldots, x_n$ are all different and $y_r = f(x_r)$ for $0 \leq r \leq n$, then the approximating function of order n is just the polynomial $y = f(x)$ itself.

12.5. WEIGHTED APPROXIMATION

Sections 12.1 through 12.4 are concerned with finding approximate solutions to equations $Ax = y$ and fitting functions to data that minimize distance and length. Since it is the distance $\|Ax - y\|$ from Ax to y, and then the length $\|x\|$ of x, that we minimize, we now ask:

What answer do we get if we weight things differently by using, instead, distances and lengths defined as $\|Ax - y\|_P = \|P(Ax - y)\|$ and $\|x\|_Q = \|Qx\|$ in terms of invertible $P \in M_m(\mathbb{R})$ and $Q \in M_n(\mathbb{R})$?

EXAMPLE

In Section 12.1 we found the shortest approximate solution $x = \begin{bmatrix} 254.90 \\ 401.96 \\ 656.86 \end{bmatrix}$ (vector of materials needed) to an equation $Ax = y$, where $A = \begin{bmatrix} 2 & 3 & 5 \\ 0 & 4 & 4 \\ 0 & 1 & 1 \end{bmatrix}$ (matrix of units of materials needed to manufacture 1 unit of product) and $y = \begin{bmatrix} 5000 \\ 4000 \\ 2000 \end{bmatrix}$ (vector of products to be manufactured). Suppose that a careful analysis of cost of materials shows that we should minimize $\|x\|_Q = \|Qx\|$, rather than $\|x\|$, where $Q = \begin{bmatrix} 1 & 1 & 0 \\ 0 & 1 & 0 \\ 0 & 0 & 2 \end{bmatrix}$. The entries of Q reflect cost of the materials, taking into account cost-related interdependencies among them. Suppose also that an analysis of costs associated with manufacturing too many or two few products shows that we should minimize $\|Ax - y\|_P = \|P(Ax - y)\|$, rather than $\|Ax - y\|$, where $P = \begin{bmatrix} 1 & 0 & 0 \\ 0 & 1 & 0 \\ 0 & 0 & 3 \end{bmatrix}$. What do we then come up with for x?

To answer our question, we assume that we have been given invertible matrices $P \in M_m(\mathbb{R})$ and $Q \in M_n(\mathbb{R})$, which we refer to as *weighting matrices*. We then use the standard inner products and lengths in $\mathbb{R}^{(m)}$ and $\mathbb{R}^{(n)}$ to define new ones in terms of P and Q:

Definition. Let $y, z \in \mathbb{R}^{(m)}$ and $w, x \in \mathbb{R}^{(n)}$. Then $\langle y, z \rangle_P = (Py, Pz)$, $\|y\|_P = \|Py\|$, $\langle w, x \rangle_Q = (Qw, Qx)$ and $\|x\|_Q = \|Qx\|$.

Regarding an $m \times n$ matrix A as a linear transformation from the inner product space $\mathbb{R}^{(n)}$ with inner product $\langle w, x \rangle_Q$ to the inner product space $\mathbb{R}^{(m)}$ with inner product $\langle y, z \rangle_p$, the counterpart of our definition of approximate inverse is

Definition. The *weighted approximate inverse* $A^\sim$ of A with respect to the weighting matrices $P \in M_m(\mathbb{R})$ and $Q \in M_n(\mathbb{R})$ is the function from $\mathbb{R}^{(m)}$ to $\mathbb{R}^{(n)}$ such that for each $y \in \mathbb{R}^{(m)}$, $A^\sim y = x$ if and only if the following conditions are satisfied:

1. $\|Ax - y\|_P$ is minimal (as small as possible).
2. For all x such that (1) is true, $\|x\|_Q$ is minimal.

Then we have

Theorem 12.5.1. The weighted approximate inverse of A with respect to weighting matrices $P \in M_m(\mathbb{R})$ and $Q \in M_n(\mathbb{R})$ is $A^\sim = Q^{-1}(PAQ^{-1})^- P$, where $(PAQ^{-1})^-$ is the approximate inverse of PAQ^{-1}.

Proof: Each of the following conditions is equivalent to the next:

1. $A^\sim y = x$;
2. $\|Ax - y\|_P$ is minimal and, subject to this, $\|x\|_Q$ is minimal;
3. $\|PAQ^{-1}Qx - Py\|$ is minimal and, subject to this, $\|Qx\|$ is minimal;
4. $(PAQ^{-1})^- Py = Qx$;
5. $Q^{-1}(PAQ^{-1})^- Py = x$.

So (1) and (5) are equivalent and we see that $A^\sim = Q^{-1}(PAQ^{-1})^- P$. ∎

Assume for the moment that the columns of A are linearly independent. Then $A'A$ is invertible and we can easily calculate $A^- = (A'A)^{-1}A'$. Similarly, the matrix $(PAQ^{-1})'PAQ^{-1}$ is invertible, and we can calculate

$$(PAQ^{-1})^- = (Q'^{-1}A'P'PAQ^{-1})^{-1}Q'^{-1}A'P' = Q(A'P'PA)^{-1}Q'Q'^{-1}A'P'$$
$$= Q(A'P'PA)^{-1}A'P'.$$

To find $A^\sim$, we use the formula $A^\sim = Q^{-1}(PAQ^{-1})^- P$ of Theorem 11.9.1 and calculate

$$A^\sim = Q^{-1}(PAQ^{-1})^- P = Q^{-1}Q(A'P'PA)^{-1}A'P'P$$
$$= (A'P'PA)^{-1}A'P'P.$$

This gives us

Corollary 12.5.2. If the columns of an $m \times n$ matrix A are linearly independent, then the weighted approximate inverse of A corresponding to weighting matrices $P \in \mathbb{R}^{(m)}$ and $Q \in \mathbb{R}^{(n)}$ is $A^\sim = (A'P'PA)^{-1}A'P'P$.

Why doesn't Q appear in the formula for $A^\sim$ in the corollary? Since the columns of A are linearly independent, $\|Ax - y\|_P$ is minimal for only one x, so that the condition that $\|x\|_Q$ be minimal never has a chance to be used.

When the columns of A are linearly dependent, we can find $A^\sim = Q^{-1}(PAQ^{-1})^- P$ by first finding $(PAQ^{-1})^-$ using the methods of Section 12.3.

EXAMPLE

In our example of $A = \begin{bmatrix} 2 & 3 & 5 \\ 0 & 4 & 4 \\ 0 & 1 & 1 \end{bmatrix}$, $y = \begin{bmatrix} 5000 \\ 4000 \\ 2000 \end{bmatrix}$, and weighting matrices $P = \begin{bmatrix} 1 & 0 & 0 \\ 0 & 1 & 0 \\ 0 & 0 & 3 \end{bmatrix}$ and $Q = \begin{bmatrix} 1 & 1 & 0 \\ 0 & 1 & 0 \\ 0 & 0 & 2 \end{bmatrix}$, we solve $Ax = y$ for x, taking the weightings into account, by letting $x = A\tilde{\ }y$. To get $A\tilde{\ }y$, we first form the matrix

$$PAQ^{-1} = \begin{bmatrix} 1 & 0 & 0 \\ 0 & 1 & 0 \\ 0 & 0 & 3 \end{bmatrix} \begin{bmatrix} 2 & 3 & 5 \\ 0 & 4 & 4 \\ 0 & 1 & 1 \end{bmatrix} \begin{bmatrix} 1 & -1 & 0 \\ 0 & 1 & 0 \\ 0 & 0 & \frac{1}{2} \end{bmatrix} = \begin{bmatrix} 2 & 1 & \frac{5}{2} \\ 0 & 4 & 2 \\ 0 & 3 & \frac{3}{2} \end{bmatrix}$$

and find its approximate inverse $\begin{bmatrix} 2 & 1 & \frac{5}{2} \\ 0 & 4 & 2 \\ 0 & 3 & \frac{3}{2} \end{bmatrix}^-$. Then

$$A\tilde{\ }y = Q^{-1}(PAQ^{-1})^- Py = \begin{bmatrix} 1 & -1 & 0 \\ 0 & 1 & 0 \\ 0 & 0 & \frac{1}{2} \end{bmatrix} \begin{bmatrix} 2 & 1 & \frac{5}{2} \\ 0 & 4 & 2 \\ 0 & 3 & \frac{3}{2} \end{bmatrix}^- \begin{bmatrix} 1 & 0 & 0 \\ 0 & 1 & 0 \\ 0 & 0 & 3 \end{bmatrix} \begin{bmatrix} 5000 \\ 4000 \\ 2000 \end{bmatrix}.$$

PROBLEMS

NUMERICAL PROBLEMS

1. Find the weighted approximate inverse of $\begin{bmatrix} 2 & 1 \\ 0 & 4 \\ 0 & 3 \end{bmatrix}$ with respect to the weighting matrices $P = \begin{bmatrix} 1 & 0 & 0 \\ 0 & 1 & 0 \\ 0 & 0 & 3 \end{bmatrix}$ and $Q = \begin{bmatrix} 1 & 1 \\ 0 & 1 \end{bmatrix}$, using Corollary 12.5.2.

MORE THEORETICAL PROBLEMS

Easier Problems

2. If A is invertible, show that $A\tilde{\ } = A^- = A^{-1}$.

Middle-Level Problems

3. If Q is a real unitary matrix, use Theorems 12.3.6 and 12.5.1 to show that $A\tilde{\ } = (PA)^- P$.
4. If P and Q are real unitary matrices, use Theorems 12.3.6 and 12.5.1 to show that $A\tilde{\ } = A^-$.
5. If P is a real unitary matrix and the columns of A are linearly independent, use Theorem 12.3.6 and Corollary 12.5.2 to show that $A\tilde{\ } = A^-$; that is, neither of the two weighting matrices P and Q has any effect.

CHAPTER
13

Linear Algorithms (Optional)

13.1. INTRODUCTION

Now that we have the theory and applications behind us, we ask:

How can we instruct a computer to carry out the computations?

So that we can get into the subject enough to get a glimpse of what it is about, we restrict ourselves to a single computational problem—but one of great importance—the problem of solving a matrix-vector equation

$$Ax = y$$

for x exactly or approximately, where the entries of A, x, and y are to be real.

To instruct a computer to find x for a given A and y,

we simply find a mathematical expression for x and devise an unambiguous recipe or step by step process for computing x from the expression.

We call such a process an *algorithm*.

How do we choose a particular algorithm to compute x? Many factors may be involved, depending on the uses that will be made of the algorithm. Here we want to be able to solve $Ax = y$ without knowing anything in advance about the size or nature of the matrix A or vectors y that we are given as input. So we take the point of view that

1. We want the algorithm to work for all rectangular matrices A and right-hand-side vectors y, and we want it to indicate whether the solution is exact or approximate.

Sec. 13.1]   Introduction   459

2. We want it to use as little memory as possible.
3. We want the computation to take as little time as possible.
4. We want the algorithm to be simple and easy to translate into language understandable by a computer.

The algorithm that plays the most central role in this chapter, the *row reduction algorithm* discussed in Section 13.3, may come about as close to satisfying all four of the foregoing properties as one could wish. It is simple and works for all matrices. At the same time it is very fast and since it uses virtually no memory except the memory that held the original matrix, it is very efficient in its use of memory. This algorithm, which row reduces memory representing an ordinary matrix into memory representing that matrix in a special *factored format*, has another very desirable feature. It is *reversible*; that is, there is a corresponding *inverse* algorithm to *undo row reduction*. This means that we can restore a matrix in its factored format back to its original unfactored format without performing cumbersome multiplications.

To solve the equation $Ax = y$ exactly, we use the row reduction algorithm to replace A in memory by three very special matrices L, D, U whose product, in the case where no row interchanges are needed, is $A = LDU$. If r is the rank of A, L is an $m \times r$ matrix which is essentially lower triangular with 1's on the diagonal (its rows may have to be put in another order to make L truly lower triangular), D is an invertible $r \times r$ matrix, and U is an echelon $r \times n$ matrix. The entries of these factors, except for the known 0's and 1's, are nicely stored together in the memory that had been occupied by A. So that we can get them easily whenever we need them, we mark the locations of the r diagonal entries of D.

EXAMPLE

When we apply the row reduction algorithm to the matrix

$$A = \begin{bmatrix} 6 & 12 & 18 & 0 \\ 2 & 9 & 6 & 10 \\ 3 & 7 & 10 & 7 \end{bmatrix},$$

we get it in its factored format,

$$\begin{bmatrix} \mathbf{6} & 2 & 3 & 0 \\ \frac{1}{3} & \mathbf{5} & 0 & 2 \\ \frac{1}{2} & \frac{1}{5} & \mathbf{1} & 5 \end{bmatrix},$$

from which we extract the matrices

$$L = \begin{bmatrix} 1 & 0 & 0 \\ \frac{1}{3} & 1 & 0 \\ \frac{1}{2} & \frac{1}{5} & 1 \end{bmatrix}, \quad D = \begin{bmatrix} \mathbf{6} & 0 & 0 \\ 0 & \mathbf{5} & 0 \\ 0 & 0 & \mathbf{1} \end{bmatrix}, \quad U = \begin{bmatrix} 1 & 2 & 3 & 0 \\ 0 & 1 & 0 & 2 \\ 0 & 0 & 1 & 5 \end{bmatrix},$$

whose product LDU is A.

Solving $Ax = y$ for x now is reduced to solving $LDUx = y$ for x. To get x, we simply solve $Lv = y$, $Dw = v$ and $Ux = w$. These equations are easy to solve (or to determine to be unsolvable) because of the special nature of the matrices L, D, U. So in this way, we get our solution x to $Ax = y$, if a solution exists.

When we are done, we can reverse the row reduction to restore the matrix A. This amounts to multiplying the factors L, D, U together in a very efficient way, getting back $A = LDU$. This is of great importance, since we may, within a small time frame, want to go back and forth many times between the unfactored and factored formats of the matrix.

After discussing this, in Sections 13.2 through 13.4 we go on to develop an algorithm for solving $Ax = y$ approximately. This algorithm also finds exact solutions, if they exist, but not as efficiently as does the algorithm for finding exact solutions. Here the algorithm replaces the matrix A by a factorization $A^- = J'K$ of its approximate inverse, where J', in the case where no row interchanges are needed, is upper triangular. Solving $Ax = y$ approximately for x now is reduced to solving $x = J'Ky$, which we solve in two steps $z = Ky$ and $x = J'z$.

Finally, we illustrate how these algorithms can be used to build a computer program for solving $Ax = y$ and finding the exact or approximate inverse of a matrix A.

PROBLEMS

NUMERICAL PROBLEMS

1. Let $A = LDU$, where

$$L = \begin{bmatrix} 1 & 0 & 0 \\ 2 & 1 & 0 \\ 3 & 4 & 1 \\ 1 & 1 & 1 \end{bmatrix}, \quad D = \begin{bmatrix} 1 & 0 & 0 \\ 0 & 2 & 0 \\ 0 & 0 & 3 \end{bmatrix}, \quad \text{and} \quad U = \begin{bmatrix} 0 & 0 & 1 & 1 & 2 & 0 & 3 \\ 0 & 0 & 0 & 0 & 1 & 4 & 2 \\ 0 & 0 & 0 & 0 & 0 & 1 & 1 \end{bmatrix}.$$

Then

(a) Solve the equation $Ax = \begin{bmatrix} 1 \\ 2 \\ 3 \\ 1 \end{bmatrix}$ for x by solving $\begin{bmatrix} 1 & 0 & 0 \\ 2 & 1 & 0 \\ 3 & 4 & 1 \\ 1 & 1 & 1 \end{bmatrix} v = \begin{bmatrix} 1 \\ 2 \\ 3 \\ 1 \end{bmatrix}$,

$\begin{bmatrix} 1 & 0 & 0 \\ 0 & 2 & 0 \\ 0 & 0 & 3 \end{bmatrix} w = v$, and $\begin{bmatrix} 0 & 0 & 1 & 1 & 2 & 0 & 3 \\ 0 & 0 & 0 & 0 & 1 & 4 & 2 \\ 0 & 0 & 0 & 0 & 0 & 1 & 1 \end{bmatrix} x = w$.

(b) Calculate x by computing A and row reducing $[A, y]$.

2. Using the factors L, D, U in problem 1 as a road map, find a row reduction of the matrix A of Problem 1 that leads to the echelon form U.

MORE THEORETICAL PROBLEMS

Easier Problems

3. If the reduction of a matrix A to its echelon form U by the method of Chapter 2 does not involve any interchanges, and if $E_1, \ldots, E_k$ denote the elementary matrices corresponding to the elementary row operations used in the reduction, show that $U = NA$, where $N = E_k \cdots E_1$ is a lower triangular matrix. Using this, show that A can be factored as $A = LDU$, where L is a lower triangular $m \times m$ matrix with 1's on the diagonal and D is a diagonal $m \times m$ matrix.

Show that the matrix

$$\begin{bmatrix} 1 & 0 & 0 & 0 \\ f & 1 & 0 & 0 \\ e & c & 1 & 0 \\ d & b & a & 1 \end{bmatrix}$$

factors as

$$\begin{bmatrix} 1 & 0 & 0 & 0 \\ f & 1 & 0 & 0 \\ 0 & 0 & 1 & 0 \\ 0 & 0 & 0 & 1 \end{bmatrix} \begin{bmatrix} 1 & 0 & 0 & 0 \\ 0 & 1 & 0 & 0 \\ e & 0 & 1 & 0 \\ 0 & 0 & 0 & 1 \end{bmatrix} \begin{bmatrix} 1 & 0 & 0 & 0 \\ 0 & 1 & 0 & 0 \\ 0 & 0 & 1 & 0 \\ d & 0 & 0 & 1 \end{bmatrix} \begin{bmatrix} 1 & 0 & 0 & 0 \\ 0 & 1 & 0 & 0 \\ 0 & c & 1 & 0 \\ 0 & 0 & 0 & 1 \end{bmatrix} \begin{bmatrix} 1 & 0 & 0 & 0 \\ 0 & 1 & 0 & 0 \\ 0 & 0 & 1 & 0 \\ 0 & b & 0 & 1 \end{bmatrix} \begin{bmatrix} 1 & 0 & 0 & 0 \\ 0 & 1 & 0 & 0 \\ 0 & 0 & 1 & 0 \\ 0 & 0 & a & 1 \end{bmatrix}.$$

13.2. THE *LDU* FACTORIZATION OF *A*

For any $m \times n$ matrix A and vector y in $\mathbb{R}^{(m)}$, solving the matrix-vector equation $Ax = y$ for $x \in \mathbb{R}^{(n)}$ by the methods given in Chapter 2 amounts to reducing the augmented matrix $[A, y]$ to an echelon matrix $[U, z]$ and solving $Ux = z$ instead. Let's look again at the reason for this, so that we can improve our methods. If M is the product of the inverses of the elementary matrices used during the reduction, a matrix that we can build and store during the reduction, then M is an invertible $m \times m$ matrix and $[MU, Mz] = [A, y]$. Since $Ux = z$ if and only if $MUx = Mz$, x is a solution of $Ux = z$ if and only if x is a solution of $Ax = y$.

From this comes something quite useful. If we reduce A to its echelon form U, gaining M during the reduction, then $A = MU$ and for any right-hand-side vector y that we may be given, we can solve $Ax = y$ in two steps as follows:

1. Solve $Mz = y$ for z.
2. Solve $Ux = z$ for x.

Putting these two steps together, we see that the x we get satisfies

$$Ax = MUx = Mz = y.$$

If no interchanges were needed during the reduction, M is an invertible lower triangular matrix. So since U is an echelon matrix, both equations $Mz = y$ and $Ux = z$ are easy to solve, provided that solutions exist. We have already seen how to solve $Ux = z$ for x by back substitution, given z. And we can get z from $Mz = y$, for a given y, using a reversed version of back substitution which we call *forward substitution*. Of course, if interchanges are needed during the reduction, we must also keep track of them and take them into account.

How do we find and store M so that $A = MU$? Let's look at

EXAMPLE

Let's take

$$A = \begin{bmatrix} 6 & 12 & 18 & 0 \\ 2 & 9 & 6 & 10 \\ 3 & 7 & 10 & 7 \end{bmatrix} \quad \text{(unfactored format for the matrix } A),$$

and row reduce it to an echelon matrix U. As we reduce A, we keep track of certain nonzero *pivot entries* a_{pq} and for each of them, we store each *multiplier* a_{tq}/a_{pq} ($t > p$) as entry (t, q) after the (t, q) entry has been changed to zero as a result of performing the operation $\text{Add}(t, p; -a_{tq}/a_{pq})$. The row operations $\text{Add}(2, 1; -\frac{1}{3})$, $\text{Add}(3, 1; -\frac{1}{2})$, $\text{Add}(3, 2; -\frac{1}{5})$ reduce A to the upper triangular matrix $V = \begin{bmatrix} 6 & 12 & 18 & 0 \\ 0 & 5 & 0 & 10 \\ 0 & 0 & 1 & 5 \end{bmatrix}$. If we write the pivot entries used during the reduction in boldface, the successive matrices encountered in the reduction are

$$\begin{bmatrix} \mathbf{6} & 12 & 18 & 0 \\ 2 & 9 & 6 & 10 \\ 3 & 7 & 10 & 7 \end{bmatrix}, \begin{bmatrix} \mathbf{6} & 12 & 18 & 0 \\ \frac{1}{3} & 5 & 0 & 10 \\ 3 & 7 & 10 & 7 \end{bmatrix}, \begin{bmatrix} \mathbf{6} & 12 & 18 & 0 \\ \frac{1}{3} & \mathbf{5} & 0 & 10 \\ \frac{1}{2} & 1 & 1 & 7 \end{bmatrix}, \begin{bmatrix} \mathbf{6} & 12 & 18 & 0 \\ \frac{1}{3} & \mathbf{5} & 0 & 10 \\ \frac{1}{2} & \frac{1}{5} & \mathbf{1} & 5 \end{bmatrix}.$$

These matrices successively displace A in memory. The upper triangular part of the last one,

$$\begin{bmatrix} 6 & 12 & 18 & 0 \\ \frac{1}{3} & 5 & 0 & 10 \\ \frac{1}{2} & \frac{1}{5} & 1 & 5 \end{bmatrix} \quad (LV \text{ factored format for } A),$$

holds $V = \begin{bmatrix} 6 & 12 & 18 & 0 \\ 0 & 5 & 0 & 10 \\ 0 & 0 & 1 & 5 \end{bmatrix}$, whereas the lower part of it holds the lower entries of the *matrix of multipliers* $L = \begin{bmatrix} 1 & 0 & 0 \\ \frac{1}{3} & 1 & 0 \\ \frac{1}{2} & \frac{1}{5} & 1 \end{bmatrix}$. We claim that $A = LV$.

Of course, it is easy to compute the product to check that $A = LV$. To see *why* $A = LV$, however, note also that if we were to apply the same operations $\text{Add}(2, 1; -\frac{1}{3})$, $\text{Add}(3, 1; -\frac{1}{2})$, $\text{Add}(3, 2; -\frac{1}{5})$ to L, we would get

$$\text{Add}(3, 2, -\tfrac{1}{5}) \, \text{Add}(3, 1, -\tfrac{1}{2}) \, \text{Add}(2, 1, -\tfrac{1}{3}) L = I,$$

which implies that L can be gotten by applying their inverses in the opposite

Sec. 13.2]   The *LDU* Factorization of *A*   463

order to *I*. So

$$L = \text{Add}(2,1;\tfrac{1}{3})\,\text{Add}(3,1;\tfrac{1}{2})\,\text{Add}(3,2;\tfrac{1}{5})I$$

and

$$LV = \text{Add}(2,1;\tfrac{1}{3})\,\text{Add}(3,1;\tfrac{1}{2})\,\text{Add}(3,2;\tfrac{1}{5})V = A.$$

Going on, we can factor $V = \begin{bmatrix} 6 & 12 & 18 & 0 \\ 0 & 5 & 0 & 10 \\ 0 & 0 & 1 & 5 \end{bmatrix}$ by taking the *matrix of pivots*

$D = \begin{bmatrix} 6 & 0 & 0 \\ 0 & 5 & 0 \\ 0 & 0 & 1 \end{bmatrix}$ and the echelon matrix $U = D^{-1}V = \begin{bmatrix} 1 & 2 & 3 & 0 \\ 0 & 1 & 0 & 2 \\ 0 & 0 & 1 & 5 \end{bmatrix}$. So we get

the factorizations $V = DU$ and $A = LDU$:

$$\begin{bmatrix} 6 & 12 & 18 & 0 \\ 0 & 5 & 0 & 10 \\ 0 & 0 & 1 & 5 \end{bmatrix} = \begin{bmatrix} 6 & 0 & 0 \\ 0 & 5 & 0 \\ 0 & 0 & 1 \end{bmatrix}\begin{bmatrix} 1 & 2 & 3 & 0 \\ 0 & 1 & 0 & 2 \\ 0 & 0 & 1 & 5 \end{bmatrix}$$

$$\begin{bmatrix} 6 & 12 & 18 & 0 \\ 2 & 9 & 6 & 10 \\ 3 & 7 & 10 & 7 \end{bmatrix} = \begin{bmatrix} 1 & 0 & 0 \\ \tfrac{1}{3} & 1 & 0 \\ \tfrac{1}{2} & \tfrac{1}{5} & 1 \end{bmatrix}\begin{bmatrix} 6 & 0 & 0 \\ 0 & 5 & 0 \\ 0 & 0 & 1 \end{bmatrix}\begin{bmatrix} 1 & 2 & 3 & 0 \\ 0 & 1 & 0 & 2 \\ 0 & 0 & 1 & 5 \end{bmatrix}.$$

Of course, we could do this directly, starting from where we left off with the

matrix $\begin{bmatrix} 6 & 12 & 18 & 0 \\ \tfrac{1}{3} & 5 & 0 & 10 \\ \tfrac{1}{2} & \tfrac{1}{5} & 1 & 5 \end{bmatrix}$. Simply divide the upper entries (entries above the

main diagonal) of *V*, row by row, by the pivot entry of the same row to get the upper entries of *U*, to get

$$\begin{bmatrix} 6 & 2 & 3 & 0 \\ \tfrac{1}{3} & 5 & 0 & 10 \\ \tfrac{1}{2} & \tfrac{1}{5} & 1 & 5 \end{bmatrix} \quad (LDU \text{ factored format for } A).$$

So not only have we factored $A = MU$, but our *M* comes to us in the factored form $M = LD$, enabling us to store it in factored form by storing *L* and *D*.

From our example we see how to find and store *M* so that $A = MU$. In fact, *M* comes to us in a factored form $M = LD$, so that $A = LDU$, and the factors *L*, *D*, *U* are stored efficiently during the reduction. In the general case, things go the same way. If no interchanges are needed, we can reduce *A* to an upper triangular matrix *V* using only the elementary row operations $\text{Add}(t, p; -a_{tq}/a)$. At the stage where we have a nonzero entry *a* in row *p* and column *q*, the *pivot entry*, and use the operation $\text{Add}(t, p; -a_{tq}/a)$, to make the (t, q) entry 0 for $t > p$, the (t, q) entry becomes available to us for *storing* the multiplier a_{tq}/a. Letting *L* be the corresponding lower triangular *matrix of multipliers*, consisting of the multipliers a_{tq}/a (with $t > p$) used in the reduction, below the diagonal, and 1's on the diagonal, we get $A = LV$. Why? In effect, to get V we are multiplying A by the elementary matrices corresponding to Add$(t, p; -a_{tq}/a)$; and we are multiplying I by their inverses, *in reverse order*, to get L. To see this, just compute the product of their inverses in reverse order, which has the same effect as writing the same multipliers in the same places, but starting from the other end and working forward. For example, if L is $\begin{bmatrix} 1 & 0 & 0 \\ 3 & 1 & 0 \\ 4 & 5 & 1 \end{bmatrix}$, it is the product

$$\begin{bmatrix} 1 & 0 & 0 \\ 3 & 1 & 0 \\ 4 & 5 & 1 \end{bmatrix} = \begin{bmatrix} 1 & 0 & 0 \\ 3 & 1 & 0 \\ 0 & 0 & 1 \end{bmatrix} \begin{bmatrix} 1 & 0 & 0 \\ 0 & 1 & 0 \\ 4 & 0 & 1 \end{bmatrix} \begin{bmatrix} 1 & 0 & 0 \\ 0 & 1 & 0 \\ 0 & 5 & 1 \end{bmatrix}.$$

So if we get $V = E_k \cdots E_1 A$, then we get $A = E_1^{-1} \cdots E_k^{-1} V = LV$. After we get $A = LV$, we go on to factor V as $V = DD^{-1}V = DU$, where D is the diagonal *matrix of pivots* whose diagonal entry in row p is 1 if row p of V is 0 and a if a is the first nonzero entry of row p of V, and where U is the echelon matrix $D^{-1}V$. Then we can rewrite the product $A = LV$ as $A = LDU = MD$, where $M = LD$. This proves

Theorem 13.2.1. If no interchanges take place in the reduction of an $m \times n$ matrix A to an echelon matrix U, then $A = LDU$, where L is the lower triangular matrix of multipliers, D is the matrix of pivots, and U is the echelon matrix.

We can further simplify the factorization $A = LDU$ by throwing away parts of the matrices that are not needed. Letting r be the rank of A, we throw away all but the first r columns of L, all but the first r rows and columns of D, and all but the first r rows of U. Then we still have $A = LDU$, but now L is a lower triangular $m \times r$ matrix with 1's on the diagonal, D is an invertible diagonal $r \times r$ matrix, and U is an $r \times n$ echelon matrix.

EXAMPLE

The product

$$LDU = \begin{bmatrix} 1 & 0 & 0 & 0 \\ 2 & 1 & 0 & 0 \\ 3 & 0 & 1 & 0 \\ 2 & 0 & 5 & 1 \end{bmatrix} \begin{bmatrix} 2 & 0 & 0 & 0 \\ 0 & 5 & 0 & 0 \\ 0 & 0 & 3 & 0 \\ 0 & 0 & 0 & 1 \end{bmatrix} \begin{bmatrix} 1 & 2 & 3 & 4 & 3 & 3 & 3 & 3 \\ 0 & 0 & 1 & 2 & 2 & 2 & 2 & 2 \\ 0 & 0 & 0 & 0 & 0 & 0 & 1 & 6 \\ 0 & 0 & 0 & 0 & 0 & 0 & 0 & 0 \end{bmatrix}$$

equals the product

$$\begin{bmatrix} 1 & 0 & 0 \\ 2 & 1 & 0 \\ 3 & 0 & 1 \\ 2 & 0 & 5 \end{bmatrix} \begin{bmatrix} 2 & 0 & 0 \\ 0 & 5 & 0 \\ 0 & 0 & 3 \end{bmatrix} \begin{bmatrix} 1 & 2 & 3 & 4 & 3 & 3 & 3 & 3 \\ 0 & 0 & 1 & 2 & 2 & 2 & 2 & 2 \\ 0 & 0 & 0 & 0 & 0 & 0 & 1 & 6 \end{bmatrix}$$

Sec. 13.2]   The *LDU* Factorization of *A*   465

because the effect of the diagonal matrix D is to multiply the rows of U by scalars, and the last column of L has no effect, since the last row of the product DU is 0.

Given the factorization $A = LDU$, whenever we are given a vector $y \in \mathbb{R}^{(m)}$, we solve $Ax = y$ for x by taking apart the problem of solving $LDUx = y$ for x as follows:

1. Solve $Lv = y$ for v by forward substitution.
2. Next, solve $Dw = v$ for w by stationary substitution.
3. Finally, solve $Ux = w$ for x by back substitution.

Putting this all back together, we then get

$$Ax = LDUx = LDw = Lv = y,$$

which shows that x is the desired solution.

When there are no interchanges, the resulting vectors v, w, x are:

1. For $p = 1$ to the rank of A:

$$v_p = y_p - \sum_{j=1}^{p-1} L_{pj} v_j \quad \text{(forward substitution)}$$

2. For $p = 1$ to the rank of A:

$$w_p = v_p / D_{pp} \quad \text{(stationary substitution)}$$

3. For $q = n$ down to 1:
   For q a column containing a (p, q) pivot entry:

$$x_q = w_p - \sum_{j=q+1}^{n} U_{pj} x_j \quad \text{(back substitution)}$$

For any other q (x_q is then an independent variable):

$$x_q = \text{any desired value}$$

EXAMPLE

Let's consider the the product

$$A = LDU = \begin{bmatrix} 1 & 0 & 0 \\ 2 & 1 & 0 \\ 3 & 0 & 1 \\ 2 & 0 & 5 \end{bmatrix} \begin{bmatrix} 2 & 0 & 0 \\ 0 & 5 & 0 \\ 0 & 0 & 3 \end{bmatrix} \begin{bmatrix} 1 & 2 & 3 & 4 & 3 & 3 & 3 \\ 0 & 0 & 1 & 2 & 2 & 2 & 2 \\ 0 & 0 & 0 & 0 & 0 & 1 & 6 \end{bmatrix}$$

of the example above and try to solve $LDUx = y$ for the values $\begin{bmatrix} 2 \\ 9 \\ 6 \\ 3 \end{bmatrix}, \begin{bmatrix} 2 \\ 9 \\ 6 \\ 4 \end{bmatrix}$ for y.

Taking $y = \begin{bmatrix} 2 \\ 9 \\ 6 \\ 3 \end{bmatrix}$ first, we try to solve $\begin{bmatrix} 1 & 0 & 0 \\ 2 & 1 & 0 \\ 3 & 0 & 1 \\ 2 & 0 & 5 \end{bmatrix} \begin{bmatrix} v_1 \\ v_2 \\ v_3 \end{bmatrix} = \begin{bmatrix} 2 \\ 9 \\ 6 \\ 3 \end{bmatrix}$. The forward substitution formula gives us $\begin{bmatrix} v_1 \\ v_2 \\ v_3 \end{bmatrix} = \begin{bmatrix} 2 \\ 9 - 2 \cdot 2 \\ 6 - 3 \cdot 2 - 0 \end{bmatrix} = \begin{bmatrix} 2 \\ 5 \\ 0 \end{bmatrix}$. Testing, we find that $\begin{bmatrix} 1 & 0 & 0 \\ 2 & 1 & 0 \\ 3 & 0 & 1 \\ 2 & 0 & 5 \end{bmatrix} \begin{bmatrix} 2 \\ 5 \\ 0 \end{bmatrix} = \begin{bmatrix} 2 \\ 9 \\ 6 \\ 4 \end{bmatrix}$. So, *for* $y = \begin{bmatrix} 2 \\ 9 \\ 6 \\ 3 \end{bmatrix}$, *there is no exact solution*. Next, let's try to solve $LDUx = \begin{bmatrix} 2 \\ 9 \\ 6 \\ 4 \end{bmatrix}$. This time, we can solve $Lv = \begin{bmatrix} 2 \\ 9 \\ 6 \\ 4 \end{bmatrix}$ by forward substitution, getting $v = \begin{bmatrix} 2 \\ 5 \\ 0 \end{bmatrix}$. We then solve $\begin{bmatrix} 2 & 0 & 0 \\ 0 & 5 & 0 \\ 0 & 0 & 3 \end{bmatrix} w = \begin{bmatrix} 2 \\ 5 \\ 0 \end{bmatrix}$, getting $w = \begin{bmatrix} 1 \\ 1 \\ 0 \end{bmatrix}$. Finally, we solve $\begin{bmatrix} 1 & 2 & 3 & 4 & 3 & 3 & 3 & 3 \\ 0 & 0 & 1 & 2 & 2 & 2 & 2 & 2 \\ 0 & 0 & 0 & 0 & 0 & 0 & 1 & 6 \end{bmatrix} x = \begin{bmatrix} 1 \\ 1 \\ 0 \end{bmatrix}$ using the back substitution formula, getting $x = \begin{bmatrix} 1 - 3 \\ 0 \\ 1 - 0 \\ 0 \\ 0 \\ 0 \\ 0 - 0 \\ 0 \end{bmatrix} = \begin{bmatrix} -2 \\ 0 \\ 1 \\ 0 \\ 0 \\ 0 \\ 0 \\ 0 \end{bmatrix}$.

PROBLEMS

NUMERICAL PROBLEMS

1. Find L, D, U for the matrices $\begin{bmatrix} 1 & 0 & 0 \\ 2 & 1 & 0 \\ 3 & 4 & 1 \end{bmatrix}, \begin{bmatrix} 1 & 0 \\ 1 & 2 \\ 0 & 0 \end{bmatrix}, \begin{bmatrix} 1 & 0 & 0 \\ 1 & 2 & 0 \\ 0 & 0 & 3 \end{bmatrix}, \begin{bmatrix} 1 & 2 & 3 \\ 2 & 1 & 3 \end{bmatrix}$.

MORE THEORETICAL PROBLEMS

Easier Problems

2. If A is invertible, show that if $A = LDU$ and $A = MEV$, where L and M are lower triangular with 1's on the diagonal, D and E are diagonal, and U and V are upper triangular with 1's on the diagonal, then $L = M$, $D = E$ and $U = V$.

3. Show that the following algorithm delivers the inverse of a lower triangular matrix with 1's on the diagonal:
   1. Perform the following row operations successively on I for each value of q from 1 to $n - 1$: For each value of p from $q + 1$ to n, Add $(p, q, -a_{pq})$.
   2. L^{-1} is the resulting matrix.

4. Show that some invertible matrices A cannot be factored as $A = LDU$, by showing that there are no values a, b, c, d such that

$$\begin{bmatrix} 0 & 1 \\ 1 & 0 \end{bmatrix} = \begin{bmatrix} 1 & 0 \\ a & 1 \end{bmatrix} \begin{bmatrix} b & 0 \\ 0 & c \end{bmatrix} \begin{bmatrix} 1 & d \\ 0 & 1 \end{bmatrix}.$$

Middle-Level Problems

5. Given the factorization $A = LDU$ of an invertible matrix A:
   (a) Shown that L, D, U are invertible.
   (b) Find algorithms for inverting $L, D,$ and U in their own memory, given at most one extra column for work space.
   (c) Give an algorithm for using the matrices L^{-1}, D^{-1}, U^{-1} to solve $Ax = y$ for x without explicitly calculating the product $A^{-1} = U^{-1}D^{-1}L^{-1}$.

Harder Problems

6. Show that if A is an $m \times n$ matrix of rank r and $A = LDU$ and $A = MEV$, where L' and M' are echelon $r \times m$ matrices, D and E are invertible diagonal $r \times r$ matrices and U and V are echelon $r \times n$ matrices, then $L = M$, $D = E$ and $U = V$.

13.3. THE ROW REDUCTION ALGORITHM AND ITS INVERSE

In Chapter 2, we gave a method for reducing a matrix to an echelon matrix. We now describe an algorithm similar to that method, but with important difference.

In order to give the algorithm, we must describe how to store an $m \times n$ matrix A in memory and how to perform and keep track of row interchanges. Of course, we must have an $m \times n$ array Memory (R, C) of real numbers in the memory of the computer, to hold the entries of A. So that we do not need to actually move entries when we perform a row or column interchange, we just keep track of the rows and columns by making and updating lists Row and Col of their rows and columns in memory. So if we load the 4×6 matrix

$$A = \begin{bmatrix} 0 & 1 & 0 & 3 & 4 & 5 \\ 1 & 3 & 4 & 5 & 5 & 5 \\ 3 & 3 & 3 & 3 & 3 & 3 \\ 2 & 2 & 2 & 2 & 2 & 2 \end{bmatrix}$$

into memory with the Row $= [1, 2, 3, 4]$ and Col $= [1, 2, 3, 4, 5, 6]$, we can keep track of Row, Col, and the entries, held in the array Memory (R, C), by the following 4×6 *matrix structure* **A**:

$$
\mathbf{A} = \begin{array}{c} \\ 1 \\ 2 \\ 3 \\ 4 \end{array} \begin{array}{c} 1 \ 2 \ 3 \ 4 \ 5 \ 6 \\ \left[\begin{array}{cccccc} 0 & 1 & 0 & 3 & 4 & 5 \\ 1 & 3 & 4 & 5 & 5 & 5 \\ 3 & 3 & 3 & 3 & 3 & 3 \\ 2 & 2 & 2 & 2 & 2 & 2 \end{array} \right] \end{array}.
$$

All we mean by this is that the matrix A that we loaded in the computer was loaded by setting up the two lists Row = $[1, 2, 3, 4]$ and Col = $[1, 2, 3, 4, 5, 6]$, and putting the entries of A into the array Memory (R, C) according to the lists Row and Col. In this case, Row and Col indicate that the usual order should be used, so the entries occur in the array Memory (R, C) in the same order as they occur in A. So giving **A** is the same as giving the lists Row and Col and the array Memory (R, C). After loading A in memory in this way, suppose that we first interchange rows 3 and 4, then rows 4 and 2, then columns 2 and 4. The matrix A undergoes the following changes:

$$
\begin{bmatrix} 0 & 1 & 0 & 3 & 4 & 5 \\ 1 & 3 & 4 & 5 & 5 & 5 \\ 3 & 3 & 3 & 3 & 3 & 3 \\ 2 & 2 & 2 & 2 & 2 & 2 \end{bmatrix}, \begin{bmatrix} 0 & 1 & 0 & 3 & 4 & 5 \\ 1 & 3 & 4 & 5 & 5 & 5 \\ 2 & 2 & 2 & 2 & 2 & 2 \\ 3 & 3 & 3 & 3 & 3 & 3 \end{bmatrix},
$$

$$
\begin{bmatrix} 0 & 1 & 0 & 3 & 4 & 5 \\ 3 & 3 & 3 & 3 & 3 & 3 \\ 2 & 2 & 2 & 2 & 2 & 2 \\ 1 & 3 & 4 & 5 & 5 & 5 \end{bmatrix}, \begin{bmatrix} 0 & 3 & 0 & 1 & 4 & 5 \\ 3 & 3 & 3 & 3 & 3 & 3 \\ 2 & 2 & 2 & 2 & 2 & 2 \\ 1 & 5 & 4 & 3 & 5 & 5 \end{bmatrix}.
$$

To make corresponding changes in the matrix structure **A**, we keep updating the lists:

Row = $[1, 2, 4, 3]$ (after interchanging rows 3 and 4)

Row = $[1, 4, 2, 3]$ (after then interchanging rows 4 and 2)

Col = $[1, 4, 3, 2, 5, 6]$ (after then interchanging columns 2 and 4).

Let's look at the matrix structure **A**, which represents A as it undergoes the corresponding transformations:

$$
\begin{array}{c} \\ 1 \\ 2 \\ 3 \\ 4 \end{array} \begin{array}{c} 1 \ 2 \ 3 \ 4 \ 5 \ 6 \\ \left[\begin{array}{cccccc} 0 & 1 & 0 & 3 & 4 & 5 \\ 1 & 3 & 4 & 5 & 5 & 5 \\ 3 & 3 & 3 & 3 & 3 & 3 \\ 2 & 2 & 2 & 2 & 2 & 2 \end{array} \right] \end{array}, \begin{array}{c} \\ 1 \\ 2 \\ 4 \\ 3 \end{array} \begin{array}{c} 1 \ 2 \ 3 \ 4 \ 5 \ 6 \\ \left[\begin{array}{cccccc} 0 & 1 & 0 & 3 & 4 & 5 \\ 1 & 3 & 4 & 5 & 5 & 5 \\ 3 & 3 & 3 & 3 & 3 & 3 \\ 2 & 2 & 2 & 2 & 2 & 2 \end{array} \right] \end{array},
$$

The Row Reduction Algorithm and Its Inverse

$$\begin{array}{c} \;\;1\;\;2\;\;3\;\;4\;\;5\;\;6 \\ \begin{array}{c}1\\4\\2\\3\end{array}\!\!\left[\begin{array}{cccccc}0&1&0&3&4&5\\1&3&4&5&5&5\\3&3&3&3&3&3\\2&2&2&2&2&2\end{array}\right], \end{array} \qquad \begin{array}{c} \;\;1\;\;4\;\;3\;\;2\;\;5\;\;6 \\ \begin{array}{c}1\\4\\2\\3\end{array}\!\!\left[\begin{array}{cccccc}0&1&0&3&4&5\\1&3&4&5&5&5\\3&3&3&3&3&3\\2&2&2&2&2&2\end{array}\right]. \end{array}$$

Having set up this labeling system, it is time to make things precise.

Definition. An $m \times n$ *matrix structure* **A** consists of Row, Col, and Memory (R, C), where Row is a $1 - 1$ onto function from $\{1,\ldots,m\}$ to itself, Col is a $1 - 1$ onto function from $\{1,\ldots,n\}$ to itself, and Memory (R, C) is an $m \times n$ array of real numbers.

When we write Row $= [1, 4, 2, 3]$, we mean that Row is the mapping Row$(1) = 1$, Row$(4) = 2$, Row$(2) = 3$, Row$(3) = 4$ from $\{1, 2, 3, 4\}$ to itself. Similarly, writing Col $= [1, 4, 3, 2, 5, 6]$ means that Col is the mapping from $\{1, 2, 3, 4, 5, 6\}$ to itself such that Col$(s) = t$, where s is in position t in the list $[1, 4, 3, 2, 5, 6]$. So since 4 is in position 2, Col$(4) = 2$.

The 4×6 matrix structure consisting of Row $= [1, 4, 2, 3]$, Col $= [1, 4, 3, 2, 5, 6]$, and on 4×6 array Memory (R, C) is just

$$\mathbf{A} = \begin{array}{c} \;\;1\;\;4\;\;3\;\;2\;\;5\;\;6 \\ \begin{array}{c}1\\4\\2\\3\end{array}\!\!\left[\begin{array}{cccccc}0&1&0&3&4&5\\1&3&4&5&5&5\\3&3&3&3&3&3\\2&2&2&2&2&2\end{array}\right]. \end{array}$$

Matrices get put in, or taken from, matrix structures according to

Definition. The $m \times n$ matrix A corresponding to the matrix structure **A** consisting of the lists Row, Col, and the $m \times n$ array Memory (R, C) is the $m \times n$ matrix A whose (r, s) entry A_{rs} is given by the formula

$$A_{rs} = \text{Memory}(\text{Row}(r), \text{Col}(s)).$$

For example, the 4×6 matrix A corresponding to the 4×6 matrix structure **A** described earlier is the matrix

$$A = (A_{rs}) = (\text{Memory}(\text{Row}(r), \text{Col}(s)),$$

which can easily be read from **A** when we write it out to look at:

$$\mathbf{A} = \begin{array}{c} \;\;1\;\;4\;\;3\;\;2\;\;5\;\;6 \\ \begin{array}{c}1\\4\\2\\3\end{array}\!\!\left[\begin{array}{cccccc}0&1&0&3&4&5\\1&3&4&5&5&5\\3&3&3&3&3&3\\2&2&2&2&2&2\end{array}\right], \end{array} \qquad A = \left[\begin{array}{cccccc}0&3&0&1&4&5\\3&3&3&3&3&3\\2&2&2&2&2&2\\1&5&4&3&5&5\end{array}\right].$$

For example A_{43} = Memory(Row(4), Col(3)) = Memory(2, 3) is read from the matrix structure by going to the row of memory marked 4 (which is row 2 of memory) and to the column of memory marked 3 (which is column 3 of memory) and getting the entry $A_{43} = 4$ in that row and column.

Of course, our objective in all of this been to represent $m \times n$ matrices by $m \times n$ matrix structures and row and column interchanges on $m \times n$ matrices by corresponding operations on $m \times n$ matrix structures. We now do the latter in

Definition. To interchange rows (respectively, columns) p and q of an $m \times n$ matrix structure **A** consisting of Row, Col, and Memory, just interchange the values of Row(p) and Row(q) [respectively, Col(p), Col(q)].

In our example above, we interchanged rows 4 and 2 when Row was Row = [1, 2, 4, 3]. The result there was that the values Row(4) = 3 and Row(2) = 2 of Row = [1, 2, 4, 3] were interchanged, resulting in the new list Row = [1, 4, 2, 3], the new values 2 and 3 for Row(4) and Row(2) having been obtained by interchanging the old ones, 3 and 2.

We can now give the row reduction algorithm. This algorithm is similar to the method given in Chapter 2 for reducing a matrix to an echelon matrix, but we've made some important changes and added some new features:

1. Our operations are performed on an $m \times n$ matrix structure rather than an $m \times n$ matrix, to make it easy to perform them and keep track of row interchanges.

2. Where "0" occurs there, we now say "less in absolute value than Epsilon" (where Epsilon is a fixed small positive value which depends on the computer to be used).

3. Instead of looking for the "first nonzero value if any" in the rest of a given column, we look for the "first value that is the largest in absolute value" in rest of that column.

4. As we reduce to the echelon matrix U, we keep track of the pivot entries and use the freed memory on and below them to store the entries of D and L. In particular, we record the number of pivot entries in the variable Rank A.

Of these changes, (2) and (3) lead to increased *numerical stability*. In other words, these changes are important if we prefer not to divide by numbers so small as to lead to serious errors in the computations. The others enable us to construct, store, and retrieve the factors, L, D, and U of A. They also enable us to reverse the algorithm and restore A.

Of course, the operations on the entries A_{rs} of A performed in this algorithm are really performed as operations on Row, Col, and the array Memory(R, C), the correspondence of entries being

$$A_{rs} = \text{Memory}(\text{Row}(r), Col(s)).$$

This algorithm does not involve column interchanges and, in fact, neither to the other algorithms considered in this chapter. So,

henceforth we take Col *to be the identity list* Col(s) = s *and we do not label columns of a matrix structure.*

Algorithm to row reduce an $m \times n$ matrix A to an echelon matrix U

Starting with (p, q) equal to $(1, 1,)$ and continuing as long as $p \leq m$ and $q \leq n$, do the following:

1. Get the first largest (in absolute value) (p', q) entry

$$a = A_{p'q} = \text{Memory}(\text{Row}(p'), \text{Col}(q))$$

of A for $p' \geq p$.

2. If its absolute value is less than Epsilon, then we decrease the value of p by 1 (so later, when we increase p and q by 1, we try again in the same row and next column), but otherwise we call it a pivot entry and we do the following:

    (a) We record that the (p, q) entry is the pivot entry in row p; we do this using a function PivotList by setting PivotList $(p) = q$.

    (b) If $p' > p$, we interchange rows p and p' (by interchanging the values of Row (p) and Row (p')).

    (c) For each row t with $t > p$, we perform the elementary row operation

    $$\text{Add}(t, p; -A_{tq}/a) \quad (\text{add } -A_{tq}/a \text{ times row } p \text{ to row } t),$$

    where A_{tq} is the current (t, q) entry of A; in doing this, we do not disturb the area in which we have already stored multipliers; we then record the operation by writing the multipler A_{tq}/a as entry (t, q) of A; (since we know that there should be a 0 there, we lose no needed information when we take over this entry as storage for our growing record).

    (d) We perform the elementary row operation

    $$\text{Multiply}(p; 1/a) \quad (\text{divide the entries of row } p \text{ by } a);$$

    we then record the operation by writing the divisor $d_p = a$ as entry (p, q) of A; (since we know there should be a 1 there, we lose no needed information when we take over this entry for our growing record).

3. We increase the values of p and q by 1 (on to the next row and column ...).

After all this has been done, we record that row $p - 1$ was the last nonzero row by setting the value Rank A equal to $p - 1$.

This algorithm changes a matrix structure representing A in unfactored format to a matrix structure representing A in LDU factored format. Since we keep track of the pivots and the number $r = \text{Rank } A$ of pivots, we can get the entries of the echelon matrix U, the diagonal matrix D, and the lower triangular matrix L:

1. L is the $m \times r$ matrix whose (p, q) entry is

$$A_{p\text{PivotList}(q)} = \text{Memory}(\text{Row}(p), \text{Col}(\text{PivotList}(q)))$$

for $p > q$, 1 for $p = q$ and 0 for $p < q$;

2. D is the $r \times r$ diagonal matrix with (q, q) entry

$$A_{q\text{PivotList}(q)} = \text{Memory}(\text{Row})(q), \text{Col}(\text{PivotList}(q)))$$

for $1 \leq q \leq r$;

3. U is the $r \times n$ echelon matrix whose (p, q) entry is

$$A_{p\text{PivotList}(q)} = \text{Memory}(\text{Row}(p), \text{Col}(\text{PivotList}(q)))$$

for $p < q$, 1 for $p = q$ and 0 for $p > q$.

The $A = LDU$ factorization of the Section 13.2 is then replaced by a factorization $A = PLDU$, where P is the *permutation matrix corresponding to the list* Row, defined by

4. P is the $m \times m$ matrix whose (p, q) entry is 1 if $p = \text{Row}(q)$ and 0 otherwise.

EXAMPLE

Let $A = \begin{bmatrix} 3 & 7 & 10 & 7 \\ 6 & 12 & 18 & 0 \\ 2 & 9 & 6 & 10 \end{bmatrix}$ be represented by the matrix structure

$$\mathbf{A} = \begin{matrix} \mathbf{1} \\ \mathbf{2} \\ \mathbf{3} \end{matrix} \begin{bmatrix} 3 & 7 & 10 & 7 \\ 6 & 12 & 18 & 0 \\ 2 & 9 & 6 & 10 \end{bmatrix} \quad \text{(unfactored format for } A\text{)},$$

with Row $= [1, 2, 3]$. Then the row operations Interchange$(1, 2)$, Add$(2, 1; -\frac{1}{2})$, Add$(3, 1; -\frac{1}{3})$, Interchange$(2, 3)$, Add$(3, 2; -\frac{1}{5})$ reduce the matrix structure $\mathbf{A}$ to

$$\mathbf{V} = \begin{matrix} \mathbf{3} \\ \mathbf{1} \\ \mathbf{2} \end{matrix} \begin{bmatrix} 0 & 0 & 1 & 5 \\ 6 & 12 & 18 & 0 \\ 0 & 5 & 0 & 10 \end{bmatrix},$$

which represents the upper triangular matrix

$$V = \begin{bmatrix} 6 & 12 & 18 & 0 \\ 0 & 5 & 0 & 10 \\ 0 & 0 & 1 & 5 \end{bmatrix}.$$

How do we get this, and what is the multiplier matrix? Writing the pivots in boldface, as in the earlier example, the successive matrices encountered in the reduction are:

$$\begin{matrix} \mathbf{1} \\ \mathbf{2} \\ \mathbf{3} \end{matrix} \begin{bmatrix} 3 & 7 & 10 & 7 \\ 6 & 12 & 18 & 0 \\ 2 & 9 & 6 & 10 \end{bmatrix}, \quad \begin{matrix} \mathbf{2} \\ \mathbf{1} \\ \mathbf{3} \end{matrix} \begin{bmatrix} 3 & 7 & 10 & 7 \\ 6 & 12 & 18 & 0 \\ 2 & 9 & 6 & 10 \end{bmatrix}, \quad \begin{matrix} \mathbf{2} \\ \mathbf{1} \\ \mathbf{3} \end{matrix} \begin{bmatrix} \frac{1}{2} & 1 & 1 & 7 \\ 6 & 12 & 18 & 0 \\ 2 & 9 & 6 & 10 \end{bmatrix},$$

Sec. 13.3] The Row Reduction Algorithm and Its Inverse

$$2\begin{bmatrix} \frac{1}{2} & 1 & 1 & 7 \\ 1 & 6 & 12 & 18 & 0 \\ 3 & \frac{1}{3} & 5 & 0 & 10 \end{bmatrix}, \quad 3\begin{bmatrix} \frac{1}{2} & 1 & 1 & 7 \\ 1 & 6 & 12 & 18 & 0 \\ 2 & \frac{1}{3} & 5 & 0 & 10 \end{bmatrix}, \quad 3\begin{bmatrix} \frac{1}{2} & \frac{1}{5} & 1 & 5 \\ 1 & 6 & 12 & 18 & 0 \\ 2 & \frac{1}{3} & 5 & 0 & 10 \end{bmatrix}.$$

Now **A** has been reduced to

$$\mathbf{V} = \begin{matrix} 3 \\ 1 \\ 2 \end{matrix}\begin{bmatrix} 0 & 0 & 1 & 5 \\ 6 & 12 & 18 & 0 \\ 0 & 5 & 0 & 10 \end{bmatrix}$$

and A to

$$V = \begin{bmatrix} 6 & 12 & 18 & 0 \\ 0 & 5 & 0 & 10 \\ 0 & 0 & 1 & 5 \end{bmatrix};$$

and the matrix structure of multipliers is

$$\mathbf{L} = \begin{matrix} 3 \\ 1 \\ 2 \end{matrix}\begin{bmatrix} \frac{1}{2} & \frac{1}{5} & 1 \\ 1 & 0 & 0 \\ \frac{1}{3} & 1 & 0 \end{bmatrix},$$

with corresponding matrix

$$L = \begin{bmatrix} 1 & 0 & 0 \\ \frac{1}{3} & 1 & 0 \\ \frac{1}{2} & \frac{1}{5} & 1 \end{bmatrix}.$$

The matrix P is obtained by listing rows 1, 2, 3 of the identity matrix as the rows 3, 1, 2 of memory, that is, P is the matrix

$$P = \begin{bmatrix} 0 & 0 & 1 \\ 1 & 0 & 0 \\ 0 & 1 & 0 \end{bmatrix}$$

obtained from the matrix structure

$$\mathbf{L} = \begin{matrix} 3 \\ 1 \\ 2 \end{matrix}\begin{bmatrix} \frac{1}{2} & \frac{1}{5} & 1 \\ 1 & 0 & 0 \\ \frac{1}{3} & 1 & 0 \end{bmatrix}$$

by removing the row labels and the multipliers.

Suppose that we now perform the product of

$$P = \begin{bmatrix} 0 & 0 & 1 \\ 1 & 0 & 0 \\ 0 & 1 & 0 \end{bmatrix}, \quad L = \begin{bmatrix} 1 & 0 & 0 \\ \frac{1}{3} & 1 & 0 \\ \frac{1}{2} & \frac{1}{5} & 1 \end{bmatrix}, \quad V = \begin{bmatrix} 6 & 12 & 18 & 0 \\ 0 & 5 & 0 & 10 \\ 0 & 0 & 1 & 5 \end{bmatrix}.$$

For LV we get $LV = \begin{bmatrix} 6 & 12 & 18 & 0 \\ 2 & 9 & 6 & 10 \\ 3 & 7 & 10 & 7 \end{bmatrix}$. Multiplying this by $P = \begin{bmatrix} 0 & 0 & 1 \\ 1 & 0 & 0 \\ 0 & 1 & 0 \end{bmatrix}$ then rearranges the rows, giving us

$$\begin{bmatrix} 3 & 7 & 10 & 7 \\ 6 & 12 & 18 & 0 \\ 2 & 9 & 6 & 10 \end{bmatrix} = A.$$

If we want to factor V further as DU, we continue with the matrix structure where we left off,

$$\begin{array}{c} 3 \\ 1 \\ 2 \end{array} \begin{bmatrix} \frac{1}{2} & \frac{1}{5} & 1 & 5 \\ 6 & 12 & 18 & 0 \\ \frac{1}{3} & 5 & 0 & 10 \end{bmatrix} \quad (LV \text{ factored format for } A),$$

and reduce it further so that it contains all three factors L, D, U as in the example above. The only change in the matrix structure is that the upper entries of V are changed to the upper entries of U, by dividing them row by row by the pivots:

$$\begin{array}{c} 3 \\ 1 \\ 2 \end{array} \begin{bmatrix} \frac{1}{2} & \frac{1}{5} & 1 & 5 \\ 6 & 2 & 3 & 0 \\ \frac{1}{3} & 5 & 0 & 2 \end{bmatrix} \quad (LDU \text{ factored format for } A).$$

Of course, since $A = PLV$ and $V = DU$, we can get A back as before as $A = PLDU$.

This reconstruction of A from L, D, U and the permutation matrix corresponding to the list Row built during reduction works the same way for any matrix A, so we have

Theorem 13.3.1. Suppose that A is represented by the matrix structure **A** with Row $= [1, \ldots, m]$, which is reduced to its LDU format with list Row updated during reduction to record the affect of interchanges. Then A is obtained by performing the product $PLDU$, where P, L, D, U are as described above.

If we have used the reduction algorithm to change a matrix structure representing A in unfactored format to a matrix structure representing A in LDU factored format, we can reverse the algorithm and return the matrix structure to matrix format by

Algorithm to undo row reduction of A to an echelon matrix U

Starting with $p = $ Rank A, and continuing as long as $p \geq 1$, do the following:

1. Let $q = $ PivotList (p) and let d_p be the current (p, q) entry of A.
2. Multiply row p of U by d_p, ignoring the storage area on and below the pivot entries.
3. For each row $t > p$ of A, perform the elementary row operation Add $(t, p; m_{tq})$, where

Sec. 13.3]   The Row Reduction Algorithm and Its Inverse   475

m_{tq} is the multiplier stored in row t below the pivot d_p of column q. (When doing this, set the current (t, q) entry to 0.

4. Decrease p by 1.

Rearrange the rows by setting Row = $[1, \ldots, n]$.

EXAMPLE

The matrix structures encountered successively when the row reduction algorithm is applied to the matrix A in the above example are:

$$\begin{array}{c}1\\2\\3\end{array}\begin{bmatrix}3&7&10&7\\6&12&18&0\\2&9&6&10\end{bmatrix} \quad \begin{array}{c}2\\1\\3\end{array}\begin{bmatrix}3&7&10&7\\6&12&18&0\\2&9&6&10\end{bmatrix} \quad \begin{array}{c}2\\1\\3\end{array}\begin{bmatrix}\tfrac{1}{2}&1&1&7\\6&12&18&0\\2&9&6&10\end{bmatrix}$$

PivotList (1) = 1

$$\begin{array}{c}2\\1\\3\end{array}\begin{bmatrix}\tfrac{1}{2}&1&1&7\\6&12&18&0\\\tfrac{1}{3}&5&0&10\end{bmatrix} \quad \begin{array}{c}2\\1\\3\end{array}\begin{bmatrix}\tfrac{1}{2}&1&1&7\\6&2&3&0\\\tfrac{1}{3}&5&0&10\end{bmatrix} \quad \begin{array}{c}3\\1\\2\end{array}\begin{bmatrix}\tfrac{1}{2}&1&1&7\\6&2&3&0\\\tfrac{1}{3}&5&0&10\end{bmatrix}$$

PivotList (2) = 2

$$\begin{array}{c}3\\1\\2\end{array}\begin{bmatrix}\tfrac{1}{2}&\tfrac{1}{5}&1&5\\6&2&3&0\\\tfrac{1}{3}&5&0&10\end{bmatrix} \quad \begin{array}{c}3\\1\\2\end{array}\begin{bmatrix}\tfrac{1}{2}&\tfrac{1}{5}&1&5\\6&2&3&0\\\tfrac{1}{3}&5&0&2\end{bmatrix} \quad \begin{array}{c}3\\1\\2\end{array}\begin{bmatrix}\tfrac{1}{2}&\tfrac{1}{5}&1&5\\6&2&3&0\\\tfrac{1}{3}&5&0&2\end{bmatrix}$$

PivotList (3) = 3

Rank $A = 3$

The same matrix structures are encountered in reverse order when the algorithm to undo row reduction of A is applied. Since Rank $A = 3$, the algorithm starts with $p = 3$, $q = $ PivotList $(3) = 3$ and $d_3 = 1$. It then goes to $p = 3 - 1 = 2$, $q = $ PivotList $(2) = 2$, $d_2 = 5$ and multiplies the rest of row 2 by 5. It sets the $(3,2)$ entry $\tfrac{1}{5}$ to 0 and performs Add $(3, 2; \tfrac{1}{5})$ (ignoring the entries still holding multipliers). Finally, it goes to $p = 2 - 1 = 1$, $q = $ PivotList $(1) = 1$, $d_1 = 6$ and multiplies the rest of row 1 by 6. It sets the $(2, 1)$ entry $\tfrac{1}{3}$ to 0 and performs Add $(2, 1; \tfrac{1}{3})$. Then it sets the $(3, 1)$ entry $\tfrac{1}{2}$ to 0 and performs Add $(3, 1; \tfrac{1}{2})$. Finally, it resets Row = $[1, 2, 3]$.

PROBLEMS

NUMERICAL PROBLEMS

1. Find L, D, U, and Row for the matrix $\begin{bmatrix}0&0&3&0&0\\0&4&0&0&0\\5&0&0&0&0\\0&0&0&0&6\\0&0&0&7&0\end{bmatrix}$. What are the values of Pivot List (p) $(p = 1, 2, 3, 4, 5)$?

MORE THEORETICAL PROBLEMS

Easier Problems

2. If the matrix A is symmetric, that is, $A = A'$, and no interchanges take place when the row reduction algorithm is applied to A, show that in the resulting factorization $A = LDU$, L is the transpose of U.

3. In Problem 2, show that the assumption that no interchanges take place is necessary.

13.4. BACK AND FORWARD SUBSTITUTION. SOLVING $Ax = y$

Now that we can use the row reduction algorithm to go from the $m \times n$ matrix A to the matrices L, D, U and the list Row, which was built from the interchanges during reduction, we ask: How do we use P, L, D, U, and Row to solve $Ax = y$? By Theorem 13.3.1, $A = PLDU$. So we can break up the problem of solving $Ax = y$ into parts, namely solving $Pu = y$, $Lv = u$, $Dw = v$, and $Ux = w$ for u, v, w, x. The only thing that is new here is solving $Pu = y$ for u. So let's look at this in the case of the preceding example. There we have Row $= [3, 1, 2]$ and P is the corresponding permutation matrix $P = \begin{bmatrix} 0 & 0 & 1 \\ 1 & 0 & 0 \\ 0 & 1 & 0 \end{bmatrix}$. So, solving

$$\begin{bmatrix} 0 & 0 & 1 \\ 1 & 0 & 0 \\ 0 & 1 & 0 \end{bmatrix} \begin{bmatrix} u_1 \\ u_2 \\ u_3 \end{bmatrix} = \begin{bmatrix} y_1 \\ y_2 \\ y_3 \end{bmatrix}$$

for $\begin{bmatrix} u_1 \\ u_2 \\ u_3 \end{bmatrix}$, we get

$$\begin{bmatrix} u_1 \\ u_2 \\ u_3 \end{bmatrix} = \begin{bmatrix} y_2 \\ y_3 \\ y_1 \end{bmatrix} = \begin{bmatrix} y_{\text{Row}(1)} \\ y_{\text{Row}(2)} \\ y_{\text{Row}(3)} \end{bmatrix}.$$

What this means is that we need make only one alteration in our earlier solution of $Ax = y$, namely, replace $\begin{bmatrix} y_2 \\ y_3 \\ y_1 \end{bmatrix}$ by $\begin{bmatrix} y_{\text{Row}(1)} \\ y_{\text{Row}(2)} \\ y_{\text{Row}(3)} \end{bmatrix}$. So, we now have the

Algorithm for solving $Ax = y$ for x

Use the row reduction algorithm to get the matrices L, D, U, and the list Row. Given a particular y in the column space of A, do the following:

1. Solve $Lv = y$ by the *forward substitution formula*

$$v_p = y_{\text{Row}(p)} - \sum_{j=1}^{p-1} L_{pj} v_j$$

for $p = 1$ to the rank of A.

2. Solve $Dw = v$ by the formula

$$w_p = v_p / D_{pp}$$

for $p = 1$ to the rank (A).

3. Solve $Ux = v$ by the *back substitution equations*:

$$x_q = w_p - \sum_{j=q+1}^{n} U_{pj} x_j \quad \text{if column } q \text{ contains a pivot entry; or}$$

$$x_q = 1 \quad \text{if column } q \text{ contains no pivot entry}$$

for $1 \leq q \leq n$.

When y is not in the column space of A, the above algorithm for solving $Ax = y$ exactly cannot be used. Instead, we can solve $Ax = y$ approximately by solving the normal equation $A'Ax = A'y$ exactly by the above algorithm. Since $A'y$ is in the column space of $A'A$, by our earlier discussion of the normal equation, this is always possible. So we have

Algorithm for solving $Ax = y$ for x approximately:

Use the algorithm for solving $A'Ax = A'y$ for x exactly.

If we need to solve $Ax = y$ approximately for many different vectors y, it is more efficient first to find A^- and then to use it to get $x = A^- y$ for each y. In the next section, we give an algorithm for finding A^- for any A.

We now turn to the important special case when the columns of A are linearly independent. In this case, we can solve the normal equation $A'Ax = A'y$ and find A^- efficiently by a simple algorithm involving the Gram–Schmidt Orthogonalization Process described in Section 4.3.

From the columns $v_1, \ldots, v_k$ of A, the Gram–Schmidt Orthogonalization Process gives us orthogonal vectors $w_1, \ldots, w_k$, where

$$w_s = v_s - \frac{(v_s, w_{s-1})}{(w_{s-1}, w_{s-1})} w_{s-1} - \cdots - \frac{(v_s, w_1)}{(w_1, w_1)} w_1$$

or

$$v_s = w_s + \frac{(v_s, w_{s-1})}{(w_{s-1}, w_{s-1})} w_{s-1} + \cdots + \frac{(v_s, w_1)}{(w_1, w_1)} w_1$$

for $1 \leq s \leq k$. Letting $u_1 = (1/|w_1|)w_1, \ldots, u_k = (1/|w_k|)w_k$, $b_{ss} = |w_s|$ and setting

$$b_{rs} = \frac{(v_s, w_r)}{(w_r, w_r)}|w_r|$$

for $r < s$, $1 \leq s \leq k$, we can rewrite this as

$$v_s = b_{1s}u_1 + \cdots + b_{s-1\,s}u_{s-1} + b_{ss}u_s$$

for $1 \leq s \leq k$. Letting Q be the m × k matrix whose columns are the orthonormal vectors $u_1,\ldots,u_k$ and R be the $k \times k$ matrix whose (r, s) entry is b_{rs} for $r \leq s$ and 0 for $r > s$, these equations imply that

$$A = QR$$

(Prove!). This is the so-called *QR factorization* of A as product of a matrix Q with orthonormal columns and an invertible upper triangular matrix R. (We leave it as an exercise for the reader to show that there is only one such factorization of A). So *applying the Gram–Schmidt Orthogonalization process to the columns of A to get orthonormal vectors* $u_1,\ldots,u_k$ *in the above manner gives us the QR factorization* $A = QR$.

For example, if $A = \begin{bmatrix} 1 & 2 \\ 3 & 4 \end{bmatrix}$, we get the orthogonal vectors

$$w_1 = \begin{bmatrix} 1 \\ 3 \end{bmatrix}$$

$$w_2 = \begin{bmatrix} 2 \\ 4 \end{bmatrix} - \frac{1 \cdot 2 + 3 \cdot 4}{1 \cdot 1 + 3 \cdot 3}\begin{bmatrix} 1 \\ 3 \end{bmatrix} = \begin{bmatrix} \frac{3}{5} \\ -\frac{1}{5} \end{bmatrix}$$

whose lengths are $|w_1| = \sqrt{10}$ and $|w_2| = (\frac{1}{5})\sqrt{10}$. From these, we get the orthonormal vectors $u_1 = (1/\sqrt{10})\begin{bmatrix} 1 \\ 3 \end{bmatrix}$ and $u_2 = (1/\sqrt{10})\begin{bmatrix} 3 \\ -1 \end{bmatrix}$. The equations

$$v_s = b_{1s}u_1 + \cdots + b_{ss}u_s \ (1 \leq s \leq 2)$$

are then

$$\begin{bmatrix} 1 \\ 3 \end{bmatrix} = \sqrt{10}\,u_1$$

$$\begin{bmatrix} 2 \\ 4 \end{bmatrix} = (\tfrac{7}{5})\sqrt{10}\,u_1 + (\tfrac{1}{5})\sqrt{10}\,u_2,$$

the matrices Q, R are $Q = (1/\sqrt{10})\begin{bmatrix} 1 & 3 \\ 3 & -1 \end{bmatrix}$ and $R = \sqrt{10}\begin{bmatrix} 1 & \frac{7}{5} \\ 0 & \frac{1}{5} \end{bmatrix}$ and the *QR*

Sec. 13.4] Back and Forward Substitution. Solving $Ax = y$

factorization of $\begin{bmatrix} 1 & 2 \\ 3 & 4 \end{bmatrix}$ is

$$\begin{bmatrix} 1 & 2 \\ 3 & 4 \end{bmatrix} = \left(\frac{1}{\sqrt{10}} \begin{bmatrix} 1 & 3 \\ 3 & -1 \end{bmatrix} \right) \left(\sqrt{10} \begin{bmatrix} 1 & \frac{7}{5} \\ 0 & \frac{1}{5} \end{bmatrix} \right).$$

Given the QR factorization $A = QR$ for a matrix A with independent columns, the normal equation

$$A'Ax = A'y$$

can be solved for x easily and efficiently. Replacing A by QR in the normal equation $A'Ax = A'y$, we get $R'Q'QRx = R'Q'y$. Since R and R' are invertible and the columns of Q are orthonormal, this simplifies to

$$Rx = Q'y$$

and

$$x = R^{-1}Q'y$$

(Prove!). Since R is upper triangular and invertible, the equation $Rx = Q'y$ can be solved for x using the back substitution equations

$$b_{qq}x_q = z_q - \sum_{j=q+1}^{k} b_{qj}x_j$$

where z_q is entry q of $Q'y$ ($1 \leq q \leq k$). Moreover, since R is invertible, *there is only one such solution x*. So x is the shortest approximate solution to $Ax = y$, from which it follows that $A^- = R^{-1}Q'$.

Computing the inverse R^{-1} can be done very easily, since R is upper triangular. One simply finds the columns $c_1, \ldots, c_k$ of R^{-1} as the solutions c_s to the equations $Rc_s = e_s$ (column s of the $k \times k$ identity matrix) using back substitution equations:

$$b_{qq}c_{qs} = 0 - \sum_{j=q+1}^{k} b_{qj}c_{js} \qquad \text{for } q \neq s$$

$$b_{ss}c_{ss} = 1 - \sum_{j=q+1}^{k} b_{sj}c_{js}$$

for each s with $1 \leq s \leq k$. As it turns out, $c_{js} = 0$ for $j > s$. So the above back substitution equations simplify to the equations

$$b_{qq}c_{qs} = - \sum_{j=q+1}^{s} b_{qj}c_{js} \qquad \text{for } q < s$$

$$b_{ss}c_{ss} = 1$$

$$c_{qs} = 0 \qquad \text{for } q > s$$

for $1 \leq s \leq k$. We leave it as an exercise for the reader to prove directly that the entries c_{qs} of the inverse R^{-1} satisfy these equations and are completely determined by them.

We summarize all of this by formulating the following algorithms.

Algorithm for solving $Ax = y$ for x approximately when the columns of A are linearly independent:

1. Use the QR factorization $A = QR$ to get the equation $Rx = Q'y$ (which replaces the normal equation $A'Ax = A'y$).
2. Solve $Rx = Q'y$ for x by the back substitution equations

$$b_{qq}x_q = z_q - \sum_{j=q+1}^{k} b_{qj}x_j$$

where z_q is entry q of $Q'y$ for $1 \leq q \leq k$.

Algorithm for finding R^{-1} for an invertible upper triangular matrix R:

Letting b_{rs} denote the (r, s) entry of an invertible upper triangular $k \times k$ matrix R, the entries c_{rs} of R^{-1} are determined by the back substitution equations

$$b_{rr}c_{rs} = -\sum_{j=r+1}^{s} b_{rj}c_{js} \qquad \text{for } r < s$$

$$b_{ss}c_{ss} = 1$$

$$c_{rs} = 0 \qquad \text{for } r > s$$

for $1 \leq s \leq k$.

Algorithm for finding A^- when the columns of A are linearly independent:

1. Use the QR factorization $A = QR$ to get Q and R.
2. Find R^{-1} by the above algorithm.
3. Then $A^- = R^{-1}Q'$.

PROBLEMS

NUMERICAL PROBLEMS

1. Suppose that the row reduction algorithm, applied to A, gives us $L = \begin{bmatrix} 1 & 0 & 0 \\ 2 & 1 & 0 \\ 3 & 0 & 1 \\ 2 & 0 & 5 \end{bmatrix}$,

Sec. 13.4]   Back and Forward Substitution. Solving $Ax = y$ 481

$$D = \begin{bmatrix} 2 & 0 & 0 \\ 0 & 5 & 0 \\ 0 & 0 & 3 \end{bmatrix}, \quad U = \begin{bmatrix} 1 & 2 & 3 & 4 & 3 & 3 & 3 \\ 0 & 0 & 1 & 2 & 2 & 2 & 2 \\ 0 & 0 & 0 & 0 & 0 & 1 & 6 \end{bmatrix},$$

and Row $= [2, 3, 4, 1]$. Then solve $Ax = y$ (or show that there is no solution) for

the vectors $\begin{bmatrix} 1 \\ 0 \\ 1 \\ 1 \end{bmatrix}, \begin{bmatrix} 1 \\ 0 \\ 1 \\ 5 \end{bmatrix}$, by modifying the discussion in the related example in Section 11.2.

2. Find the inverse of $\begin{bmatrix} 2 & 3 & 4 \\ 0 & 5 & 6 \\ 0 & 0 & 7 \end{bmatrix}$ by the method described in this section.

3. Find the QR decomposition of the matrix $\begin{bmatrix} 1 & 3 \\ 2 & 4 \end{bmatrix}$ and compare it with the example given in this section.

4. Using the QR decomposition, find the approximate inverse of the matrix $\begin{bmatrix} 1 & 2 \\ 0 & 0 \\ 3 & 4 \end{bmatrix}$.

MORE THEORETICAL PROBLEMS

Easier Problems

5. For an invertible upper triangular $k \times k$ matrix R with entries b_{rs}, show that the entries c_{rs} of R^{-1} satisfy the equations

$$b_{qq}c_{qs} = -\sum_{j=q+1}^{s} b_{qj}c_{js} \qquad \text{for } q < s$$

$$b_{ss}c_{ss} = 1$$

$$c_{qs} = 0 \qquad \qquad \text{for } q > s$$

for $1 \leq s \leq k$, and that they are completely determined by them.

Middle-Level Problems

6. Suppose that vectors $v_1, \ldots, v_k \in F^{(m)}$ are expressed as linear combinations

$$v_s = b_{1s}u_1 + \cdots + b_{ks}u_k \qquad (1 \leq s \leq k)$$

of vectors $u_1, \ldots, u_k \in F^{(m)}$. Letting A be the $m \times k$ matrix whose columns are $v_1, \ldots, v_k$, Q be the $m \times k$ matrix whose columns are $u_1, \ldots, u_k$, and R be the $k \times k$ matrix whose (r, s) entry is b_{rs} for $1 \leq r, s \leq k$, show that $A = QR$.

Harder Problems

7. Show that if $QR = ST$, where Q and S are $m \times k$ matrices whose columns are orthonormal and R and T are invertible upper triangular $k \times k$ matrices, then $Q = S$ and $R = T$.

13.5. APPROXIMATE INVERSE AND PROJECTION ALGORITHMS

In Chapter 12 we saw how to find the approximate solution x to an equation $Ax = y$, where A is an $m \times n$ real matrix. To do this efficiently for each of a larger number of different y, we should first get A^- and then compute x as $x = A^- y$. How do we get A^-? In principle, we can get A^- by using the methods of Chapter 12 to calculate each of its columns $A^- e_s$ (where e_s is column s of the identity matrix) as an approximate solution x_s to the equation $Ax_s = e_s$. However, there are more efficient methods, which are based on Theorem 12.3.5 and diagonalization of a symmetric matrix. Unfortunately, however, these methods are also somewhat complicated.

To avoid the complications, we have worked out an efficient new method for finding A^- which uses only elementary row operations. This method, a variation of the method in Chapter 3 for finding the inverse of an invertible matrix, is based on two facts. The first of these is that the matrix $\text{Proj}_{A'\mathbb{R}^{(m)}}$ is just $J'J$ where J is gotten by row reducing $A'A$ to an orthonormalized matrix in the sense of

Definition. An *orthonormalized matrix* is an $m \times n$ matrix J satisfying the following conditions:

1. Each nonzero row of J has length 1.
2. Any two different nonzero rows of J are orthogonal.

We can always row reduce a matrix to an orthonormalized upper triangular matrix. How? First, reduce it to an echelon matrix. Then orthonormalize the nonzero rows in the reverse order $r, \ldots, 1$ by performing the following row operations on A for each value of k from r down to 1:

1. Multiply $(k; 1/u_k)$, where u_k is the current length of row k.
2. For each value of q from $k-1$ down to 1, Add $(q, k; -v_{kq})$, where v_{kq} is the inner product of the current rows k and q.

How do we show that $\text{Proj}_{A'\mathbb{R}^{(m)}}$ equals $J'J$? First, we need some preliminary tools.

Definition. A matrix $P \in M_n(\mathbb{R})$ is a *projection* if $P = P'$ and $P^2 = P$.

Theorem 13.5.1. If P is a projection, then $P = \text{Proj}_{P\mathbb{R}^{(n)}}$.

Proof: Let's denote the column space of P by W. Let $v \in \mathbb{R}^{(n)}$ and write $v = v_1 + v_2$, where $v_1 \in W$, $v_2 \in W^\perp$. Then $v_1 = Pu$ fo some u, so that $Pv_1 = P^2u = Pu = v_1$. Thus $Pv_1 = v_1$. Letting u now represent an arbitrary element of $\mathbb{R}^{(n)}$, Pu is

Sec. 13.5]   Approximate Inverse and Projection Algorithms   483

in W, so that Pu and v_2 are orthogonal. This implies that

$$0 = (Pu)'v_2 = u'P'v_2 = u'Pv_2.$$

But then Pv_2 is orthogonal to u for all $u \in \mathbb{R}^{(n)}$, which implies that $Pv_2 = 0$. It follows that $Pv = Pv_1 + Pv_2 = v_1 + 0 = v_1 = \text{Proj}_W(v)$ for all v, that is, $P = \text{Proj}_W(v)$. ∎

By Theorem 11.4.3, the nullspaces of $A'A$ and A are equal. So since the matrices $A'A$ and A both have n columns and the rank plus the nullity adds up to n for both of them, by Theorem 7.2.6, the ranks of $A'A$ and A are equal. Using this, we can prove something even stronger, namely

Theorem 13.5.2. The column spaces of $A'A$ and A' are the same.

Proof: Certainly, the column space of $A'A$ is contained in the column space of A'. Since the dimensions of the column spaces of $A'A$ and A' are the ranks of $A'A$ and A, respectively, and since these are equal as we just saw, it follows that the column spaces of $A'A$ and A' are equal. ∎

We now can prove two theorems that give us row reduction algorithms to compute $\text{Proj}_{A'\mathbb{R}^{(m)}}$, the nullspace of A and A^-.

Theorem 13.5.3. Let A be a real $m \times n$ matrix. Then for any orthonormalized matrix J that is row equivalent to A, $\text{Proj}_{A'\mathbb{R}^{(m)}} = J'J$ and the columns of $I - J'J$ span the nullspace of A.

Proof: Since J is an orthonormalized matrix, we get $(JJ')J = J$. But then $J'JJ'J = J'J$. Since $(J'J)' = J'J$ and $(J'J)^2 = J'J$, $J'J$ is a projection and $J'J = \text{Proj}_{J'J\mathbb{R}^{(n)}}$. Since J and A are row equivalent, J' and A' have the same column spaces. So, by Theorem 13.5.2, $J'J$, J', A' have the same column spaces. But then $J'J = \text{Proj}_{J'J\mathbb{R}^{(n)}} = \text{Proj}_{A'\mathbb{R}^{(m)}}$. It follows that the columns of $I - J'J$ span the nullspace of A, since:

1. The nullspace of A is $(A'\mathbb{R}^{(m)})^\perp$ (Prove!).
2. $(A'\mathbb{R}^{(m)})^\perp = (J'J\mathbb{R}^{(n)})^\perp = (I - J'J)\mathbb{R}^{(n)}$ since $J'J$ is a projection (Prove!). ∎

Lemma 13.5.4. Let $P \in M_n(\mathbb{R})$ satisfy the equation $PP' = P'$. Then P is a projection.

Proof: Since $P' = PP'$, $P = P'$. But then $P = P' = PP' = PP = P^2$, so P is a projection. ∎

Theorem 13.5.5. Let A be an $m \times n$ real matrix, and let M be an invertible $n \times n$ matrix such that $J = MA'A$ is orthonormalized. Then $A^- = J'MA'$.

Proof: Let $B = J'MA'$. We claim first that

1. $BA = J'J = \text{Proj}_{A'\mathbb{R}^{(m)}}$
2. $BAB = B$
3. $AB = \text{Proj}_{A\mathbb{R}^{(n)}}$.

Since $BA = J'MA'A = J'J$, and since $A'A$ and A' have the same column space by Theorem 13.5.2, (1) follows from Theorem 13.5.3. For (2), we first use the equation $JJ'J = J$ from the proof of Theorem 13.5.3 to get the equation $J'JJ' = J'$. Then $BAB = J'JB = J'JJ'MA' = J'MA' = B$. For (3), we first show that AB is a projection. By Lemma 13.5.4 it suffices to show that $(AB)(AB)' = (AB)'$, which follows from the equations

$$AB(AB)' = (AJ'MA')(AM'JA') = AJ'(MA'A)M'JA'$$
$$= AJ'JM'JA' = (AJ'J)M'JA' = AM'JA' = (AB)'.$$

Here we use the fact that since $J'JA' = A'$, by Theorem 13.5.3, $AJ'J = A$. (Prove!) Finally, the column space of A contains $AB\mathbb{R}^{(m)}$, which in turn contains $ABA\mathbb{R}^{(n)}$ $A(BA\mathbb{R}^{(n)})$, which in turn is $A(A'\mathbb{R}^{(m)})$ by (1). Since AA' and A have the same column space by Theorem 13.5.2, it follows that all these spaces are actually equal; that is,

$$A\mathbb{R}^{(n)} = AB\mathbb{R}^{(m)} = ABA\mathbb{R}^{(n)} = A(BA\mathbb{R}^{(n)}) = A(A'\mathbb{R}^{(m)}) = A\mathbb{R}^{(n)}.$$

So A and AB have the same column spaces. But then the projection AB is just $AB = \text{Proj}_{A\mathbb{R}^{(n)}}$.

To show that $B = A^-$, let $x = By$. Then from (3) we get that $Ax = ABy = \text{Proj}_{A\mathbb{R}^{(n)}} y$, and from (1) and (2) we get that $\text{Proj}_{A'\mathbb{R}^{(m)}} x = BAx = BABy = By = x$. So x is the shortest solution to $Ax = \text{Proj}_{A\mathbb{R}^{(n)}} y$ and $x = A^- y$. Since this is true for all y, we get that $B = A^-$. ∎

From these theorems and Theorem 12.2.2, we get the following methods, each of which builds toward the later ones:

Algorithm to compute the projection $\text{Proj}_{A'\mathbb{R}^{(m)}}$ for a real $m \times n$ matrix A

1. Row reduce $A'A$ to an orthonormalized matrix $J = MA'A$.
2. Then $\text{Proj}_{A'\mathbb{R}^{(m)}}$ is $J'J$.

Algorithm to compute the nullspace of a real $m \times n$ matrix A

3. Then the columns of $I - J'J$ span the nullspace of A.

Algorithm to compute the approximate inverse of a real $m \times n$ matrix A

4. Then A^- is $J'K$, where $K = MA'$.

Algorithm to compute the projection $\text{Proj}_{A'\mathbb{R}^{(n)}}$ for a real $m \times n$ matrix A

5. Then $\text{Proj}_{A'\mathbb{R}^{(m)}}$ is AA^-.

Algorithm to find all approximate solutions of $Ax = y$

6. The shortest approximate solution to $Ax = y$ is $x = A^- y$, which we have by (4).
7. Every approximate solution is $x + w$, where $w \in (I - J'J)\mathbb{R}^{(n)}$, by (3).

It is instructive to look at some examples.

EXAMPLES

1. To compute the projection of $\mathbb{R}^{(2)}$ onto the column space of $\begin{bmatrix} 1 \\ 2 \end{bmatrix}$, row reduce $\begin{bmatrix} 1 \\ 2 \end{bmatrix}[1,2] = \begin{bmatrix} 1 & 2 \\ 2 & 4 \end{bmatrix}$ to its orthonormalized echelon form

$$J = \begin{bmatrix} 1/\sqrt{5} & 2/\sqrt{5} \\ 0 & 0 \end{bmatrix}.$$

Then the projection is $J'J = \begin{bmatrix} \frac{1}{5} & \frac{2}{5} \\ \frac{2}{5} & \frac{4}{5} \end{bmatrix}$.

2. Let's find A^- for $A = \begin{bmatrix} 2 & 3 \\ 0 & 4 \\ 0 & 1 \end{bmatrix}$. Since

$$A'A = \begin{bmatrix} 2 & 0 & 0 \\ 3 & 4 & 1 \end{bmatrix} \begin{bmatrix} 2 & 3 \\ 0 & 4 \\ 0 & 1 \end{bmatrix} = \begin{bmatrix} 4 & 6 \\ 6 & 26 \end{bmatrix},$$

we row reduce $\begin{bmatrix} 4 & 6 & 2 & 0 & 0 \\ 6 & 26 & 3 & 4 & 1 \end{bmatrix}$ to the echelon form $\begin{bmatrix} 1 & 1.5 & .5 & 0 & 0 \\ 0 & 1 & 0 & \frac{4}{17} & \frac{1}{17} \end{bmatrix}$.

We then apply the operation Add $(1, 2, -v_{21})$ where v_{21} is the inner product 1.5 of $(1, 1.5)$ and $v = (0, 1)$, getting $\begin{bmatrix} 1 & 0 & .5 & -\frac{6}{17} & -\frac{1.5}{17} \\ 0 & 1 & 0 & \frac{4}{17} & \frac{1}{17} \end{bmatrix}$. Since $J = I$,

A^- is $IK = K = \begin{bmatrix} .5 & -\frac{6}{17} & -\frac{1.5}{17} \\ 0 & \frac{4}{17} & \frac{1}{17} \end{bmatrix}$, the same answer that we got when we computed A^- directly in Section 12.2.

Example 2 illustrates the fact that in the case where the columns of A are linearly independent, the row reduction is the same as reduction of $A'A$ to reduced echelon form I, with the operations applied to the augmented matrix. In this case, $A^- = K$.

EXAMPLE

Let's next find A^- for the matrix $A = \begin{bmatrix} 2 & 3 & 5 \\ 0 & 4 & 4 \\ 0 & 1 & 1 \end{bmatrix}$, to see what happens when the columns of A are linearly dependent. Since

$$A'A = \begin{bmatrix} 2 & 0 & 0 \\ 3 & 4 & 1 \\ 5 & 4 & 1 \end{bmatrix} \begin{bmatrix} 2 & 3 & 5 \\ 0 & 4 & 4 \\ 0 & 1 & 1 \end{bmatrix} = \begin{bmatrix} 4 & 6 & 10 \\ 6 & 26 & 32 \\ 10 & 32 & 42 \end{bmatrix},$$

we row reduce $\begin{bmatrix} 4 & 6 & 10 & 2 & 0 & 0 \\ 6 & 26 & 32 & 3 & 4 & 1 \\ 10 & 32 & 42 & 5 & 4 & 1 \end{bmatrix}$ successively to the matrices

$$\begin{bmatrix} 2 & 3 & 5 & 1 & 0 & 0 \\ 6 & 26 & 32 & 3 & 4 & 1 \\ 10 & 32 & 42 & 5 & 4 & 1 \end{bmatrix}, \begin{bmatrix} 2 & 3 & 5 & 1 & 0 & 0 \\ 0 & 17 & 17 & 0 & 4 & 1 \\ 0 & 17 & 17 & 0 & 4 & 1 \end{bmatrix},$$

$$\begin{bmatrix} 2 & 3 & 5 & 1 & 0 & 0 \\ 0 & 17 & 17 & 0 & 4 & 1 \\ 0 & 0 & 0 & 0 & 0 & 0 \end{bmatrix}, \begin{bmatrix} 2 & 3 & 5 & 1 & 0 & 0 \\ 0 & 1 & 1 & 0 & \frac{4}{17} & \frac{1}{17} \\ 0 & 0 & 0 & 0 & 0 & 0 \end{bmatrix},$$

To avoid further fractions, we orthogonalize $(2, 3, 5)$ and $(0, 1, 1)$ directly, by the operation Add $(1, 2; -4)$, where 4 was chosen as the inner product 8 of $(2, 3, 5)$ and $(0, 1, 1)$ divided by the inner product 2 of $(0, 1, 1)$ and $(0, 1, 1)$. We then get

$$\begin{bmatrix} 2 & -1 & 1 & 1 & -\frac{16}{17} & -\frac{4}{17} \\ 0 & 1 & 1 & 0 & \frac{4}{17} & \frac{1}{17} \\ 0 & 0 & 0 & 0 & 0 & 0 \end{bmatrix}.$$

To normalize the orthogonal vectors $(2, -1, 1)$, $(0, 1, 1)$ to vectors of length 1, we apply the operations Multiply $(1, 1/\sqrt{6})$ and Multiply $(2, 1/\sqrt{2})$, getting

$$\begin{bmatrix} .8165 & -.4082 & .4082 & .4082 & -.3842 & -.0961 \\ 0 & .7071 & .7071 & 0 & .1664 & .0416 \\ 0 & 0 & 0 & 0 & 0 & 0 \end{bmatrix}.$$

So $A^- = J'K$ is the matrix

$$\begin{bmatrix} .8165 & 0 & 0 \\ -.4082 & .7071 & 0 \\ .4082 & .7071 & 0 \end{bmatrix} \begin{bmatrix} .4082 & -.3842 & -.0961 \\ 0 & .1664 & .0416 \\ 0 & 0 & 0 \end{bmatrix},$$

which we multiply out to

$$A^- = \begin{bmatrix} .3333 & -.3137 & -.0784 \\ -.1667 & .2745 & .0686 \\ .1667 & -.0391 & -.0098 \end{bmatrix}.$$

Sec. 13.5]   Approximate Inverse and Projection Algorithms

We can check to see if this does what it is supposed to do, by checking whether

$A^- \begin{bmatrix} 5000 \\ 4000 \\ 2000 \end{bmatrix}$ is the approximate inverse $\begin{bmatrix} 254.90 \\ 401.96 \\ 656.86 \end{bmatrix}$ which we calculated in Section 11.5. We find that

$$A^- \begin{bmatrix} 5000 \\ 4000 \\ 2000 \end{bmatrix} = \begin{bmatrix} .3333 & -.3137 & -.0784 \\ .1667 & .2745 & .0686 \\ .1667 & -.0391 & -.0098 \end{bmatrix} \begin{bmatrix} 5000 \\ 4000 \\ 2000 \end{bmatrix} = \begin{bmatrix} 254.90 \\ 402.96 \\ 657.86 \end{bmatrix},$$

which is the same.

In using the algorithms discussed above, note that:

1. To find approximate solutions x to $Ax = y$ for each of a large number of y, the most efficient way to proceed, using the algorithm above, is probably to keep A^- in its factored form $A^- = J'K$ and get $x = A^- y$ for each y by getting first $v = Ky$, then $x = J'v$. Here J' is lower triangular and the number of nonzero rows of K equals the rank of A. (Prove!)
2. The algorithm requires only enough memory to hold the original matrix A and the matrix $A'A$. Then $J'J$ will occupy the memory that had held $A'A$, and K occupies the memory that had held A.
3. We can carry out the multiplication $J'K$ to get A^- without using additional memory, by multiplying J' times K column by column until done. As a result of doing this, A will be replaced by A^- in the memory that had held it.
4. We can recover A^- by repeating the process, replacing A^- by $(A^-)^-$. Then the resulting matrix is just A (see Problem 4).
5. When memory is limited, we can assume without loss of efficiency that $m \geq n$, since A^- is the transpose of A'^- (see Problem 5).

PROBLEMS

NUMERICAL PROBLEMS

1. Compute A', A'^-, A^-, $(A^-)^-$ for the following matrices.

    (a) $\begin{bmatrix} 1 & 0 & 1 & 1 \\ 1 & 2 & 0 & 1 \end{bmatrix}$.

    (b) $\begin{bmatrix} 1 & 2 & 2 \\ 2 & 2 & 2 \\ 3 & 2 & 2 \end{bmatrix}$.

    (c) $\begin{bmatrix} 1 & 2 & 3 \\ 2 & 2 & 3 \\ 3 & 2 & 1 \end{bmatrix}$.

MORE THEORETICAL PROBLEMS

Easier Problems

2. Show that if $B = A^-$, then $ABA = A$, $BAB = B$, $(AB)' = AB$, $(BA)' = BA$.

Middle-Level Problems

3. Show that for any $m \times n$ matrix A, there is exactly one $n \times m$ matrix B such that $ABA = A$, $BAB = B$, $(AB)' = AB$, $(BA)' = BA$, namely $B = A^-$.

Harder Problems

4. Using Problem 3, show that $(A^-)^- = A$.
5. Using Problem 3, show that $(A^-)' = (A')^-$.
6. Using Problem 3, show that $(A'A)^- = A^- A'^-$.
7. Using Problem 3, show that $J^- = J'$.

Very Hard Problems

8. Using Problem 3, show that if A is given in its LDU factored format $A = LDU$, where A is $m \times n$ of rank r, L' is echelon $r \times m$, D is nonsingular $r \times r$, and U is echelon $r \times n$, then $A^- = U^- D^- L^-$.
9. Show that when A^- is factored in the form $A^- = J'K$ as described above, only the first r rows of K are nonzero where $r = r(A)$.

13.6. A COMPUTER PROGRAM FOR FINDING EXACT AND APPROXIMATE SOLUTIONS

Having studied the basic algorithms related to solving equations, we now illustrate their use in a program for finding exact and approximate solutions to any matrix equation.

The program is written in TURBO PASCAL* for use on microcomputer. It can easily be rewritten in standard PASCAL by bypassing a few special features of TURBO PASCAL used in the program.

This program enables you to load, use and save *matrix files*. Here, a matrix file is a file whose first line gives the row and column degrees of a matrix, as positive integers, and whose subsequent lines give the entries, as real numbers. For example, the 2×3 matrix

$$A = \begin{bmatrix} 1.000 & -2.113 & 4.145 \\ 3.112 & 4.212 & 5.413 \end{bmatrix}$$

is represented by the contents of the matrix file

```
    2        3
1.000    -2.113    4.145
3.112     4.212    5.413
```

* TURBO PASCAL is a trademark of Borland International.

Sec. 13.6] Computer Program for Exact, Approximate Solutions

To load a matrix file, the program instructs the computer to get the row and column degrees m and n from the first line, then to load the entries into an $m \times n$ matrix structure. So, in the example, it gets the degrees 2 and 3 and loads the entries, as real numbers, into a 2×3 matrix structure.

While running the program, you have the following options, which you can select any number of times in any order by entering the commands L, S, W, E, A, D, U, I, X and supplying appropriate data or choices when prompted to supply it:

L Load a matrix A from a matrix file
   (You will be asked to supply the file name for a matrix file. If you wish, you can supply CON, the special file name for the keyboard, in which case the computer loads A as you enter it from the keyboard—with two positive integers m and n on the first line, and n real numbers on each of the next m lines.)

S Save a matrix A to a matrix file
   You will be asked to supply the file name. If you wish, you can supply LPT1, the special file name for the printer, in which case the computer prints A on the printer.)

W Display a window of entries of the current matrix A

E Solve $Ax = y$ for x exactly, if x exists, and approximately if not
   (You will be asked to enter y's for which x's are desired until you indicate that you are done)

A Find the shortest approximate solution x to $Ax = y$ for any given matrix A
   (You will be asked to enter y's for which x's are desired until you indicate that you are done)

D Decompose A, given in matrix format, into its LDU factored format

U Undo the LDU decomposition to recover A, given in LDU factored format, into its matrix format

I Find the approximate inverse of A, which is the exact inverse when A is invertible

X Exit from the program

PROBLEMS

Note: In Problems 1 to 9, you are asked to write various procedures to add to the above program. As you write these procedures, integrate them into the programs and expand the menu accordingly.

1. Write a procedure for computing the inverse of an invertible upper or lower triangular matrix, using the back substitution equations of Section 13.4.
2. Write a procedure that uses the *LDU* factorization and the procedure in Problem 1 to compute the inverse of any invertible matrix A in its factored format $A^{-1} = U^{-1}D^{-1}L^{-1}$.
3. Write a procedure for determining whether the columns of a given matrix A are linearly independent.
4. Write a procedure for determining whether a given vector is in the column space of the current matrix.

5. Write a procedure for computing the matrices Q and R of the QR factorization of a matrix A with linearly independent columns.
6. Write a procedure for computing the approximate inverse of A using the QR factorization of A, when the columns of A are linearly independent, by solving the equations

$$Ax_1 = e_1, \ldots, Ax_n = e_n$$

approximately.

7. Write a procedure for computing, for a given matrix A, a matrix N whose columns form a basis for the nullspace of A.
8. Write a procedure to compute the matrices $\text{Proj}_{A(\mathbb{R}^{(n)})}$ and $\text{Proj}_{A'(\mathbb{R}^{(m)})}$.
9. Write a set of procedures and add to the menu an option EDIT to enable the user to create and edit matrices. EDIT shoud enable you to:
   (a) Use the cursor keys to control what portion of a large current matrix is displayed on the screen. Alternately, control what portion of a large current matrix is displayed on the screen by specifying a row and a column.
   (b) Use the cursor keys to move the cursor to a desired entry displayed on the screen.
   (c) Change the entry under the cursor.
   (d) Copy to a matrix file all or part of the current matrix defined by specifying ranges for rows and columns.
   (e) Copy from a matrix file into a specified portion of the current matrix.
10. Rewrite the program LinAlg in standard Pascal.

Computer Program

```
PROGRAM LinAlg;    {Version 1.0.           An instructional aid.}

{************************* DECLARATIONS ******************************}

CONST
  Epsilon = 0.1E-6; MaxDegree = 16;Format = 3;

TYPE
  IntegerVector   = ARRAY  [1..MaxDegree            ]  OF INTEGER;
  RealVector      = ARRAY  [1..MaxDegree            ]  OF REAL;
  RealMatrix      = ARRAY  [1..MaxDegree,1..MaxDegree ] OF REAL;
  StringType      = STRING [80];

VAR
  TurnedOn                  {Turns on the program loop} : BOOLEAN;
  RowDegA,ColDegA,RankA,PivotRowA,RowDegB,ColDegB       : INTEGER;
  ZeroList, IdentityList, RowListA, ColListA, PivotListA,
  InvRwListA , InvClListA                               : IntegerVector;
  MatrixA, MatrixB                                      : RealMatrix;
  MatrixEntry                                           : REAL;
  v,w,x,y     {Solve Lv=y, Dw=v, Ux=w so Ax=LDUx=LDw=Lv=y} : RealVector;
  InputFile,OutputFile                                  : TEXT;
  InputFileN, OutputFileN, AnySt                        : StringType;

{*********************** LDU DECOMPOSITION OF A **************************}

PROCEDURE InterchangeRow (p,q: INTEGER);

{ Interchange rows p and q in both matrices A and B. }

VAR  CopyOfRow: INTEGER;
BEGIN
  CopyOfRow:=RowListA[q]; RowListA[q] := RowListA[p]; RowListA[p]:=CopyOfRow;
END;

PROCEDURE   AddRow(p,q :INTEGER;u:REAL;TruncationColA,ClDgB:INTEGER);

{ Add u times row q to row p in both matrices A and B, skipping the storage
  area in A determined by TruncationColA. }

VAR  s : INTEGER;
BEGIN
  FOR s := TruncationColA TO ColDegA DO        { Skipping the storage area, }
    MatrixA [RowListA[p],ColListA[s]]               { add u times row q to row p. }
      := MatrixA[RowListA[p],ColListA[s]] + u*MatrixA[RowListA[q],ColListA[s]];
  FOR s := 1 TO ClDgB DO                  { Operate in parallel on MatrixB. }
    MatrixB [RowListA[p],ColListA[s]]
      := MatrixB[RowListA[p],ColListA[s]] + u*MatrixB[RowListA[q],ColListA[s]];
END;
```

492 Computer Program

```
PROCEDURE GetPivotRow (p,q : INTEGER; VAR PivotRow: INTEGER);

{ Get the PivotRow, when at the stage of row p and column q. }

VAR  r:INTEGER;
BEGIN
  PivotRow := p;   { We record our best guess, at the outset, for PivotRow. }
  FOR r := p+1 TO RowDegA DO
    BEGIN          { We then look for a row with a bigger entry of column q. }
      IF (ABS(MatrixA [RowListA[r],ColListA[q]])
      > ABS(MatrixA [RowListA[PivotRow],ColListA[q]]))
      THEN PivotRow := r;      { If we find one, we update our last guess. }
    END;
END;

PROCEDURE ReduceAndStoreMultipliers (VAR p,q: INTEGER; VAR PivotEntry: REAL;
                          ClDgB: INTEGER; ReduceUtoDU, StoreL: BOOLEAN);

{ For PivotEntry in row p and column q, use the multipliers to reduce the
  subsequent rows and, if StoreL, store them below the PivotEntry as entries
  of L. }

VAR  t: INTEGER; Multiplier: REAL;
BEGIN
  FOR t := p+1 TO RowDegA DO
    BEGIN
      Multiplier := MatrixA[RowListA[t],ColListA[q]]/PivotEntry;
      AddRow(t,p,-Multiplier,q,ClDgB);           { Reduce MatrixA, MatrixB. }
      IF StoreL THEN MatrixA [RowListA[t],ColListA[q]] := Multiplier
      ELSE  MatrixA [RowListA[t],ColListA[q]] := 0;
    END;

{ If ReduceUtoDU, further reduce the factorizations A = LU, B = LV to
  A = LDU, B = LDV. }

  IF ReduceUtoDU THEN begin
    FOR t := q+1 TO ColDegA DO
      MatrixA [RowListA[p],ColListA[t]]
        := MatrixA[RowListA[p],ColListA[t]]/PivotEntry;
    FOR t := 1 TO ClDgB DO
      MatrixB [RowListA[p],ColListA[t]]
        := MatrixB[RowListA[p],ColListA[t]]/PivotEntry;
  END;
END;

PROCEDURE DoLDUdecompose (ClDgB: INTEGER; ReduceUtoDU, StoreL: BOOLEAN);

{ Starting with PivotRowA = 1, PivotColA = 1, keep getting the next Pivots
  and decompose Matrix A to its LDU factored format. }

VAR  PivotRow, PivotColA :INTEGER; VAR PivotEntry:REAL;
BEGIN
  IF PivotRowA<>0 THEN         { Do Nothing - the matrix is in LDU Format. }
    ELSE BEGIN                   { Begin to put the matrix in LDU Format. }
```

```
    WRITELN('DoLDUdecompose...');
    RowListA:=IdentityList; ColListA:=IdentityList;
    PivotRowA:=1; PivotColA:=1;            { Start in upper left hand corner }
    WHILE ((PivotRowA<=RowDegA) AND (PivotColA<=ColDegA)) DO BEGIN
      GetPivotRow(PivotRowA,PivotColA,PivotRow);      { and get PivotRow. }
      PivotEntry := MatrixA [RowListA[PivotRow],ColListA[PivotColA]];
      IF (ABS(PivotEntry) < Epsilon) THEN PivotRowA:=PivotRowA-1 {Try again.}
      ELSE                              { PivotColA is a PivotColumn. }
      BEGIN
        PivotListA[PivotRowA]:=ColListA[PivotColA];{Save PivotColA for UnDo.}
        IF PivotRow <> PivotRowA THEN InterchangeRow(PivotRowA,PivotRow);
        ReduceAndStoreMultipliers (PivotRowA,PivotColA,PivotEntry,ClDgB,
                            ReduceUtoDU, StoreL);
      END;
      PivotRowA:=PivotRowA+1;PivotColA:=PivotColA+1; { Move on to the next. }
    END;
    PivotRowA:=PivotRowA-1;RankA:=PivotRowA; { Maintain PivotRowA for UnDo. }
  END;
END;

PROCEDURE RetrieveMultipliersAndRebuild (VAR p,q:INTEGER;ClDgB:INTEGER;
                                    ReduceUtoDU:Boolean);

{ Retrieve the stored Multipliers from L, to use to UnDo the decomposition
  of Matrix A to its LDU factored format and rebuild A. }

VAR  t:INTEGER;  Multiplier, PivotEntry:REAL;
BEGIN
    q:=PivotListA[p];                         { Get the pivot column q. }
    PivotListA[p]:=0 ;                        { Undo PivotListA too. }
    PivotEntry:=MatrixA[RowListA[p],ColListA[q]];    {Undo U in Matrix A. }
    IF ReduceUtoDU THEN BEGIN                 { to Undo DU to U. }
      FOR t := q+1 TO ColDegA DO
        MatrixA [RowListA[p],ColListA[t]]
          := MatrixA[RowListA[p],ColListA[t]]*PivotEntry;
      FOR t := 1 TO ClDgB DO                  { Undo in Matrix B too. }
        MatrixB [RowListA[p],ColListA[t]]
          := MatrixB[RowListA[p],ColListA[t]]*PivotEntry;
    END;
    FOR t:= RowDegA DOWNTO p+1 DO                       { Undo U to A. }
      BEGIN
        Multiplier := MatrixA[RowListA[t],ColListA[q]];{UnDo StoreMultipliers.}
        MatrixA [RowListA[t],ColListA[q]] := 0;
        AddRow(t,p, Multiplier,q,ClDgB);                { UnDo Reduce A. }
      END;
END;

PROCEDURE UnDoLDUdecompose(ClDgB:INTEGER; ReduceUtoDU:Boolean);

{ UnDo the decomposition of Matrix A to its LDU factored format, restoring
  it to its unfactored format A. }

VAR  i, PivotCol, Row, Col: INTEGER; PivotEntry: REAL;
```

494   Computer Program

```
BEGIN              { Note that if PivotRowA already 0, the procedure is idle. }
  IF PivotRowA <> 0 THEN WRITELN('UnDoLDUdecompose...');
  WHILE PivotRowA>=1 DO        { Starting with PivotRowA, get PivotCol }
  BEGIN                        { and rebuild in it; undo PivotRowA.   }
    RetrieveMultipliersAndRebuild (PivotRowA,PivotCol,ClDgB,ReduceUtoDU);
    PivotRowA:=PivotRowA-1;                           { Undo the PivotRow. }
  END;
  FOR i:=1 TO RowDegA DO RowListA[i]:=i;                { Undo RowListA. }
END;

{********************** APPROXIMATE INVERSE OF A *************************}

{ FOR EFFICIENCY: Modify according to the outline given in the chapter
  LINEAR ALGORITHMS, Section 5. }

PROCEDURE ApproximateInverse;

{ Start with Matrix A and replace it with its approximate inverse. }

VAR MatrixC: RealMatrix; RowDegC, ColDegC: INTEGER; r,s,t: INTEGER;

PROCEDURE TransposeAtoB(MatA: RealMatrix; VAR MatB: RealMatrix;
          RDegA, CdegA: INTEGER; VAR RdegB, CdegB: INTEGER);

{ Put the transpose of A in B. }

VAR i,j:INTEGER;
BEGIN
  RdegB := CdegA; CdegB := RdegA;
  FOR i := 1 TO RdegB DO
    FOR j := 1 TO CdegB DO
      MatB[i,j] := MatA[RowListA[j],i];     { Sometimes, RowList is needed. }
END;

PROCEDURE BtimesBprimeToA(MatB: RealMatrix;    VAR MatA      : RealMatrix;
                   RDegB,CDegB: INTEGER;VAR RdegA,CdegA: INTEGER);

{ Put B times its transpose in A. }

VAR  r,s,t :INTEGER;
BEGIN
  RdegA:=RdegB;CdegA:=RdegB;
  FOR r:= 1 TO RdegB DO
  FOR t:= 1 TO RdegB DO
    BEGIN
      MatA[r,t] := 0.0;
      FOR s:= 1 TO CdegB DO MatA[r,t]:=MatA[r,t]+MatB[r,s]*MatB[t,s];
    END
END;

PROCEDURE OrthogonalizeUp(RkA:INTEGER);

{ Orthogonalize the nonzero rows of A in reverse order by the Gram-Schmidt
  process. }

VAR dt,lngth : REAL; i,j,k : INTEGER;
```

```
FUNCTION Dot(i,k:INTEGER):REAL;

BEGIN
  dt:=0.0;
  FOR j:= 1 TO ColDegA DO dt
    :=dt + MatrixA[RowListA[i],j]*MatrixA[RowListA[k],j];
  Dot:=dt;
END;

FUNCTION Length(i:INTEGER):REAL;

BEGIN
  Lngth:=Dot(i,i); length:=sqrt(lngth);
END;

BEGIN {* OrthogonalizeUp *}
  FOR i:=RkA DOWNTO 1 DO
  BEGIN
    lngth:=Length(i);
    FOR j:= 1  TO ColDegA DO MatrixA[RowListA[i],j]
      :=MatrixA[RowListA[i],j]/lngth;
    FOR j:= 1  TO ColDegB DO MatrixB[RowListA[i],j]
      :=MatrixB[RowListA[i],j]/lngth;
    FOR k:= i-1 DOWNTO 1 DO AddRow(k,i,-Dot(i,k),1,ColDegB);
  END;
END;

BEGIN   {* ApproximateInverse *}
  IF PivotRowA<>0 THEN UnDoLDUdecompose(0,True);  { UnDo LDU decomposition. }
  WRITELN('ApproximateInverse...');
  TransposeAtoB(MatrixA,MatrixB, { Could be avoided by loading Aprime in B. }
     RowDegA,ColDegA,RowDegB,ColDegB);              { Get(A,Aprime). }
  BtimesBprimeToA(MatrixB,MatrixA,
             RowDegB,ColDegB,RowDegA,ColDegA);    { Get(AprimeA,Aprime). }
  DoLDUdecompose(ColDegB, False, False);    { LU decompose to get (J1,K1). }
  OrthogonalizeUp(RankA);                { Go on to get(J,K) from (J1,K1). }
  TransposeAtoB(MatrixA,MatrixC,
             RowDegA,ColDegA,RowDegC,ColDegC);  { Get Jprime from (J,K). }
  ColDegA:=ColDegB;   { Give MatrixA its new column degree, that of Aprime. }
  FOR r:=1 TO RowDegA DO    { Get JprimeK in MatrixA from Jprime and (J,K). }
    BEGIN
      FOR t:=1 TO ColDegA DO
        BEGIN
          MatrixA[r,t]:=0.0;
          FOR s:= 1 TO {ColDegA}RowDegA DO
            MatrixA[r,t]:=MatrixA[r,t]+MatrixC[r,s]*MatrixB[RowListA[s],t];
        END;
    END;                          { The approximate inverse is JprimeK. }
  FOR r:=1 TO RowDegA DO BEGIN RowListA[r]:=r; PivotListA[r]:=r;END;
  PivotRowA:=0;  { Return A to matrix format now that it is a matrix again. }
END;
```

{****************** SOLVE Ax=y EXACTLY or APPROXIMATELY ******************}

PROCEDURE Multiply(VAR MatAminus:RealMatrix;RowDg,ColDg:INTEGER;
                    VAR y:RealVector);

{ Solve Ax = y by multiplying y by the approximate inverse of A to get x. }

VAR i,j:INTEGER;
BEGIN
  FOR i:= 1 TO RowDg DO
     BEGIN x[i]:=0; FOR j:=1 TO ColDg DO x[i]:=x[i]+MatAminus[i,j]*y[j]; END;
END;

PROCEDURE ForwardSubstitute(VAR MatL:RealMatrix; VAR y,v:RealVector);

VAR p,j:INTEGER;
BEGIN
  FOR p:=1 TO RowDegA DO
    BEGIN
      v[p]:=y[RowListA[p]];
      FOR j:=1 TO p-1   DO v[p]:=v[p]-MatL[RowListA[p],j]*v[j]
    END
END;

PROCEDURE BackSubstitute(VAR w,x:RealVector;VAR MatU:RealMatrix);

VAR p,q,j:INTEGER;
BEGIN
  FOR j:=1 TO ColDegA DO x[j]:= 0; { Decide on values of independent x[j]. }
                                  { If j is pivot column, x[j] redefined later. }
  FOR p:= RankA DOWNTO 1 DO
    BEGIN
      q := PivotListA[p];
      x[q] := w[p];
      FOR j:=q+1 TO ColDegA DO x[q]:=x[q]-MatU[RowListA[p],j]*x[j]
    END
END;

PROCEDURE EnterVector (VAR v:RealVector;RwDgA:INTEGER);

{ Prompt user to enter the vector v. }

VAR i : INTEGER;
BEGIN WRITELN;
  FOR i := 1 TO RwDgA DO
    BEGIN WRITE (' Enter y[',i:1,'] = '); READLN (v[i]); END;
  WRITELN;
END;

PROCEDURE WriteVector(z:RealVector;ClDgA:INTEGER);

{ Write the vector z on the screen. }

VAR j : INTEGER;

```
BEGIN
  FOR j:=1 TO ClDgA DO
    BEGIN WRITE('        x[',j:1,'] = '); WRITELN(z[j]:6:2) END;
  WRITELN;
END;

PROCEDURE GetRightHandSide(SolveItExactly:BOOLEAN;RwDgA,ClDgA:INTEGER);

{ Call EnterVector to get right hand side vector y from user. Solve
  Ax = y exactly using ForwardSubstitute and BackSubstitute;  or
  approximately using Multiply. }

VAR i: Integer; Ch: Char;
BEGIN
  Ch := 'Y'; WRITELN('  Right Hand Side Vector y: ');
  WHILE UPCASE(Ch) ='Y' DO BEGIN                 { to get y and solve for x. }
    EnterVector(y,RwDgA);
    IF SolveItExactly THEN
      BEGIN       { to solve Lv = y, Dw = v, Ux = w so LDUx = LDw = Lv = y. }
        ForwardSubstitute(MatrixA,y,v);                             { Get v. }
        FOR i:=1 TO RankA DO w[i]:=v[i]/MatrixA[RowListA[i],i];     { Get w. }
        BackSubstitute(w,x,MatrixA);                                { Get x. }
      END
    ELSE                                 { we are solving it approximately. }
    Multiply(MatrixA,ClDgA,RwDgA,y);  { where Matrix A holds inverse of A. }
    WRITELN('Final Solution Vector x: ');
    WriteVector(x,ClDgA);
    WRITELN ('Would you like to enter a different vector b? (Y/N):');
    READ (KBD,Ch); WRITELN(UPCASE(Ch));
  END
END;

PROCEDURE SolveApproximately;

{ Solve Ax = y for its shortest approximate solution x. }

VAR Ch: Char;
BEGIN
  ApproximateInverse;                    { Put the inverse of A in Matrix A. }
  GetRightHandSide(False,ColDegA,RowDegA);{ Get y and deliver x until done. }
  WRITELN ('Restore the matrix? (Y/N):'); READ (KBD,Ch);
  IF UPCASE(Ch)='N' THEN {Do Nothing} ELSE ApproximateInverse;
END;

PROCEDURE SolveExactly;
VAR Singular:Boolean; Ch: Char;
BEGIN
  Singular:=False;
  IF RowDegA <> ColDegA THEN Singular:=True ELSE DoLDUdecompose(0,True,True);
  IF RowDegA<>RankA THEN Singular:= True;
  IF NOT Singular THEN BEGIN                          { to solve exactly. }
    GetRightHandSide(True,RowDegA,ColDegA);
    WRITELN ('Restore to matrix format? (Y/N):');
    READ (KBD,Ch); WRITELN(UPCASE(Ch));
  IF UPCASE(Ch)='N' THEN {Do Nothing} ELSE UnDoLDUdecompose(0,True);
END
```

```
    ELSE BEGIN                          { to solve approximately or exit to menu. }
      WRITELN('Matrix singular. Solve Approximately? (Y/N) ');
      Read(KBD,Ch); IF UPCASE(Ch) = 'Y' THEN SolveApproximately;
    END;
END;

{*************************** WRITEMATRIX ****************************}

PROCEDURE WAIT; BEGIN WRITELN('Press a key ...');Repeat until KeyPressed END;

PROCEDURE InvertList( VAR RwList,ClList: IntegerVector; m, n:INTEGER);

{ Invert the row list. }

VAR i: Integer; AnyList: IntegerVector;
BEGIN   FOR i:= 1 TO m DO AnyList[RowListA[i]]:=i; RwList:= AnyList;
        FOR i:= 1 TO n DO AnyList[ColListA[i]]:=i; ClList:= AnyList; END;

PROCEDURE WriteMatrix (Mat:RealMatrix;m,n:INTEGER);

{ Writes RowListA, ColListA and matrix to screen. }

VAR
  i,j : INTEGER;
BEGIN
  InvertList(InvRwListA,InvClListA,m,n); WRITELN; WRITE('      ');
  FOR j:= 1 TO n DO WRITE('C',InvClListA[j]:2,'   ');WRITELN;WRITELN;
  FOR I:= 1 TO m DO
    BEGIN
      WRITE('R',InvRwListA[i]:2,' ');
      FOR j:= 1 TO n DO WRITE(Mat[i ,j]:6:2);
      WRITELN; WRITELN;
END;END;

PROCEDURE Window(Msg:StringType);

{ Display a message and write from the matrix to a window on the screen. }

BEGIN
  ClrScr; WRITELN(Msg); WRITELN('RowListA......Entries in memory:');
  WriteMatrix(MatrixA,RowDegA,ColDegA); Write('Rank = ',RankA,'   ');
  IF PivotRowA=0 THEN WRITELN('Matrix Format') else WRITELN('LDU Format');
  Wait;
END;

{************************ INPUT/OUTPUT FILES ****************************}

PROCEDURE ReadFileMatrix ;

{ The input Matrix A is read into memory. }

VAR  i , j : INTEGER;
```

Computer Program

```
BEGIN
  RowListA:=IdentityList;ColListA:=IdentityList; PivotRowA:=0;{ A is matrix.}
  READLN (InputFile, RowDegA , ColDegA );
  IF RowDegA>MaxDegree Then RowDegA:=MaxDegree;
  IF ColDegA>MaxDegree Then ColDegA:=MaxDegree;
  FOR i := 1 TO RowDegA DO BEGIN
    FOR j := 1 TO ColDegA DO  READ (InputFile, MatrixA [i,j]);
    READLN (InputFile);
  END;
END;

PROCEDURE WriteFileMatrix ;

{ The Matrix B is written to file memory. }

VAR  i , j : INTEGER;
BEGIN
  WRITELN (OutputFile, RowDegA:10 ,ColDegA:10 );
  FOR i := 1 TO RowDegA DO BEGIN
    FOR j := 1 TO ColDegA DO
      WRITE (OutputFile, MatrixA [RowListA[i],ColListA[j]]:3*Format:Format);
    WRITELN (OutputFile);
  END;
END;

PROCEDURE GetMatrix;

{ Gets degrees, entries. }

VAR
  FileExists : BOOLEAN;
BEGIN
  Write('Loading '+InputFileN+' (or enter new File Name): ');READLN (AnySt);
  IF NOT(AnySt = '') THEN InputFileN:=AnySt;
  ASSIGN (InputFile,InputFileN);
  {$I-}   RESET (InputFile); {$I+};
  FileExists := (IOresult = 0);
  IF FileExists THEN BEGIN
     ReadFileMatrix; CLOSE (InputFile); RankA:=0; Window(InputFileN+' :   ')
  END
  ELSE BEGIN WRITELN('NO '+InputFileN+' exists'); Wait END;
END;

PROCEDURE PutMatrix;

{ Gets filename and writes degrees, entries to file. }

VAR  i,j  : INTEGER; AnySt:StringType;
BEGIN
  Write('Saving '+OutputFileN+' (or enter new File Name): '); READLN (AnySt);
  IF NOT(AnySt = '') THEN OutputFileN:=AnySt;          { Get file name. }
  ASSIGN (OutputFile,OutPutFileN);
  REWRITE (OutPutFile);
  IF pivotRowA <>0 THEN UnDoLDUdecompose(0,True);    { First make A a matrix. }
  WriteFileMatrix ;  CLOSE (OutputFile);             { Then save it.          }
END;
```

Computer Program

```
{********************* INITIALIZATION OF VARIABLES **********************}

PROCEDURE Initialize;

VAR i: Integer;
BEGIN
  TurnedOn := True;                         { Turns on the main program loop. }
  MatrixA[1,1]:=1; RowDegA:=1;ColDegA:=1;PivotRowA:=0;RankA:=0; { Matrix A. }
  FOR i:= 1 TO MaxDegree DO BEGIN IdentityList[i]:=i; ZeroList[i]:=0; END;
  RowListA:=IdentityList; InvRwListA:=IdentityList; ColListA:=IdentityList;
  PivotListA:=ZeroList;
  RowDegB:=0; ColDegB:=0;                                    { Matrix B. }
  InputFileN:='M1.MAT ';OutputFileN:='N1.MAT';               { Files   . }
END;

{***************************** MENU *************************************}

PROCEDURE MenuSelections;

BEGIN
  ClrScr;
  WRITELN('< To enter a matrix from console, use CON as the LOAD file >');
  WRITELN('< To print a matrix, use LPT1 as the SAVE file             >');
  WRITELN('< Please enter the upper case letter of your selection     >');
  WRITELN;
  WRITELN('    Loadmatrix:'+InputFileN+'           Savematrix:'+OutputfileN);
  WRITELN('    Decompose/Undecompose.         Window.');
  WRITELN('    Exactly solve/Approximately solve Ax=y for x.   ');
  WRITELN('    Invert exactly or approximately.                        eXit');
END;

Procedure Menu (VAR TurnedOn: Boolean);
VAR Answer: Char;
BEGIN
  MenuSelections;
REPEAT Read(KBD,Answer); WRITELN(UPCASE(Answer));GotoXY(WhereX,WhereY-1);
UNTIL UPCASE(Answer) in   ['L','S','W','E','A','D','U','I','X'];
  CASE Answer of
'L','l': GetMatrix ;
'S','s': PutMatrix ;
'W','w': Window('CurrentEntries');
'E','e':BEGIN ClrScr; SolveExactly;                                   END;
'A','a':BEGIN ClrScr; SolveApproximately;                             END;
'D','d':BEGIN ClrScr; DoLDUdecompose(0,True,True); Window('Decomposed.');END;
'U','u':BEGIN ClrScr; UnDoLDUdecompose(0,True); Window('UnDecomposed.'); END;
'I','i':BEGIN ClrScr; ApproximateInverse;
  IF ((RowDegB=ColDegB)AND(ColDegB=RankA)) THEN Window(' Exact Inverse: ')
  ELSE Window(' Approximate inverse: ');                              END;
'X','x': TurnedOn :=False ;
  END;END;

{************************* MAIN PROGRAM *********************************}

BEGIN  Initialize; REPEAT Menu(TurnedOn) UNTIL Not TurnedOn; ClrScr  END.
```

Index

A

A History of Mathematics, 408
A Maggot, 408
Absolute value of a complex number, 49
Action of a matrix or transformation on a vector, 35, 98
Add $(r, s; u)$, 80, 257
Addition
  of matrices, 3, 91, 273
  of a multiple of a row or column, 80, 257
Adjacent objects, interchange of, 222–224
Adjoint
  classical, 253
  Hermitian, 116, 354
Algorithm
  for finding A^- when the columns of A are linearly independent, 480
  for finding R^{-1} for an invertible upper triangular matrix R, 480
  for solving $Ax = y$ for x approximately, 477
  for solving $Ax = y$ approximately, when the columns of A are linearly independent, 480
  for solving $Ax = y$ exactly, using the LDS Factorization, 476
  to compute the approximate inverse of a real $m \times n$ matrix A, 484
  to compute the nullspace of a real $m \times n$ matrix A, 484
  to compute the projection onto the column space of A for a real $m \times n$ matrix A, 484
  to compute the projection onto the column space of A' for a real $m \times n$ matrix A, 484
  to find all approximate solutions of $Ax = y$, 485
  to find the inverse of a lower triangular matrix, 467
  to row reduce an $m \times n$ matrix A to an echelon matrix U, 443, 444, 448
  to undo row reduction of A to an echelon matrix U, 474
Approximate inverse of an $m \times n$ matrix, 443, 444, 448
  factorization of, 456, 487
  weighted, 456
Approximate solution to a system of linear equations, 434, 440
  shortest, 441, 443
Approximating functions, 450
Approximation, 450
  first-order, 451
  functional, 450
  general nth order, 451
  second-order, 452
  weighted, 455
Associative law, 5, 8, 15, 30, 96
  for mappings and functions, 30
  for products of matrices, 96
  for sums of matrices, 96
Augmented matrix, 88, 135, 146

B

Back substitution, 86, 476
Bases, all possible, 141

Index

Basis, 137–145
  canonical, 151
  of $F^{(n)}$, 157
  orthonormal, 178, 319–329
  standard, 151
  for a subspace, 166
  for a vector space, 308
Beginning of a branch, 423
Block matrices, 275
  elementary, 278
Block multiplication, 275
Borland International, 488
Boyer, Carl, 408
Branches in incidence diagrams, 423

C

Cancellation law, for multiplication, 5, 8
Canonical basis of $F^{(n)}$, 151
Canonical forms, 376
Cartesian plane, 35
Cayley, Arthur, 38
Cayley–Hamilton Theorem, 38, 39, 62, 184, 270
Change of basis, 140, 153
  of $F^{(n)}$, 140
  matrix of, 140, 153
Characteristic polynomial, 38, 39, 266, 346
  of a linear transformation, 346
  of a matrix, 39, 266
Characteristic root, 127, 183, 185, 267, 346
Characteristic vector, associated with a given scalar, 127, 346
Class, similarity, of a matrix, 155
Classical adjoint, 253
Closed, with respect to products, 94
Coefficient matrix, 69, 87
Cofactors, 246
  expansion by, 247
Column operations, 233
Column space of a matrix, 285
Column vector, 35, 98
Combinations, linear, 94, 131, 135
  trivial, 131
Commutative law for addition and multiplication, 15, 94, 96
Commuting matrices, 5
Complement, orthogonal, 173, 178, 315, 325, 353
Complex conjugate, 49
Complex conjugation, 304
Complex number, 2, 45
  absolute value of, 49
  argument of, 41
  conjugate of, 49
  equality of, 46
  i, 46
  length of, 50
  one, a, 47
  polar form of, 51
  product of, 46
  square root of, 48
  sum of, 46
Component-wise operation, 35
Composition, 30, 37, 105
  of functions, 30
  of matrices, 37, 105
Computations, 458
Computers, 488
Conjugate of a complex number, 49
Continuous real-valued functions, 296
Convergence, of a power series in a matrix, 393
Coordinates, 35, 36, 99
  change of, 36
  transformation of, 36
Cosine, of a matrix, 398
Cramer, Gabriel, 26
Cramer's Rule, 26, 243–245

D

De Moivre, Abraham, 52
De Moivre's Theorem, 52
Dependent variable, 64
Derivative of the exponential function, 401
Derivative of a vector function, 400
Descartes, René, 35
Determinant, of a matrix, 21, 22, 214
  characterization of, 280
Determinants, product rule for, 262, 279, 281
Diagonal, main, 11
Diagonal matrix, 6, 92
Diagonalization
  of a Hermitian matrix, 62, 195
  of a symmetric matrix, 196
Differential equation, 400, 430
  homogeneous, 430
  matrix, 400
  vector space of solutions to, 430
Dimension
  of a subspace, 167
  of a vector space, 167, 310

Direct sum of subspaces, 172, 325
  orthogonal, 173
Displacements at nodes in a mechanical structure, 423
Distributive law, 8, 15
Dot product, 55
Double summation, 10
Dummy variable or index in a summation, 10, 126

E

Echelon matrix, 79, 83
  reduced, 287
Eigenvalue, see Characteristic root
Eigenvector, see Characteristic vector
Electrical network model, 427
Electrical potential, 427
Elementary matrices, 256
Elementary row and column operations, 76, 79, 80
Elimination of variables, 72, 74
End of a branch, 423
Energies of the physical state, 193
Entry, 3, 11, 91
Equal matrices, 2–3, 91
Equations of curves, 414–415
  of circles, 415
  of lines, 414
Equivalent systems, 74, 77
Existence and uniqueness theorem
  for solutions to a homogeneous linear differential equation of order n, 432
  for solutions to a matrix differential equation, 405–406
Expansion, of the determinant of a matrix by cofactors, 274
Exponent, 12, 32
Exponential of A, 393–394
Exponential function e^{tT}, 396–398
  computation of, 396
Exponents, law of, 32, 95
External currents, 428

F

Fibonacci, Leonardo, 408
  numbers, 408
  sequence, 408
Field F, 91
First-order approximation, 451
Forward substitution, 476

Fourier series, 316
Fowles, John, 408
Free variable, see Independent variable
Function, identity, 30
Functional approximation, 405
Fundamental Theorem of Algebra, 41, 51
  over $\mathbb{R}$, 209

G

Gauss, Karl Friedrich, 51
General functional approximation of order n, 451
Generalized characteristic subspace of a matrix, 381
Generalized nullspace of a matrix, 381
Generating set of vectors
  minimal, 307
  for a subspace, 307
Golden mean, 408
Golden section, 408
Gram–Schmidt orthogonalization process, 176, 321
Growth, in population, 402

H

Hamilton, William Rowan, 38
Hamiltonian, 193
Hermite, Charles, 116
Hermitian adjoint, 116, 354
  of a linear transformation, 354
  of a matrix, 116
Hermitian linear transformation, 351, 354
Hermitian matrix, 60, 192–199
  diagonalization of, 62, 195
Homogeneous systems of linear equations, 80, 282–291
Homomorphism Theorem, 362
Homomorphisms of vector spaces, 300–307

I

Identity, 94
  function or mapping, 31
  matrix, 5, 91
Image of an element under a function or mapping, 29
Imaginary part of a complex number, 46
Incidence diagrams, 423
  connected, 424
Incidence models, 423

504 Index

Independent variable, 64, 88
Induction, mathematical, 199
   going by, 200
Inductive definition, 214
Inner product, 55, 124, 314
   of column vectors, 55, 124
   in an inner product space, 314
Inner product space, 314
Interchange, 76, 224, 239, 257, 470
   of adjacent objects, 224
   of rows or columns in a matrix structure, 470
   of rows or columns of a matrix, 224, 239
Interchange (r, s), 80, 81, 257
Inverse, 6, 31, 95
   of a 1-1 onto function, 31
   of a linear transformation, 339
   of a matrix, 6, 95
Invariant subspace, 366
Invertible matrix, 6, 95, 145
Isomorphism, transfer principles for, 301, 310, 327
Isomorphism of vector spaces, 300–307, 307–313

J

Jordan, Camille, 376
Jordan canonical form, 376

K

Kernel of a homomorphism, 305
Kirchoff's Laws, 425

L

LDU decomposition, see LDU factorization
LDU factorization, of an $m \times n$ matrix, 462–467
   echelon matrix U, 464
   matrix of multipliers L, 462
   matrix of pivots D, 462
   uniqueness of, 467
Leading entry of a row, 83
Least squares methods, 434
Legendre, Adrian Marie, 324
Legendre polynomials, 324
Length of a vector, 58, 127, 318
Leonardo of Pisa, see Fibonacci, Leonardo
Liber Abaci, 408
Linear algorithms, see Algorithm

Linear combination of vectors, 131, 295
   trivial, 131
Linear equations, systems of, 62–89, 282
Linear transformation, 2, 35–36, 102, 282, 330, 331
   characteristic polynomial of, 346
   characteristic root of, 346
   characteristic vector of, 346
   determinant of, 346
   from one space to another, 372
   Hermitian, 351, 354
   Hermitian adjoint of, 354
   inverse of, 339
   invertible, 339
   matrix of, 342, 372
   negative of, 335
   product of, 337
   scalar multiple of, 334
   skew-Hermitian of, 354
   sum of, 334
   trace of, 346
   unitary, 209–211, 354
   vector space of, 335
   zero, 335
Linearity conditions, 35–36, 283
Linearly dependent set of vectors, 133
Linearly independent set of vectors, 133, 311
List of rows and columns in memory, 467
Local prices, 423

M

Main diagonal, 69
Manufactured products, how many units to make, 438
Mappings, 29–34
   associative law for, 30
   composition of, 30
   identity, 31
   product of, 30
Markov, Andrei Andreevich, 416
Markov processes, 416
Mathematical induction, 199
Matrices
   commuting, 94
   difference of, 3, 91
   equality of, 2–3, 90–91
   over $\mathbb{C}$, 2, 90
   over F, 91
   as mappings, 35, 98, 106
   over $\mathbb{R}$, 2, 90
   product of, 4, 93, 106

Matrices (*Continued*)
  row equivalent, 83
  similar, 44, 154
  sum of, 3, 91, 273
Matrix
  2×2, 2, 3
  addition, 3, 52, 91, 273
  augmented, *see* Augmented matrix
  of coefficients, 62–79, 282
  of the change of basis, 140, 153
  column space of, 35, 98, 285
  corresponding to a given matrix structure, 469, 471
  cosine of, 398
  diagonal, 6, 92
  differential equation, 400
  echelon, 79, 83
  equation, 62, 69, 282, 458
  exponential of, 393–394
  files, 488
  Hermitian, 60, 192–199
  identity, 5, 91
  invertible, 6, 95, 145
  of a linear transformation, 106, 149
  $m \times n$, 67, 283
  multiplication, 4, 273
  $n \times n$, 90
  negative of, 3–4, 91–92
  nilpotent, 187
  nullity of, 181, 287
  nullspace of, 287
  rank of, 88, 180, 287, 288–289
  rectangular, 67, 273
  reduced echelon, 287
  row space of, 288
  scalar, 6, 92
  similarity class of, 155
  sine of, 398
  singular, 6
  skew-Hermitian, 120, 121
  skew-symmetric, 20
  square, 90
  subtraction, 3, 52
  symmetric, 19, 120
  trace of, 16, 109
  transpose of, 9, 19, 116
  unipotent, 187
  unitary, 58, 127, 160
  units, 94
  upper triangular, 7, 69
  which satisfies a polynomial, 39
  zero, 3, 94

Matrix differential equation, 400
Matrix structure, for representing a matrix in memory, 468
Matrix–vector equations, algorithmic study of, 458–490
Memory, 458, 469
Method of least squares, 474, 497
Method to
  determine whether a matrix is invertible, 146
  determine whether $Ax = y$ has a solution, 88
  determine whether given vectors are linearly independent, 133
  express a given vector as a linear combination of given vectors, when possible, 135
  row reduce A to echelon form, 84
  solve $Ax = y$, 87, 284
  solve linear equations by back substitution, 72, 86
Minimum polynomial, 43, 184
Minor submatrix, 215
Minors, expansion by, 216
Models
  electrical network, 427
  incidence, 423
  transportation, 427
Multiplication, 4, 93, 99, 106, 337
  of linear transformations, 337
  of matrices, 4, 93, 106
  of a matrix by a scalar, 93
  of a row or column by a scalar, 99
  of a vector by a scalar, 99
Multipliers
  matrix of, 462
  in row reduction, 462
Multiply (r, s), 80–81, 257

N

Nearest vector to a subspace, 435
Negative
  of a linear transformation, 335
  of a matrix, 3–4, 91–92
  of a vector, 99
Nilpotent matrix, 187
Nodes, in incidence and transition diagrams, 423
Normal equation, 446
Normal matrix, 60, 356
Nullity, of a matrix, 180, 287–289

Nullspace of a matrix, 287
  generalized, 381
Numerical stability, 470

O

Ohm's Law, 428
One-to-one mapping, 30, 34
Orthogonal complement, 173, 178, 325, 353
Orthogonal direct sum, 173, 325–326
Orthogonal vectors, 56, 125, 319
Orthogonalization, process of, 176, 321
Orthonormal set or basis, 178, 319
Orthonormalized $m \times n$ matrix, 482

P

PASCAL, standard, 490
Permutation matrix corresponding to row list, 472
Pivot entries, in row reduction, 462
Pivot list, 471
Polynomial, characteristic, 38, 39, 266, 346
Polynomial, minimum, 43, 184
Population
  growth in, 402
  of rabbits, 408
Power series, in a matrix, 393
Potential
  difference, 429
  electrical, 427
  vector, 428
Power
  of a function, 32
  of a matrix, 12, 95
Predators, 402
Preservation
  of an inner product or length, 59, 327
  of structure, 300, 327
Prey, 402
Price
  difference, 425
  vector, 426
Product rule, for the determinant function, 262, 279, 281
Product
  of functions, 30
  of linear transformations, 337
  of matrices, 4, 93, 273
  of a matrix and a vector, 35, 98
  of a scalar and a matrix, 93
  of a scalar and a vector, 99, 292
Projection, 33, 303, 435, 482
  onto the column space of A, 446
  onto the column space of A', 446
  matrix, 482
  of a vector on a subspace, 435
  of a vector on the span of mutually orthogonal vectors, 437
  onto the x-axis, 33
  onto the y-axis, 33
Pseudoinverse of a matrix, 444
Pure imaginary, a, 46

Q

QR factorization of A, 478
Quadratic formula for complex numbers, 48

R

Rabbit population, 408
Rank of a matrix, 88, 180, 287, 288, 289
Real numbers, 2
Real part
  of a complex number, 46
  of a complex solution to a matrix differential equation, 403
Resistance to flow along a transportation route, 425
Row equivalent matrices, 83, 287
Row operations, 221
Row reduction algorithm, 459, 467
  inverse of, 474
Row space of a matrix, 288

S

Scalar, 4
Scalar matrix, 6, 92
Schwarz, Hermann Amandus, 317
Schwarz Inequality, 317
Second-order approximation, 452
Shape of a matrix, 273
Shift operator, 303
Sign change of the determinant, 222–224, 239
Similar matrices, 44, 154
Similarity class of a matrix, 155
Sine, of a matrix, 398

Singular matrix, 6
Skew-Hermitian linear transformation, 345
Skew-Hermitian matrix, 60-61
Solution, to a matrix differential equation, 400
Solution, to a system of linear equations, 64-89, 282-291, 458-490
Span of a set of vectors, 134, 312
Square root of a complex number, 48
State, in a Markov process, 416, 420
Steady state, in a Markov process, 416
Subspace, 164-176
  basis of, 166
  proper, 164, 297
  spanned by given vectors, 134, 312
  trivial, 164
  of vectors orthogonal to a subspace, 173, 178, 325, 353
Subspaces
  direct sum of, 172-173
  mutually orthogonal, 173
  sum of, 163-164
Substitution
  back, 476
  forward, 476
Sum
  of linear transformations, 334
  of matrices, 3, 52, 91, 273
  of subspaces, 163-164, 173
  of vectors in $F^{(n)}$, 35, 99
Summation, 10
  double, 10
  dummy index or variable, 10, 126
Symmetric matrix, 19, 120
System of equations, see System of linear equations
System of linear equations, 62-89, 282

T

Time-study experiment, 451
Total consumption, 425
Total production, 425
Trace
  of a linear transformation, 346
  of a matrix, 16, 109
Trace function, 109, 346
Transfer principle for isomorphisms, 301, 310, 327
  of inner product spaces, 327
  of vector spaces, 301, 310

Transformations, linear, see Linear transformations
Transition in a Markov process
  diagram for, 420
  matrix for, 420
  probabilities for, 416
  regular, 421
Transportation model, 427
Transportation routes, 427
Transpose, of a matrix, 9, 19, 116
Triangle inequality, 51, 130, 318
Triangularization
  of a complex matrix, 199
  of a real matrix, 208
Triangular matrix
  lower, 7
  upper, 7
TURBO PASCAL, 488 n.

U

Unipotent matrix, 187
Unit vector, 158
Unitary linear transformation, 160, 354
Unitary matrix, 58, 127, 160

V

Variable, dependent and independent, 64, 88
Vector
  acted on by a linear transformation, 35, 98
  length of, 58, 127, 318
  multiplied by scalar, 99, 314
  negative of, 3-4, 91, 314
  price, 426
Vector analysis, 36
Vectors
  mutually orthogonal, 173
  orthogonal, 125
  span of, 134, 312
  sum of, 35, 99
Vector spaces, 3, 101, 292-329
  dimension of, 167, 310
  finite dimensional, 298
  of functions, 293
  infinite dimensional, 311
  of linear combinations of a set of vectors, 295
  of linear transformations, 335

Vector spaces (*Continued*)
   of n times differentiable complex-valued functions, 430
   of polynomials, 294
   of solutions to a homogeneous differential equation, 430
Vector spaces, isomorphic, 305

W

Wronskian, 313, 406

X

x-axis, projection onto, 33

Y

y-axis, projection onto, 33

Z

Zero matrix, 3, 94